Lecture Notes in Computer Science 4024

Commenced Publication in 1973
Founding and Former Series Editors:
Gerhard Goos, Juris Hartmanis, and Jan van Leeuwen

Susanna Donatelli   P.S. Thiagarajan (Eds.)

Petri Nets
and Other Models
of Concurrency –
ICATPN 2006

27th International Conference on Applications and Theory
of Petri Nets and Other Models of Concurrency
Turku, Finland, June 26-30, 2006
Proceedings

 Springer

Volume Editors

Susanna Donatelli
Università degli Studi di Torino, Dipartimento di Informatica
Corso Svizzera 185, 10149 Torino, Italy
E-mail: susi@di.unito.it

P.S. Thiagarajan
National University of Singapore, School of Computing
3 Science Drive 2, Singapore 117543, Singapore
E-mail: thiagu@comp.nus.edu.sg

Library of Congress Control Number: 2006926657

CR Subject Classification (1998): F.1-3, C.1-2, G.2.2, D.2, D.4, J.4

LNCS Sublibrary: SL 1 – Theoretical Computer Science and General Issues

ISSN   0302-9743
ISBN-10   3-540-34699-6 Springer Berlin Heidelberg New York
ISBN-13   978-3-540-34699-9 Springer Berlin Heidelberg New York

Springer is a part of Springer Science+Business Media

springer.com

© Springer-Verlag Berlin Heidelberg 2006
Printed in Germany

Typesetting: Camera-ready by author, data conversion by Scientific Publishing Services, Chennai, India
Printed on acid-free paper   SPIN: 11767589   06/3142   5 4 3 2 1 0

Preface

This volume consists of the proceedings of the 27th International Conference on Applications and Theory of Petri Nets and Other Models of Concurrency (ICATPN 2006). This series of conferences provides a forum for presenting the current state of research on Petri nets and related approaches to concurrent systems. Both applications and theoretical developments are represented. Shorter presentations introducing novel tools or substantial enhancements to existing tools are also encouraged. Further, a range of invited talks that survey related domains are presented.

ICATPN 2006 was co-located with the 6th International Conference on Applications of Concurrency to System Design (ACSD 2006). The two conferences had common satellite events and invited speakers as well as coordinated programs to enable participants to benefit from both conferences.

The ICATPN 2006 conference — as well as ACSD 2006 — was organized by the Department of Computer Science, Åbo Academi University, Turku, Finland. We would like to heartily thank the Organizing Committee, chaired by Johan Lilius, for the considerable effort invested to bring off the two conferences smoothly. Detailed information about ICATPN 2006 and the related events can be found at http://www.cs.abo.fi/atpn2006/.

This year we received 93 submissions from authors from 26 different countries. The Program Committee selected 22 contributions classified as: theory papers (10 accepted), application and theory papers (3 accepted), application papers (3 accepted) and tool papers (6 accepted). We thank all the authors who submitted papers. We wish to thank the Program Committee members and other reviewers (whose names appear on page VII) for their careful and timely evaluation of the submissions before the Program Committee meeting in Turin, Italy. Special thanks are due to Martin Karusseit, University of Dortmund, for his technical support with the Online Conference Service. Finally, we wish to express our gratitude to the four invited speakers, Ralf-Johan Back, Javier Campos, Ekkart Kindler and Jianli Xu, who chose to have their papers appear in this volume.

As usual, the Springer LNCS Team provided high quality support in the preparation of this volume. Last but not least, our heartfelt thanks are due to Phan Thi Xuan Linh and Yang Shaofa who put in a great deal of work toward the compilation of these proceedings.

April 2006   Susanna Donatelli and P.S. Thiagarajan

Organization

Steering Committee

Wil van der Aalst, The Netherlands
Jonathan Billington, Australia
Jörg Desel, Germany
Susanna Donatelli, Italy
Serge Haddad, France
Kurt Jensen, Denmark (chair)
H.C.M. Kleijn, The Netherlands
Maciej Koutny, UK

Sadatoshi Kumagai, Japan
Tadao Murata, USA
Carl Adam Petri, Germany (honorary member)
Lucia Pomello, Italy
Wolfgang Reisig, Germany
Grzegorz Rozenberg, The Netherlands
Manuel Silva, Spain

Organizing Committee

Jerker Björkqvist
Robert Gyllenberg
Tiina Haanila

Johan Lilius (chair)
Lionel Morel
Xinrong Zhou

Tool Demonstration

Jerker Björkqvist (chair)

Program Committee

Jonathan Billington, Australia
Didier Buchs, Switzerland
Nadia Busi, Italy
Gianfranco Ciardo, USA
Jose Manuel Colom, Spain
Philippe Darondeau, France
Susanna Donatelli, Italy
  (co-chair, applications)
Giuliana Franceschinis, Italy
Boudewijn Haverkort,
  The Netherlands
Xudong He, USA
Kees M. van Hee, The Netherlands
Monika Heiner, Germany
Jane Hillston, UK
Kunihiko Hiraishi, Japan

Petr Jančar, Czech Republic
Gabriel Juhás, Germany
Maciej Koutny, UK
Lars Michael Kristensen, Denmark
Johan Lilius, Finland
Madhavan Mukund, India
Wojciech Penczek, Poland
Laure Petrucci, France
Lucia Pomello, Italy
Laura Recalde, Spain
Karsten Schmidt, Germany
P.S. Thiagarajan, Singapore (co-chair, theory)
Toshimitsu Ushio, Japan
Rudiger Valk, Germany
Francois Vernadat, France

Referees

Slim Abdellatif
Baver Acu
Alessandra Agostini
Alessandro Aldini
Ashok Argent-Katwala
Eric Badouel
Kamel Barkaoui
Marek Bednarczyk
Simona Bernardi
Luca Bernardinello
Gérard Berthelot
Bernard Berthomieu
Jerker Björkqvist
Jean-Paul Bodeveix
Luciano Bononi
Jeremy Bradley
Roberto Bruni
Benoit Caillaud
Ang Chen
Christine Choppy
Lucia Cloth
Salem Derisavi
Raymond Devillers
Jean Fanchon
Carlo Ferigato
Mamoun Filali
Jana Flochova
Blaise Genest
Stephen Gilmore

Andreas Glausch
Monika Heiner
Keijo Heljanko
Loïc Hélouët
Jarle Hulaas
David Hurzeler
Wojtek Jamroga
Agata Janowska
Pawel Janowski
Jorge Julvez
Victor Khomenko
Hanna Klaudel
Jetty Kleijn
Fabrice Kordon
Matthias Kuntz
Sebastien Lafond
Niels Lohmann
Marco Loregian
Robert Lorenz
Roberto Lucchi
Levi Lucio
Matteo Magnani
Cristian Mahulea
Jose M. Martinez
Peter Massuthe
Hiroshi Matsuno
Sebastian Mauser
Samia Mazouz
José Merseguer

Toshiyuki Miyamoto
Madhavan Mukund
Wojciech Nabialek
Igor Nai Fovino
Atsushi Ohta
Luca Padovani
Wieslaw Pawlowski
Luis Pedro
Elisabeth Pelz
Giovanni Michele Pinna
Denis Poitrenaud
Agata Pólrola
Franck Pommereau
Ivan Porres
Wolfgang Reisig
Anne Remke
Matteo Risoldi
Diego Rodriguez
Christian Stahl
Tatsuya Suzuki
Maciej Szreter
Shigemasa Takai
Dragos Truscan
Kohkichi Tsuji
Robert Valette
Enrico Vicario
Jozef Winkowski
Satoshi Yamane
Gianluigi Zavattaro

Table of Contents

Tool Papers

Invariant Based Programming

Ralph-Johan Back

Abo Akademi University, Turku, Finland
backrj@abo.fi
www.abo.fi/~backrj

Abstract. Program verification is usually done by adding specifications
and invariants to the program and then proving that the verification conditions are all true. This makes program verification an alternative to or
a complement to testing. We study here an another approach to program
construction, which we refer to as *invariant based programming,* where we
start by formulating the specifications and the internal loop invariants for
the program, before we write the program code itself. The correctness of
the code is then easy to check at the same time as one is constructing it.
In this approach, program verification becomes a complement to coding
rather than to testing. The purpose is to produce programs and software
that are correct by construction. We present a new kind of diagrams,
nested invariant diagrams, where program specifications and invariants
(rather than the control) provide the main organizing structure. Nesting
of invariants provide an extension hierarchy that allows us to express
the invariants in a very compact manner. We study the feasibility of
formulating specifications and loop invariants before the code itself has
been written. We propose that a systematic use of figures, in combination with a rough idea of the intended behavior of the algorithm, makes
it rather straightforward to formulate the invariants needed for the program, to construct the code around these invariants and to check that
the resulting program is indeed correct.

1 Introduction

Program construction proceeds through a sequence of rather well-established
steps: *Requirement analysis* (understand the problem domain and work out a
specification of the problem), *design* (work out an overall structure for the program), *implementation* (code the program in the chosen programming language),
testing (check that the program works as intended), *debugging* (correct the errors
that testing has revealed), and *documentation* (provide a report on the program
for users, for maintenance and as a basis for later extensions and modifications).
Figure 1 illustrates the traditional design-implement-test-debug cycle and the
feedback loops in it.

Testing increases our confidence in the correctness of the program, but does
not prove its correctness. To establish correctness, we need a precise mathematical specification of what the program is intended to do, and a mathematical
proof that the implementation satisfies the specification. This kind of software
verification is both difficult and time consuming, and is presently not considered

S. Donatelli and P.S. Thiagarajan (Eds.): ICATPN 2006, LNCS 4024, pp. 1–18, 2006.
© Springer-Verlag Berlin Heidelberg 2006

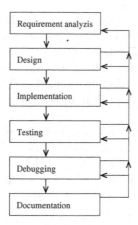

Fig. 1. The programming process

cost effective in practice. Moreover, verification would not replace testing in the above work flow, because most programs are not correct to start with, and need to be debugged before one can even attempt to verify them. This is one reason why modern verification tools like ESC Java [13], JML [6] and Spec# [5] are focused more on debugging than on complete verification of program correctness.

Verifying a simple program is usually broken down in the following steps:

1. provide the *pre-* and *postconditions* for the program,
2. provide the necessary *loop invariants* for the program,
3. compute the *verification conditions* for the program, and
4. *check* that each verification condition is indeed true.

Each of these steps have their own difficulties. Formulating pre- and postconditions requires that we have some formal or semi-formal theory of the problem domain. It can also be difficult to identify all the conditions implied by the informal problem statement. The loop invariants must be inferred from the code, and they are often quite difficult to formulate and to make complete enough so that the verification conditions can be proved. Computing verification conditions can be tedious by hand, but this step can be easily automated. Finally, proving that the verification conditions are correct can sometimes be difficult, but for most parts it is quite straightforward. The verification conditions usually consist of a large number of rather trivial lemmas that an *automatic theorem prover* can verify (provided it has a good understanding of the domain theory). The rest of the lemmas are more difficult, and have to be proved by hand or by an *interactive theorem prover* with some help and guidance from the user.

The work flow above is known as *a posteriori* verification of software, and is known to be cumbersome. An alternative approach, originally propagated by Dijkstra, is therefore to construct the program and the correctness proof hand in hand. This is known as the *correct by construction* approach to verification [7]. In other words, verification is done in the coding step rather than after (or as part of) the testing step. This means that each subunit of the program is specified before it is coded, and it is checked for correctness immediately after it has been written. Writing pre- and postconditions for the program explicitly, as well as loop invariants, is a considerably help in understanding the program and avoids a large number of errors from being introduced in the first place. Combined with *stepwise refinement* [8], this approach allows the reliable construction of quite large programs.

In this paper, we will propose to move correctness concerns to an even earlier place, to the design phase and immediately following the requirement analysis. This means that not only pre- and postconditions but also loop invariants (and class invariants in object-oriented systems) are written before the code itself is written, as a continuation of the design. This will require that the invariants are expressed in the framework and language of the design phase (figures, formulas, texts and diagrams), rather than in the framework and language of the coding phase (a programming language).

We refer to this approach as *invariant based programming*. The idea itself is not new, similar ideas were proposed in the 70's by John Reynolds, Martin van Emden, and myself, in different forms and variations. Dijkstra's later work on program construction also points in this direction [9], he emphasizes the formulation of a loop invariant as a central step in deriving the program code. Basic for all these approaches is that the loop invariants are formulated before the program code is written. Eric Hehner was working along similar lines, but chose relations rather than predicates as the basic construct. Figure 2 illustrates the differences in work flow between these three approaches.

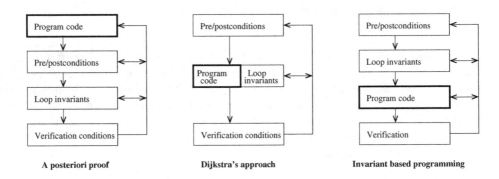

A posteriori proof   **Dijkstra's approach**   **Invariant based programming**

Fig. 2. Three different approaches to verification

The idea that program code should be enhanced with program invariants is a repeating theme in software engineering. It becomes particularly pressing to include program invariants when we want to provide mechanized support for program verification, because synthesizing program invariants from the code seems to be too hard to mechanize efficiently. The idea that the program invariants should be formulated before the code itself is written has not, however, received much attention.

Reynolds original approach [16] was to consider a program as a transition diagram, where the invariants were the nodes and program statements were transitions between the nodes. Invariants become program labels in the code, and the transitions between invariants are implemented with goto's. Reynolds starts by formulating the invariants hand in hand with the transitions between the invariants. The need to formulate invariants early on also highlights the problem of describing the invariants in an intuitive way. For this purpose, Reynolds proposes a visual way of describing properties of arrays. Reynolds work served as the basis for my own early attempt at invariant based programming, where I studied the method in case studies, and proposed multiple-exit statements as programming construct for invariant based programming [2, 3, 4]. Van Emden considered programming as just a way of writing verification conditions [17].

Hehner's approach is in spirit very similar to invariant based programming, but comes with different building blocks [12]. He describes imperative programs using tail recursion. This means that he is calling parameterless recursive procedures rather than jumping to labels. Semantically, he is working with input-output specifications (relations) rather than with invariants (predicates). This approach nicely supports stepwise refinement in a non-structured programming language, but the basic paradigm for constructing programs is different from invariant based programming.

I hope to show in this paper that, by a collection of proper enhancements to these early ideas on invariant based programming, it is possible to construct programs that are proved correct with approximately the same amount of work that we today spend on programs that are just tested for correctness. The main issues that we will tackle in the sequel are:

- what is a suitable notation for writing invariant based programs,
- what is a suitable methodology for constructing invariant based programs (in particular, how do find the invariants before the code has been written),
- how do we check the correctness of invariant based programs, and
- what are the specific challenges and opportunities with invariant based programming.

We will address each of these issues in the remaining sections of this paper.

2   Nested Invariant Diagrams

The traditional programming approach requires that the program code is kept as simple and intuitive as possible. The programmer has to have a good understanding of the structure of the code and on how the code works, because he understands the code in terms of its possible execution sequences. This is the motivation for the *structured programming* principle [8], which emphasize single-entry-single-exit control structures in order to keep the code simple and disciplined. This makes it easier to build, understand, extend and modify the code.

When one tries to make the program code as simple as possible, there is a danger that the invariants become more complicated. This is because the invariant has to fit the code, and there is no guarantee that simple code leads to simple invariants. Essentially, we have two possible ways of structuring the program, using either control or invariants as the primary structure and the other as the secondary structure, which is adapted to the primary structure. These two structures need not coincide. When we are constructing invariants first, then we should focus on structuring the invariants so that they are as simple as possible to build, understand, extend and modify. We will look at this issue next.

Consider the simple summation program in Figure 3, described as a flowchart. The program adds up the first n integers, and shows the result in the variable *sum*.

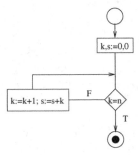

Fig. 3. Simple flowchart

Figure 4 shows the same program on the left, but we now added the pre- and postconditions and the loop invariant. Let us introduce the term *situation* for a condition that describes a collection of states. Here, the precondition, the postcondition and the loop invariant are all situations. We notice that a termination function is also needed, to show that the computation does not loop forever.

The program on the right in the same figure is also equivalent to the previous one, but now we have deemphasized the program statements by writing them as annotations on the arrows between the situations. We write both the condition for traversing an arrow (the *guard* [g]) and the effect of traversing the arrow (an *assignment statement*) on the arrow.

We are looking for a way to impose some structure on the situations described here. A situation describes a set of states (the set of states that satisfies the condition for the situation). The typical thing that we do with a situation is that we strengthen it by adding new constraints. This means that we identify a subset of the original situation where the new constraints are satisfied. Thinking of situations as sets of states, we can use Venn diagrams to describe this strengthening of situations. Figure 5 shows this nesting of situations for our example program.

We have here omitted conditions that are implicit because of nesting. For instance, $n \geq 0$ is written only once in the outermost situation, but it will hold in all nested situations. The situations are thus expressed more concisely with nesting. The diagram also shows more clearly how the different situations are related.

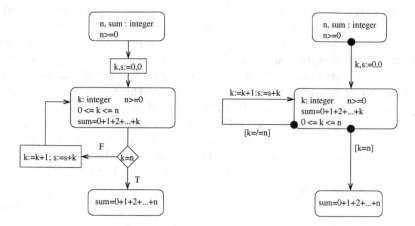

Fig. 4. Flow chart and invariants

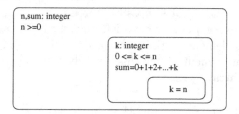

Fig. 5. Nested situations

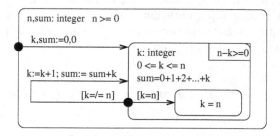

Fig. 6. Nested invariant diagram

Figure 6 shows the same program as before, but with the situations nested. The transitions are exactly as before. We will refer to diagrams of this kind as *(nested) invariant diagrams* (or *situation diagrams*).

We also need a way to indicate the termination function. We write the termination function inside the box in the right upper corner of the invariant. It shows that the termination function must be bounded from below ($n - k \geq 0$) in the indicated situation, and that the termination function $n - k$ must be decreased before re-entering this situation. This description of the original flow chart is equivalent to the previous ones, the difference is just in the presentation, which emphasizes the structure of situations rather than the structure of code.

Nested invariant diagrams are similar to *state charts* [11, 10]. Both are essentially extensions of state transition diagrams. However, the interpretation and intended use is different. State charts are intended to specify the control flow in reactive systems, without any concern for correctness, whereas invariant diagrams specifically address the correctness issue of algorithms. A state chart is usually seen as describing some specific aspect of a larger software system, i.e., it is a form of abstraction, whereas invariant diagrams describe the whole program. Our formalism can be extended with features from state charts that are deemed useful (like product states, and signals/procedure calls), but we will concentrate here on the bare essentials.

3   Constructing Invariant Based Programs

We next show how to construct invariant based programs with the help of an example. We consider the simplest possible sorting program, *selection sort*. The array is sorted by moving a cursor from left to right in the array. At each stage we find the smallest element to the right of the cursor, and exchange this element with the cursor element. After this, we advance the cursor, until we have traversed the whole array.

Formulating the problem. We assume that $Sorted(A, i, j)$ means that the array elements are non-decreasing in the (closed) interval $[i, j]$, that $Partitioned(A, i)$ means that every element in array A below index i is smaller or equal to any element in A at index i or higher, and that $Permutation(A, A0)$ means that the elements in array A form a permutation of the elements in array $A0$.

Let us first identify the initial situation and the required final situation. Figure 7 shows how we can visualize the problem.

We assume that

$$n : integer \land A : array\, 1 : n\, of\, integer \land n \geq 1 \land Permutation(A, A_0)$$

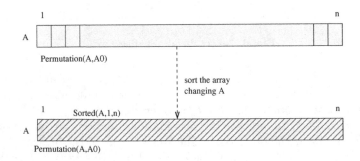

Fig. 7. Visualizing the specification

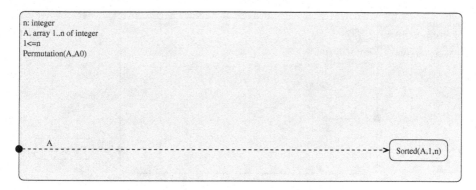

Fig. 8. Initial and final situation

holds initially. The final situation requires that in addition

$$Sorted(A, 1, n)$$

holds. We describe this using the nested invariant diagram in Figure 8.

Here the dashed arrow indicates the computation that we want to define. The annotation A on the arrow shows that only program variable A (among the variables introduced so far) may be changed.

Increasing sortedness. The next step is to identify an intermediate situation (an invariant). The most plausible one is that in the intermediate situation, part of the array has been already sorted, and that none of the remaining elements are smaller than any of the elements in the sorted part. This is illustrated in Figure 9.

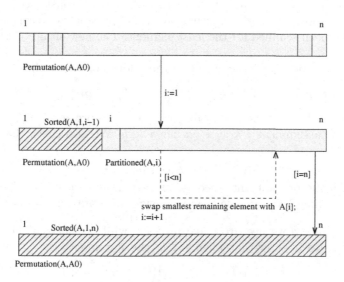

Fig. 9. Sorting with invariant

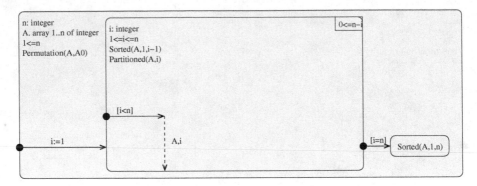

Fig. 10. Intermediate situation

The intermediate situation can be characterized by adding the following conditions to the initial situation

$$i : integer \land 1 \leq i \leq n \land Sorted(A, 1, i - 1) \land Partitioned(A, i)$$

This gives us the invariant diagram in Figure 10.

It is quite easy to check that the initial assignment $i := 1$ will establish this intermediate situation. It is also easy to check that the condition $i = n$ will imply the final situation. Hence, the only thing that remains is to figure out how to make progress towards this condition (termination function $n - i$) while maintaining the intermediate situation. This is to be done by only changing A and i.

Finding smallest element. Finding the smallest remaining element indicates that we need to scan over all the remaining elements, so we obviously need a loop here also. We add a fourth innermost situation, where part of the unsorted elements have already been scanned for the least element. The new situation is shown in Figure 11.

The new situation is characterized by the additional constraints

$$k, j : integer \land i < n \land i \leq k \leq j \leq n \land A[k] = min\{A[h] | i \leq h \leq j\}$$

We check that this situation is established from the previous intermediate situation by the assignment $j, k := i, i$ when $i < n$. We also check that if $j = n$, then

$$A[i], A[k] := A[k], A[i]; \ i := i + 1$$

will establish the previous intermediate situation, as indicated in the diagram. The corresponding invariant diagram is shown in Figure 12.

Final step. Finally, we need to make progress while preserving the second invariant. We need to show that when $j < n$, the statement

$$j := j + 1; \ if \ A[j] < A[k] \ then \ k := j \ fi$$

preserves the second invariant. This is also easily checked.

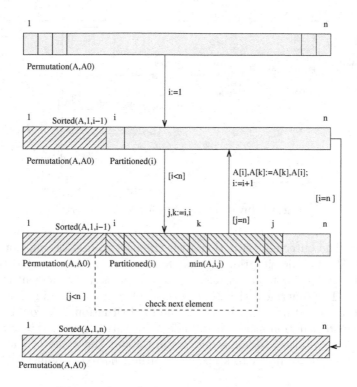

Fig. 11. Sorting program with two invariants

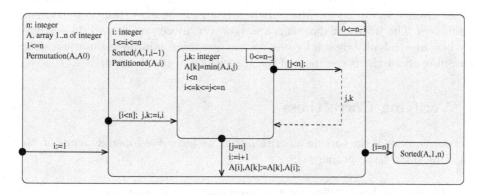

Fig. 12. Second invariant

The inner loop will eventually terminate because $n - j$ is decreased but is bounded from below. The outer loop will terminate because $n - i$ is decreased and is bounded from below.

This concludes our derivation of the sorting algorithm. Figure 13 shows the invariant diagram for the final program that we have derived.

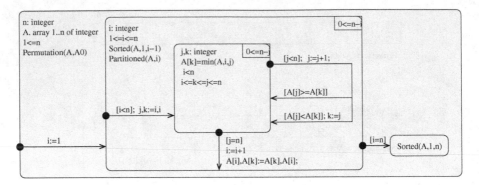

Fig. 13. Invariant diagram for selection sort

Discussion. The outermost situation gives the background assumptions for the algorithm. The middle situation is what holds when we have sorted the array up to $i-1$, but have not yet started to scan for the smallest element in the rest of the array. The innermost situation holds while we are scanning for the least element. The second nested situation is the final situation. We could also have nested the final situation inside the middle invariant. However, that would have indicated that we also had some information about the value of i at exit. As this is not needed, we prefer to keep this invariant nested only inside the initial situation.

We used figures quite extensively in the derivation of the invariant diagram. The figures allow us to read out the logical formulation of the invariants in a rather straightforward way. Most of the design of the algorithm is done using the figures. The invariant diagrams are, however, more precise and do not have any bias due to badly chosen figures, so they are more suitable for mathematical reasoning about the correctness of the transitions in the diagram.

4 Verifying Correctness

The correctness of the sorting algorithm that we have developed depends on the following three basic properties:

- The diagram is *consistent*, in the sense that the situations are preserved by the transitions,
- the diagram is *terminating*, in the sense that there are no infinite execution paths, and
- the diagram is *non-blocking*, in the sense that execution will eventually reach a final situation.

Let us consider these issues in more detail. Consistency is the central property that we require. It means that all the verification conditions are satisfied. The verification conditions can be constructed in different ways, using backward propagation (weakest preconditions) or forward propagation (strongest postconditions). A quite intuitive way is to take the diagram and turn all assignments into single assignment statement, by introducing a new primed version of a program variables whose value is changed. Thus, x stands for the original value of the program variable, x' stands for the value after the first assignment to x, x'' stands for the value after the second assignment to x, and so on. Figure 14 shows the selection sort diagram adapted in this way.

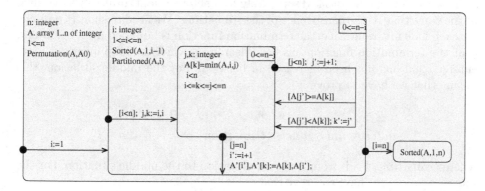

Fig. 14. Single assignment statements

Consider as an example the transition from innermost invariant to the middle invariant. The verification condition that we need to prove is shown in Figure 15.

$$
\begin{aligned}
\text{(outer)} \quad & n \in integer \wedge A \in array\ 1..n\ of\ integer \wedge Permutation(A, A_0) \wedge 1 \leq n \\
\text{(mid)} \quad & \wedge\ i \in integer \wedge 1 \leq i \leq n \wedge Sorted(A, 1, i-1) \wedge Partitioned(A, i) \\
\text{(inner)} \quad & \wedge\ j, i \in integer \wedge A[k] = min(A, i, j) \wedge i < n \wedge i \leq k \leq j \leq n \\
\text{(trans.)} \quad & \wedge\ j = n \wedge i' = i + 1 \wedge A'[i'] = A[k] \wedge A'[k] = A[i'] \wedge \\
& \wedge\ (\forall h \neq k, i' \cdot A'[h] = A[h]) \\
& \Rightarrow \\
\text{(outer)} \quad & n \in integer \wedge A' \in array\ 1..n\ of\ integer \wedge Permutation(A', A_0) \wedge 1 \leq n \\
\text{(mid)} \quad & \wedge\ i' \in integer \wedge 1 \leq i' \leq n \wedge Sorted(A', 1, i'-1) \wedge Partitioned(A', i)
\end{aligned}
$$

Fig. 15. Verification condition

Note that line 5 in the verification condition is the frame condition for array assignment, i.e., array elements that are not explicitly changed remain the same.

In this case, there is quite a lot to be proved. However, proving each fact at a time is a very efficient way of finding errors, either in the transition or in the situations. The amount of checking that has to be done can be decreased quite drastically by omitting those facts that cannot be wrong. Thus, the transition from the inner situation back to itself in case $A[j] \geq A[k]$ requires us to only prove the following facts under the given assumptions:

$$j' \in integer \wedge A[k] = min(A, i, j') \wedge k \leq j' \leq n$$

In principle we should prove that all facts in $outer \wedge mid \wedge inner$ hold after the transition, but the other facts all state properties about program variables that are unchanged by this transition.

Termination can be checked separately for each invariant, or it can be verified at the same time as the consistency of the transitions. We need to show two things for each loop invariant: that the termination function is bounded from below and that the termination function has been decreased when the loop invariant is re-entered, and has not been increased in between. For the middle situation, this means that we have to prove

$$outer \wedge mid \Rightarrow 0 \leq n - i$$
$$outer \wedge mid \wedge inner \wedge transition_1 \Rightarrow n - i' < n - i$$

where $transition_1$ leads from the inner situation to the middle situation. For the inner situation, we have to prove that

$$outer \wedge mid \wedge inner \Rightarrow 0 \leq n - j$$
$$outer \wedge mid \wedge inner \wedge transition_2 \Rightarrow n - j' < n - j$$

where $transition_2$ leads from the inner situation back to the inner situation. In addition, we need to check that the termination function $n - i$ is not increased in any other transition in the diagram, and similarly for the termination function $n - j$.

Finally, we also need to show that the execution does not get blocked at an intermediate situation in the diagram. In our case, we need to show that the final situation $Sorted(A, 1, n)$ is eventually reached. Execution can get stuck at an intermediate situation if there is no enabled transition out of the situation. There are two transitions from the inner situation, one which is enabled when $j < n$ and the other which is enabled when $j = n$. Similarly, for the middle situations, there are two transitions, one enabled when $i < n$ and the other enabled when $i = n$. Hence, we need to prove that

$$outer \wedge mid \wedge inner \Rightarrow j < n \vee j = n$$
$$outer \wedge mid \Rightarrow i < n \vee i = n$$

Both of these statements are easily seen to be true.

In fact, we could also have a transition that blocks in the middle of its execution, if we reach a branching point where none of the guards are enabled. In our example, the transition from the inner situation back to the inner situation has a conditional branching point, but the two guards together exhaust all possibilities $(A[i] < A[k] \vee A[i] \geq A[k])$, so execution will always proceed past this point.

The diagram has been designed so that the verification condition can be read directly from the diagram. Correctness can thus be checked by careful inspection of the diagram. In manual proofs, it is easy to add dashes to the program variables that are updated in a transition, either by hand in the diagram, or just mentally. In a computer supported environment, the single assignment version of the diagram can be seen as just another view of the diagram.

5   Invariant Based Programs vs Ordinary Programs

Let us finally consider some of the main differences between invariant based programming and programming as it is ordinarily carried out.

Correctness notion. There is a difference in philosophy between ordinary programs and invariant based programs, in particular concerning the notion of correctness. The traditional notion of correctness states that if a program is started in an initial state that satisfies the precondition of the program, then the program must terminate in a final state that satisfies the postcondition of the program.

For an invariant based program, this requirement is strengthened, because an initial state can be in any situation. In particular, this means that we require that correctness also holds for initial states in interior situations that may not even be reachable by a legal execution from a state in the given initial situation. The correctness of an invariant based program is also stronger in that it is not sufficient that the final situations are satisfied upon termination, all interior situations must also be satisfied whenever execution reaches them.

This means that we can have an invariant based program that is correct in the traditional sense, but not correct as an invariant based program. The program does compute the intended input-output relation, but the invariants are chosen badly, so that consistency does not hold. We can in this case either argue that the program should be considered correct, because it does solve the given problem, or that it should be considered incorrect, because it does not solve the problem in the way the programmer intended the problem to be solved. The programmers intentions are written down in the form of the invariants of the program.

The correctness requirement for invariant based programs is in a way stronger than needed. However, as soon as a program has at least one loop, an inductive argument is needed for the correctness proof. If this argument takes the form of a proof with invariant assertions, one will in fact end up establishing the correctness of the program as an invariant based program. In other words, the usual proof technique establishes a stronger correctness property for a program than what is required by the traditional notion of correctness. This is a consequence of program verification just being an inductive proof, and it is quite common for the induction hypothesis to be stronger than the lemma that we are proving.

Establishing correctness as a side effect. Standard programming methods produce code that needs to be tested and debugged. It encourages a trial and error approach to program construction, and can only achieve higher reliability through a considerable amount of effort spent on testing. Verification can in principle be done a posteriori, but this is usually not done. The reasons for this are partly psychological: the problem has already been solved (kind of), there is already something that works. The additional verification effort is quite laborious, and the market situation is such that one can get away with a reasonable number of errors in the code.

With invariant based programming, we get the invariants for free, as part of the programming process. Checking that the invariants are preserved by the code is done continuously, as part of the program construction process. Writing down the invariants does require an additional effort, but on the other hand, it helps in understanding and solving the programming problem and leads to much higher reliability of the program code.

Role of figures. The work flow for constructing invariant based programming is based on using figures as stepping stones to formulate the invariants. Any programmer who is solving a problem like the one above will (or at least should) draw the kind of figures we have shown, to get a feeling for how the program behaves. These figures are preserved as the invariants in invariant based programming. In ordinary programming, they are usually lost in subsequent development steps. By elevating these figures to a more distinguished position in program construction, we are more likely to preserve them and keep them up to date. The role of figures here is similar to the role of figures in physics and engineering: it establishes the central constants and variables that are needed to express the problem and the solution, and shows how they are related to each other.

Smaller grain of modularity. Invariant based programming provides a grain of modularity smaller than procedures, the usual smallest units of modularity in imperative programming languages. We can *reason* about invariant based programs in a very local fashion. We can check each situation and its transitions one by one, ignoring other situations for the moment. This locality of reasoning is the payoff of using the stronger notion of correctness for invariant based programs.

We also *construct* an invariant based program in a local fashion, as illustrated in the example above. The situations are introduced one by one, as are the transitions that connect the situations to each other. Similarly, we *change and fix* an invariant based program in a local fashion.

Only termination requires an global view of the program: one needs to check that each possible cycle in the diagram decreases some termination function.

Programming methodology. The minimum requirement for an invariant based program is that it is consistent. The program does not have to terminate or be free of deadlocks. There may be deadlocks in the program because we have not (yet) covered all possible cases for some internal situation, i.e., there are cases for which we still need to provide transitions. However, the program constructed thus far is consistent (although incomplete). Similarly, we may have a consistent program, but we have yet to tackle the termination of the program, which may require redefinition of some invariants or transitions.

An invariant based program is constructed by a sequence of successive development steps, where each step preserves the consistency of the previous step, while deadlock freedom and termination may vary during development. Each step will add, modify or remove some situations or transitions in the diagram.

This approach requires that we carefully check the consistency of each transition when it is introduced. Leaving the consistency checks to a later stage in program development will only accumulate errors in the program and make the consistency checking very laborious. It will also decrease the motivation for carrying out consistency checks at all, because too many interdependent things then need to be considered and changed. Consistency checks can be done at different levels of rigor, but a good rule of thumb is to use the same rigor as one would use for checking a normal mathematical lemma.

6   Conclusions and Work in Progress

We have in this paper argued for an alternative approach to constructing simple algorithms, where we start by constructing the preconditions, postconditions and intermediate invariants of the program (the situations) before writing any code. The program code is then constructed in the form of transitions that allow us to move from one situation to another, and checking that the invariants are preserved by these transitions. This allows us to construct a program and its correctness proof at the same time, in a sequence of successive consistency preserving enhancements and modifications to the program. We have argued that the careful use of figures makes it quite straightforward to find the right invariants for the program, once a basic understanding of how the algorithm should work is at hand. We have provided a diagrammatic notation (nested invariant diagrams) that provides an intuitive way of describing invariant based programs, and have shown how to structure a program using nested invariants in this way. The (nested) invariant diagram provides a concise and compact overview of the whole program, and shows the basic structure of the program.

We are presently working on extending this approach in a number of different directions. The notations and formalisms presented here are very simple, and are only intended to convey a basic understanding of invariant based programming. We are presently extending the approach to larger units of construction, like procedures, data modules, classes and processes.

We have tried out the invariant based programming on a collection of classical first year CS programming course algorithms, where it seems to work quite well. We have also worked on a collection of more demanding pointer manipulation algorithms, including list manipulation and marking algorithms. We are presently working on a somewhat larger system, with a collection of 10 - 12 classes (a text editor), to see how well this approach works in an object oriented framework. The experiences from these experiments are very good so far. Tool support is highly desirable, but it is not necessary, one can get quite a lot of work done manually.

We are also building toop support for the approach. The *Socos environment* supports invariant based programing by providing a graphical diagram editor for nested invariant diagrams. It computes the verification conditions automatically for all transitions, and sends them to either an automatic prover (presently *Simplify* [14]) or to an interactive prover (presently *PVS* [15]). Those assertions that could not be verified automatically are shown on demand, as a list of possible semantic errors in the program. An item on this list can indicate one of three things: the invariants are not quite correct, or the transition is not quite correct, or then both are correct but the automatic verifier was too weak to prove the assertion. This approach to checking invariant based programs is described in more detail in [1].The constructed program can also be expressed in an existing programming language (presently *Python* [18]), and executed automatically. We provide a debugging mode for execution, where the invariant assertions are checked during run time, and those assertions which are false are highlighted.

Another issue that we are studying is teaching invariant based programming to novices. Our experiences with teaching small groups indicates that the method is as easy to learn for complete novices as it is for experts in programming methodology. The main obstacle for novices is how to express the invariants in a logical language and how to prove rigorously that the verification conditions are correct. However, this is essentially background knowledge that programmers would need to learn anyway, in order to understand and develop software systems, and it can be taught as a course on practical logic in a CS curriculum.

A good test of how well these ideas work in practice would be to teach invariant based programming as a first programming course in high school. This course could be seen as a mathematics or a physics course (but simpler than the usual math and physics course). The course would teach the basic principles for constructing programs that work correctly. The students in such a course can convince themselves that their programs work correctly, and get the satisfaction of building something solid, rather than the uneasy feeling of undetected errors that they get with todays trial and error approach to programming.

References

1. Ralph Back and Magnus Myreen. Tool support for invariant based programming. In *Proceedings of the 12th Asia-Pacific Software Engineering Conference*, Taipei, Taiwan, December 2005.
2. Ralph-Johan Back. Program construction by situation analysis. Research Report 6, Computing Centre, University of Helsinki, Helsinki, Finland, 1978.
3. Ralph-Johan Back. Exception handling with multi-exit statements. In H. J. Hoffmann, editor, *6th Fachtagung Programmiersprachen und Programmentwicklungen*, volume 25 of *Informatik Fachberichte*, pages 71–82, Darmstadt, 1980. Springer-Verlag.
4. Ralph-Johan Back. Invariant based programs and their correctness. In W. Biermann, G Guiho, and Y Kodratoff, editors, *Automatic Program Construction Techniques*, number 223-242. MacMillan Publishing Company, 1983.

18   R.-J. Back

5. Mike Barnett, K. Rustan M. Leino, and Wolfram Schulte. The spec-sharp programming system: An overview. In *CASSIS 2004 Proceedings*, 2004.
6. Lilian Burdy, Yoonsik Cheon, David R. Cok, Michael D. Ernst, Joseph R. Kiniry, Gary T. Leavens, K. Rustan M. Leino, and Erik Poll. An overview of jml tools and applications. *Software Tools for Technology Transfer*, 7(3), June 2005.
7. E. W. Dijkstra. A constructive approach to the problem of program correctness. *BIT*, 8:174 – 186, 1968.
8. E. W. Dijkstra. Notes on structured programming. In Ole-Johan Dahl, C.A.R Hoare, and E.W Dijkstra, editors, *Structured Programming*. Academic Press, New York, 1972.
9. E. W. Dijkstra. *A Discipline of Programming*. Prentice-Hall, 1976.
10. Martin Fowler. *UML Distilled*. Addison Wesley, 1999.
11. D. Harel. State charts: a visual formalism for complex systems. *Science of Computer Programming*, 8:231–274, 1987.
12. E. Hehner. Do considered od: a contribution to the programming calculus. *Acta Informatica*, 11:287 – 304, 1979.
13. K. Rustan M Leino and Gregg Nelson. An extended static checker for modula-3. In *Proceedings of the 7th International Conference on Compiler Construction*, volume 1383 of *Lecture Notes in Computer Science*, pages 302–305, April 1998.
14. Greg Nelson. *Techniques for Program Verification*. PhD thesis, Stanford University, 1980.
15. Sam Owre, Natarajan Shankar, and John Rushby. Pvs: A prototype verification system. In *CADE 11*, Saratoga Springs, NY, 1992.
16. J. C. Reynolds. Programming with transition diagrams. In D. Gries, editor, *Programming Methodology*. Springer Verlag, Berlin, 1978.
17. M. H. van Emden. Programming with verification conditions. *IEEE Transactions on Software Engineering*, SE-5, 1979.
18. Guido Van Rossum and Fred L. Jr. Drake. *The Python Tutorial - An Introduction to Python*. Network Theory Ltd., 2003.

On the Integration of UML and Petri Nets in Software Development*

Javier Campos and José Merseguer

Departamento de Informática e Ingeniería de Sistemas
Universidad de Zaragoza
C/María de Luna, 1, 50018 Zaragoza, Spain
{jcampos, jmerse}@unizar.es

Abstract. Software performance engineering deals with the consideration of quantitative analysis of the behaviour of software systems from
the early development phases in the life cycle. This paper summarizes
in a semiformal and illustrative way our proposal for a suitable software
performance engineering process. We try to integrate in a very pragmatic
approach the usual object oriented methodology —supported with UML
language and widespread CASE tools— with a performance modelling
formalism, namely stochastic Petri nets. A simple case study is used to
describe the whole process. More technical details should be looked up
in the cited bibliography.

1 Introduction

The design and implementation of complex systems is a difficult engineering
task. In the last years the modelling, validation, performance evaluation and
implementation of such systems has been usually tackled with the help of formal
models.

Petri nets (PNs) [1] is an adequate formal paradigm to support the whole life-
cycle engineering of a complex discrete event system. They have been used for the
modelling and evaluation of flexible manufacturing systems [2], multiprocessor
architectures [3], communication systems [4], and also for the writing of reliable
and efficient concurrent programs [5].

Software systems are complex systems, probably the most complex construction tasks that humans undertake. Functional requirements of software applications are obviously important, but they are not the only concern. Performance
objectives are also important: the degree to which a software system meets its
objectives for timeliness is important in many cases and it is critical in some
real-time applications.

Being software engineering a relatively young discipline, the importance of the
use of well established methodologies, even formal methods, languages and tools
has already been detected and assumed. Unfortunately, performance objectives

---

* This work has been developed within the projects TIC2003-05226 of the Spanish Ministry of Science and Technology and IBE2005-TEC-10 of the University of
Zaragoza.

S. Donatelli and P.S. Thiagarajan (Eds.): ICATPN 2006, LNCS 4024, pp. 19–36, 2006.

are still not usually included in first stages of the software life cycle. Being performance requirements critical to the success of today's software systems, many final software products fail to meet those requirements. The usual practice, as nicely summarized in [6], consists on *make it run, then make it run right, and finally make it run fast.* But this practice is in many cases too expensive, because fixing performance problems can oblige to modify the initial design.

Just to illustrate the previous statement with a single famous crisis, we recall the Denver airport baggage system story. It was planned for the United Airlines terminal, but during the development it was enlarged to support all the airport terminals but without considering the new system's workload. As a result of the inadequate performance characteristics of the system, the airport was opened 16 months later with a loss of hundreds of million dollars [7].

Therefore, many researchers defend the principle that performance should be included in the software design process from the very beginning. This principle is one of the main goals of the *International Workshop on Software and Performance*, a forum for researchers interested in the intersection of software engineering and performance evaluation that started in 1998 and has already held five editions. The research field that deals with the goal of building software with predictable performance by specifying and analysing quantitative behaviour from the early development phases of a system throughout its entire life cycle has been coined with the term *Software Performance Engineering* (SPE) [8].

The Unified Modelling Language (UML) [9] combined with an object oriented methodology, such as [10], is nowadays the most widely used approach in the software engineering community. Thus, most of Computer Aided Software Engineering (CASE) tools support OO methodology and use UML as the design language.

Since performance goals are not included in the usual practice of software engineers, we think that there exists a need of integrating performance modelling with the existing software development methodologies and tools. Markov models, queueing networks, stochastic process algebras and stochastic PNs are probably the best studied performance modelling paradigms. Among all of them, we bet on PNs due to its special adequacy to model parallel and distributed systems, its mathematical simplicity, its modelling generality, its adequacy for expressing all basic semantics of concurrency, its locality both of states and actions, its graphical representation, its well-developed qualitative and quantitative analysis techniques, and the existence of analysis tools.

This paper tries to present in a semiformal and illustrative way, some of our experience in the process of integration of performance modelling within software development process. In section 2 we summarize the main phases of a suitable software performance engineering process based on UML and stochastic PNs. Section 3 is devoted to a more detailed presentation of the process by means of a simple case study. A basic client for checking mail from a server using POP3 protocol is described. The system is modelled using UML language. In particular, use case diagram, statecharts diagrams, sequence diagram, activity diagram and the deployment diagram are considered. Each diagram is annotated with timing information according to the UML *Profile for Schedulability, Performance and Time* [11] of the Object Management Group (OMG). For each of the annotated UML diagram, the highlights of the translation into a corresponding stochastic PN is presented. Also some examples of performance figures that can be derived from the analysable model are presented. The paper ends with some concluding remarks in section 4.

2   Software Performance Process

Several works have been proposed to combine UML and a performance modelling formalism to analyze quantitative aspects of software systems. All of them share some basic principles. Then it could be argued that there exists a widely accepted *process* among the SPE community:

- The behaviour and architecture of the system is described by a (set of) UML diagram(s), that make up the *system design*.
- This UML design is annotated according to a standard OMG profile. Then gaining the *annotated design*.
- The annotated design is converted into a *performance modelling formalism*.
- A qualitative analysis will be carried out, if the formal model allows it.
- The formal model is analyzed using quantitative analysis techniques already developed for the target formalism.

Now, we reveal some key aspects of the process.

Use cases (UCs) use to be the starting point to describe the system behaviour. They are used to specify the requirements of a system, subsystem or class and for the specification of their functionality. Sequence diagrams (SDs) or activity diagrams (ADs) are the common choice to detail UCs. SDs specify a set of partially ordered messages, each one defining a communication mechanism as well as the roles to be played by the sender and the receiver. Thus SDs represent patterns of interaction between objects. ADs represent internal control flow of processes. SDs and ADs are useful to conduct performance-based scenario analysis. Few approaches introduce the statecharts (SCs) to complement them. SCs, as ADs, are used to describe the behaviour of a model element. The software architecture completes the *system design* by means of the deployment diagram (DD), then modeling the distribution of the software components in the hardware platform.

Workloads, utilizations, response times or throughput characterize the performance view of the system design, then gaining an *annotated design*. The UML profile for schedulability, performance and time specification (UML-SPT) [11] is the widely used OMG standard to annotate them. The OMG QoS profile [12] is also used for these purposes. Actually, the convergence of both profiles is expected.

Three paradigms have been mainly used as *performance modeling formalisms* in the SPE process: Stochastic Petri nets (SPNs), stochastics process algebras (SPA) and (layered) queuing networks (LQN). Several methods have been proposed to translate the UML syntax into the syntax of the target formalism.

Among the translations using SPNs as target are [13, 14, 15], using SPA [16], using QN [17, 18] and LQN [19, 20]. Such methods use different approches and technologies: Customization Rules [21], XSL transformations [19], algebra-based transformations [22], direct formalization [23] or Graph Grammars [20]. For a survey of such translations see [24].

Tools are essential for SPE. The preferred approach is to integrate performance aspects into existing CASE tools. Figure 1 depicts the OMG framework to develop performance evaluation tools following the *SPE process*.

During the last decades researchers developed techniques to analyze performance models. They were implemented in analysis tools [25, 26, 27] that the engineer uses to analyze the resulting models. Moreover these analysis tools are used as the core of the CASE tools [28, 29, 30, 31, 32] that follow the proposal in Figure 1.

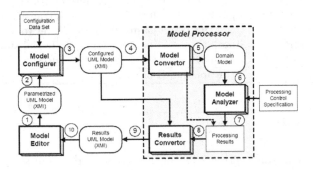

Fig. 1. OMG architectural framework for SPE tools

The authors have developed some work in the SPE field following the process so far outlined. Now, we recall how each process step is considered in the proposal here recalled:

- Concerning the UML diagrams, we consider: Use cases, sequence diagrams, state machines, activity diagrams and the deployment diagram [33].
- They are annotated according to the UML-SPT. Table 1 summarizes the subset of annotations currently supported.
- SPNs are the target of the translation methods proposed in [34] for SMs, in [35] for SDs and in [36] for ADs.
- ArgoSPE [37] provides tool support for the proposal.

This proposal is illustrated, in section 3, through the development of a case study. It is worth to notice that today a big SPE challenge is about how to integrate the bunch of UML diagramatic notations and the performance formalisms. So, our approach has evolved and now fits in the PUMA framework [32]. It defines an intermediate performance model, the core scenario model (CSM) [38], to encompass this challenge. Other works complement the PUMA approach [39].

Table 1. Performance annotations

Stereotype	Tag	Diagram	Kind*; Type	Comment
PAcontext	-	SD,SC,AD	-;-	Performance model
PAclosedLoad	PArespTime	SD	pred.;time-value	Execution time
		SC		Object lifetime
		AD		Execution time
	PApopulation	SC	-; nat-number	Number of instances
PAstep	PAdemand	SD,SC,AD	assm.; time-value	Msg. transm. delay
		SC,AD		Activity duration
	PArespTime	SC	pred.;time-value	State exec. time
		AD		Trans. exec. time
	PAprob	SD	-; real	Condit. msg; Branch
		SC,AD		Guarded transition
	PArep	SD	-; nat-number	Iterated msg
	PAsize	SD,SC,AD	assm.; nat-numb.	Msg size
	PAthroughput	SC	pred.; -	Transition throughput
PAcommun.	PAspeed	DD	assm.; Kbytes	Network speed
PAinitialCond.	PAinitialState	SC	bool	Initial state
GRMcode	-	DD	-	Resident classes
(*)predicted or assumed				

3 Case Study

The process outlined in section 2 has been applied in different software domains. For example in [40] to test the design of a mobile agents system or in [41] to study the QoS of fault tolerant systems distributed over Internet. In this paper, we recall a case study [42] that was useful to show the feasibility of the proposal. Now, we come back to that work to illustrate some aspects of the SPE process.

The case study models a basic mail client. Here we will focus in the first Use Case (UC) showed in Figure 2: checking mail from a server using the POP3 [43] protocol.

The behavior of the referred client is rather intuitive for this UC. The SC in Figure 3 depicts the behaviour of a mail client, it can be useful to get a first view of the general system behaviour. The client tries to establish a TCP connection with the server via port 110. If it succeeds (reception of a greeting message), both (client and server) begin the authentication (authorization) phase. The client sends the username and his/her password through a USER and PASS

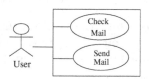

Fig. 2. Use Case view of the 'mail client' model

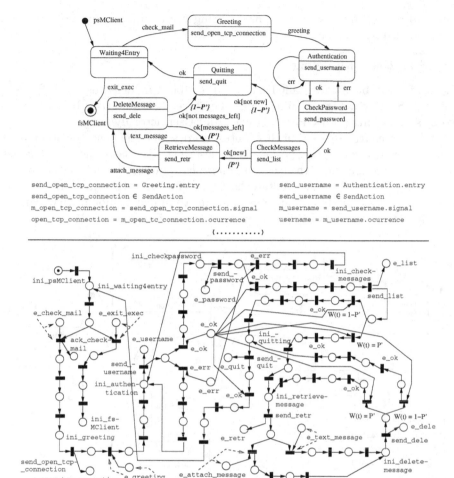

send_open_tcp_connection = Greeting.entry
send_open_tcp_connection ∈ SendAction
m_open_tcp_connection = send_open_tcp_connection.signal
open_tcp_connection = m_open_tc_connection.ocurrence

send_username = Authentication.entry
send_username ∈ SendAction
m_username = send_username.signal
username = m_username.ocurrence

(............)

Fig. 3. Statechart and resulting LGSPN for the dynamics of the class ClientHost

command combination. For the sake of simplicity, usage of the APOP command has not been contemplated here.

If the server has answered with a positive status indicator ("+OK") to both messages, then the POP3 session enters the transaction state (phase). Otherwise (e.g., the password doesn't match the one specified for the username), it returns to the beginning of the authorization phase.

In the transaction phase, the client checks for new mail using the LIST command. If there is any, the client obtains every e-mail by means of the RETR and DELE commands. It must be noted that, for simplicity, potential errors have not been considered here; thus, no negative status messages ("-ERR") are modelled.

Once every e-mail has been downloaded, the mail client issues a QUIT command to end the interaction. This provokes the POP3 server to enter the update state and release any resource acquired during the transaction phase. The protocol is ended with a goodbye (”+OK”) message.

3.1   Statechart Modeling

The mail client behavior (MailClient class) has been modeled using an SC, see Figure 3. Similarly, Figure 4 illustrates the server host and user behaviors via two SCs describing the POP3Server class and User actor dynamics.

UML statecharts are used in this context to describe the behavior of a model element, concretely classes. Specifically, it describes possible sequences of states and actions through which the element can proceed during its lifetime as a result of reacting to discrete events. A statechart maps into a UML state machine that differs from classical Harel state machines in a number of points that can be found in [44]. Studies about their semantics can be found in [34, 45].

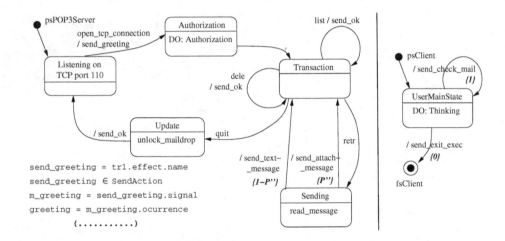

Fig. 4. Statecharts for the dynamics of the classes ServerHost and User (actor)

A state in a statechart diagram is a condition during the life of an object or an interaction during which it satisfies some condition, performs some action, or waits for some event. See state Waiting4Entry in Figure 3, where the mail client is waiting either for the event check_mail or the event exit_exec. An event is a noteworthy occurrence that may trigger a state transition [44]. A simple transition is a relationship between two states indicating that an object in the first state will enter the second state.

The approach taken to translate a statechart into an Labelled Generalized Stochastic Petri Net [46] (LGSPN) consists in the translation of each element in the metamodel of the state machines into an LGSPN subsystem. Just to take the flavor of the translation, Figure 5 depicts the LGSPNs subsystems obtained by the translation of a simple state.

An important aspect of the translation concerns with the activities. Since they consum processing time, then they will be translated as timed transitions. The

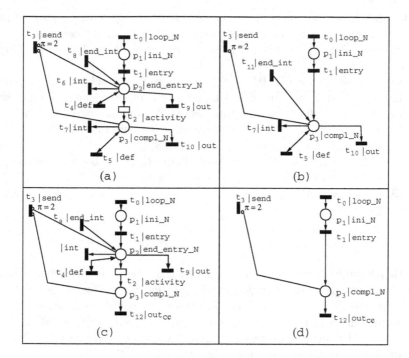

Fig. 5. Different labelled "basic" systems for a simple state N: (a) with activity and no immediate outgoing transition; (b) no activity and no immediate outgoing transition; (c) with activity and immediate outgoing transition; (d) no activity and with immediate outgoing transition

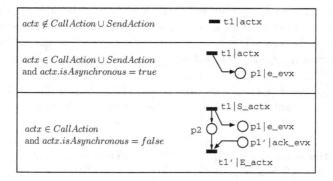

Fig. 6. Translation of the different types of actions

delay of such transitions, i.e. the mean of a exponentially distributed ramdom variable, is calculated as the inverse of such processing time. The translation of the actions is given in Figure 6. The translation of the other elements in the metamodel can be found in [23, 34].

Not-event-driven decisions are modelled using guards with its success probability. Concretely, a combination of guards and events has been used in some SC transitions. That probability will be represented in the Petri net using an immediate transition.

In order to obtain an LGSPN ($\mathcal{LS}_{sc}$) for a given statechart sc, the subsystems that represent its metamodel elements are composed using a composition operator defined in [46]. Basically, this operator composes two LGSPNs into a third one by fusing the places and transitions with equal labels. The details of the composition method can be found also in [23, 34].

Figure 3 shows the LGSPN for the MailClient SC. It has been obtained applying the transformation method for SCs given in [23] and briefly recalled here. The LGSPNs for the ServerHost and the User are not given to avoid a cumbersome presentation.

3.2   Activity Diagram

An in-depth study of our mail client showed that the activity Authenticate, associated to the state Authorization in the SC for ServerHost (Figure 4), was relevant to the system performance. Moreover, it would be necessary a more accurate modelling of its behaviour to complete the system description. To fill this gap, the SPE process proposes to model the actions performed within the states of the SC, concretely by using the activity diagram (AD).

Thus, we used an AD (see Figure 7) to model the behaviour of the Authenticate activity. Although it may be more useful in cases where there is not such a strong external event dependence (e.g., 'internal' operations). The activity could have been described extending the SC but, in general, ADs provide some additional expresiveness for certain tasks.

Activity diagrams (ADs) represent UML activity graphs and are just a variant of UML state machines (see [44]), in fact, a UML activity graph is a specialization of a UML state machine. The main goal of ADs is to stress the internal control flow of a process in contrast to statecharts, which are often driven by external events.

Considering the fact that ADs are suitable for internal flow process modeling, they are relevant to describe activities performed by the system, usually expressed in the statechart as *doActivities* in the states. The AD will be annotated with the information to model routing rates and the duration of the basic actions. See the annotations PAprob and PArespTime in Table 1.

Moreover, the AD will be annotated with the size of the messages, PAsize tag, when it models event-driven behaviour. The closed load will be modelled attaching to the initial state the PApopulation tag. Table 1 summarizes the annotations proposed for the AD.

According to the UML specification, most of the states in an AD should be an action or subactivity state, so most of the transitions should be triggered by the ending of the execution of the entry action or activity associated to the state. Since UML is not strict at this point, then the elements from the SMs package could occasionally be used with the interpretation given in [34].

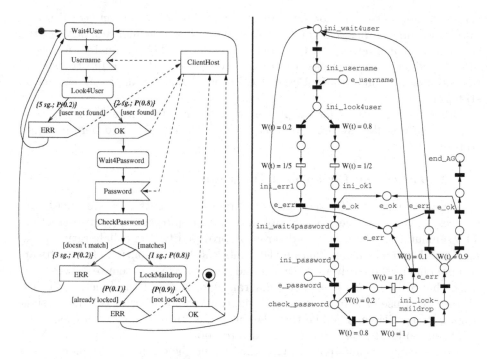

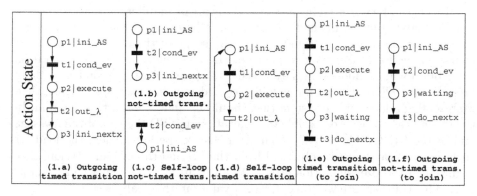

Fig. 7. Activity Diagram for POP3ServerHost::Authenticate and resulting LGSPN

Fig. 8. Translation of the Action States in ADs

As far as this issue is concerned, our decision is not to allow other states than action, subactivity or call states, and thus to process external events just by means of call states and control icons involving signals, i.e. signal sendings and signal receipts. As a result of this, events are always deferred (as any event is always deferred in an action state), so an activity will not ever be interrupted when it is described by an AD.

The performance model obtained from an AD in terms of LGSPNs as proposed in [36] can be used with performance evaluation purposes with two goals:

A) just to obtain some performance measures of the model element they describe or B) to compose this performance model with the performance models of the statecharts that use the activity modeled in order to obtain a final performance model of the system described by the referred statecharts. The full translation process can be found in [36], Figure 8 offers its flavour by depicting the action states translation.

Other interpretations for the AD propose it as a high level modeling tool, that of the workflow systems, but the SPE approach presented here does not consider this role.

3.3   Sequence and Deployment Diagrams

A sequence diagram (SD) describes a communication pattern performed by instances playing the roles to accomplish a specific purpose, i.e. an interaction. The semantics of the sequence diagram is provided by the collaboration package [44]. The deployment diagram (DD) models the distribution of the software components in the hardware platform and the operative system resources.

In this SPE process a sequence diagram should detail the functionality expressed by a use case in the use case diagram, by focusing in the interactions among its participants. While the DD will be useful to deploy each class (that has been modeled with a SC) in a hardware node.

From the performance point of view the SD and the DD have relevant elements and constructions. They will be annotated, according to Table 1. In the following these elements and constructs are explained.

Objects can be executed in the same hardware or in different ones in the case of distributed systems. In the first case, it can be assumed that the time spent to send the message is not significant in the scope of the modeled system. Of course the actions taken as a response of the message can spend computation time, but it will be modeled in the statechart diagram. For the second case, those messages *travelling through the net*, it is considered that they spend time, then representing a load for the system that should be annotated in this diagram. In the second case, it is also possible to annotate in the SD the size of the message and in the DD the bit rate at which the network operates. Then, the load for each message can be easily obtained from the SD and the DD annotations. See Figure 9 for the size annotation. Figure 10 annotates the bit rate with a variable $TR. This transfer rate will take different values in the analysis step.

A *condition* can be attached to each message in the diagram, representing the possibility that the message could be dispatched. Even multiple messages can leave a single point each one labeled by a condition. From the performance point of view it can be considered that routing rates are attached to the messages.

A set of messages can be dispatched multiple times if they are enclosed and marked as an *iteration*. This construction also has its implications from the performance point of view.

In [35], the translation process of a given sequence diagram *sd* into its corresponding LGSPN $\mathcal{LS}_{sd}$ can be found.

Finally, we use SDs to obtain performance analytical measures in a certain context of execution. Figure 9 shows an example of interaction between both server and client. Some results for this particular scenario will be obtained in the next subsection.

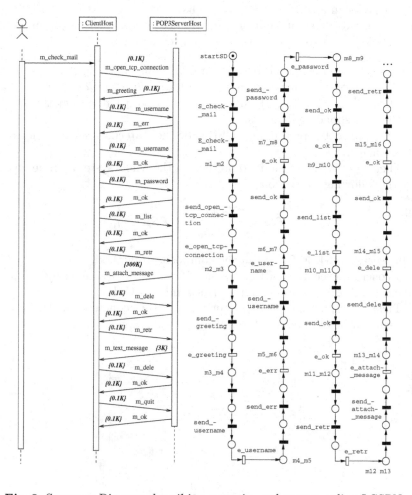

Fig. 9. Sequence Diagram describing scenario, and corresponding LGSPN

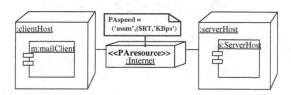

Fig. 10. Deployment Diagram

3.4 Analysis

Once the final LGSPN models are obtained (following the composition rules detailed in [34, 35]) performance estimates can be computed. These figures can be related to either the whole system behavior (somehow unrestricted) or the system behavior in a concrete scenario (thus adjusted to certain restrictions).

Therefore, two different kind of performance models can be obtained using the translations surveyed in the previous subsections.

A. Supose a system described by a set of UML statecharts $\{sc_1, \ldots, sc_k\}$ and a set of activity diagrams refining some of their *doActivities*, $\{ad_1, \ldots, ad_l\}$.

Then $\{\mathcal{LS}_{sc_1}, \ldots, \mathcal{LS}_{sc_k}\}$ and $\{\mathcal{LS}_{ad_1}, \ldots, \mathcal{LS}_{ad_l}\}$ represent the LGSPNs of the corresponding diagrams.

$\mathcal{LS}_{sc_i-ad_j}$ will represent an LGSPN for the statechart sc_i and the activity diagram ad_j.

Then a performance model representing the whole system can be obtained by the following expression:

$$\mathcal{LS} = \overset{i=1,\ldots,k j=1,\ldots,l}{\underset{Labels}{||}} \mathcal{LS}_{sc_i-ad_j}$$

The works in [34, 36, 23] detail the composition method.

B. If a concrete execution of the system $\mathcal{LS}$ in [A] is described by a sequence diagram sd, $\mathcal{LS}_{sd}$ represents its corresponding LGSPN.

Then a performance model representing this concrete execution of the system can be obtained by the following expression:

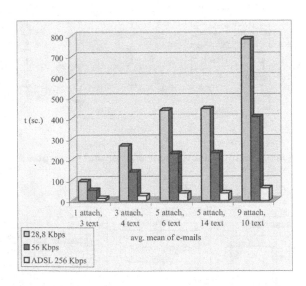

Fig. 11. Effect of number of mails and attached file proportion on the downloading time for different connection speed

$$\mathcal{LS}_{execution} = \mathcal{LS} \underset{Labels}{||} \mathcal{LS}_{sd}$$

The works in [35, 47] detail the composition method.

Figure 11 shows the effect of the number of mails and different proportion of them with attached files on the downloading time for different connection speeds. Bars labelled with "a attach - b text" correspond with the case where the average number of downloaded mails is $a + b$ and an average of a of them have attached files.

The graph in Figure 12 represents the effective transfer rate of the client when checking mail (maximum transfer rate: 56 Kbps). Note that higher amounts of data minimize the relative amount of time spent by protocol messages. The analysis has been taken considering the whole system behavior (that is, using the net obtained by composition of the ones corresponding to the SCs and the AD).

Meanwhile, the graph in Figure 13 represents the time cost of executing the interaction illustrated in figure 9 in function of different attach file sizes and maximum network speeds. The analysis has been taken using the SD to construct the net for the constrained case [35]. In general, SDs can be extremely useful to check the behaviour of the system for a particular use case. Moreover, analysts may use them to model test conditions in an easy way.

In another paper included in this volume [37], ArgoSPE tool is presented. It implements the features explained here for the software performance modelling process. So, the system is modeled as a set of UML diagrams, annotated according to the UML-SPT, which are translated into LGSPN. The UML diagrams used to obtain a performance model by means of ArgoSPE are those considered in our process: statecharts, activity diagrams and sequence diagrams. The class and the implementation diagrams (components and deployment) are used to collect some system parameters (system population or network speed).

ArgoSPE has been implemented as a set of Java modules, that are plugged into the open source ArgoUML CASE tool [48]. It follows the software architecture

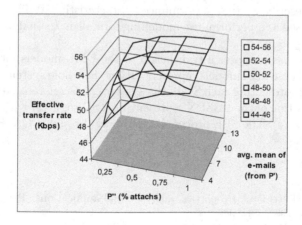

Fig. 12. Effective transfer rate of the client when checking mail

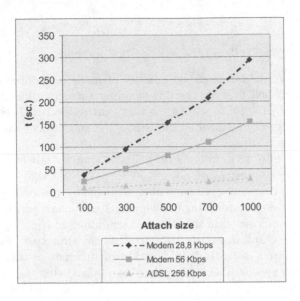

Fig. 13. Some analytical results for the presented case study

proposed in the UML-SPT, see Figure 1. ArgoSPE has been used to model and analyze software fault tolerant systems [41] and mobile agents software [40].

4   Conclusions

This paper has presented the application of a SPE process through a case study. Such SPE process has as relevant features the use of some UML diagrams and stochastic Petri nets. The UML behavioral diagrams (statecharts, activity and sequence) togheter with the deployment diagram allow to model system functionality and describe system performance characteristics. While the stochastic Petri nets are used as a performance modelling formalism to analyze quantitative aspects of the system.

The SPE process proposes a translation of the UML models into the performance model. These final models represent either the whole system (composition of the LGSPN representing each SC in the system) or a concrete execution of the system (composition of the LGSPN representing a SD togheter with the LGSPNs of the involved SCs). The SPE process has support through a CASE tool.

References

1. Murata, T.: Petri nets: Properties, analysis, and applications. Proceedings of the IEEE **77**(4) (1989) 541–580
2. DiCesare, F., Harhalakis, G., Proth, J.M., Silva, M., Vernadat, F.: Practice of Petri Nets in Manufacturing. Chapman & Hall, London (1993)

3. Ajmone Marsan, M., Balbo, G., Conte, G.: Performance Models of Multiprocessor Systems. MIT Press, Cambridge, Massachussetts (1986)
4. Billington, J., Diaz, M., Rozenberg, G., eds.: Application of Petri Nets to Communication Networks. Number 1605 in Lecture Notes in Computer Science, Advances in Petri Nets. Springer-Verlag (1999)
5. Ajmone Marsan, M., Balbo, G., Conte, G., Donatelli, S., Franceschinis, G.: Modelling with Generalized Stochastic Petri Nets. John Wiley Series in Parallel Computing - Chichester (1995)
6. Smith, C., Williams, L.: Software Performance Engineering. In: UML for Real: Design of Embedded Real-Time Systems. Kluwer (2003)
7. Gibbs, W.: Trends in computing: Software's chronic crisis. Scientific American **271**(3) (1994) 72–81
8. Smith, C.U.: Performance Engineering of Software Systems. The Sei Series in Software Engineering. Addison–Wesley (1990)
9. Booch, G., Jacobson, I., Rumbaugh, J.: The Unified Modeling Language. Addison Wesley (1999)
10. Jacobson, I., Christenson, M., Jhonsson, P., Overgaard, G.: Object-Oriented Software Engineering: A Use Case Driven Approach. Addison-Wesley (1992)
11. Object Management Group http:/www.omg.org: UML Profile for Schedulabibity, Performance and Time Specification. (2005)
12. Object Management Group http:/www.omg.org: UML Profile for Modeling Quality of Service and Fault Tolerant Characteristics and Mechanisms. (2004)
13. Bondavalli, A., Dal Cin, M., Latella, D., Majzik, I., Pataricza, A., Savoia, G.: Dependability analysis in the early phases of UML-based system design. International Journal of Computer Systems Science & Engineering **16**(5) (2001) 265–275
14. Pettit IV, R., Gomaa, H.: Modeling Behavioral Patterns of Concurrent Software Architectures Using Petri Nets. In: 4th Working IEEE / IFIP Conference on Software Architecture (WICSA), Oslo, Norway, IEEE Computer Society (2004) 57–68
15. Saldhana, J., Shatz, S.: UML Diagrams to Object Petri Net Models: An Approach for Modeling and Analysis . In: Twelfth International Conference on Software Engineering and Knowledge Engineering, Chicago, IL, USA, Knowledge Systems Institute (2000) 103–110
16. Canevet, C., Gilmore, S., Hillston, J., Kloul, L., Stevens, P.: Analysing UML 2.0 activity diagrams in the software performance engineering process. In: Proceedings of the 4th International Workshop on Software Performance (WOSP'04), Redwood Shores, California, USA, ACM (2004) 74–78
17. Cortellessa, V., Mirandola, R.: Deriving a queueing network based performance model from UML diagrams. In: Proceedings of the Second International Workshop on Software and Performance (WOSP2000), Ottawa, Canada, ACM (2000) 58–70
18. Smith, C.U., Williams, L.: Performance Solutions. Addison–Wesley (2002)
19. Gu, G., Petriu, D.: XSLT transformation from UML models to LQN performance models. In: Proceedings of the Third International Workshop on Software and Performance (WOSP2002), Rome, Italy, ACM (2002) 25–34
20. Petriu, D., Shen, H.: Applying the UML performance profile: Graph grammar-based derivation of LQN models from UML specifications. In Field, T., Harrison, P.G., Bradley, J., Harder, U., eds.: Computer Performance Evaluation, Modelling Techniques and Tools 12th International Conference, TOOLS 2002. Volume 2324 of Lecture Notes in Computer Science., London, UK, Springer (2002) 159–177

21. Baresi, L., Pezzè, M.: On formalizing UML with high-level Petri nets. In Agha, G., De Cindio, F., Rozenberg, G., eds.: Concurrent Object-Oriented Programming and Petri Nets. State of the Art. Volume 2001 of Advances in Petri Nets. Lecture Notes in Computer Science, (LNCS). Springer-Verlag, Heidelberg (2001) 276–304

22. Gu, G., Petriu, D.: From UML to LQN by XML algebra-based model transformations. In: Proceedings of the Fifth International Workshop on Software and Performance (WOSP2005), Palma, Spain, ACM (2005) 99–110

23. Merseguer, J.: Software Performance Engineering based on UML and Petri nets. PhD thesis, University of Zaragoza, Spain (2003)

24. Balsamo, S., Di Marco, A., Inverardi, P., Simeoni, M.: Model-based performance prediction in software development: a survey. IEEE Transactions on Software Engineering 30(5) (2004) 295–310

25. The GreatSPN tool (http://www.di.unito.it/~greatspn)

26. The TimeNET tool (http://pdv.cs.tu-berlin.de/~timenet/)

27. The Möbius Tool (http://www.mobius.uiuc.edu/)

28. ArgoSPE: (http://argospe.tigris.org)

29. Argo Performance: (http://argoperformance.tigris.org)

30. Cortellessa, V., Gentile, M., Pizzuti, M.: Xprit: An xml-based tool to translate uml diagrams into execution graphs and queueing networks. In: 1st International Conference on Quantitative Evaluation of Systems (QEST 2004). (2004) 342–343

31. Marzolla, M., Balsamo, S.: UML-PSI: the UML Performance SImulator. (In: 1st International Conference on Quantitative Evaluation of Systems (QEST 2004)) 340–341

32. Woodside, M., Petriu, D., Petriu, D., Shen, H., Israr, T., Merseguer, J.: Performance by unified model analysis (PUMA). In: Fifth International Workshop on Software and Performance (WOSP'05), Palma, Spain, ACM (2005) 1–12

33. Merseguer, J., Campos, J.: Exploring roles for the UML diagrams in software performance engineering. In: Proceedings of the 2003 International Conference on Software Engineering Research and Practice (SERP'03), Las Vegas, Nevada, USA, CSREA Press (2003) 43–47

34. Merseguer, J., Bernardi, S., Campos, J., Donatelli, S.: A compositional semantics for UML state machines aimed at performance evaluation. In Giua, A., Silva, M., eds.: Proceedings of the 6th International Workshop on Discrete Event Systems, Zaragoza, Spain, IEEE Computer Society Press (2002) 295–302

35. Bernardi, S., Donatelli, S., Merseguer, J.: From UML sequence diagrams and statecharts to analysable Petri net models. In: Proceedings of the Third International Workshop on Software and Performance (WOSP2002), Rome, Italy, ACM (2002) 35–45

36. López-Grao, J., Merseguer, J., Campos, J.: From UML activity diagrams to stochastic Petri nets: Application to software performance engineering. In: Proceedings of the Fourth International Workshop on Software and Performance (WOSP'04), Redwood City, California, USA, ACM (2004) 25–36

37. Gómez Martínez, E., Merseguer, J.: ArgoSPE: Model-based software performance engineering. In: 27th International Conference on Application and Theory of Petri Nets and Other Models Of Councurrency, LNCS (2006) In this volume.

38. Petriu, D., Woodside, M.: A metamodel for generating performance models from uml designs. In: Proc. UML 2004. Volume 3273 of LNCS., Lisbon, Portugal, Springer-Verlag (2004) 41–53

39. Grassi, V., Mirandola, R., Sabetta, A.: From design to analysis models: a kernel language for performance and reliability analysis of component-based systems. In: Proceedings of the Fifth International Workshop on Software and Performance (WOSP'05). (2005) 25–36

40. Merseguer, J., Campos, J., Mena, E.: Analysing internet software retrieval systems: Modeling and performance comparison. Wireless Networks: The Journal of Mobile Communication, Computation and Information **9**(3) (2003) 223–238

41. Bernardi, S., Merseguer, J.: QoS assesment of fault tolerant applications via stochastics analysis. IEEE Internet Computing (2006) To appear.

42. López-Grao, J., Merseguer, J., Campos, J.: Performance engineering based on UML and SPNs: A software performance tool. In: Proceedings of the Seventeenth International Symposium On Computer and Information Sciences (ISCIS XVII), Orlando, Florida, USA, CRC Press (2002) 405–409

43. Myers, J., Rose, M.: RFC 1725: Post Office Protocol - version 3 (1994)

44. Object Management Group http:/www.omg.org: OMG Unified Modeling Language Specification. (2003) version 1.5.

45. Domínguez, E., Rubio, A., Zapata, M.: Dynamic semantics of UML state machines: A metamodelling perspective. Journal of Database Management **13** (2002) 20–38

46. Donatelli, S., Franceschinis, G.: PSR Methodology: integrating hardware and software models. In Billington, J., Reisig, W., eds.: Proceedings of the 17th International Conference on Application and Theory of Petri Nets. Volume 1091 of Lecture Notes in Computer Science., Osaka, Japan, Springer (1996) 133–152

47. Bernardi, S.: Building Stochastic Petri Net models for the verification of complex software systems. PhD thesis, Dipartimento di Informatica, Università di Torino (2003)

48. ArgoUML project: (http://argouml.tigris.org)

Component Tools: Integrating Petri Nets with Other Formal Methods

Ekkart Kindler, Vladimir Rubin, and Robert Wagner

Department of Computer Science, University of Paderborn, Germany
{kindler, vroubine, wagner}@upb.de

Abstract. The field of *formal methods* provides all kinds of powerful techniques for the specification, design, verification, validation, and ramp-up of systems. Petri nets, in different versions and "levels", are among those techniques that have successfully been used in various fields of application and for different engineering tasks.

During the full development process of a system, different engineers have different tasks. Each engineering task has its specific purpose and, therefore, some version of Petri net or some formal method is better suited for this task than others. In order to take full advantage of formal methods, engineers need to switch between different techniques.

Unfortunately, different techniques have different underlying formalisms and notations, they use different concepts and methods, and they are supported by different and, in many cases, incompatible tools. Therefore, most applications of formal methods are restricted to one technique, formalism, and tool – though using several techniques in combination would have many benefits.

The *Component Tools* project aims at easing the application and the integration of several formal methods with different underlying formalisms, notations, and tools for all kinds of application areas. To this end, Component Tools supports the definition of components with different underlying formal models in different notations and on different levels of abstraction for different purposes. Moreover, Component Tools supports the definition of transformations to different tools. This way, an engineer can use these components for designing, verifying, and validating a system with support from formal methods and their tools under a uniform graphical user interface – without even knowing the details of the underlying techniques and formalisms.

Initially, Component Tools was inspired by the use of different versions of Petri nets during the development process. But, it turned out that the concepts of Component Tools are much more general and can be used for integrating other formal methods. In this paper, we outline the basic idea, the concepts, and the main ingredients of Component Tools by using a simplified example from the area of flexible manufacturing systems, where the models are Petri nets on different levels of abstraction.

Keywords: Petri nets, formal methods, integration, components, transformation, system engineering.

S. Donatelli and P.S. Thiagarajan (Eds.): ICATPN 2006, LNCS 4024, pp. 37–56, 2006.
© Springer-Verlag Berlin Heidelberg 2006

1   Introduction

The field of *formal methods* provides quite powerful techniques for the specification, design, implementation, verification, validation, and ramp-up of all kinds
of systems. Each technique has its strengths and weaknesses and is better suited
for some engineering tasks than another. There is no single technique best suited
for all tasks during the design and development of a system. In order to take full
advantage of formal methods, a system engineer must apply different techniques
from the field of formal methods, which, typically, use different formalisms, notations, and tools – let alone different principles and philosophies. Since engineers
are, typically, not experts in all necessary techniques of formal methods, they
cannot switch back and forth between different formalisms. Even if they can,
it will require modelling the same system in different notations over and over
again, which is expensive, time consuming and error prone. Therefore, the high
potential of formal methods is never fully exploited.

In order to improve this situation, *Component Tools* supports the definition
of components for a particular application area along with a set of models and
transformations to external tools that support the different engineering tasks in
that area. We call such a definition a *component library*. Component Tools along
with a component library definition, implements a tool with a uniform graphical
user interface for system engineering in a chosen application area; this *component
tool* supports the chosen formal methods for all engineering tasks in the chosen
application area. The basic idea is to define *components* that are equipped with
ports. Different *instances* of such components can be *connected* at these ports.
This way, an engineer can construct a system such as a manufacturing system
from these components. Along with each component definition, there will be one
or more models of the behaviour of this component, which can be in different
notations and on different levels of abstraction or can cover different aspects of
the behaviour of the component. In addition, there will be *transformations* that
define how the models of the different instances are combined into an overall
model in order to apply a particular formal method. On this model, an *external
tool* for the particular formal method can be started and the results of that tool
can be transformed back to the visual representation of Component Tools. In
order to support the bidirectional transformation, we use *triple graph grammars*
(TGGs) [18].

In addition to the system models, many formal methods need input or provide
output that specifies legal or illegal behaviour. In many cases, this behaviour
refers to executions or runs of the models. Unfortunately, each formalism comes
with its own version of runs, which makes it difficult to exchange and to compare
them among different tools. In order to cope with this problem, Component
Tools defines a general concept of a run, which we call a *scenario*: A scenario
uses the concepts from the component library only; therefore, it is independent
from the underlying formalisms. A scenario can be transformed back and forth
between runs of the different formalisms. This way, runs resp. scenarios can
be exchanged among different tools and different tasks during the development
process.

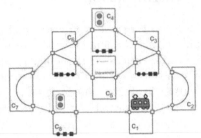

Fig. 1. The computer controlled toy-train   **Fig. 2.** The construction plan

In this paper, we discuss the basic idea and the concepts of Component Tools by the help of a simple toy-train example, which is introduced in Sect. 2 – as a simplified version of an application from the area of flexible manufacturing systems. By the help of this example, we explain how a combination of different techniques could be used during the development and ramp-up of a flexible manufacturing system. The definition of components, the component library, is discussed in Sect. 3, and the transformations as well as the integration of tools for particular design or development tasks is discussed in Sect. 4.

2   The Example

In this section, we briefly introduce our example of a computer-controlled toy-train. In our research project, this toy-train serves as a simplified version of a case study from the area of flexible manufacturing systems. Figure 1 shows a photograph of the setup where the toy-train is controlled by a Petri net [11].

Here, this toy-train serves as a running example for explaining the purpose, the concepts, and the use of Component Tools. But, Component Tools is much more general and can be used for many other application areas – actually Component Tool's focus is on the definition of component libraries and tools for all kinds of application areas that use prefabricated components.

2.1   The Construction Plan

Figure 2 shows a simple construction plan for our toy-train example, in which only a part of the system from Fig. 1 is used. The construction plan consists of four different kinds of *components*: *straight tracks* (c_1, c_5), *curved tracks* (c_2, c_7), *switches* (c_3, c_6), and *signal units* (c_4, c_8). These components are combined by *connections*, which are indicated by arcs between the corresponding *ports*. There are one-to-one connections from white square ports to white circle ports, which represent the mechanical joints of the tracks of the toy-train.

In our toy-train example, we assume that the trains are running all the time and can be stopped only at the signal units by switching the signal units to stop. This setup is inspired by a mono-rail system from flexible manufacturing

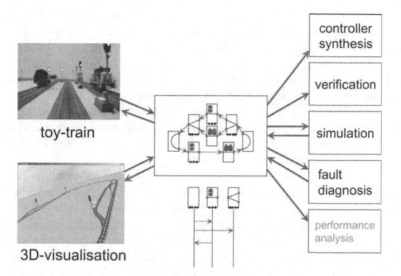

Fig. 3. Engineering tasks

systems, which is called *Montrac*. In order to control the route of the trains, the switch components and the signal units are equipped with *actuators*: additional ports for changing the direction of the switch and for changing the signal states between stop and go. These ports are shown as black squares at the corresponding components. They correspond to electric plugs, which can be connected to some controller, which is not shown in Fig. 2. Moreover, there are *sensors* for sensing the current direction of a switch and for sensing the presence of a train at some signal unit. The ports corresponding to these sensors are represented as black circles.

2.2 Some Engineering Tasks

One engineering task during the development of a flexible manufacturing system is to design the construction plan of the *plant*. Though Component Tools supports this task by a particular view (see Sect. 3.5), we do not deal with this task here; rather, we assume that the construction plan is given already. The other engineering tasks start from that point.

Figure 3 gives an overview on some tasks. The first task, could be to design a *controller* for the plant: the specification could be the states of the components that should be avoided, which we call *error states*. Given the construction plan and the error states of the components, a *controller synthesis* algorithm could construct a controller for this plant – ensuring that the plant will never end up in an error state. This controller could then by refined and extended by the engineer. Once, the controller is finished, the engineer could *verify* the correctness of the controller by some model checker. And he could *simulate* and *animate* the behaviour in a 3D-visualisation.

Once, he is sure that the plant and controller behave correctly, he could start building the plant and begin the ramp-up phase. In this phase, some of the used hardware components might have some faults, resulting in unexpected behaviour – even though the controller was verified and behaves correctly. The reason is that the hardware does not behave correctly because of some faults. This could be detected by some *fault diagnosis* tool. To this end, Component Tools could record the behaviour of the plant and use it as input for the fault diagnosis. Based on the recorded failure behaviour and some fault models, the fault diagnosis algorithm can identify the faulty component – again by using techniques from model checking.

Once the faulty components are replaced by non-faulty components the engineer might gradually increase the work-load of the plant. In order to further improve the throughput, he might use some techniques for *performance* analysis.

Currently, tools for supporting the above design tasks from Component Tools are being implemented in bachelor and master theses. Here, we cannot go into the details of these techniques. The only important observation is that the different tasks need different models and need to be on different levels of abstraction. For using model checking, the models need to be quite simple due to the state explosion problem. For the simulation, the models need to be equipped with further information on the form and shape of the objects. For fault-analysis, the models need to be equipped with further behaviour, which occurs when certain faults are present. And, for performance analysis, there needs to be some timing information.

Moreover, there needs to be some interaction between different tools. For example, the fault diagnosis needs the information on the observed behaviour of the plant, which is recorded by Component Tools. This can be represented as a *scenario*. This scenario can be passed to the fault-diagnosis tool, which extends this scenario so that it includes the position at which a fault occurs. This scenario can then be animated by the 3D-visualisation for presenting the cause of a fault to the engineer (see Sect. 2.5 for more details).

2.3   The Models

In order to support these and other tasks, there must be some models for the components and, in particular, for the dynamic behaviour of the components. Component Tools is completely independent from any formalism, notation, or syntax and from any formal method. In our example, we use Petri nets for a very simple model of the dynamic behaviour of the components. Figure 4 shows the simple behaviour models for the signal unit and the switch units, where the shaded squares indicate Petri net transitions that can be controlled by an attached controller. With these models, we can simulate the system and by applying some formal methods, we could synthesise a simple controller for our toy-train (plant).

Actually, there can be many more models for each component, which serve different purposes and can be used in different tasks during the development, implementation, and ramp-up process of the plant and the controller. For example,

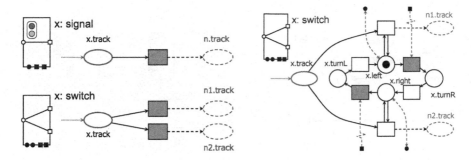

Fig. 4. Components: Simple models   **Fig. 5.** Components: Detailed model

there could be models that define the physical shape of the trains, the tracks and slightly extended Petri net models in order to animate the behaviour in a 3D-visualisation. This simulation and visualisation could help in the validation of the controller. Another example are more detailed models of the behaviour that cover all kinds of intermediate states so that a more detailed controller can be designed avoiding all kinds of illegal intermediate states. An example of such a more detailed Petri net model of a switch is shown in Fig. 5.

In order to support fault diagnosis during the ramp-up process of a new plant, there can be even more detailed models that also cover possible faults in the hardware components. These fault-models can be used by some techniques from model checking for detecting hardware faults from observed failure behaviour.

In this paper, we cannot not go into the details of the underlying formal methods and the used formalisms and notations that are employed for the different development tasks. The crucial point, here, is that we need different models on different levels of abstraction in, possibly, different formalisms and notations, which support different design tasks.

2.4   The Transformations

During the design and implementation of a new plant and controller, there are many different tasks in different stages of the development process that can be supported by some technique from the field of formal methods. Component Tools supports the definition of such tasks and the integration of tools supporting these tasks. To this end, the different models of the components will be used and transformed into formalisms that are understood by some tool supporting this formal method, and the results can be transformed back to the construction plan in order to visualise them to the engineer.

For example, from the simple Petri net models shown in Fig. 4, similar models for the other components, and the construction plan from Fig. 2, we could generate the Petri net model shown in Fig. 6. Note that these transformations are not programmed, but are defined in a declarative way by *triple graph grammars* (TGGs), which will be discussed in more detail in Sect. 4. The main benefit is that TGGs support a back-and-forth translation, which easily allows results obtained by one method to be used by others. For example, a counter-example,

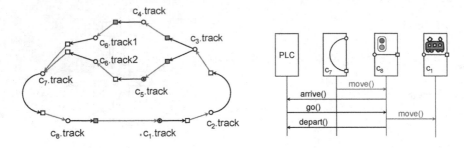

Fig. 6. A generated model   **Fig. 7.** A scenario

i. e. an execution that shows that a particular property is not valid, obtained by some model checker, could be passed to a tool that animates this execution in a 3D-environment. This way, an engineer virtually sees what goes wrong. In combination with a fault diagnosis tool, this could be also used to visualise the possible hardware faults in a 3D-model of the plant.

2.5   The Scenarios

Unfortunately, each formal method or notation has its own notion of runs or executions. Component Tools defines a concept of run in terms of the objects present in the construction plan: a *scenario* shows some components and their interaction along their connections. Figure 7 shows an example of such a scenario: In this scenario, a train moves from the track component c_7 to the signal component c_8; then a sensor at the signal component informs the controller (PLC) on this arrival, the controller, switches the signal to go, and the train starts moving again; by the sensor of the signal component, the controller is informed when the train leaves the signal component.

Since scenarios use the concepts from the construction plan only, the runs from the different underlying models can be transformed back to a scenario. This way, runs can be exchanged among different tasks. For example, a faulty behaviour can be recorded, sent to the fault diagnosis task; the fault diagnosis identifies the faulty component and adds the position where the fault occurs for the first time. Then, this trace can be visualised to the engineer.

3   Component Libraries

In order to use Component Tools in some new application area with some prefabricated components, it needs some input.

The most important input is the definition of a *component library*: the main constituents of a component library are the definitions of different *components*. Then, a construction plan is built from *instances* of these components which are joined via *connections* at the *ports*.

The other input for Component Tools is the definition of the engineering *tasks* and *transformations* to the used formalism.

In this section, we will discuss the component library and the different concepts that need to be defined. The definition of tasks will be discussed in Sect. 4.

3.1 Ports

Since the same port type can occur in different components, ports need to be defined first. A component library defines any number of ports, each of which corresponds to a physical, electrical or mechanical jack or joint. Note that, actually, the defined ports are *port types* – but we do not dwell on this issue here.

In addition to a unique *name*, each port type defines its particular *appearance*, i. e. its shape, line colour and fill colour. In our example, we used white circles and squares for the mechanical connections of tracks, and we used black squares and circles for electrical jacks. Moreover, each definition of a port type has a *description* of this particular jack or joint that informally defines its function and purpose. For example, a description could refer to its technical specification.

Clearly, the information on the appearance will be exploited when editing and displaying the components. The description of a port type can be used as a tool tip when the cursor is over such a port in order to help the engineer understand the purpose of this particular port type.

3.2 Connections

In a component library, there may be any number of *connections* resp. *connection types*. Again, each connection type has a unique *name* and an *appearance*, which defines the arrow heads and the colour of the arc, when connecting two ports. Again, there will be a *description* for this particular connection type.

More importantly, a component library comprises a *connection paradigm*, which defines how ports may be connected by particular connections and in which direction the connections may run. Moreover, the connection paradigm can define fan-in and fan-outs of connections in order to restrict the number of incoming and outgoing connections at some ports. In particular, the connection paradigm can restrict connections at ports to one-to-one connections.

Component Tools will use the connection paradigm in order to guarantee that the construction plan never violates any structural restrictions implied by the components in the particular application domain.

3.3 Components

The definition of *components* resp. *component types* is the most important part of a component library. Each component type definition consists of a unique *name*, a definition of its *appearance*, i. e. its size, fill and line colour, and a *description* of its purpose and behaviour.

Moreover, each component has a list of *port definitions*. Each port definition, again, consists of a unique *name*, a reference to a port type, a *description*, and a *position* of this particular port. Note that it is not necessary to define the possible connections to a port of a component because this is already defined for the port types by the connection paradigm.

In each component type definition, there can be a list of *parameter definitions*, which, again, consist of a *name* and some data type defining the range of values of this parameter. In our example, the component type straight track has a parameter "number", which defines the number of trains that are on that particular track. Moreover, there are parameters defining the start and end points of the track.

When a new instance of a component type is created in a construction plan, the values for all the parameters of this component type must be provided.

3.4 Models

In order to apply formal methods, each component is equipped with one or more models defining the behaviour or other aspects of the component (which, possibly, depends on the values of the parameters of the particular instance).

In order to deal with models, a component library defines a set of model types that are associated with each component type. Basically, this is a finite list of *names* equipped with a reference to some *format* in which these models are represented. And, again, there is a *description* of the purpose of this particular type of model. Then, each component type is equipped with all the models defined in the required representation.

These models will be used later for generating the models for the formal methods, which will be discussed in Sect. 4.

3.5 Views

Up to now, we have discussed the basic structure of a component library. Once all these concepts are defined in a component library, a *project editor* can be used for editing a construction plan from the defined components.

At last, we discuss a concept that supports different kinds of users involved in the same project. For example, there could be electrical engineers, mechanical engineers, and computer scientists working on the same project. All of them have different points of views and a different focus. In order to allow the different groups of users to focus on those aspects that are important for them, a component library may define different *views*.

Each view, basically, consists of a *name*, a *description* of its purpose or the class of intended users as well as a set of ports, connections, and components that are visible in that view. Moreover, each view can define a different appearance for the ports, connections, and components.

When an engineer selects a specific view on the project, he sees only the parts defined in that view. And the parts appear as defined for that particular view. The underlying construction plan, however, is the same for all engineers in order to avoid inconsistent views.

These different views can be all viewed and edited with the standard project editor of Component Tools. But, it is also possible to add a *dedicated editor* for a particular view[1]. For example, there could be an editor in which the physical

---

[1] We use Eclipse plug-ins and extension points for implementing these concepts. But, these implementational details are not discussed in this paper.

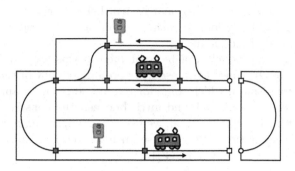

Fig. 8. The geometry view of the construction plan

extensions of the components are exactly visualised. In fact, the size and the position of the component instances defines the value of the corresponding parameters. Also the position of the ports could be changed and connectible ports could be connected by snapping to each other. Figure 8 shows how such a *geometry view* of the very same construction plan of our example could look like. Note, that this view does not show the ports for the electrical wiring; it shows the ports for the mechanical connections only, since it is the mechanical engineers view.

3.6 Extensions

Note that there are all kinds of conceptual refinements and extensions when defining component libraries, which are not yet implemented, but could be useful in order to properly support the development tasks. For example, components are displayed as rectangles in the current version. Future versions could support other freely definable shapes such as polygons or dedicated images; in some views there could be even some vertexes that could be freely moved in order to define some parameter. Moreover, there could be all kinds of icons associated with the different components.

Up to now, all components are defined in XML documents. Future versions of Component Tools could also support the user in defining his own components in a hierarchical way from components of the library. The concept itself is not difficult; still it is not yet implemented.

Up to now, there is only a very simple mechanism for defining the connection paradigms: a list of pairs that maybe connected by this connection along with a simple fan-in and fan-out. A future version could use a much more powerful mechanism; the exact mechanism that suits the needs of typical components and connections, however, has yet to be identified.

4   Tasks and Transformations

Up to this point, a component library can be used for building construction plans from its components and connections and different engineers can work on them in their favourite views. The models along with their description might be useful in understanding the behaviour of each component; but, strictly speaking, the models do not have a function at all.

The purpose of the models will be explained in this section. Basically, the models will be used for constructing some overall model out of the construction plan in some formalism on which some formal method can be applied. In order to support transformations back and forth between the construction plan and the tools supporting a formal method, we use *triple graph grammars* (*TGGs*) for defining the transformations [18]. We use an interpreter for actually executing the transformation defined by a TGG [13, 6].

4.1   Triple Graph Grammars

Here, we cannot discuss TGGs in full detail. Rather, we will explain the idea of TGGs by the help of our example.

Figure 9 shows a TGG rule that captures the translation of an instance of a signal unit to the corresponding Petri net. This graph resembles UML's object diagrams, where circle nodes are used only for emphasising the correspondence to the respective ports of the component and to the places of the Petri net. The Petri net model was shown in Fig. 4 already. Basically, the part below the dashed line shows an instance of the signal unit with its two ports called in and out on the left-hand side, and it shows the corresponding Petri net model on the right-hand side. In the middle part, it shows the *correspondence* or mapping of the elements of the component to the elements of the Petri net. For example, all elements of the Petri net correspond to the signal unit, whereas the place of the Petri net corresponds to the in port of the component, and the transition of the Petri net corresponds to the out port of the component. Minus syntactic sugar, the part below the dashed line of the TGG rule in Fig. 9 is a more formal presentation of the Petri net model corresponding to the component along with the mapping of the ports (cf. Fig. 4).

Next, we explain the meaning of Fig. 9 as a TGG rule. It represents a graph grammar rule. As all grammars, it consists of two parts: a left-hand side and a right-hand side, where the parts that do not change occur on both sides. The

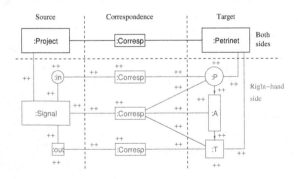

Fig. 9. TGG rule for a signal unit instance

top part above the dashed line of Fig. 9 belongs to both sides of the rule, and the part below the dashed line belongs to the right-hand side only. Such a rule means that once there is a graph as shown above the line, the part below the line can be added. This is the reason, why all edges and nodes in the lower part have a label ++. In fact, it is these labels which makes the nodes right-hand side nodes only. The horizontal dashed line is not part of the TGG rule. So, this rule, in combination with similar rules for the other components, can generate a set of component instances in the construction plan (called project) along with the corresponding Petri net models. Though the translation will not be executed this way, conceptually, we can assume that whenever a component is added to the project on the *source* side of the TGG, the corresponding Petri net elements will be generated on the *target* side along with correspondence objects that link the nodes on the source side with nodes on the target side. This way, the TGG rules define a transformation from a construction plan, i.e. a project on the source side, to a Petri net on the target side.

Up to this point, the connections have not yet been considered. In our simple example, we need only one TGG rule for dealing with the connections. This rule is shown in Fig. 10. Again, the new parts (the ones occurring on the right-hand side of a graph grammar only) in this rule are labelled with ++. This rule can be read in the following way: When a connection from some out port to some in port is inserted to the project, a Petri net arc is inserted to the Petri net between the transition and place corresponding to these ports.

Altogether, one TGG rule for each component and its model and the single rule for the connection precisely define a translation of the construction plan to a Petri net model. For example, the construction plan from Fig. 2 along with the TGG rules for the simple Petri net models would translate to the Petri net shown in Fig. 6 – except for the initial marking, which will be discussed shortly.

When explaining the TGG rules, we assumed that both sides, the construction plan and the Petri net, are created at the same time. But, this is not necessary; we used this point of view only for explaining the idea and the general principle of TGGs. In practise, we assume that we have a construction plan, and we try to map a part of the project with the source part of the TGG rule and than generate the missing part in the target part (Petri net) and the correspondence part. The idea of such an interpreter of TGGs was discussed in [13, 6].

As mentioned already, the above TGG rules do not yet generate an initial marking for our Petri net. The reasons is that we did not consider the parameters of the component that define the number of the trains on that component. Figure 11 shows a rule that transforms the parameter number to a marking of the corresponding place. With this rule added, we would obtain the Petri net from Fig. 6 for the construction plan from Fig. 2. But, this rule can do more. Suppose, some analysis or simulation tool changes the marking of a place in the Petri net. Then, we can apply the rule backward, and translate the changed marking back to the component instance and change the parameter number accordingly. In principle, we could even change the Petri net and change the construction plan

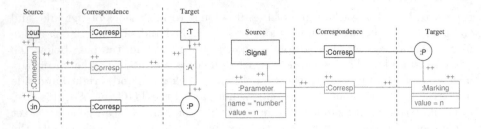

Fig. 10. TGG rule for a connection   **Fig. 11.** TGG rule for initial markings

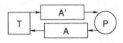

Fig. 12. Meta model for Petri nets

accordingly. But, since not all Petri nets correspond to a construction plan, we run in problems with more complicated changes.

Altogether, TGGs are an appropriate means for defining transformations among different models, and for actually performing these transformations in both directions. In order to make this work, we need to define the nodes and the associations between these nodes that may occur in the source model, in the target model, and in the correspondence part of the TGGs, which is a *meta model* of the source (the Component Tools project) and the target model (the Petri net). Figure 12 shows an example of the meta model for Petri nets, which resembles a UML class diagram. The node types occurring in the rules of a TGG actually refer to such meta models[2].

4.2   Tasks

In order to complete the definition of a component tool based on a component library, we need to define how the tool supports the different tasks of the engineer. We call this a *task definition*.

A task definition, basically, consists of two parts: a TGG transformation along with the corresponding meta models and an *external tool* that will be started on the transformed model. Technically, an external tool is a class implementing a tool interface with a method for passing the model to the external tool and a method to start the tool on that model. Even more technically, this class must be installed as a plug-in in the Eclipse platform, but we do not go into these details here.

The external tool could do anything on the transformed model; it could modify it, analyse it save it or start third party tools on it. Actually, we distinguish

---

[2] Here we discuss the conceptual part only. For technical reasons, we will need some mapping from Java objects implementing a meta model to the objects of the meta model. These *model adapters* will be discussed in Sect. 5.

two kinds of tasks. *Attached tasks* are those tasks that stay connected to the original model (via the TGG correspondences) and all modifications made on the transformed model will be transformed back to the construction plan and can be seen in the Component Tool project or geometry editor. *Detached tasks* do not stay connected to the original models once they are started.

Clearly, the attached tools are the more attractive ones since they show their results or effects in the construction plan of the engineer. So, the engineer does not need to adjust to a different formalism or notation for viewing the results. In some cases, the engineer might want to interact with the tool directly, in which case it might be used as a detached tool.

4.3   The Example Revisited

Altogether, a component library along with the transformations and task definitions defines a tool for a particular application area. With these definitions Component Tools implements a *component tool*: With this component tool, an engineers can use the components of the component library for editing construction plans and different engineers can have their particular view on these construction plans. The component tool will provide buttons for each defined task so that the engineers can start the corresponding transformations and external tools, and – for attached tasks – see the result in the construction plan. The different models and formalisms for each component can be inspected by the engineer, but it is not necessary for him to see them and to know the formalisms at all.

For our toy-train example, we have equipped the different components with simple Petri net models and the transformations discussed above. The external tool started on the transformed model is the *Petri Net Kernel* (PNK) [21] along with the PNVis tool [12], which simulates the Petri net and animates the behaviour in a 3D-visualisation as shown in Fig. 13. All the user must do for starting this visualisation is providing a construction plan as shown in Fig. 2 and then press the button of the corresponding task. What is more, while the 3D-animation is running the number of the corresponding shuttles at the particular instances of the components would change in the construction plan accordingly.

Other tasks, working on the very same construction plan, can be easily added. Right now, external tools for the controller synthesis, and for generating code for PLC controllers for flexible manufacturing systems, and fault analysis during the ramp-up process of such plants are being implemented in master thesis projects [17, 24].

5   Implementation and Research Issues

In the previous sections, we have discussed Component Tools on a conceptual level and from the perspective of potential users of Component Tools. Actually, there are two kinds of users, the first type would be someone defining a component tool for a specific application area using some specific formal methods; the other type would be a user of that component tool. In this section, we discuss

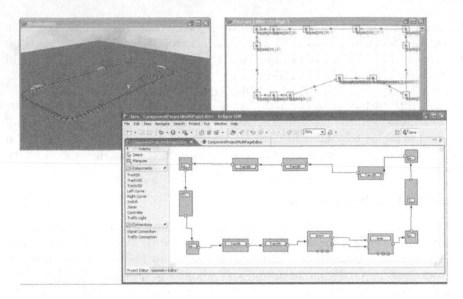

Fig. 13. PNVis running as a task of a component tool

some implementation issues of Component Tools and give an overview on the state of the implementation and some future extensions.

A first prototype of Component Tools was implemented based on the Eclipse platform [8]. This prototype covered the basic concepts of the component library, but did not cover transformations and did not support tasks. Basically, the implementation consisted of an editor for construction plans. The component library, its ports, connections, components was defined in some XML files.

In a one-year master's project with a group of ten students [6], this first prototype was extended into a more complete implementation. In particular, the concepts of views and dedicated editors for some views were added. Moreover, the concepts of transformations and tasks were implemented, which included an algorithm that interprets a TGG for transforming the corresponding models.

This TGG interpreter works on Java implementations of the underlying meta models. In order to access these Java implementations, the Java implementation must be mapped to the meta model of the TGGs. To this end, we developed the concept of an *model adapter* that must be implemented for each Java implementation of a particular model. With the help of these adapters, a TGG interpreter can transform the source model into the target model and vice versa [13].

The concept of scenarios was added in another bachelor thesis [16], which will serve as a basis for implementing a play-in/play-out concept similar to the one proposed by Harel and Marelly [10].

Altogether, the current implementation of Component Tools covers the example discussed in this paper. A first version of it will be made available in April 2006 – this version does not yet cover scenarios. A release of a version covering scenarios and play-in/play-out of scenarios is scheduled for July 2006.

Some interesting features, however, are still missing, which will be added in the near future by some bachelor and master theses:

Library and TGG editors

Up to now, the component library is defined by several XML files. For practical use, it is important to have a graphical editor for defining a component library and, in particular, the triple graph grammars defining the transformations. With such an editor, the component library could even be extended dynamically by user defined components and components could be defined from other components.

Visualising analysis information

Up to now, the result of the tools can be displayed in the construction plan only by changing some of its parameters. In order to make the visualisation more flexible, we need to develop a more sophisticated concept. For example, some ports, connections, or components could be high-lighted and equipped with some text explaining the results (see [7] for more details).

Generic adapter

Up to now, a new adapter must be implemented whenever a new Java implementation of a source or target model is needed for some new tool or component model, which needs some programming effort. The general idea of Component Tools is that defining a new component library, new transformations, and tasks does not require any (or at least not much) programming effort. Therefore, we are thinking of implementing a language for describing this mapping and implementing a *generic model adapter* that takes this description as a parameter and then serves as a model adapter for the described mapping.

An other idea, which is currently investigated, is a combination of TGGs with the Eclipse Modeling Framework (EMF); this will allow the use of TGGs with EMF models without any adapter [9].

Mapping scenarios to runs

As mentioned earlier, scenarios are formalism independent representations of runs. Up to now, the transformation of a run in some formalism to a scenario needs to implemented by hand; usually this is not difficult, but still needs some time, and one must make sure that the transformation is faithful. We believe that the transformation of the runs could specified jointly with the TGG transformations of the models, which would save some work and guarantee consistency between the two transformations. But, the exact concepts need to be worked out in detail.

In addition to the above ideas, there are many long-term ideas on how Component Tools could be equipped with "cool features". These, however, are beyond the scope of this paper.

Concerning theory, there is much work that needs to be done. Actually, Component Tools is only a framework for integrating different formal methods. The theory for the different techniques and, in particular, for integrating different models still needs much more attention. For example, we need methods that guarantee that the different models in different formalisms are consistent with each other. Or we need theory for combining models on different levels of abstraction. For example, we would like to combine models on different levels of abstraction in order to avoid the state explosion problem in model checking: One component of the construction plan that should be verified in detail could be represented with a detailed model, whereas all the other components could use a more abstract model. This way, the overall model has less states and could still be verified by model checking. Generating such a combined model is no problem. The question, however, is the validity of the results. This needs careful design of the models on the different levels of abstraction and theory that helps verifying the validity of the combined models – or at least sufficient conditions for the correctness of some verification results. Therefore, Component Tools is only a start for productively using and integrating theory still to come.

6   Related Work

In this section, we give an overview of the related work and tools and systems that inspired our work.

Tools. First, we discuss typical representatives of tools for the design, simulation, and visualisation of plants from components.

Tool suites such as *Simulink* of The MathWorks, $SimOffice^{TM}$ of MSCSoftware, $AutoMod^{TM}$ [2] and *eMPower* [19] are used for simulation and provide an environment for building the models using the libraries of components. They support modelling in different areas, such as airport industry, manufacturing constructions, and logistics. The *LONTROL* tool [5] is focused on control engineering for material handling, automation, and assembly logistics. It uses the library of components for building the models and functional components to control the system. The research tool $d^3FACT\ INSIGHT$ [4] is built to support the analysis of simulation models. The focus of this tool lies on the possibility to create and simulate complex models based on components in collaborative work. The SEA Environment [15] presents a methodology for the design, analysis, and simulation of embedded real-time systems using different modelling paradigms and tools.

All of the tools discussed above use components in a similar way to Component Tools. The difference is in the purpose. Basically, all of the above tools support simulation and visualisation. The focus of Component Tools is on analysis and verification and the integration of new formal methods.

Model Transformations. Due to the success of the Model-Driven Architecture (MDA) [14], model transformation is now the focus of many research activities. This leads to many different approaches for model transformation – each for a special purpose and within a particular domain with its own requirements. Here, we cannot give a complete discussion (see [3] for a survey).

Maybe, one of the best-known approaches for model transformation is XSLT [23]. It is used for the transformation of models represented as XML documents via the XMI specification. However, the description of the transformation is done textually in a highly procedural form. Hence, the specification of a transformation is not very user friendly.

Another class of transformation approaches comprises graphical transformation languages which are based on graph grammars and graph transformations. These approaches operate on graphs representing the data structures which have to be transformed. The transformation is executed by searching a pattern in the graph and applying an action which transforms the pattern to a new data structure. Examples for model transformation approaches based on graph grammars and graph transformation include VIATRA [20], GreAT [1], and UMLX [22]. Common to all mentioned approaches is that the transformation must be specified for each transformation direction separately. Hence, it is not well suited for the specification of bidirectional transformations as required by our approach. This is why we use TGGs [18].

7   Conclusion

In this paper, we discussed the basic ideas and concepts of Component Tools. The objective of Component Tools is to combine a bunch of different formal models, formal methods, and tools supporting them under a uniform graphical user interface, which defines a component tool. With such a component tool, an engineer can use the component library and all the associated task without even knowing the underlying formalisms. What is more, there is only one composition mechanism for constructing systems from component libraries; it is up to the transformations defined along with the component library to map these compositions to the composition mechanism provided by the underlying formal method. Moreover, there is a single concept of scenario, which allows an engineer exchanging specifications and analysis results among different task – in particular analysis results can be visualised to the engineer.

Though we discussed these ideas by the help of a simple toy-train example only, the scope of Component Tools is much broader. The concepts of Component Tools have been inspired by a project working on techniques that support engineers in the area of flexible manufacturing systems in the task of designing PLC code and in identifying hardware faults during the ramp-up phase of a new flexible manufacturing system.

Up to now, the different models of a component library are completely independent of each other. It is an interesting (but formalism-dependent) task to investigate conditions that guarantee that one model refines another or that different models are consistent to each other so that the results of different tasks based on different models can be combined. This, however, is challenging and interesting future research.

Acknowledgement. We would like to thank all the students that participated in our one year master's project 'Component Tools', which helped to advance the concepts of Component Tools and implemented the prototype.

References

1. A. Agrawal, G. Karsai, Z. Kalmar, S. Neema, F. Shi, and A. Vizhanyo, "The design of a simple language for graph transformations," *Journal in Software and System Modeling*, 2005, in review.
2. Brooks Automation Corporate, "AutoMod Suite," http://www.automod.com/products/products.asp.
3. K. Czarnecki and S. Helsen, "Classification of model transformation approaches," in *OOPSLA'03 Workshop on Generative Techniques in the Context of Model-Driven Architecture*, 2003.
4. W. Dangelmaier, B. Mueck, C. Laroque, and K. R Mahajan, "d3fact insight: A simulation-tool for multiresolution material flow models," in *Simulation in Industry - 16th European Simulation Symposium (ESS2004) SCS - Europe*, I. Lipovszki, György; Molnár, Ed., 2004, pp. 17–22.
5. FASTEC GmbH, "Lontrol," http://www.fastec.de/.
6. A. Gepting, J. Greenyer, E. Kindler, A. Maas, S. Munkelt, C. Pales, T. Pivl, O. Rohe, V. Rubin, M. Sanders, A. Scholand, C. Wagner, and R. Wagner, "Component tools: A vision of a tool," in *Proc. of the 11th Workshop on Algorithms and Tools for Petri Nets (AWPN)*, ser. Tech. Rep. tr-ri-04-251, Paderborn, Germany, September 2004, pp. 37–42.
7. H. Giese, E. Kindler, F. Klein, and R. Wagner, "Reconciling scenario-centered controller design with state-based system models," in 4th *International Workshop on Scenarios and State Machines: Models, Algorithms and Tools (SCESM'05), Satellite event of ICSE '05*, May 2005.
8. J. Greenyer, "Maintaining and using component libraries for the design of material flow systems: Concept and prototypical implementation," Bachelor thesis, Department of Computer Science, University of Paderborn, October 2003.
9. J. Greenyer, "A study of technologies for model transformation: Reconciling TGGs with QVT," Master thesis, Department of Computer Science, University of Paderborn, July 2006 (in preparation).
10. D. Harel and R. Marelly, "Come let's play: Scenario-based programming using LSCs and the Play-engine," Springer-Verlag, 2003.
11. E. Kindler and F. Nillies: "Petri nets and the real world," in: *AWPN workshop 2005*, proceedings, Berlin, Germany, K. Schmidt, C. Stahl, Ed., September 2005, pp. 19–24.
12. E. Kindler and C. Páles, "3D-visualization of Petri net models: Concept and realization," in *Application and Theory of Petri Nets 2004, 25th International Conference*, ser. LNCS, J. Cortadella and W. Reisig, Eds., vol. 3099. Springer, June 2004, pp. 464–473.
13. E. Kindler, V. Rubin, and R. Wagner, "An adaptable TGG interpreter for in-memory model transformation," in *Proc. of the Fujaba Days 2004*, Darmstadt, Germany, September 2004, pp. 35–38.
14. OMG, *Model Driven Architecture*, http://www.omg.org/mda/.
15. C. Rust, J. Stroop, and J. Tacken, "The Design of Embedded Real-Time Systems using the SEA Environment," in *Proc. of the 5th Annual Australasian Conference on Parallel And Real-Time Systems (PART '98), Adelaide, Australia*, September 1998.
16. D. Schmelter, "Play-In und Play-Out von Szenarien von Component-Tools," Bachelor thesis (in German), Department of Computer Science, University of Paderborn, November 2005.

17. A. Scholand, "Steuerungssynthese für komponentenbasierte Systeme unter partieller Sichtbarkeit," Master thesis (in German), Department of Computer Science, University of Paderborn, April 2006 (in preparation).
18. A. Schürr, "Specification of graph translators with triple graph grammars," in *Graph-Theoretic Concepts in Computer Science, 20th International Workshop, WG '94*, ser. LNCS, E. W. Mayr, G. Schmidt, and G. Tinhofer, Eds., vol. 903, Herrsching, Germany, June 1995, pp. 151–163.
19. Tecnomatix Technologies Ltd., "eMPower Solutions," http://www.tecnomatix.com/.
20. D. Varró, G. Varró, and A. Pataricza, "Designing the automatic transformation of visual languages," *Science of Computer Programming*, vol. 44, no. 2, pp. 205–227, August 2002.
21. M. Weber and E. Kindler, "The Petri Net Kernel," in *Petri Net Technologies for Modeling Communication Based Systems*, ser. LNCS, H. Ehrig, W. Reisig, G. Rozenberg, and H. Weber, Eds.   Springer, 2003, vol. 2472, pp. 109–123.
22. E. D. Willink, "UMLX: A graphical transformation language for MDA," in *MDAFA'03*.   Entschede, Netherlands: University of Twente, September 2003, pp. 13–24.
23. W3C, "XSL Transformations (XSLT) Version 1.0," November 1999. [Online]. Available: http://www.w3.org/TR/xslt
24. M. Zarbock, "Ermittlung von Fehlerursachen in Prduktionsanalgen," Master thesis (in German), Department of Computer Science, University of Paderborn, July 2006 (in preparation).

Using Colored Petri Nets and Tools to Support Mobile Terminal and Network Software Development

Jianli Xu

Nokia Research Center
P.O. Box 407, FIN-00045, NOKIA GROUP, Finland
jianli.xu@nokia.com

Abstract. In this paper we report our experience in several research projects that use Colored Petri Nets in modeling and analysis of the software systems of Nokia products. These research projects use both formal language - Colored Petri Nets and semi-formal language – UML to describe, model and analyze different software systems. Due to the particular limitations, formal methods have been only used to model and analyze a certain property or a few strongly related properties, or a certain critical part of a big system at a time. This short paper reports our experience in applying formal methods in industry settings. We mainly address our problems in applying formal methods in supporting our product software development, such as system complexity, model complexity, the appropriate abstraction levels of a formal model, tracing the problem from abstract formal model to real software and vice versa, tool support, etc.

1 Introduction

System modeling and analysis play an important role inside Nokia in improving the quality of both its system design and the final products. Modeling and model analysis are becoming a common practice for software architects and designers of Nokia since the wide adoption of UML. Due the obvious reasons, UML alone cannot meet our requirements in modeling and analysis to guarantee important system properties, especially those run-time/behavioral properties. For a long time we have been making big effort to introduce formal methods into the product software development practice. Many projects have been carried out in trying out different formal techniques in system modeling and analysis, and testing the possibility of integrating formal methods and tools in the current software development process.

In this short paper, we only describe some projects inside our own research group. We try to share with the readers our experience of using formal methods in modeling and analysis of software systems and our expectations for the research community. As the formal modeling language we use Colored Petri Nets (CPN), a visual, both action and state-based specification formalism, that is suitable for modeling concurrent activities and flows in complex systems [1][2][3]. CPN has formal execution semantics that make it possible to simulate the behavior specified by CPN models. The strong tool support [4][5], especially in model construction and simulation, and

S. Donatelli and P.S. Thiagarajan (Eds.): ICATPN 2006, LNCS 4024, pp. 57–64, 2006.

long time cooperation with the CPN group at Aarhus University have greatly helped us in those projects.

In this paper we first briefly introduce four research projects conducted since late 90's at Nokia that used CPN as one of the main modeling languages, discuss the results and lessons learnt from each projects. Then in the last part of the paper we try to summarize and share with practitioners our experience of using formal methods in industry research projects, and convey our expectations to the research community.

2   Some CPN Application Projects at Nokia

The size of projects described in this section varies from 2 person months to 1.5 person years. There were 1 to 4 people directly involved in a project. Two projects, the first one and the third one, were carried out only by Nokia research engineers. The other two projects were cooperation projects between Nokia Research Center and the CPN group at University of Aarhus, Denmark.

2.1   Modeling and Analyzing the Execution Architecture of Mobile Phone Software

The modeling work of this project was done during the early design phase of a new mobile phone product family. The goal of the project was to evaluate the initial software architecture design against the system requirements [6]. The execution/run-time architecture was modeled with CPN and analyzed with the Design/CPN tool.

This was our first CPN project. Before we started to create the CPN model, we had spent about one and a half person months to prepare for the real work, including learning CPN and SML, learning how to use Design/CPN tool, and doing small case studies with the Design/CPN tool. The modeling work started immediately after the principal structure of the new software architecture had been outlined and continued along with the component architecture design and detailed system design. The CPN modeling iterated over three main steps: creating or modifying the model, simulating, and analyzing.

The CPN model of the new architecture shares the system structure which was described using UML class diagrams. The decomposition of the model into parts follows the decomposition of the system and mapping from the entities of the model to the entities of the system is clear. This means that the CPN model of the new architecture directly assists in designing and configuring different system variants. Since this CPN model shares the system structure, it is easy to see how well the model conforms to the system design. This increases the confidence that the model can give correct analysis and predictions of system properties.

The behavior properties of the new architecture were analyzed mainly with model simulation that was supported by the Design/CPN tool. The CPN model was annotated with timing parameters and memory consumption parameters of all system operations been modeled. The behavior properties include the task switch time, message delay and message buffer usage at the best and worst case. The CPN model had been used to:

- specify and verify the task control mechanism
- specify and verify the task communication mechanism
- evaluate different task divisions and allocations
- simulate typical use cases
- estimate the message buffer usage and message delays

The CPN model allowed us to analyze both the time and space performance of the new software architecture. Our experience shows that a special purpose model can be constructed for large systems also and this model predicts system properties accurately enough to be useful. By using the CPN model we evaluated alternative communication mechanisms object to different task allocations and task configuration and we found potential problems in specific design alternatives.

Due to the large size of the model and the limitation of processing power and memory on the computer where we run the Design/CPN tool, we could not analyze the model as a whole with the state space tool of Design/CPN. The state space analysis was performed only on several important sub-models, hence only some local properties were verified.

This project demonstrates that formal architectural models can be built for industrial systems and general-purpose formal modeling techniques can be applied to architectural modeling. Formal modeling can provide quantitative results to be used as a guide in developing the system further. Existing tools have limited analytical power but simulation can be used in practice. The CPN execution architecture model also demonstrates the important benefit of architectural modeling. Structural similarity between different views helps in keeping the design and model coherent.

Our main problem during this project was the lack of documentation on how to model software architectures using formal techniques. We had to figure out how to model different constructs, where to abstract, how to model the input to the system and how to monitor the behavior. The approach used in this project was also applied in the later projects, and they turned out to be a lot easier and much more effective than the first case.

2.2 Modeling Feature Interactions in Mobile Phones

We studied the problems of feature interactions in the user interface (UI) software of Nokia mobile phones in this project [7]. The mobile phone UI software is designed to support the most frequent and important user tasks flexibly and smoothly, hence it enables many features to interplay and be active at the same time. The dependencies or interplay of features are called feature interactions and range from simple usage dependencies to more complex combinations of several independent behaviors.

The research work of this project was motivated by the fact before the project started, that feature and feature interactions were not systematically documented at Nokia and often the most complex feature interactions were not fully understood before the features were implemented. In the design and development of features, focus was often on the behavior and appearance of individual features. Based on our positive experience with the previous project, we believed CPN can help us to create both a static graphical description of features and allow simulation of feature interactions in an interactive environment. The targets of the project ware:

- To identify and document typical patterns of feature interactions.
- To develop a systematic methodology for describing feature interactions.
- To provide an environment for interactive exploration and simulation of the feature interactions for demonstrational or analytical purpose.

This project was a joint project between Nokia Research Center and the CPN group at the University of Aarhus. The project lasted for a year from November 2000 to November 2001 with a total amount of 15 person months of effort. The project group consisted of two research engineers from Nokia Research Center and three people from the CPN group, hence it contains both application domain experts and experts in formal methods and tools (CPN and Design/CPN).

Before the joint project started, initial work was done at Nokia Research Center to experiment with the use of CPN for modeling of feature interactions. Through the initial experiment, the research engineers at Nokia thought CPN and Design/CPN tools were very promising in modeling of mobile phone UI features and feature interactions, however, we found it difficult to develop by ourselves an UI oriented approach which enables the interaction between the CPN model and UI designers/developers in their own languages. Hence a joint project was setup with the CPN group to have CPN and tool experts involved directly in the work. When the joint project started, one researcher from the CPN group worked full time at Nokia Research Center for six months. Other project members from the CPN group provided guidance and technical support on the modeling work.

The CPN model had been constructed in a period of six months and had been constructed in several iterative steps with more and more elaborated models each presented to the UI designers/developers. In order to present the CPN model to users not familiar with Petri Nets, the CPN model was extended with domain-specific graphics at a very early stage. The graphics was developed and extended in parallel with the underlying CPN model.

The core of our approach is an executable behavioral model of the underlying UI architecture and the individual features. The feature interaction patterns were modeled using explicit behavioral models of features and interactive graphical simulation. A categorization of feature interactions was produced and has been used in UI specifications of Nokia mobile phones. The tool that we use makes it possible to add domain specific graphics for visualization and interaction purposes. A GUI that animates the real mobile phone UI was created as the front-end of the CPN model. A user of the model can play different scenarios by just clicking the buttons on the screen and see the current state of the system from the animated phone display. The system internal actions are executed via automatic model simulation. For all the played scenarios the tool we used can generate corresponding message sequence charts of the simulated feature interactions.

We also experimented with the CPN model of feature interaction to see if automatic analysis can be done to detect all possible feature interactions. Using state space analysis tool of Design/CPN initial attempts had been made to evaluate the possibility of performing automatic state space analysis on the CPN model. The state spaces of different combinations of selected features had been generated (not for all feature combinations and not for the full CPN model), possible feature interactions and deadlock situations were detected using queries of two properties expressed in temporal logic on the state space generated. The CPN model of feature interaction is suitable for automatic analysis, however, we expect the state space of the full CPN model to be very large and can not be well handled by the current tools.

This project has improved our knowledge about features and feature interactions of mobile phone UI and has influenced the change of design practice of features in new products. The feature interaction patterns and their categorizations have been used as the template for mobile phone UI specifications at Nokia.

2.3 Evaluating System Performance and Reliability Properties

In this project we also used CPN to evaluate key properties of software architecture of the call record management subsystem of Nokia DX200 switch system [9]. The CPN model focused on reliability and performance properties of the subsystem been modeled. This is a one person year project conducted by Nokia research engineers at Nokia Research Center using standard Design/CPN tools.

Reliability and performance, which are amongst the most important properties of telecommunication systems, are often bound up with each other. Fault-tolerant mechanisms built to provide high reliability usually slow down the system or reduce system throughput. In both practice and research reliability modeling and performance modeling are often separated, this fact makes it very difficult to evaluate the co-relations between this two properties. In this project we attempted to create architecture behavior model that addresses both reliability and performance properties of the architecture solution been modeled. We aimed at developing sound and balanced solutions for the reliability and performance properties of the system and make the system cost effective.

First an architectural behavior model was created which covers the key functions of the system and with sufficient details of the fault-tolerant mechanisms. Then this model was used to analyze the fault-tolerant mechanisms under various software and hardware failure cases. Finally the model was extended with performance factors and the effect of the fault-tolerant mechanisms on system performance had been checked through simulation. An alternative architectural solution proposed to improve the original system was also modeled and analyzed in the same way. The performance and reliability properties of the two solutions were analyzed and compared through extensive model simulations of the two CPN models using the simulation tool of Design/CPN. Simulation results revealed unexpected system properties that had not been discovered through other methods.

The model analysis results provided useful feedback to the design of new products of high speed switching, and helped in finding an ideal solution with high performance, low cost and reasonable reliability that meets the industry standards.

The capability of tool support is still the main barrier to the application of formal methods. In this project, due to the size of the CPN model, the amount of simulation data and the limit of the processing power of the computer where we run the Design/CPN tool, we could not analyze the entire model with the state space analysis tool. The formal analysis (state space based) was performed only on several selected important sub-models, hence only some local properties were verified.

2.4 Estimation of Memory Usage of Mobile Phone Application Software

This project was the most recent one using formal method in our research group, it is also a joint research project with the CPN group. Two researchers from the CPN group and two research engineers from Nokia are involved in the project. The total effort of the project is less than one person year.

In the project we developed a tool which mobile phone software engineers at Nokia can use during specification of software architectures to obtain estimates of worst-case memory usage of interacting software components [8]. The tool was created on top of the current CPN Tools [5]. A user of the tool specifies a set of scenarios showing interactions of interest. The scenarios are specified as UML sequence diagrams in IBM Rational Rose; relevant classes, components, and messages are annotated with estimates of memory usage. All possible interleaving of the individual events of the given set of scenarios are captured in the state space of a Colored Petri Nets model called the formal analysis engine.

Formal analysis carried out in CPN Tools produces a path in the state space, which corresponds to a scenario that at some point has the worst-case memory usage. The analysis results are presented for the user in an Excel spreadsheet. The use of CPN tools is hidden from the user and the analysis tools can be used together with the software design tool, such as IBM Rational Rose [10]. CPN Tools is automatically launched from Rose, and Excel is automatically launched from CPN Tools. The formal analysis engine of the tool is a generic CPN model, although it captures the properties of the scheduler of the underlying mobile phone operating system, it is not the case that a new CPN model is created for each new set of scenarios that must be analyzed. The CPN model is shared by all applications of the tool. It is only the initial marking that is different for different sets of sequence diagrams to be analyzed.

This project is a little different than the earlier ones. Although all the earlier projects produced promising results, it has turned out to be difficult to transfer the results of the projects from research to broad application in real software development. The main reason for this may be that the common characteristic of the three earlier projects is that the approach used in the projects requires software designers/developers to create CPN models. When we recommend to the managers responsible for real development projects that they should use a formal method like CPN, we believe that the design rationales for the memory estimation technique are arguments that can be used to increase the chance that a transfer from research to development will happen.

The tool relies on generation of a state space. Consequently, state explosion is a potential general problem. Currently the tool typically analyses just a few scenarios, and each scenario consists of relatively few events. Therefore, the state spaces are often quite small. On the other hand, if the technique proves to be useful, engineers may demand more from it. In this way, it may be necessary to address the state explosion problem in the future.

3   Observations and Conclusions

Our experience from the projects described above shows that formal modeling and analysis techniques can be used on software systems of reasonable size. By the cooperation with the university and having research engineers with formal methods back-ground, this kind of project can be carried out successfully in the industry research setting. However, it is still not in the stage of using formal methods in large scale in business and production units of companies.

Nine years ago when we had our first formal method application project – "modeling and analyzing the execution architecture of mobile phone software", we noticed that from an industrial perspective architectural modeling using formal methods was not yet a mature field. In order to make architectural modeling a practical tool for software developers, additional support was required in several forms [6]:

- Modeling techniques should be classified according to the reasoning they can support.
- Model templates or examples should exist for different architectural constructs allowing analysis of different attributes.
- Models should be composable so that the composite model will correctly predict the properties of the composed system.
- Model-checking techniques should be more practical. Models can be partitioned to overcome computational limitations but no guidance exists on how to do it.
- Guidelines on how to model different architectural constructs should exist. Currently the results depend entirely on the personal skills of the architect.

But up to now the situation has not been fundamentally improved. In project "evaluating system performance and reliability properties", we developed a general approach of creating run-time formal property models of software architecture [9], and try to make it applicable to a wide range of software systems. We still need more case studies to test and improve the approach.

Other reasons that prevent the application of formal methods from being a common practice in industry are:

- The cost of training engineers to use formal methods are too high;
- There are few industrial tools that support formal method in software development;
- Limitation of current formal techniques in handling large complex system, such as state explosion, lack of different levels of abstraction, model composition;
- Most of the formal languages are far away from the actual design and implementation languages used in software development;
- It is difficult to integrate formal modeling and analysis into the current software development process.

One possible solution to the problems is to provide appropriate tool support for formal modeling and analysis. The tool should be able to hide formal models from the end user and can perform the formal analysis automatically. We also need tools that can generate (automatically or semi-automatically) formal analytical models from system design (for example UML design model) or implementation (source code). Another expectation from industry is being able to generate the implementation from the formal specification models of a system we are going to build, this may not be very practical, but the resent research [11] [12] make us see the light in the end of the tunnel.

References

1. K. Jensen. "Coloured Petri Nets. Basic Concepts, Analysis Methods and Practical Use. Volume 1, Basic Concepts". Monographs in Theoretical Computer Science. Springer-Verlag, 1997. ISBN:3-540-60943-1.
2. K. Jensen. "Coloured Petri Nets. Basic Concepts, Analysis Methods and Practical Use. Volume 3, Practical Use". Monographs in Theoretical Computer Science. Springer-Verlag, 1997.
3. L.M. Kristensen, S. Kristensen, and K. Jensen, "The practitioner's guide to Coloured Petri nets". International Journal of Software Tools for Technology Transfer 2(2): 1998, pages 98-132. Springer.
4. Design/CPN Online. http://www.daimi.au.dk/designCPN/
5. CPN Tools. http://www.daimi.au.dk/CPNtools/
6. J. Xu and J. Kuusela. "Analyzing the execution architecture of mobile phone software with coloured Petri nets". International Journal of Software Tools for Technology Transfer 2(2): 1998, pages 132-143. Springer.
7. L. Lorentsen, A.-P. Tuovinen, and J. Xu. "Modelling of Features and Feature Interactions in Nokia Mobile Phones using Coloured Petri Nets". In Proc. of 23rd Petri Nets Conf., volume 2360 of LNCS, pages 294–313, Adelaide, Australia, 2002. Springer.
8. J.B. Jørgensen, S. Christensen, A.-P. Tuovinen and J. Xu. "Tool Support for Estimating the Memory Usage of Mobile Phone Software". To be published in International Journal of Software Tools for Technology Transfer. Springer.
9. J. Xu. "Evaluating and Balancing Reliability and Performance Properties of Software Architecture Using Formal Modeling Techniques". To be published in the proceedings of the 30th NASA/IEEE Workshop on Software Engineering, 25-27 April, 2006, Colombia, USA. IEEE Computer Science Press.
10. IBM Rational Rose. www-136.ibm.com/developerworks/rational/products/rose
11. J.B. Jørgensen, C. Bossen. "Executable Use Cases: Requirements for a Pervasive Health Care System", IEEE Software, 2004.
12. W. van der Aalst, J.B. Jørgensen, K.B. Lassen. "Let's Go All the Way: From Requirements Via Colored Workflow Nets to a BPEL Implementation of a New Bank System", 13th Intl. Conf. on Cooperative Information Systems (CoopIS), Springer LNCS, 2005.

Compensation in Workflow Nets

Baver Acu and Wolfgang Reisig

Humboldt-Universität zu Berlin
Institut für Informatik
Unter den Linden 6
D-10099 Berlin
{acu, reisig}@informatik.hu-berlin.de

Abstract. We present a formal model to specify *compensation* of workflows: Each acyclic workflow net W (as defined by v.d. Aalst) is canonically extended to a net W^+, representing the potential occurrence of failures, and the compensation of so-far executed actions. We show that the construction is compositional and correct, i.e. meets the expected properties.

1 Introduction

The concept of *compensation* has been introduced in the area of database transactions. A transaction is a unit of work which makes consistent changes on a database. It ensures consistency by holding its resources until it completes. In case of a failure, a trasaction undoes everything by reversing its updates. A long running transaction (LRT) as its name implies is a transaction which may take long time to complete. A LRT holding its resources for a long time may cause deadlocks or increase the rate of cancellations. To alleviate this problem, a LRT is defined as a *Saga* [5], i.e. as a sequence of sub-transactions. Each subtransaction completes its work, i.e. releases its resources, before being sure that overall transaction completes. The effects of completed sub-transactions may be used by other transactions and hence in case of a failure an undo may not always be possible. An undo of such a work may only be possible, in a way by paying a compensation. Therefore to undo completed works, every sub-transaction is associated with a compensating transaction. When a Saga fails, the effects of already completed sub-transactions are undone by scheduling their associated compensating transactions in the reverse order of their completetion.

As Workflow systems are widely used to implement large, scalable long running business processes, reliability and consistency issues became more prominent. Such large scale systems cannot be left in inconsistent states in case of failures. A suitable mechanism for reliability and consistency is to apply to workflows the resource holding methods that are well established in database transactions. However, the application of this method leads workflows to suffer from the same problems as they occur in LRTs. Compensation mechanism can be used as a means of recovery to keep workflows in consistent states upon failures. To this end, individual workflow tasks have been defined together with

S. Donatelli and P.S. Thiagarajan (Eds.): ICATPN 2006, LNCS 4024, pp. 65–83, 2006.

their compensating counter part. Upon a failure, the already executed tasks are compensated in the reverse order of their completion.

Our aim in this paper is to formally model the compensation behavior of acycli workflows. As a model for workflows we take Van der Aalst's Workflow Nets (WF-nets) [1]. We refine the basic task model of WF-nets to include compensation behavior. We apply the same technique to basic workflow patterns as defined in [2]. We show the semantics of tasks to construct workflows with compensation capability, i.e., Compensational Workflow nets (CWF-nets) are compositional. We show that the constructed CWF-nets meet the intended compensational behavior of workflows.

The rest of this paper is organized as follows: Section 2 and Section 3 give a short background of the compensation mechanism and Workflow nets. In Section 4, we describe how we refine a WF-net task model to include compensational behavior. In Section 5, we explain the semantics of the tasks and show that our construction is compositional: Composition of compensational tasks distributes over the constructing components. In Section 6, we define Compensational Workflow nets formally and show the construction in fact meets the expected properties: A compensational workflow net either completes, or properly compensates all executed tasks in the reverse partial order. Section 7 finishes this paper with conclusions and future work.

2 Recovery by Compensation

The next sections describe the use of compensation mechanism in Database managements systems (DBMS) and Workflow management systems (WFMS).

2.1 Compensation in DBMS

The notion of compensation emerged as a solution for the atomicity problem of long running transactions. In Database management systems, every transaction is required to meet the *ACID* principle [4]. This is a short notation for *Atomicity, Consistency, Isolation* and *Durability*. This principle requires every transaction

- either to complete successfully(commit) or leave irregularly (abort) and erase all the effects as if it never happened (*Atomicity*),
- to produce effects which obey the underlying database constraints, like the integrity constrains of data stored in a database (*Consistency*),
- to run isolated from other concurrent transactions so that they will not affect the result of the transaction (*Isolation*), and
- to persist its effects so that they will stay permanently on the stable storage (*Durability*).

Applying the ACID principle to a LRTs means holding resources during its entire duration. Holding resources for a long duration, however, causes deadlocks and increases the rate of transaction aborts. One of the solutions to tackle this problem is to define a LRT as a sequence of sub-transactions, called *Saga* [5]. Each sub-transaction in itself preserves ACID'ity. This sequence is allowed to be interleaved with other long running transactions to allow more concurrency. However, interleaving sub-transactions, causes intermediate results to be visible to other concurrent transactions, before the saga finishes. This may cause inconsistent states if the saga aborts. To amend inconsistency, the Saga method defines a *compensating transaction* for each sub-transaction. This compensating transaction semantically undoes the effects of a committed transaction. In case of a failure, the compensating transactions of all completed transactions are called in the reverse order of their completion. That means, given a saga $T_1, T_2, ..., T_n$ with their compensating transactions $C_1, C_2, ..., C_{n-1}$ either the sequence

$$T_1, T_2...T_n \text{ or the sequence } T_1, T_2...T_j, C_j, ...C_1 \tag{1}$$

for some $1 \leq j < n$ occurs. In a saga every sub-transaction has a compensating counter part except the last one. The Saga method assumes that completion of the last step finishes a saga and no compensation will be available anymore.

A compensating transaction does not necessarily have to be the reverse of its associated transaction. For example, a transaction withdrawing money from a bank account can be compensated with its exact reverse of depositing the same amount. However, a flight reservation may not be compensated by simply canceling the reservation. Canceling the flight reservation may require a charge to be paid.

2.2   Compensation in WFMS

A *workflow* is a computerized facilitation or automation of a business process in whole or part [8]. A workflow taking a long time to complete resembles to a LRT. The difference is that a workflow may consist of partially ordered high level programs or human tasks instead of sequential database operations. Reliable execution of workflows can be guaranteed using the similar approach as in LRTs by holding the resources until the workflow finishes. A more flexible approach is to apply the Saga method to workflows. A workflow itself can be seen as a saga, since a workflow may consist of several substeps, which are called *workflow tasks*. Individual tasks can be assumed as atomic sub-transactions and every workflow task can be associated its *compensating task*. A compensating task is defined for a workflow task to semantically undo its effects. A task with a compensation capability is called *compensational task*. When a workflow fails, the tasks already completed are compensated by calling their compensating tasks. In a workflow, the partial order relation $\prec$ over a the set of tasks $T_1, T_2, ..., T_n$ together with their compensating tasks $C_1, C_2, ..., C_n$ is defined as follows:

- $T_i \prec T_j$ if T_i occurs before T_j,
- $T_i \prec C_i$, a compensating task should occur after its original task,
- $C_j \prec C_i$ if and only if $T_i \prec T_j$.

This partial order ensures that sequential tasks are compensated in the reverse order of their completion but in addition, it allows concurrent tasks to be compensated in parallel.

In this paper, we assume that in workflows every task has an associated compensating task including the last one. This enables compensating an entire workflow. Compensating an entire workflow is particularly important when a workflow is used to implement a single task of another workflow. To guarantee that a workflow can no longer be compensated, we append an additional task called *Complete* to the workflow. That means, execution of a workflow can be either of the following partially ordered sets:

$$\{T_1, T_2, ..., T_n, Complete\} \text{ or } \{T_1, T_2, ..., T_j, C_j, ..., C_1\} \qquad (2)$$

for some $1 \leq j \leq n$.

There have been also attempts to reduce the number of compensations, by partial recoveries [6] or to change the order of compensating tasks according to time, control or data dependency [3]. Partial recovery strategies require a correctness definition based on application semantics. They are mostly used in the context of forward recoveries. In this paper we assume workflows recovering completely using backward recovery and the order of compensation depending on control flow only.

3   Workflow Nets

Workflow nets (WF-nets) have been introduced by Van der Aalst [1] as a means to specify workflows:

Definition 1 (WF-net). *A Petri net $N = (P,T,F)$ is a WF-net iff :*

  i. *N has two special places: i and o s.t. $^\bullet i = \emptyset$ and $o^\bullet = \emptyset$. i and o are called source and sink places respectively.*
 ii. *If we add a transition t^* to N with $^\bullet t^* = \{o\}$ and $t^{*\bullet} = \{i\}$, then the resulting Petri net is strongly connected.*

(A Petri net is strongly connected if and only if for every pair of nodes, i.e. places and transitions, x and y, there is a path leading from x to y.)

Many workflow patterns can be modeled easily by Petri nets [2]. Figure 1 depicts the Petri net models for AND-split, AND-join, OR-split, and OR-join. Figure 2 shows an example WF-net, W_1 which is constructed using these basic patterns. W_1 has a source place i and a sink place o.

In WF-nets, a task is modeled by a transition. Input places of a transition represent preconditions and output places represent post conditions. A single basic task is reasonably modeled by a Petri net with one transition:

Definition 2 (Basic workflow task). *A Petri net $N = (P,T,F)$ is a basic workflow task iff*

  i. *N is connected,*
 ii. *T has a single transition element, t,*
iii. *input and output places of t are disjoint, i.e. $^\bullet t \cap t^\bullet = \emptyset$.*

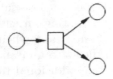

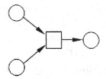

(a) AND-split (Parallelism)

(b) AND-join (Synchronous join)

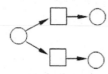

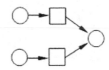

(c) OR-split (Non-determinism)

(d) OR-join (Asynchronous join)

Fig. 1. Basic workflow patterns

A transition with tokens in its input places corresponds to a task with its pre-conditions satisfied. Firing of the transition means execution of the task. In this paper, we model a workflow task with a transition whose input and output places are disjoint, i.e. an acyclic task, and restrict ourselves to workflow nets constructed from the patterns shown in Fig.1.

The WF-net W_1 of Fig. 2 specifies the control flow among six workflow tasks. Initially having a token in source place i, the task represented by transition t_1 is performed first. Following the completion of task t_1, both tasks t_2 and t_3 are started concurrently. After the execution of t_2, a decision is made between the tasks t_4 and t_5. Execution of task t_6 starts only when both t_3 and either of t_4 or t_5 complete. Completion of task t_6 also completes the workflow execution.

An execution of a workflow can be simulated by a run of its WF-net model. A run of a WF-net can be represented as a sequence $M_0 \xrightarrow{t_1} M_2 \xrightarrow{t_2} \dots$ of firings $M_{i-1} \xrightarrow{t_i} M_i$ with M_0 the initial marking. However, an interleaved run does not capture the concurrency among its actions. Runs of nets in which independent occurrence of actions are explicitly shown can be represented by *occurrence nets* [7]. They are special acyclic nets with unbrached places:

Definition 3 (Occurrence net). *A net* $K = (P_K, T_K, F_K)$ *is called an occurrence net iff*

i. *for each* $p \in P_K$ $|{}^\bullet p| \leq 1$ *and* $|p^\bullet| \leq 1$,
ii. *the transitive closure of* F_k^+, *frequently written* $<_K$, *of* F_k *is irreflexive (i.e.,* $x_1 F_k x_2 F_j \dots F_k x_n$ *implies* $x_1 \neq x_n$),
iii. *for each* $x \in (P \cup T)$, $\{y | x <_K y\}$ *is finite.*

Two elements x, $y \in (P \cup T)$ *are concurrent iff neither* $x <_K y$ *nor* $y <_K x$.

Each transition of an occurrence net corresponds to an occurrence of an action. Occurrence of this action is represented by the transition's labeling. Repeated

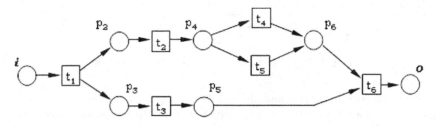

Fig. 2. An example Workflow net W_1

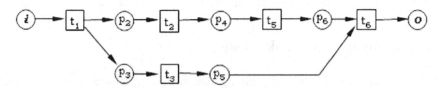

Fig. 3. A concurrent run of the WF-net W_1

occurrence of the same action is represented by different transitions with the equal labels. Similarly each place in an occurrence net corresponds to an occurrence of a local state. Formally a concurrent run of a net is defined as:

Definition 4 (Concurrent Run). *Let N be a net, let K be an occurrence net and let $l : K \rightarrow N$ be an element labeling of K. K is a concurrent run of N iff*

i. concurrent elements of K are differently labeled,
ii. for each $t \in T_K, l(t) \in T_N$, $l({}^\bullet t) = {}^\bullet l(t)$ and $l(t^\bullet) = l(t)^\bullet$.

In WF-nets we assume that every transition has the *progress* property and therefore we use the term *run* to mean *maximal run*. Figure 3 shows a concurrent run of W_1 initially with a single token in place i. In this concurrent run, the non-deterministic choice at p_4 is made for the transition t_5. In fact, W_1 has another concurrent run in which t_4 occurs instead of t_5. Both concurrent runs of W_1 terminate with a single token in place o.

A proper termination definition for WF-nets is defined by the soundness notion. It is introduced by Van der Aalst [1] as a correctness criterion for WF nets.

Definition 5 (Sound). *A WF-net $N=(P,T,F)$ is sound iff:*

i. For every marking M reachable from marking i, there exists a firing sequence leading from marking M to marking o.
ii. The marking o is the only marking reachable from marking i with at least one token in place o.
iii. For each transition $t \in T$, there is a marking M reachable from marking i which enables t.

Soundness requires that a workflow initially having a token in source place i always terminates with one token in sink place o and all other places containing

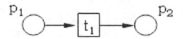

Fig. 4. Petri net model of a basic workflow task

no tokens. In addition, there should not be any dead transitions, i.e. for each transition t it should be possible to reach to a state (from the marking i) where t is enabled. The WF-net W_1 is a sound WF-net. With a single token in place i, W_1 always terminates with a single token in place o. In addition, for each transition in W_1 there is a reachable marking which enables it.

4   Compensational Task Model

In WF-nets, tasks are modeled assuming their ordinary behaviors, i.e., without failures and compensations. Each task behaves as follows: a task which is *ready to start, executes* and then *finishes*. The Petri net model given in Fig. 4 specifies this behavior. Place p_1 represents the state in which the task is ready to start. Transition t_1 represents the task execution and the place p_2 is the state where task is successfully executed.

In a workflow where failures and compensations may occur, the model of each task must also capture the occurrence of its failure as well as its compensation. A compensational task:

1. may execute successfully and activates its compensating pair (*Normal behavior*),
2. may fail during its execution (*Failure behavior*) or,
3. it may be compensated after a successful execution due to a failure in workflow (*Compensation behavior*).

Our aim is to refine the basic workflow task model in Fig. 4 step by step to include these additional behaviors.

The model in Fig. 4 already represents the normal behavior of a compensational task except activating the compensating task(Its activation will be discussed later). The representation of failure behavior requires an additional transition. Figure 5a extends Fig. 4 by the corresponding transition t_1^a. We call this transition as *failure transition*. The place p_1' obviously represents the *failed state*. Failures occur during the execution of tasks. As we model a task execution as a (instantaneous) transition occurrence, we see only the outcome of a task in the model: The task either completes successfully or fails.

Next, we are going to implement compensational behavior. Compensating a workflow means to compensate each of its tasks. In the Petri net model we therefore introduce a *compensating transition* t' for each transition t that models a task. t' must be executed upon a failure occurring *after* t. Therefore we have extended the task model in Fig. 5a as in Fig. 5b. Transition t_1' represents the compensating task. Places p_2' and p_3' represent the *beginning* and *end of the*

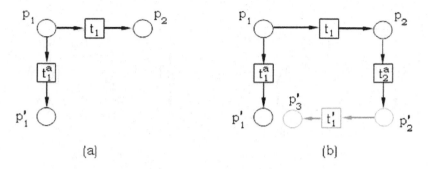

Fig. 5. Extending basic task with failure (a) and compensation behavior (b)

compensating task. We have drawn the compensating task in grey color to distinguish it from the original task model. Failure after an execution is specified by the failure transition t_2^a.

A final requirement remained is activating the compensating task after a successful execution. A task should not be compensated if it is not executed. This is particularly important when workflows make non-deterministic choices among their tasks. We implement this requirement using a place, p_1^a, between the transitions t_1 and t_1'. Figure 6a shows this implementation. Firing of transition t_1 creates a token in p_1^a, and consequently contributes to the enabling of the compensating transition t_1'. We call this place *activation place*, since it activates the compensating transition. The resulting model in Fig. 6a covers all possible behaviors of a compensational task. This model, in addition to successfully finished state, p_2, has two failure states p_1' and p_3'. p_1' is reached due to a failure in t_1, whereas p_3' is reached due to compensating t_1. Nevertheless, both yield the same effect, i.e. a failure. Merging these two places into a single one simplifies the model and preserves its semantics. Figure 6b depicts the final model of a compensational task. Given a token in place p_1, the following transition sequences simulate all possible behaviors of a compensational task:

1. $\{p_1\} \xrightarrow{t_1} \{p_2, p_1^a\}$ (Normal behavior)
2. $\{p_1\} \xrightarrow{t_1^a} \{p_1'\}$ (Failure behavior)
3. $\{p_1\} \xrightarrow{t_1} \{p_2, p_1^a\} \xrightarrow{t_2^a} \{p_2', p_1^a\} \xrightarrow{t_1'} \{p_1'\}$ (Compensation behavior)

In the final model, the compensating task model is structurally equal to the original task model except its inverted flow relation. This is justified by the fact that a compensating task always operates in the reverse order of its original task. Therefore given a basic workflow task, its compensating task model can be defined by its *inverse copy* :

Definition 6 (Copy, inverse copy). *Let $N = (P, T, F)$ be a net.*

 i. *A copy of N is $N' = (P', T', F')$ where $p' \in P'$ iff $p \in P$, $t' \in T'$ iff $t \in T$, and $(x', y') \in F'$ iff $(x, y) \in F$.*
 ii. *An inverse copy of N is defined by $N^* = (P', T', (F')^{-1})$.*

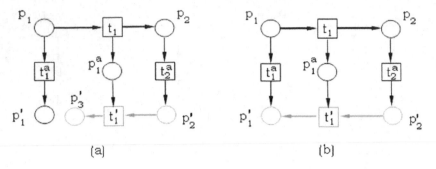

Fig. 6. Compensational task with activation place (a) and single failure place (b)

All copies (and all inverse copies) of a net N are unique up to isomorphism. As we don't distinguish isomorphic nets, we speak about "the copy N'" and "the inverse copy N^*" of a net N. As indicated by the above example, the compensating task of a basic workflow task N can be modeled by N^*. Consequently, a compensational task is modeled by a basic workflow task, its inverse copy, its failure transitions and activation places. Structurally we call such nets *paired nets*. Formally a paired net is defined as follows:

Definition 7 (Paired net). *Let* $N = (P, T, F)$ *be a net and let* T^a *and* P^a *be two sets s.t. for each* $p \in P$ *there is a unique* $t^p \in T^a$ *and for each* $t \in T$, *there is a unique* $p^t \in P^a$. *A paired net of* N *is defined by* $\overline{N} = (\overline{P}, \overline{T}, \overline{F})$ *where*

   i. $\overline{P} = P \cup P' \cup P^a$,
  ii. $\overline{T} = T \cup T' \cup T^a$ *and*
 iii. $\overline{F} = F \cup (F')^{-1} \cup F^a$ *where*
      $F^a = \{(p, t^p) | \ p \in P \ and \ t^p \in T^a\} \cup \{(t^p, p') | \ t^p \in T^a \ and \ p' \in P'\} \cup$
      $\{(t, p^t) | \ t \in T \ and \ p^t \in P^a\} \cup \{(p^t, t') | \ p^t \in P^a \ and \ t' \in T'\}.$

Analogously to copies and inverse copies, all paired nets $\overline{N}$ of a net N are isomorphic; hence we speak about "the paired net $\overline{N}$" of a net N. As an example, the net shown in Fig. 6b is a paired net of Fig. 4. For each p_i we write t_i^a instead of t^{p_i}. As indicated by this example, the compensational task model of a basic workflow task N is defined by the paired net $\overline{N}$.

    An important property of paired nets is that there is no path which connects an element in an inverse copy net to an element in its original net. For a compensational task model it means after firing a compensating transition t', it is not possible to fire transition t again.

Lemma 1. *Let* $N = (P, T, F)$ *be a net, let* $\overline{N}$ *be the paired net of* N *and let* $p \in (P \cup T)$. *There is no* $q' \in (P' \cup T')$ *s.t.* $q'(\overline{F})^+ p$.

Proof. Given $\overline{F} = (F \cup (F')^{-1} \cup F^a)$, the flow relations F and $(F')^{-1}$ do not define any arcs between the elements of two sets $(P \cup T)$ and $(P' \cup T')$ (see Definition 6). Therefore, the only way to proceed from any $q' \in (P' \cup T')$ to any

$p \in (P \cup T)$ is through an arc of F^a. The construction of F^a (see Definition 7) does not define any element (q', t^q) or (t^p, p), hence it is not possible to have $q'(\overline{F})^+p$.   □

We have applied the same modeling technique to other basic workflow patterns shown in Fig. 1. Compensational models for AND-split, AND-join, OR-split, and OR-join are depicted in Fig. 7.

5   Compositional Semantics of Compensational Tasks

Compositional semantics specifies the procedure of how to create large systems from building blocks. In our case, it describes how compensational tasks should be combined to create a compensational workflow. For WF-nets, compositional semantics of tasks is quite simple. A task after its execution initiates the start of its successor task. That means, the end of one task is the start of another task. This can be modeled with Petri nets by merging the end place of the first task with the start place of the second task. Figure 8a shows this composition.

If two basic workflow tasks given in Fig. 8a represent two workflow tasks T_1 and T_2 as in (2) together with the partial order relation ($\prec$) given in Sect. 2.2, the execution of this composition should yield the set $\{T_1, T_2\}$ with $T_1 \prec T_2$. With a token in place p_1, this behavior can be shown by the following run:

1. $\{p_1\} \xrightarrow{t_1} \{p_2\} \xrightarrow{t_2} \{p_3\}$

Composition of two compensational tasks analogously requires that the end of one task initiates the start of another. Additionally, in compensational workflows, failure or compensation of the second task should initiate the compensation of the first task. This additional requirement can be implemented in our compensational task model by merging the second task's failure transition and failure place with the first task's failure transition and begin place of compensating task. Figure 8b shows this composition between two compensational tasks. Elements with corresponding indices i are merged, i.e. in Fig. 8b for $i = 2$ the places p_2, p_2' and transition t_2^a.

Structurally, composition is done by joining the end of the first task with the start of the second task. In terms of Petri nets, composition is the union of the two nets representing two tasks. Obviously composition is defined as:

Definition 8 (Composition). Let $N_1 = (P_1, T_1, F_1)$ and $N_2 = (P_2, T_2, F_2)$ be two nets. The composition of N_1 and N_2 is defined by

i. $N_1 \oplus N_2 = (P_1 \cup P_2, T_1 \cup T_2, F_1 \cup F_2)$.

Fig. 8b represents two workflow tasks T_1 and T_2 with their compensating tasks C_1 and C_2 as in (2). The following transition sequences represent all possible executions of this composition w.r.t. the partial order ($\prec$) given in Sect. 2.2:

1. $\{p_1\} \xrightarrow{t_1^a} \{p_1'\}$ (*Abort*; nothing executed)
2. $\{p_1\} \xrightarrow{t_1} \{p_1^a, p_2\} \xrightarrow{t_2^a} \{p_1^a, p_2'\} \xrightarrow{t_1'} \{p_1'\}$ ($T_1 \prec C_1$)

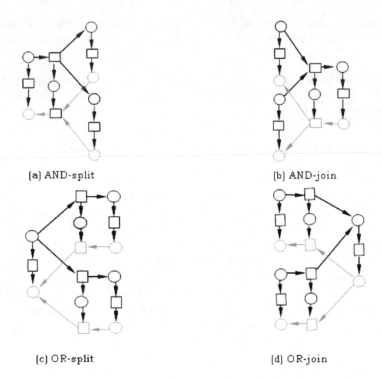

(a) AND-split

(b) AND-join

(c) OR-split

(d) OR-join

Fig. 7. Basic workflow patterns extended to include compensation behavior

3. $\{p_1\} \xrightarrow{t_1} \{p_1^a, p_2\} \xrightarrow{t_2} \{p_1^a, p_2^a, p_3\}$ $(T_1 \prec T_2)$

4. $\{p_1\} \xrightarrow{t_1} \{p_1^a, p_2\} \xrightarrow{t_2} \{p_1^a, p_2^a, p_3\} \xrightarrow{t_3^a} \{p_1^a, p_2^a, p_3'\} \xrightarrow{t_2'} \{p_1^a, p_2'\} \xrightarrow{t_1'} \{p_1'\}$
   $(T_1 \prec T_2 \prec C_2 \prec C_1)$

Another approach to the composition of two compensational tasks is first composing two basic workflow tasks and then generating its compensational model. In other words, given the composed two basic workflow task in Fig. 8a, we can generate its compensational model as in Fig. 8b. These two construction mechanism yield the same net. In other words, the construction mechanism of paired nets distributes over composition:

Theorem 1. *Given basic workflow tasks N_1 and N_2, the following holds:*

$$\overline{N}_1 \oplus \overline{N}_2 = \overline{(N_1 \oplus N_2)}.$$

Proof. The proof of the theorem follows from the application of Definition 7 and 8 to both sides of the equation. $\square$

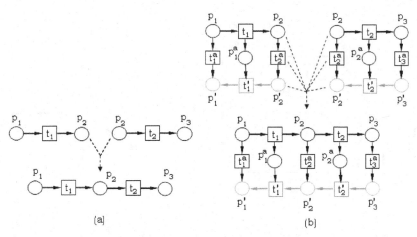

Fig. 8. Compositional semantics of basic and compensational task models

6   Specifying Compensation of WF-Nets

The following sections introduce Compensational Workflow nets (CWF-Nets). Section 6.1 gives the formal definition of CWF-nets. Sections 6.2 and 6.3 define failure and recovery notions.

6.1   Compensational Workflow Nets (CWF-Nets)

The compensational task model and its compositional semantics enable us to extend a given WF-net so that it covers the potential failures of the workflow as well as their compensation. We call such nets *Compensational Workflow nets* (CWF-nets). The following informal steps describe how to construct a CWF-net W^+ from a given WF-net W:

1. Decompose W into its basic workflow tasks.
2. Construct a compensational task model for each basic workflow task.
3. Compose compensational tasks according to the compositional semantics.
4. Append the *completion task* of W to the resulting net.

The sole purpose of the completion task is to indicate that workflow has completed and will not compensate.

Definition 9 (Completion task). *Let W be a WF-net with sink place o and let $N = (P, T, F)$ be a basic workflow task. N is a* completion task *of W iff*

i. $P = \{o, c\}$, $T = \{\checkmark\}$ *and* $F = \{(o, \checkmark), (\checkmark, c)\}$.

Figure 9 shows the CWF-net W_1^+ constructed from W_1 of Fig. 2. Realize that W_1^+ is composed of the paired net of W_1 and the completion task. This can be explained by the formal construction of CWF-nets. Given a WF-net W composed of basic workflow tasks $N_1, N_2, ..., N_n$ and its completion task N_c, the CWF-net W^+ of W can be conceived in various ways:

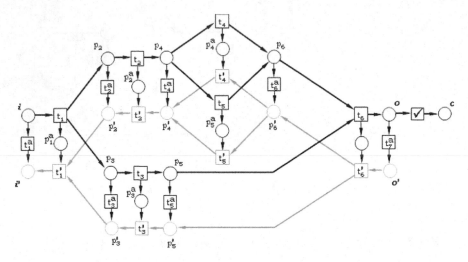

Fig. 9. Compensational model of the WF-net W_1

$$W^+ = (\overline{N}_1 \oplus \overline{N}_2 \oplus ... \oplus \overline{N}_n) \oplus N_c$$

$$W^+ = \overline{(N_1 \oplus N_2 \oplus ... \oplus N_n)} \oplus N_c \text{ by Theorem 1}$$

$$W^+ = \overline{W} \oplus N_c.$$

Definition 10 (Compensational workflow net). *Let W be a WF-net and let N_c be a completion task of W. The* Compensational workflow net (CWF-net) *W^+ of W is defined as:*

$$W^+ =_{def} \overline{W} \oplus N_c.$$

The source place of W^+ is identical with the source place i of W. i' and c are the places of W^+ with empty postsets. We call i' the *failure sink place* and c the *successful sink place.*

Lemma 2. *Let W be a WF-net with source place i. Then i is also the source place of W^+. Furthermore, W^+ has i' and c as the only sink places.*

Proof. Let $W = (P, T, F)$ be a WF-net and let N_c be its completion task. Given $W^+ = \overline{W} \oplus N_c$; Since N_c has a single sink place c, $\overline{W}$ should have a single source place i and single sink place i'. Place i was the single source place of W and we know from paired net definition (see Definition 7) that there is no transition $t \in \overline{T}$ s.t. $(t, i) \in F^a$ or $(t, i) \in (F')^{-1}$. Therefore, i is still a source place. For all other places $p^t \in P^a$ there is a transition $t \in T$ s.t. $(t, p^t) \in F^a$ and $p' \in P'$ there is a transition $t^p \in T^a$ s.t. $(t^p, p') \in F^a$. Hence i is the only source place in $\overline{W}$. Next, we'll show that i' is the only sink place of $\overline{W}$. For all places $p \in P$ there is a transition $t^p \in T^a$ s.t. $(p, t^p) \in F^a$ and for all places $p^t \in P^a$ there is a transition $t \in T$ s.t. $(t, p^t) \in F^a$. Since each place $p' \in P'$ is connected with $(F')^{-1}$, i' is the only sink place in P' and hence only sink place in $\overline{W}$. □

Since a WF-net and its inverse copy have similar structures, we would like to show CWF-nets graphically in a compact and simple way. Figure 10 shows the compact graphical representation of the CWF-net W_1^+. In this representation, W_1 is shown in the foreground and its inverse copy W_1^* in the background. We haven't shown the failure transitions and the activation places. We simply assume them *between* these two nets.

6.2   Failures in CWF-Nets

In accordance with WF-nets, a CWF-net initially has a single token in its source place i. If a CWF-net does not fire any failure transitions, the tokens follow the control flow of its WF-net. However a CWF-net has additional runs including failure and compensating transition occurrences. A *failure* in a concurrent run of a CWF-net is the concurrent occurrence of one or more failure transitions. A failure is called an *absolute failure* if it occurs in all concurrent branches. We call a run with a failure a *failure run*.

Definition 11 (Failure, absolute failure, failure run). *Let W be a WF-net, let $K = (P_K, T_K, F_K)$ be a concurrent run of W^+, let $TA = \{t|\ t \in T_K$ and $l(t) \in T^a\}$ be the set of failure transition occurrences of W^+ and let $TN = \{t|\ t \in T_K$ and $l(t) \in T\}$ be the set of transition occurrences of W.*

  i. *A set $f \subseteq TA$ is called a* failure *of K iff the elements of f are pairwise concurrent and no $x \in TA \backslash f$ is concurrent to all $y \in f$.*
 ii. *A failure f is called an* absolute failure *iff no $x \in TN$ is concurrent to all $y \in f$.*
iii. *A concurrent run K is called a failure run iff K has some failure $f \neq \emptyset$.*

Figure 11 shows a failure run of W_1^+ with a failure $f = \{t_4^a, t_5^a\}$. Since no transition of W_1 occurs concurrent to f, it is also an absolute failure. The places on the dotted lines C_1 and C_2 represent the markings of W^+ before and after the failure, respectively. We have drawn the element occurrences of W_1^* in grey color since they represent compensating task occurrences.

   In a workflow, a failure of a task is expected to cause all concurrently executing tasks to fail, i.e., it causes an absolute failure. In our compensational workflow net model, we didn't specify such an explicit termination. An explicit termination assumes no message delays between concurrent branches. In distributed environments where workflow tasks reside on different sites, message delays are more realistic. Not implementing such an explicit termination allows us to simulate message delays and to gain a simpler model. Instead, we rely on the soundness notion of WF-nets for an absolute failure.

   The soundness notion requires that a workflow starting with a single token in source place i always terminates with a single token in sink place o, with all the other places empty. This requirement implies that every parallel branch (AND-split) of a WF-net to be eventually synchronized (AND-join). As we mentioned earlier, in a CWF-net W^+ initially having a single token in place i tokens follow the control flow of WF-net W. An occurrence of a failure transition simply removes a token

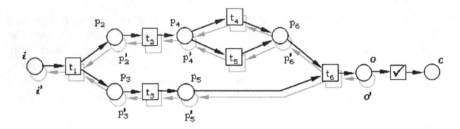

Fig. 10. A compact graphical representation of the CWF-net W_1^+

from the net W and places it to the net W^*. A token disappearing from the net W disables a synchronization which consequently disables other synchronizations. The tokens in parallel branches of W will eventually, by the progress assumption, will fire a failure transition as the only alternative. Figure 12 shows the CWF-net W_1^+ at a marking where in the upper branch a failure transition occurred. The token in place p_3 will eventually realize at place p_5 that transition t_6 cannot be fired and therefore fires a failure transition.

Theorem 2. *Let W be a sound WF-net. Each failure run of W^+ has an absolute failure.*

Proof. Let $W = (P, T, F)$ be a sound WF-net, let $K = (P_K, T_K, F_K)$ be a failure run of W^+, let f be a failure and let $TN = \{t|\ t \in T_K \text{ and } l(t) \in T\}$ be the set of transition occurrences of W in K. Assume there is an element $x \in TN$ which is concurrent to each $y \in f$. Soundness notion implies that elements following x and y will be synchronized at a transition $t \in TN$, where $x, y <_K t$. Let $v, z \in {}^\bullet t$ be two elements s.t. v and z occurs after x and y respectively, i.e., $x <_K v$ and $y <_K z$. We know from flow relation of paired nets (see Definition 7) that for each transition $t \in T$, input places of t are in P, which implies $l(v), l(z) \in P$. From the same definition we also know that output places of each transition $t^p \in T^a$ are in P', which implies $l(y)^\bullet \subset P'$. By Lemma 1, we know that an element $p \in P$ is not reachable from an element $q' \in P'$. In other words, z cannot occur after y. Therefore there cannot be a transition $x \in TN$ which is concurrent to each $y \in f$.   □

In a CWF-net generated from a sound WF-net, a failure in a concurrent run is not only an absolute failure but also the only failure. An absolute failure simply empties a WF-net and places all the tokens into its inverse copy net. The structure of a paired net prevents the tokens reaching to the WF-net again (see Lemma 1).

Theorem 3. *Let W be a sound WF-net, let K be a failure run of W^+ and let f be a failure. Then f is the only failure of K.*

Proof. Let $W = (P, T, F)$ be a sound WF-net, let $K = (P_K, T_K, F_K)$ be a failure run of W^+, let f be a failure and let $TA = \{t|\ t \in T_K \text{ and } l(t) \in T^a\}$

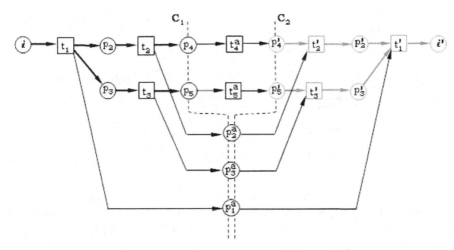

Fig. 11. A failure run of the CWF-net W_1^+

be the set of failure transition occurrences of W^+. Assume f is not the only failure of K. This implies by Definition 11(i) that there is an element $y \in TA$ which occurs after (or before) an element $x \in f$: $x <_K y$. From the definition of paired net(see Definition 7), for each $t^p \in T^a$, ${}^\bullet(t^p) \subset P$, hence occurrence of y requires an element v s.t. $l(v) \in P$ and occurs after x: $x <_K v <_K y$. From the same definition we know also that for each $t^p \in T^a$, $(t^p)^\bullet \subset P'$ which implies $l(x)^\bullet \subset P'$. By Lemma 1, we know that $l(v)$ is not reachable from $l(x)$, hence y cannot occur in K. $\qquad\square$

In the failure run of Fig. 11, $f = \{t_4^a, t_5^a\}$ is the only failure since no failure transition occurs before or after it.

6.3   Recovery in CWF-Nets

A workflow experiencing a failure is required to compensate its completed tasks in the reverse order of their completion. In terms of failure runs that means the run occurring after an absolute failure should be the inverse copy of the run occurring before an absolute failure. A failure run satisfying this requirement is called *completely recovered*.

Definition 12 (Cut, prefix, suffix). *Let $K = (P_K, T_K, F_K)$ be an occurrence net.*

   i. A set $C \subseteq P_K$ is called a cut of K if the elements of C are pairwise concurrent and no $x \in P_K \backslash C$ is concurrent to each $y \in C$.

  ii. A Prefix of a cut C is the net induced from the largest subset $C_{\leftarrow}$ of K s.t. for each $x \in C_{\leftarrow}$, there is a $y \in C$ s.t. $x <_K y$.

 iii. A Suffix of a cut C is the net induced from the largest subset $C_{\rightarrow}$ of K s.t. for each $x \in C$, there is a $y \in C_{\rightarrow}$ s.t. $x <_K y$.

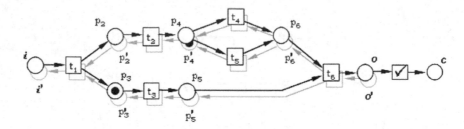

Fig. 12. The CWF-net W_1^+ in a failure state

Definition 13 (Completely recovered). *Let W be a WF-net, let K be a failure run of W^+, let f be an absolute failure and let C_1 and C_2 be two cuts of K s.t. for each $x \in f$, ${}^\bullet x \subseteq C_1$ and $x^\bullet \subseteq C_2$.* [1] *K is called* completely recovered *iff $C_{1\leftarrow} = (C_{2\rightarrow})^*$.*

The failure run of Fig. 11 is a completely recovered run. The markings C_1 and C_2 correspond to the cuts before and after an absolute failure. The activation place occurrences are shared by both cuts. They belong neither to the prefix nor to the suffix of the cuts. As we can see from Fig. 11, the run occurred after the cut C_2 (drawn in grey) is the inverse copy of the run occurred before the C_1, hence completely recovering.

A CWF-net generated from a sound acyclic WF-net has always completely recovered failure runs. In a failure run of a CWF-net W^+, an absolute failure moves a token from a place in W to its copy place in W^*. The only way for W^* to have an inverse copy run is to fire the copy of the same transitions occurred in W. An activation place in a CWF-net contributes to enabling a copy transition, if the corresponding original transition occurs in W. Since W^* has an inverse flow relation of W, W^* will generate an inverse copy run after an absolute failure. In this paper, we excluded cycles in WF-nets. In a failure run, a transition occurring n times requires a compensating transition occurring exactly n times. This semantics cannot be captured within the low level Petri net model.

Definition 14 (Acyclic net). *Let $N = (P, T, F)$ be a net and let $p \in (P \cup T)$. N is an* acyclic net *iff $\neg(p F^+ p)$.*

Theorem 4. *Let W be a sound acyclic WF-net. Each failure run of W^+ is completely recovered.*

Proof. Let $W = (P, T, F)$ be a sound acyclic WF-net, let K be a failure run of W^+, let f be a failure and let C_1 and C_2 be the cuts s.t. for each $x \in f$, ${}^\bullet x \subseteq C_1$ and $x^\bullet \subseteq C_2$. By Theorem 3 we know that f is the only failure of K. Since f is the only failure, for each element $x \in C_{1\leftarrow}$, $l(x) \in W$. By Theorem 2 we know

---

[1] Here we overloaded the subset ($\subset$) operation for nets. The subset operation holds for two nets if it holds between the corresponding place, transition and flow relation sets.

also that f is an absolute failure. Therefore, given a marking M of W enabling each failure transition occurred in K, an absolute failure creates a marking M' of W^*. For $C_{2\rightarrow}$ to be inverse copy of $C_{1\leftarrow}$, W^* with an initial marking M' should fire the copy of the same transitions occurred in $C_{1\leftarrow}$. Since W is acyclic, we don't have to consider multiple occurrences of the same transition. In a paired net (see Definition 7) for each $t \in T$ and $t' \in T'$, there is a place $p^t \in P^a$ s.t. $^\bullet p^t = \{t\}$ and $p^{t\bullet} = \{t'\}$. A transition t' can occur in a concurrent run if and only if transition t occurs before. Therefore, W^* can only fire copy of the transitions that W already fired. Since W^* fires these transitions using $(F')^{-1}$, we can conclude that $C_{2\rightarrow}$ is an inverse copy of $C_{1\leftarrow}$. $\qquad\qquad\square$

A completely recovered run terminates with a single token in the failure place i'. Similarly a run without a failure completes with a single token in the successful sink place c. In case of successful completion, together with the token in c, all the activation places contain a token. We didn't explicitly clean those tokens. Here two modelling approaches can be followed. In the first approach, the tokens in activation places and in place c can be left there and accepted as a final marking. In this case the marking is interpreted as "the workflow has finished its execution and executed the tasks indicated by the activation places". The second approach is to clean the tokens in activation places, which can be interpreted as "the workflow has finished its execution and released its remembered compensations". The token cleaning procedure can be implemented for a CWF-net W^+ by attaching the copy of W*, $(W^*)' = (P'', T'', (F'')^{-1})$, to the activation places with an additional arc from the transition ($\checkmark$) to the place o''. A run of this net upon a successful completion cleans all the tokens.

7 Conclusions and Future Work

In this paper, we developed an approach to specify the compensation of workflows in case of failures. We refined the basic task model of WF-nets to include compensation behavior. We showed that our construction is compositional. We also show that a CWF-net generated from a sound acyclic WF-net meets the expected properties, i.e., either completes successfully or compensates everything executed so far. We didn't present the case where WF-nets contain arbitrary loops and span over multiple organizations. Specifying compensation of cyclic WF-nets remains as our next work. Effects of failures and compensations in inter-organizational workflows still require investigation.

References

1. Aalst, W.M.P. van der: *The Application of Petri Nets to Workflow Management.* The Journal of Circuits, Systems and Computers, **8(1)**:2166, 1998.
2. Aalst, W.M.P. van der, Hofstede, A.H.M. ter: *Workflow Patterns: On the Expressive Power of (Petri-net-based) Workflow Languages.* In K. Jensen, editor, Proceedings of the Fourth Workshop on the Practical Use of Coloured Petri Nets andCPN Tools (CPN 2002), volume **560** of DAIMI, pages 1-20, Aarhus, Denmark, August 2002.

3. Du, W., Davis, J., Shan, M.C.: *Flexible specification of workflow compensation scopes.* GROUP 1997: 309–316.
4. Gray, J., Reuter, A.: *Transaction Processing: Concepts and Techniques.* Morgan Kaufmann, 1993.
5. Garcia-Molina, H., Salem, K.: *Sagas.* In SIGMOD 87: Proceedings of the 1987 ACM SIGMOD international conference on Management of data, pages 249–259. ACM Press, 1987.
6. Leymann, F.: *Supporting Business Transactions Via Partial Backward Recovery In Workflow Management Systems.* BTW 1995: 51–70
7. Reisig, W.: *Elements of Distributed Algorithms: Modelling and Analysis with Petri Nets.* Springer-Verlag, Berlin, Heidelberg, 1998.
8. WFMC: *Workflow Management Coalition, The Workflow Reference Model (**TC00-1003**).* Technical Report, Workflow Management Coalition, Hempshire, UK, 1995.

The Synthesis Problem of Netcharts*

Nicolas Baudru and Rémi Morin

Laboratoire d'Informatique Fondamentale de Marseille
Université de Provence, 39 rue Joliot-Curie, F-13453 Marseille cedex 13, France

Abstract. A netchart is basically a Petri net whose places are located at some
process and whose transitions are labeled by message sequence charts (MSCs).
Two recent papers showed independently that any globally-cooperative high-level
MSC corresponds to the behaviors of some communicating finite-state machine
— or equivalently a netchart. These difficult results rely either on Thomas' graph
acceptors or Zielonka's construction of asynchronous automata. In this paper we
give a direct and self-contained synthesis of netcharts from globally-cooperative
high-level MSCs by means of a simpler unfolding procedure.

1 Introduction

Message Sequence Charts (MSCs) are a popular model often used for the documentation of telecommunication protocols. They profit by a standardized visual and textual
presentation (ITU-T recommendation Z.120 [12]) and are related to other formalisms
such as sequence diagrams of UML. An MSC gives a graphical description of communications between processes. It usually abstracts away from the values of variables and the
actual contents of messages. Yet this formalism can be used at an early stage of design
to detect errors in the specification [11]. In this direction, several studies have already
brought up methods and complexity results for the model-checking and implementation
of MSCs viewed as a specification language [1, 2, 3, 5, 6, 8, 9, 10, 15, 16, 17, 18].

Collections of MSCs are often specified by means of high-level MSCs (HMSCs).
The latter can be seen as directed graphs labeled by component MSCs. However such
specifications may be unrealistic because this formalism allows to specify sets of MSCs
that correspond to no communicating finite-state machine. Furthermore *it is undecidable whether a HMSC describes an implementable language*. In [17], Mukund et al. introduced a new formalism for specifying collections of MSCs: *Netcharts* can be seen as
HMSCs with some distributed control whereas HMSCs require implicitly some global
control over processes in the system. Basically a netchart is a Petri net whose places are
labeled by processes and whose transitions are labeled by MSCs. This new approach
benefits from a graphical description, a formal semantics, and an appropriate expressive power: As opposed to HMSCs, *netcharts describe precisely all implementable
languages* and it is actually easy to derive an equivalent communicating finite-state
machine from a netchart. It follows that it is undecidable whether a HMSC is equivalent to some netchart.

Many model-checking problems are undecidable with general HMSCs. For this reason subclasses of HMSCs have been investigated in the literature, in particular *globally-
cooperative* HMSCs [8]. Logical and algebraic characterizations of these HMSCs were

---

* Supported by the ANR project SOAPDC.

S. Donatelli and P.S. Thiagarajan (Eds.): ICATPN 2006, LNCS 4024, pp. 84–104, 2006.

established in [16] and various related verification techniques are now available [9]. Recently two papers showed that globally-cooperative HMSCs describe implementable languages [5, 9]. These works extend a seminal result by Henriksen et al. who showed that all regular sets of MSCs are implementable [10]. In [5] Bollig and Leucker apply the theory of graph acceptors [21] to prove that any set of MSCs definable in existential MSO logic is implementable. In [9] Genest et al. apply Zielonka's theorem [22] to prove that any existentially-bounded recognizable set of compositional MSCs is implementable. Both studies are rather difficult and quite technical. In the particular case of finitely generated and recognizable sets of MSCs [16], both results imply that any globally-cooperative HMSC describes an implementable set of MSCs, i.e. it corresponds to some netchart. *The aim of this paper is to present a direct, self-contained, and simpler implementation technique to transform a globally-cooperative HMSC into an equivalent netchart.* The translation from netcharts into communicating finite-state machines is rather simple to define but quite tedious to handle in detailed proofs. We adopt in this paper the formalism of netcharts in order to simplify the presentation of our construction. Besides netcharts were at the origine of our first intuitions.

The paper is organized as follows. In Section 1 we recall the basic definitions of MSCs, Petri nets, and netcharts. Next Section 2 presents the semantics of a netchart as the set of MSCs that correspond to the behaviors of some underlying Petri net. Section 3 introduces the notion of HMSC regarded as an automaton labeled by MSCs. We define there a simple but naive transformation of HMSCs into netcharts. In some cases this transformation leads to a netchart whose behaviors differ from those of the given HMSC. Our strategy is motivated by an example that shows that it is sufficient to *unfold* the given HMSC in order to ensure that the naive transformation into netcharts preserves the semantics. Section 4 presents in details our unfolding algorithm of globally-cooperative HMSCs together with some simple but crucial properties of the resulting structure. Finally Section 5 explains why the naive transformation preserves the behaviors when it is applied to the unfolding of any globally-cooperative HMSC.

Our unfolding algorithm proceeds inductively on the number of communication types involved in the given HMSC by defining a family of globally-cooperative HMSCs called triangles and boxes. A triangle corresponds intuitively to a partial unfolding that represents only part of the behaviors starting from a given node of the HMSC. The role of boxes is to complete triangles by connecting copies of triangles with missing edges.

Admittedly this unfolding resembles an algorithm designed recently in [4] in the framework of Mazurkiewicz traces [7] to build asynchronous automata of polynomial size in terms of the number of states from asynchronous systems. However it is often quite difficult to transfer results or techniques from Mazurkiewicz trace theory to the framework of MSCs (see e.g. [2, 9, 10]) because communication no longer means synchronisation. The unfolding procedure presented here differs from the one used in [4] in several aspects: The induction proceeds over communication types, not component basic MSCs; the termination of the construction of boxes relies essentially on the hypothesis that loops of globally-cooperative HMSCs have a connected communication graph whereas [4] unfolds asynchronous systems with possible unconnected loops and termination is there obvious; last but not least, the present unfolding algorithm is *exponential* in the number of nodes of the given HMSC.

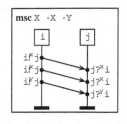

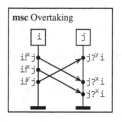

Fig. 1. FIFO MSC   **Fig. 2.** Non-FIFO MSC

2 Background

Message sequence charts (MSCs) are defined by several recommendations that indicate how one should represent them graphically [12]. Examples of MSCs are given in Figures 1 and 2 in which time flows top-down. In this paper we regard MSCs as particular labeled partial orders (or pomsets) following a traditional trend of modeling concurrent executions [7, 13, 20].

A *pomset* over an alphabet Σ is a triple $t = (E, \preccurlyeq, \xi)$ where $(E, \preccurlyeq)$ is a finite partial order and ξ is a mapping from E to Σ. A pomset can be seen as an abstraction of an execution of a concurrent system. In this view, the elements e of E are *events* and their label $\xi(e)$ describes the basic action of the system that is performed by the event $e \in E$. Furthermore, the order $\preccurlyeq$ describes the causal dependence between events.

An *order extension* of a pomset $t = (E, \preccurlyeq, \xi)$ is a pomset $t' = (E, \preccurlyeq', \xi)$ such that $\preccurlyeq \subseteq \preccurlyeq'$. A *linear extension* of t is an order extension that is linearly ordered. It corresponds to a sequential view of the concurrent execution t. Linear extensions of a pomset t over Σ can naturally be regarded as words over Σ. By $\mathrm{LE}(t) \subseteq \Sigma^*$, we denote the set of linear extensions of a pomset t over Σ.

2.1 Basic Message Sequence Charts

We present here a formal definition of basic MSCs. The latter appear as particular pomsets over some alphabet $\Sigma_{\mathcal{I}}^{\Lambda}$ that we introduce first. Let $\mathcal{I}$ be a finite set of processes (also called *instances*) and Λ be a finite set of messages. For any instance $i \in \mathcal{I}$, the alphabet $\Sigma_i^{\Lambda} = \Sigma_{!,i}^{\Lambda} \cup \Sigma_{?,i}^{\Lambda}$ is the disjoint union of the set of *send actions* $\Sigma_{!,i}^{\Lambda} = \{i!^x j \mid j \in \mathcal{I} \setminus \{i\}, x \in \Lambda\}$ and the set of *receive actions* $\Sigma_{?,i}^{\Lambda} = \{i?^x j \mid j \in \mathcal{I} \setminus \{i\}, x \in \Lambda\}$. The alphabets Σ_i^{Λ} are disjoint and we put $\Sigma_{\mathcal{I}}^{\Lambda} = \bigcup_{i \in \mathcal{I}} \Sigma_i^{\Lambda}$. Given an action $a \in \Sigma_{\mathcal{I}}^{\Lambda}$, we denote by $\mathrm{Ins}(a)$ the unique instance i such that $a \in \Sigma_i^{\Lambda}$, that is the particular instance on which each occurrence of action a takes place.

For any pomset $(E, \preccurlyeq, \xi)$ over $\Sigma_{\mathcal{I}}^{\Lambda}$ we denote by $\mathrm{Ins}(e)$ the instance on which the event e occurs: $\mathrm{Ins}(e) = \mathrm{Ins}(\xi(e))$. We say that f *covers* e and we write $e {-\!\!\!<} f$ if $e \prec f$ and $e \prec g \preccurlyeq f$ implies $g = f$. We say that two events e and f are two *matching events* and we write $e \rightsquigarrow f$ if e is the n-th send event $i!^x j$ and f is the n-th receive event $j?^x i$: In other words, we put $e \rightsquigarrow f$ if there are two instances i and j and some message $x \in \Lambda$ such that $\xi(e) = i!^x j, \xi(f) = j?^x i$ and $\mathrm{Card}\{e' \in E \mid \xi(e') = i!^x j \wedge e' \preccurlyeq e\} = \mathrm{Card}\{f' \in E \mid \xi(f') = j?^x i \wedge f' \preccurlyeq f\}$.

DEFINITION 2.1. *A basic* message sequence chart (MSC) *over the set of messages* Λ *is a pomset* $M = (E, \preccurlyeq, \xi)$ *over* $\Sigma_{\mathcal{I}}^{\Lambda}$ *that fulfills the four following conditions:*

M_1: $\forall e, f \in E$: $\mathrm{Ins}(e) = \mathrm{Ins}(f) \Rightarrow (e \preccurlyeq f \vee f \preccurlyeq e)$
M_2: $\forall e, f \in E$: $e \rightsquigarrow f \Rightarrow e \preccurlyeq f$
M_3: $\forall e, f \in E$: $[e \!\prec\!\!\!-\!\!\!< f \wedge \mathrm{Ins}(e) \neq \mathrm{Ins}(f)] \Rightarrow e \rightsquigarrow f$
M_4: $\forall i, j \in \mathcal{I}, \forall x \in \Lambda, |M|_{i!^x j} = |M|_{j?^x i}.$

By M_1, events occurring on the same instance are linearly ordered: Hence non-deterministic choice cannot be described within an MSC. Property M_2 formalizes simply that the reception of any message will occur after the corresponding send event. By M_3, causality in M consists only in the linear dependency over each instance and the ordering of pairs of corresponding send and receive events. Finally, Condition M_4 requires each send event matches some receive event: The matching relation $\rightsquigarrow$ builds a one-to-one correspondence between send events and receive events. We let bMSC denote the set of all basic MSCs. Note here that if two basic MSCs share some linear extension then they are equal. We denote by $\mathrm{Ins}(M)$ the set of *active instances* of a basic MSC M: $\mathrm{Ins}(M) = \{i \in \mathcal{I} \mid \exists e \in E, \mathrm{Ins}(e) = i\}$.

In Figure 2, the basic MSC exhibits some *overtaking* of message y above two messages x. A basic MSC is called *FIFO* if it shows no overtaking, that is, the messages from one instance to another are delivered in the order they are sent (Fig. 1). Non-FIFO basic MSCs allow for scenarios that use several channels (or message types) between pairs of processes (Fig. 2).

For convenience we shall use at some point the notion of MSC *with ϵ-actions*. For each instance $i \in \mathcal{I}$ we define a new symbol ϵ_i and we put $\mathrm{Ins}(\epsilon_i) = i$. Then *a basic MSC with ϵ-actions* is simply a pomset over the extended alphabet $\Sigma_{\mathcal{I}}^{\Lambda} \cup \{\epsilon_i \mid i \in \mathcal{I}\}$ which satisfies the conditions M_1 to M_4.

2.2   Petri Nets

Let us now recall the definition of a Petri net and some usual notations. A *Petri net* is a triple $\mathcal{P} = (P, T, F)$ where P is a set of *places*, T is a set of *transitions* such that $P \cap T = \emptyset$, and $F \subseteq (P \times T) \cup (T \times P)$ is a *flow relation*. We shall use the following usual notations. For all $x \in P \cup T$, we put ${}^\bullet x = \{y \in P \cup T \mid (y, x) \in F\}$ and $x^\bullet = \{y \in P \cup T \mid (x, y) \in F\}$. Clearly, for all transitions t, ${}^\bullet t$ and $t^\bullet$ are sets of places, and conversely for all places $p \in P$, ${}^\bullet p$ and $p^\bullet$ are both sets of transitions. A *marking* $\mathfrak{m}$ of $\mathcal{P}$ is a multiset of places $\mathfrak{m} \in \mathbb{N}^P$. A transition t is *enabled* at $\mathfrak{m} \in \mathbb{N}^P$ if $\mathfrak{m}(p) \geqslant 1$ for all $p \in {}^\bullet t$. In this case, we write $\mathfrak{m}\,[t\rangle\,\mathfrak{m}'$ where the marking $\mathfrak{m}'$ is defined by $\mathfrak{m}'(p) = \mathfrak{m}(p) - 1$ if $p \in {}^\bullet t \setminus t^\bullet$, $\mathfrak{m}'(p) = \mathfrak{m}(p) + 1$ if $p \in t^\bullet \setminus {}^\bullet t$, and $\mathfrak{m}'(p) = \mathfrak{m}(p)$ otherwise.

In this paper, we consider Petri nets provided with an *initial marking* $\mathfrak{m}_{\mathrm{in}}$ and a *finite* set of final markings $\mathfrak{F}$. An *execution sequence* from $\mathfrak{m}$ to $\mathfrak{m}'$ is a word $u = t_1...t_n \in T^\star$ such that there are markings $\mathfrak{m}_0,..., \mathfrak{m}_n$ satisfying $\mathfrak{m}_0 = \mathfrak{m}$, $\mathfrak{m}_n = \mathfrak{m}'$, and $\mathfrak{m}_{k-1}\,[t_k\rangle\,\mathfrak{m}_k$ for all naturals $k \in [1, n]$. Then the sequence $s = \mathfrak{m}_0\,[t_1\rangle\,\mathfrak{m}_1...\mathfrak{m}_{n-1}\,[t_n\rangle\,\mathfrak{m}_n$ is called the *firing sequence* of u from $\mathfrak{m}$ to $\mathfrak{m}'$ and is denoted by $s = \mathfrak{m}\,[u\rangle\,\mathfrak{m}'$. If $\mathfrak{m} = \mathfrak{m}_{\mathrm{in}}$ and $\mathfrak{m}' \in \mathfrak{F}$ then the execution sequence u is called *complete*. The language $L(\mathcal{P})$ consists of all complete execution sequences of $\mathcal{P}$.

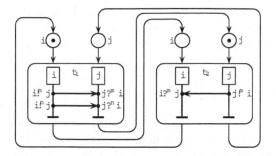 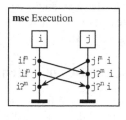

Fig. 3. A netchart $\mathcal{N}$ and a corresponding MSC

2.3   Netcharts

{A netchart is basically a Petri net whose places are labeled by instances and whose transitions are labeled by FIFO basic MSCs. Similarly to Petri nets, netcharts admit an intuitive visual representation: Examples of netcharts are given in Fig. 3, 7, 9 and 11.

DEFINITION 2.2. *A netchart over Λ consists of a Petri net $(P, T, F, \mathfrak{m}_{in}, \mathfrak{F})$ and two mappings* Ins $: P \to \mathcal{I}$ *and* $\mathcal{M} : T \to$ bMSC *such that* Ins *associates each place $p \in P$ with some instance* Ins(p) *and $\mathcal{M}$ associates each transition $t \in T$ with a FIFO basic MSC $\mathcal{M}(t)$ over the set of messages Λ. Three conditions are required for such a structure to be a netchart:*

N_1: *For each instance $i \in \mathcal{I}$, there is a single token within places located on instance i, i.e. $\sum_{\mathrm{Ins}(p)=i} \mathfrak{m}_{in}(p) = 1$.*

N_2: *For each transition $t \in T$ and each instance $i \in \mathcal{I}$ there is at most one place $p \in t^{\bullet}$ such that* Ins$(p) = i$.

N_3: *For each transition $t \in T$ and each instance $i \in \mathcal{I}$ there is at most one place $p \in {}^{\bullet}t$ such that* Ins$(p) = i$.

A netchart is called prime *if for all $t \in T$ we have* Ins$({}^{\bullet}t) =$ Ins$(t^{\bullet}) =$ Ins$(\mathcal{M}(t))$.

By N_1 the initial marking of a netchart is safe; furthermore each instance is associated with a unique initial place. Intuitively this observation extends to the semantics of netcharts: In each reachable marking a token denotes the current local state of each instance. Axiom N_2 stipulates that an instance occurs at most once in the postcondition of any transition. This condition ensures that the local state of each instance corresponds to a single token. Axiom N_3 requires that at most one place located on instance i is a precondition of a given transition. The semantics detailed below will show that transitions that do not satisfy this requirement cannot take part entirely in the behaviors of the netchart: We could remove N_3 without affecting the expressive power of netcharts.

Prime netcharts are those introduced in [17]. This additional requirement ensures in particular that ${}^{\bullet}t \cup t^{\bullet}$ is empty as soon as $\mathcal{M}(t)$ is the empty MSC. In the next section we make use of basic MSCs with ϵ-actions to extend the semantics of prime netcharts studied in [17, 3] to the relaxed setting adopted here. Noteworthy any netchart can easily be transformed into an equivalent prime one: Consequently the expressive power of these extended netcharts is the same as the prime ones. This remark simplifies the presention of our result and allows us to apply to the present setting some of the results from [3].

3   Semantics of Netcharts

In this section we fix a netchart $\mathcal{N} = ((P, T, F, \mathfrak{m}_{in}, \mathfrak{F}), \text{Ins}, \mathcal{M})$ over the set of messages Λ and define formally its behaviors. The semantics of $\mathcal{N}$ consists of FIFO basic MSCs over Λ (Fig. 3). The latter are derived from the FIFO basic MSCs that correspond to the complete execution sequences of some low-level Petri net $\mathcal{P}_{\mathcal{N}}$ (Fig. 5). Actually, the execution sequences of $\mathcal{P}_{\mathcal{N}}$ use a *refined set of messages* Λ° and the behaviors of $\mathcal{N}$ are obtained by projection of messages from Λ° onto Λ.

3.1   From MSCs to Petri Nets

The construction of the low-level Petri net $\mathcal{P}_{\mathcal{N}}$ starts with the translation of each transition $t \in T$ with component FIFO basic MSC $\mathcal{M}(t) = (E, \preccurlyeq, \xi)$ into some Petri net $\mathcal{P}_t = (P_t, T_t, F_t)$. This natural operation is depicted in Fig. 4.

This construction needs to regard each basic MSC (with ϵ-actions) $M = (E, \preccurlyeq, \xi)$ as a dag (direct acyclic graph) denoted by $(E, \prec, \xi)$. For any instance $i \in \mathcal{I}$ we let $\preccurlyeq_i$ be the restriction of $\preccurlyeq$ to events located on instance i. Then $e \prec_i f$ if e occurs immediately before f on instance i. Then the binary relation $\prec$ consists of all pairs of matching events together with all pairs of covering events w.r.t. $\preccurlyeq_i$.

DEFINITION 3.1.   *The MSC dag of a basic MSC $M = (E, \preccurlyeq, \xi)$ with possibly ϵ-actions is a labeled directed acyclic graph $(E, \prec, \xi)$ such that we have $e \prec f$ if $e \leadsto f$ or $e \prec_i f$ for some instance $i \in \mathcal{I}$.*

Clearly we can recover the basic MSC from its MSC dag. The reason for this is that $\prec \subseteq \preccurlyeq$ hence $\preccurlyeq$ is simply the reflexive and transitive closure of $\prec$. That is why we will identify a basic MSC with its corresponding MSC dag in the sequel of this paper.

We can now formalize how each component MSC $\mathcal{M}(t) = (E, \prec, \xi)$ is translated into some Petri net $\mathcal{P}_t = (P_t, T_t, F_t)$. First we add to the basic MSC $\mathcal{M}(t)$ an event labeled ϵ_i on instance i if the instance i is not active in $\mathcal{M}(t)$ while there exists a place $p \in {}^{\bullet}t$ such that $\text{Ins}(p) = i$. Note that these new events are isolated because no other event occurs on this instance.

Now the places P_t are identified with pairs from $\prec$. In particular places do not depend on possibly added events labeled ϵ_i. On the other hand the transitions T_t are

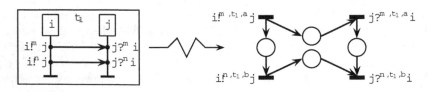

Fig. 4. From transition t_1 to Petri net $\mathcal{P}_{t_1}$

identified with some send or receive actions over the new set of messages $\Lambda^\circ = \Lambda \times T \times P_t$ or with added event labeled by ϵ_i. Formally, we put $P_t = \prec$ and

$$T_t = \{i!^{m,t,(e,f)}j, j?^{m,t,(e,f)}i \mid (e,f) \in \prec \wedge \xi(e) = i!^m j \wedge \xi(f) = j?^m i\}$$
$$\cup\{ (\epsilon_i, t) \mid i \notin \mathrm{Ins}(\mathcal{M}(t)) \wedge \exists p \in {}^\bullet t, \mathrm{Ins}(p) = i\}.$$

Note that the translation from the basic MSC $\mathcal{M}(t)$ into the Petri net $\mathcal{P}_t$ is one-to-one: We will be able to recover the basic MSC $\mathcal{M}(t)$ from the Petri net $\mathcal{P}_t$. For this, we let ρ be the mapping from T_t to E such that $\rho(i!^{m,t,(e,f)}j) = e$, $\rho(j?^{m,t,(e,f)}i) = f$ and $\rho(\epsilon_i, t) = \epsilon_i$. To complete the definition of $\mathcal{P}_t$ we choose a flow relation F_t in accordance with the causality relation $\prec$ of $\mathcal{M}(t)$: We put

$$F_t = \{(r, (e, f)) \in T_t \times P_t \mid \rho(r) = e\} \cup \{((e, f), r) \in P_t \times T_t \mid \rho(r) = f\}.$$

In the next subsection the transitions of the Petri net $\mathcal{P}_t = (P_t, T_t, F_t)$ will be connected to places of $\mathcal{N}$ by means of the following connection relation:

$$F'_t = \{(p, r) \in P \times T_t \mid p \in {}^\bullet t \wedge {}^\bullet r = \emptyset \wedge \mathrm{Ins}(\rho(r)) = \mathrm{Ins}(p)\}$$
$$\cup \{(r, p) \in T_t \times P \mid p \in t^\bullet \wedge r^\bullet = \emptyset \wedge \mathrm{Ins}(\rho(r)) = \mathrm{Ins}(p)\}.$$

3.2 Low-Level Petri Net and Its FIFO Behaviors

Now, in order to build the low-level Petri net $\mathcal{P}_\mathcal{N}$ of the netchart $\mathcal{N}$, we replace each transition $t \in T$ of $\mathcal{N}$ by its corresponding Petri net $\mathcal{P}_t$ as shown in Fig. 5.

The low-level Petri net $\mathcal{P}_\mathcal{N} = (P_\mathcal{N}, T_\mathcal{N}, F_\mathcal{N}, \mathrm{m}_{\mathrm{in}}, \mathfrak{F}_\mathcal{N})$ is built as follows. First, the set of places $P_\mathcal{N}$ collects the places of $\mathcal{N}$ and the places of all $\mathcal{P}_t$: $P_\mathcal{N} = \bigcup_{t \in T} P_t \cup P$. Second, the set of transitions collects all transitions of all $\mathcal{P}_t$: $T_\mathcal{N} = \bigcup_{t \in T} T_t$. For latter purposes we also define the map Comp that associates each transition a from $T_\mathcal{N}$ with the transition $t \in T$ such that $a \in T_t$. Thus $\mathrm{Comp}(i!^{m,t,p}j) = t$, $\mathrm{Comp}(i?^{m,t,p}j) = t$, and $\mathrm{Comp}(\epsilon_i, t) = t$. Now the flow relation consists of the flow relation F_t of each $\mathcal{P}_t$ together with the connection relations F'_t: $F_\mathcal{N} = \bigcup_{t \in T} F_t \cup F'_t$. The initial marking of $\mathcal{P}$ is the one of $\mathcal{N}$: The new places $p \in P_\mathcal{N} \setminus P$ are initially empty. Similarly a marking m of $\mathcal{P}_\mathcal{N}$ is final if the restriction of m to the places of $\mathcal{N}$ is a final marking of $\mathcal{N}$ and if all other places are empty: $\mathfrak{F}_\mathcal{N} = \{\mathrm{m} \in \mathbb{N}^P \mid \mathrm{m}_{|P} \in \mathfrak{F} \wedge \mathrm{m}_{|P_\mathcal{N} \setminus P} = 0\}$.

Any complete execution sequence $u \in L(\mathcal{P}_\mathcal{N})$ of the low-level Petri net leads from the initial marking to some final marking for which all places from $P_\mathcal{N} \setminus P$ are empty.

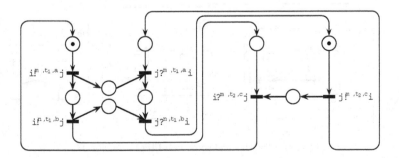

Fig. 5. The low-level Petri net $\mathcal{P}_\mathcal{N}$ associated to the netchart $\mathcal{N}$ of Fig. 3

Moreover u is actually a linear extension of a unique basic MSC over the set of extended messages Λ° that consists of triples (m, t, p).

DEFINITION 3.2. *The MSC language $\mathcal{L}_{\mathrm{fifo}}(\mathcal{P}_\mathcal{N})$ consists of the FIFO basic MSCs M such that at least one linear extension of M is a complete execution sequence of $\mathcal{P}_\mathcal{N}$.*

Interestingly, it can be easily shown that a basic MSC M belongs to $\mathcal{L}_{\mathrm{fifo}}(\mathcal{P}_\mathcal{N})$ if and only if *all* linear extensions of M are complete execution sequences of $\mathcal{P}_\mathcal{N}$. Noteworthy it can happen that a complete execution sequence of the low-level Petri net $\mathcal{P}_\mathcal{N}$ corresponds to a *non-FIFO* MSC (see e.g. [17, Fig. 5] or Fig. 7). Following [17], we focus on FIFO behaviors and neglect this kind of execution sequences in this paper.

3.3   Set of MSCs Associated to Some Netchart

Recall now that MSCs from $\mathcal{L}_{\mathrm{fifo}}(\mathcal{P}_\mathcal{N})$ may contain some events labeled by ϵ_i and use a refined set of messages Λ° that consists of triples (m, t, p) where $m \in \Lambda$, $t \in T$, and $p \in P_t$. We let $\pi^\circ : \Lambda^\circ \to \Lambda$ denote the labelling that associates each triple $(m, t, p) \in \Lambda^\circ$ with the message $m \in \Lambda$. This labelling extends to a function that maps actions from $\Sigma_\mathcal{I}^{\Lambda^\circ}$ onto actions of $\Sigma_\mathcal{I}^\Lambda$ in a natural way. Furthermore this mapping extends in the obvious way from the FIFO basic MSCs over Λ° onto the FIFO basic MSCs over Λ. Since we deal here with MSCs with possibly ϵ-actions, we ask in this paper that π° removes all actions ϵ_i, too. The semantics of the netchart $\mathcal{N}$ is defined now from the semantics of its low-level Petri net $\mathcal{P}_\mathcal{N}$ by means of the projection π°.

DEFINITION 3.3. *The MSC language $\mathcal{L}_{\mathrm{fifo}}(\mathcal{N})$ is the set of FIFO basic MSCs obtained from an MSC of its low-level Petri net by the projection π°: $\mathcal{L}_{\mathrm{fifo}}(\mathcal{N}) = \pi^\circ(\mathcal{L}_{\mathrm{fifo}}(\mathcal{P}_\mathcal{N}))$.*

EXAMPLE 3.4. Consider the netchart $\mathcal{N}_1$ depicted in Figure 7 for which the initial marking is the single final marking. Its language $\mathcal{L}_{\mathrm{fifo}}(\mathcal{N}_1)$ is the set of all basic MSCs that consist only of messages a and b exchanged from i to j in a FIFO manner. The MSC M on the right-hand side of this figure illustrates a complete execution sequence of the low-level Petri net of $\mathcal{N}$ that does not correspond to some FIFO basic MSC.

The main property of *prime* netcharts from [17] is that their MSC language can be implemented in polynomial time as the behaviors of some communicating finite-state machine. Clearly this observation extends easily to the netcharts adopted in this paper.

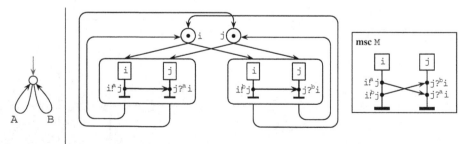

Fig. 6. $\mathcal{G}_1$ **Fig. 7.** Netchart $\mathcal{N}_1$ and some non-FIFO behavior $M \notin \mathcal{L}_{\mathrm{fifo}}(\mathcal{N})$

4   Netcharts vs. High-Level Message Sequence Charts

In this section we recall the equivalent notions of high-level MSCs (HMSCs) and MSC-graphs (MSGs). We recall also some decidability results about the respective expressive power of MSGs and netcharts. By means of three examples we introduce a naive translation of MSGs into netcharts and motivate the seek for an unfolding procedure to ensure a correct implementation of globally-cooperative MSGs as netcharts.

4.1   HMSCs and MSGs

Let us now recall how one can build high-level MSCs from basic MSCs. First, the *asynchronous concatenation* of two basic MSCs $M_1 = (E_1, \preccurlyeq_1, \xi_1)$ and $M_2 = (E_2, \preccurlyeq_2, \xi_2)$ is the basic MSC $M_1 \cdot M_2 = (E, \preccurlyeq, \xi)$ where $E = E_1 \uplus E_2$, $\xi = \xi_1 \cup \xi_2$ and the partial order $\preccurlyeq$ is the transitive closure of $\preccurlyeq_1 \cup \preccurlyeq_2 \cup \{(e_1, e_2) \in E_1 \times E_2 \mid \text{Ins}(e_1) = \text{Ins}(e_2)\}$. This concatenation allows to compose specifications in order to describe infinite sets of basic MSCs: We obtain high-level message sequence charts (HMSCs) as rational expressions or equivalently automata labeled by basic MSCs.

DEFINITION 4.1. *An MSC-graph (MSG) is a structure $\mathcal{G} = (Q, \imath, \Sigma, \longrightarrow, Q_f)$ where Q is a finite set of nodes with some initial node $\imath$ and some final nodes $Q_f \subseteq Q$, Σ is a finite subset of basic MSCs, and $\longrightarrow \subseteq Q \times \Sigma \times Q$ is a set of labeled edges.*

The semantics of MSGs is quite natural. The language associated with an MSG consists of all basic MSCs that are the product of MSCs appearing along a path from the initial node to some final node. By Kleene's theorem, a set of basic MSCs corresponds to some MSG iff it is rational, i.e. it can be built from finite sets by means of union, product, and iteration.

EXAMPLE 4.2. Let A and B be the two components MSCs of the netchart $\mathcal{N}_1$ depicted in Fig. 7. The language $\mathcal{L}_{\text{fifo}}(\mathcal{N}_1)$ corresponds to the HMSC $(A + B)^\star$ and to the MSG of Fig. 6.

We showed in [3] that it is undecidable whether the language $\mathcal{L}_{\text{fifo}}(\mathcal{N})$ of a given netchart is rational, that is, can be described by some MSG [3, Cor. 4.4]. We showed also that it is undecidable whether the language of some given MSG can be described by some netchart [3, Th. 4.7].

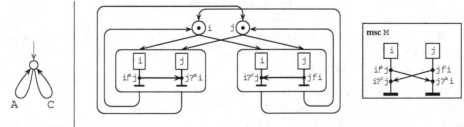

Fig. 8. $\mathcal{G}_2$

Fig. 9. Wrong implementation of $(A + C)^\star$

4.2 Globally-Cooperative MSG

Most model-checking issues related to MSGs are undecidable in general. For this reason subclasses of MSGs have been introduced in the past years. We are here interested in globally-cooperative MSGs from [8]. These MSGs correspond precisely to the ⋆-connected HMSCs from [16] and extend the class of bounded or locally-synchronized MSGs from [1, 18] by removing the requirement that the set of MSCs described by these MSGs be channel-bounded [15]. These restrictions are motivated by a similar approach in Mazurkiewicz trace theory [14, 19].

We need first to introduce the following notion. The *communication graph* $CG(M)$ of a basic MSC $M = (E, \preccurlyeq, \xi)$ is the directed graph $(\mathcal{I}, \mapsto)$ such that $(i, j) \in \mapsto$ if there is an event $e \in E$ such that $\xi(e) = i!^x j$ for some $x \in \Lambda$. An instance $i \in \mathcal{I}$ is called *active* if either $i \mapsto j$ or $j \mapsto i$ for some j. In this paper a directed graph $(\mathcal{I}, \mapsto)$ is called *connected* if the symmetric closure of its restriction to active instances is connected.

DEFINITION 4.3. *An MSG is* globally-cooperative *(for short, a gc-MSG) if for all loops*
$$q_0 \xrightarrow{M_1} q_1 \xrightarrow{M_2} \ldots \xrightarrow{M_n} q_n = q_0 \text{ the product basic MSC } M_1 \cdot M_2 \cdot \ldots \cdot M_n \text{ has a connected}$$
communication graph.

Algebraic and logical characterizations of the languages described by gc-MSGs were established in [16]. More recently two articles showed independently that these languages are implementable by communicating finite-state machines provided that one restricts to FIFO MSCs [5, 9]. On the other hand we have showed in [3, Th. 3.7] that all implementable sets of MSCs can be described by netcharts. As a consequence, *the language of any gc-MSG can be described by some netchart*. Note here that [5] relies on Thomas' graph acceptors [21] whereas [9] is based on the construction of asynchronous cellular automata [22]. Both approaches are quite involved and have high complexity costs. We give in this paper a direct, self-contained, and simpler construction that transforms any given gc-MSG into an equivalent netchart.

4.3 Naive Implementation Technique

Our method uses a translation of MSGs into netcharts illustrated by Figures 6 to 11.

DEFINITION 4.4. *Let* $\mathcal{G} = (Q, \iota, \Sigma, \longrightarrow, Q_f)$ *be an MSG. The corresponding netchart* $\widehat{\mathcal{G}}$ *is the structure* $\widehat{\mathcal{G}} = (P, T, F, \mathfrak{m}_{in}, \mathfrak{F}, \text{Ins}, \mathcal{M})$ *where*
 - $P = Q \times \mathcal{I}$ *with* $\text{Ins}(q, k) = k$,
 - $T = \longrightarrow \subseteq Q \times \Sigma \times Q$ *with* $\mathcal{M}(q_1 \xrightarrow{M} q_2) = M$,
 - *for all edges* $t = (q_1 \xrightarrow{M} q_2)$ *from* T *and all places* $(q, k) \in P$ *we have* $(q, k) \in {}^\bullet t \Leftrightarrow q = q_1$ *and* $(q, k) \in t^\bullet \Leftrightarrow q = q_2$,
 - $\mathfrak{m}_{in} = \{(\iota, k) \mid k \in \mathcal{I}\}$ *and a multiset of places* $\mathfrak{m} \in \mathbb{N}^P$ *is final if there exists a final node* $q_f \in Q_f$ *such that for each* $(q, k) \in Q$ *we have* $\mathfrak{m}(q, k) = 1$ *if* $q = q_f$ *and* $\mathfrak{m}(q, k) = 0$ *otherwise.*

EXAMPLE 4.5. Consider first again the netchart $\mathcal{N}_1$ of Fig. 7 and its two component MSCs A and B. Clearly the MSG $\mathcal{G}_1$ depicted on Fig. 6 accepts $(A + B)^\star$. It is easy to check that $\mathcal{N}_1 = \widehat{\mathcal{G}_1}$ with $\mathcal{I} = \{i, j\}$. Note here that $\widehat{\mathcal{G}_1}$ is a correct implementation of $\mathcal{G}_1$ since $\widehat{\mathcal{G}_1}$ and $\mathcal{G}_1$ both accept $(A + B)^\star$.

EXAMPLE 4.6. Consider now the netchart $\mathcal{N}_2$ of Figure 9 with its two component MSCs A and C. Then $\mathcal{N}_2$ is exactly the netchart $\widehat{\mathcal{G}_2}$ associated with the MSG $\mathcal{G}_2$ of Fig. 8. Observe here that $\mathcal{G}_2$ accepts $(A + C)^*$ whereas $\widehat{\mathcal{G}_2}$ accepts some MSC $M \notin (A + C)^*$ depicted on the right-hand side of Figure 9.

This example shows that the direct construction of the netchart $\widehat{\mathcal{G}}$ from some MSG $\mathcal{G}$ may fail to produce a correct implementation of $\mathcal{G}$. This is no surprise since we know that there are MSGs whose languages are not implementable and it is even undecidable to check implementability of MSGs. That is why we shall restrict to globally-cooperative MSGs in the next section.

Although $\mathcal{G}_2$ and $\widehat{\mathcal{G}_2}$ from Example 4.6 accept distinct languages we have in general the following useful inclusion relation.

PROPOSITION 4.7. *For any MSG $\mathcal{G}$ we have $L(\mathcal{G}) \subseteq L(\widehat{\mathcal{G}})$.*

For each node $q \in Q$ we let m_q denote the marking of the low-level Petri net of the netchart $\widehat{\mathcal{G}}$ such that $\mathsf{m}_q(p) = 1$ if $p = (q, k)$ for some instance k — that is, p is a place from the netchart $\widehat{\mathcal{G}}$ that corresponds to the node q — and $\mathsf{m}_q(p) = 0$ otherwise. We say that a firing sequence $s = \mathsf{m}\,[u\rangle\,\mathsf{m}'$ in the low-level Petri net of $\widehat{\mathcal{G}}$ is *arched* if there are two nodes q and q' in $\mathcal{G}$ such that $\mathsf{m} = \mathsf{m}_q$ and $\mathsf{m}' = \mathsf{m}_{q'}$. Noteworthy each complete execution sequence that leads the low-level Petri net of the netchart $\widehat{\mathcal{G}}$ from its initial marking to some final marking corresponds to an arched firing sequence. The next observation will be used to prove our main technical lemma.

REMARK 4.8. Let $\mathcal{G}$ be an MSG and $\mathsf{m}_q\,[u\rangle\,\mathsf{m}_{q'}$ be an arched firing sequence of the low-level Petri net of $\widehat{\mathcal{G}}$. Then u is the linear extension of some basic MSC M_u. Recall now that each transition t of the low-level Petri net of $\widehat{\mathcal{G}}$ corresponds to a transition $\mathrm{Comp}(t)$ of $\widehat{\mathcal{G}}$ which is defined as an edge $q_1 \xrightarrow{M} q_2$ from $\mathcal{G}$. If an arched firing sequence $\mathsf{m}_q\,[u\rangle\,\mathsf{m}_{q'}$ satisfies $q \neq q'$ and there is some edge a such that all transitions t that appear in u satisfy $\mathrm{Comp}(t) = a$ then a equals $q \xrightarrow{\pi^\circ(M_u)} q'$.

Let j be some instance and q some node of $\mathcal{G}$. The behavior of instance j within a firing sequence of the netchart $\widehat{\mathcal{G}}$ from m_q may be projected to a path from q in Q. Intuitively the local state and the behavior of instance j along a firing sequence corresponds to some token moving from places to places, all located at instance j, some of them corresponding to a state of $\mathcal{G}$. The idea here is simply to collect the sequence of states of $\mathcal{G}$ visited by instance j. Formally we associate inductively each firing sequence $s = \mathsf{m}_q\,[u\rangle\,\mathsf{m}'$ in the low-level Petri net of $\widehat{\mathcal{G}}$ with a path $s{\downarrow}j$ in $\mathcal{G}$ called the *projection of s on instance j* as follows:

- If s is the empty firing sequence restricted to m_q then $s{\downarrow}j = q$;
- If $s = s' \cdot f$ where $f = \mathsf{m}\,[a\rangle\,\mathsf{m}'$ then
  - $s{\downarrow}j = s'{\downarrow}j \cdot t$ if $\mathrm{Ins}(\rho(a)) = j$, $\mathrm{Comp}(a) = t$, and $\sum_{q' \in Q} \mathsf{m}'(q', j) = 1$;
  - and $s{\downarrow}j = s'{\downarrow}j$ otherwise.

4.4 Unfolding Strategy

We conclude this section by introducing our unfolding approach with the help of an example. Let $\mathcal{G}_1 = (Q_1, \imath_1, A, \longrightarrow_1, F_1)$ and $\mathcal{G}_2 = (Q_2, \imath_2, A, \longrightarrow_2, F_2)$ be two MSGs

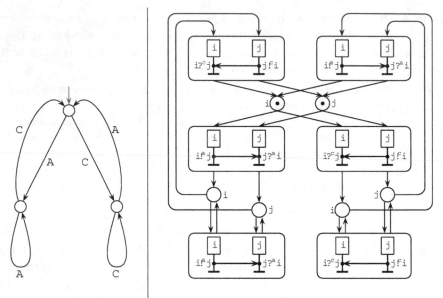

Fig. 10. MSG G_2' **Fig. 11.** Correct implementation of $(A + C)^\star$

over a subset of actions $A \subseteq \Sigma$. A *morphism* $\sigma : \mathcal{G}_1 \to \mathcal{G}_2$ *from* $\mathcal{G}_1$ *to* $\mathcal{G}_2$ is a mapping $\sigma : Q_1 \to Q_2$ from Q_1 to Q_2 such that $\sigma(\imath_1) = \imath_2$, $\sigma(F_1) \subseteq F_2$, and $q_1 \xrightarrow{a}_1 q_1'$ implies $\sigma(q_1) \xrightarrow{a}_2 \sigma(q_1')$. In particular, $L(\mathcal{G}_1) \subseteq L(\mathcal{G}_2)$. Moreover if $\mathcal{G}_2$ is globally-cooperative then $\mathcal{G}_1$ is globally-cooperative, too. A morphism $\sigma : \mathcal{G}_1 \to \mathcal{G}_2$ is called *full* if the following two requirements are satisfied: $\sigma(F_1) = F_2 \cap \sigma(Q_1)$ and for all nodes $q_1 \in Q_1$ and all actions $a \in A$, if $\sigma(q_1) \xrightarrow{a}_2 q_2'$ for some $q_2' \in Q_2$ then $q_1 \xrightarrow{a}_1 q_1'$ for some $q_1' \in Q_1$ such that $\sigma(q_1') = q_2'$. In that case we have $L(\mathcal{G}_1) = L(\mathcal{G}_2)$.

Our strategy is motivated by the following example.

EXAMPLE 4.9. We continue Example 4.6 and consider the MSG $\mathcal{G}_2'$ depicted in Figure 10. Clearly $\mathcal{G}_2'$ accepts $(A + C)^\star$ similarly to $\mathcal{G}_2$. Note here that there is an obvious full morphism from $\mathcal{G}_2'$ onto $\mathcal{G}_2$ which leads us to call informally $\mathcal{G}_2'$ an *unfolding* of $\mathcal{G}_2$. The netchart $\widehat{\mathcal{G}_2'}$ is depicted in Fig. 11. It is not difficult to check that this netchart accepts $(A + C)^\star$, too. Thus $\widehat{\mathcal{G}_2'}$ is a correct implementation of $\mathcal{G}_2$.

This example shows that in some cases it is sufficient to unfold the MSG in order to ensure that the simple translation into netcharts from Definition 4.4 yields a correct implementation. In the two next sections we show that this approach is valid for any gc-MSG.

5   Unfolding of a Globally-Cooperative MSG

In the rest of the paper we fix a *globally-cooperative* MSG $\mathcal{G} = (Q, \imath, \Sigma, \longrightarrow, F)$ where each MSC from Σ is FIFO. The aim of this section is to associate with $\mathcal{G}$ a family of MSGs called *boxes* and *triangles* which are defined inductively. The last box built by this construction will be called the *unfolding* of $\mathcal{G}$ (Def. 5.1). Boxes and triangles are associated with an initial node that may not correspond to the initial node of $\mathcal{G}$. They are associated also with a subset of MSCs $A \subseteq \Sigma$. For these reasons, for any node $q \in Q$ and any subset of actions $A \subseteq \Sigma$, we let $\mathcal{G}_{A,q}$ denote the MSG $(Q, q, A, \longrightarrow_A, F)$ where $\longrightarrow_A$ is the restriction of $\longrightarrow$ to the edges labeled by MSCs in A: $\longrightarrow_A = \longrightarrow \cap (Q \times A \times Q)$.

We shall proceed inductively on directed graphs over $\mathcal{I}$. For each directed graph $T \subseteq \mathcal{I}^2$ we let $\Sigma_T \subseteq \Sigma$ denote the subset of basic MSCs from Σ whose communication graph is included in T. For convenience we put $\mathcal{G}_{T,q} = \mathcal{G}_{\Sigma_T,q}$. We shall define the box $\square_{T,q}$ for all nodes $q \in Q$ and all subgraphs $T \subseteq \mathcal{I}^2$. The box $\square_{T,q}$ is a pair $(\mathcal{B}_{T,q}, \beta_{T,q})$ where $\mathcal{B}_{T,q}$ is an MSG over T and $\beta_{T,q} : \mathcal{B}_{T,q} \to \mathcal{G}_{T,q}$ is a morphism. Similarly, we shall define the triangle $\triangle_{T,q}$ for all nodes q and all *non-empty* subgraphs T. The triangle $\triangle_{T,q}$ is a pair $(\mathcal{T}_{T,q}, \tau_{T,q})$ where $\mathcal{T}_{T,q}$ is an MSG over Σ_T and $\tau_{T,q} : \mathcal{T}_{T,q} \to \mathcal{G}_{T,q}$ is a morphism. Since $\mathcal{G}$ is globally-cooperative (Def. 4.3), all boxes $\mathcal{B}_{T,q}$ and all triangles $\mathcal{T}_{T,q}$ are globally-cooperative, too.

The *height* of a box $\square_{T,q}$ or a triangle $\triangle_{T,q}$ is the cardinality of T. Boxes and triangles are defined inductively on the height. We first define the box $\square_{\emptyset,q}$ for all nodes $q \in Q$. Then triangles of height h are built upon boxes of height $g < h$ and boxes of height h are built upon triangles of height h. More precisely each box $\square_{T,q}$ is made of copies of triangles $\triangle_{T,q'}$. The precise construction of $\square_{T,q}$ will depend on the connectivity of the directed graph T. Moreover we shall make use of the hypothesis that $\mathcal{G}$ is globally-cooperative when defining the construction of the $\square_{T,q}$ associated with a non-connected graph T.

This family of boxes and triangles will lead us to the definition of the unfolding of $\mathcal{G}$ which is simply the box $\mathcal{B}_{T,q}$ with $T = \mathcal{I}^2$ and $q = \imath$.

DEFINITION 5.1. *The* unfolding $\mathcal{G}_{\mathrm{Unf}}$ *of* $\mathcal{G} = (Q, \imath, \Sigma, \longrightarrow, F)$ *is the box* $\mathcal{B}_{\mathcal{I}^2,\imath}$.

Along the definition of boxes we will observe that each morphism $\beta_{T,q} : \mathcal{B}_{T,q} \to \mathcal{G}_{T,q}$ is full. This is precisely the main property of boxes as opposed to triangles.

The base case of the induction deals with boxes of height 0. For all nodes $q \in Q$, the box $\square_{\emptyset,q}$ consists of the morphism $\beta_{\emptyset,q} : \{q\} \to Q$ that maps q to itself together with the MSG $\mathcal{B}_{\emptyset,q} = (\{q\}, q, \emptyset, \emptyset, F_{\emptyset,q})$ where $F_{\emptyset,q} = \{q\}$ if $q \in F$ and $F_{\emptyset,q} = \emptyset$ otherwise. More generally a node of a box or a triangle is final if it is associated with a final node of $\mathcal{G}$.

5.1 Building Triangles from Boxes

Triangles are made of boxes of lower height. Boxes are inserted into a triangle inductively along a tree-like structure and several copies of the same box may appear within a triangle. We need to keep track of this structure in order to prove properties of triangles (and boxes) inductively. This requires to distinguish between nodes inserted within different copies of different boxes or different copies of the same box. To achieve this, each node of a triangle is equipped with a *rank* $k \in \mathbb{N}$ such that all nodes with the same rank come from the same copy of the same box. For these reasons, a node of a triangle $\triangle_{T^\circ,q^\circ} = (\mathcal{T}_{T^\circ,q^\circ}, \tau_{T^\circ,q^\circ})$ is encoded as a quadruple $v = (w, T, q, k)$ such that w is a node from the box $\square_{T,q}$ with $T \subsetneq T^\circ$; moreover v is added within the k-th box inserted into the triangle in construction. By convention the node v maps to the node $\tau_{T^\circ,q^\circ}(v) = \beta_{T,q}(w) \in Q$, i.e. the insertion of boxes preserves the correspondence to the nodes of $\mathcal{G}$. Thus the morphism τ_{T°,q° of a triangle $\triangle_{T^\circ,q^\circ}$ is encoded in the data structure of its nodes. We denote by $\mathcal{B}' = \text{MARK}(\mathcal{B}, T, q, k)$ the generic process that creates a copy $\mathcal{B}'$ of an MSG $\mathcal{B}$ by replacing each node w of $\mathcal{B}$ by $v = (w, T, q, k)$.

The construction of the triangle $\triangle_{T^\circ,q^\circ}$ starts with using this marking procedure and building a copy $\text{MARK}(\square_{\emptyset,q^\circ}, \emptyset, q^\circ, 1)$ of the base box $\square_{\emptyset,q^\circ}$ which gets rank $k = 1$ and whose marked initial node $(\imath_{\square,\emptyset,q^\circ}, \emptyset, q^\circ, 1)$ becomes the initial node of $\triangle_{T^\circ,q^\circ}$. Along the construction of this triangle, an integer variable k counts the number of boxes already inserted in the triangle to make sure that all copies inserted get distinct ranks. The construction of the triangle $\triangle_{T^\circ,q^\circ}$ proceeds by successive insertions of copies of boxes according to the single following rule.

A new copy of the box $\square_{T',q'}$ is inserted into the triangle $\triangle_{T^\circ,q^\circ}$ in construction if there exists a node $v = (w, T, q, l)$ in the triangle in construction and a basic MSC $M \in \Sigma_{T^\circ}$ such that

T_1: $\beta_{T,q}(w) \xrightarrow{M} q'$ *in the MSG $\mathcal{G}_{T^\circ,q^\circ}$;*

T_2: $T \subsetneq T' \subsetneq T^\circ$ *and* $T' = T \cup CG(M)$;

T_3: *no edge labeled by M relates so far v to the initial node of some copy of*
$\square_{T',q'}$ *in the triangle in construction.*

In that case an edge labeled by M is added in the triangle in construction from v to the initial node of the new copy of the box $\square_{T',q'}$.

Note here that Condition T_2 ensures that inserted boxes have height at most $|T^\circ| - 1$. By construction all copies of boxes inserted in a triangle are related in a tree-like structure built along the application of the above rule. It is easy to implement the construction of a triangle from boxes as specified by the insertion rule above by means of a list of inserted boxes whose possible successors have not been investigated, in a depth-first-search or breadth-first-search way. Note here that if a new copy of the box $\square_{T',q'}$ is inserted and connected from $v = (w, T, q, l)$ then $T \subsetneq T'$ thus the communication graph T grows along the branches of this tree-structure. This shows that this insertion process eventually stops and the resulting tree has depth at most $|T|$. Moreover, since we start from the empty box and edges in boxes $\square_{T,q}$ carry basic MSCs from Σ_T, we get the next key property.

LEMMA 5.2. *If a word $u \in \Sigma^\star$ leads in the triangle $\triangle_{T^\circ,q^\circ}$ from its initial node to some node $v = (w, T, q, l)$ then the communication graph of u is precisely T.*

Note also that it is easy to check that the mapping τ_{T°,q° induced by the data structure builds a morphism from $\triangle_{T^\circ,q^\circ}$ to $\mathcal{G}_{T^\circ,q^\circ}$. However this morphism may not be full in some cases. The role of boxes is precisely to take care of this drawback with the help of the next notion.

DEFINITION 5.3. *Let $T^\circ \subseteq \mathcal{I}^2$ be a subgraph of $\mathcal{I}^2$ and q°, q' be two nodes of $\mathcal{G}$. The set of missing edges $\text{MISSING}(T^\circ, q^\circ, q')$ consists of all pairs (v, M) where $v = (w, T, q, l)$ is a node of $\triangle_{T^\circ,q^\circ}$ and M is a basic MSC such that*

- $\beta_{T,q}(w) \xrightarrow{M} q'$ *in the MSG $\mathcal{G}_{T^\circ,q^\circ}$;*
- $T \subsetneq T \cup CG(M) = T^\circ.$

Note here that the insertion rule T_2 for triangles forbids to insert a box $\mathcal{B}_{T^\circ,q}$ and to add an edge labeled by M from node v. This missing edge will be added into boxes of height $|T^\circ|$ in order to get a full morphism.

5.2 Building Boxes from Triangles

Boxes $\square_{T^\circ,q^\circ}$ are made of triangles $\triangle_{T^\circ,q}$ associated with the same directed graph T°. Again several copies of the same triangle are often necessary to build a box and the structure relating these triangles plays a crucial role. For this reason we adopt a similar data structure: A node w of a box $\square_{T^\circ,q^\circ}$ is a quadruple (v, T°, q, k) where v is a node of the triangle $\triangle_{T^\circ,q}$ and $k \in \mathbb{N}$. The rank k will allow us to distinguish between different copies of the same triangle. The construction of boxes uses here again an integer variable k that counts the number of triangles already inserted in the box in construction to make sure that all copies inserted get distinct ranks. On the other hand the parameter T is useless here but we keep it to get a uniform data structure.

As announced in the introduction of this section the construction of the box $\square_{T^\circ,q^\circ}$ depends on the connectivity of T°. Recall that an instance $i \in \mathcal{I}$ is *active* in the directed graph $T^\circ \subseteq \mathcal{I}^2$ if there is an edge $(i, j) \in T^\circ$ or an edge $(j, i) \in T^\circ$ for some instance $j \neq i$. Moreover a directed graph $T^\circ \subseteq \mathcal{I}^2$ is *connected* if the symmetric closure of its restriction to its active instances is connected.

We assume first that $T^\circ \subseteq \mathcal{I}^2$ is a *non-connected* directed graph and define the box $\square_{T^\circ,q^\circ}$. The definition of boxes with a connected directed graph is postponed to the next subsection. The construction of the box $\square_{T^\circ,q^\circ}$ starts with building a copy $\mathrm{MARK}(\triangle_{T^\circ,q^\circ}, T^\circ, q^\circ, 1)$ of the triangle $\triangle_{T^\circ,q^\circ}$ which gets rank $k = 1$ and whose marked initial node $(\imath_{\triangle,T^\circ,q^\circ}, \emptyset, q^\circ, 1)$ becomes the initial node of $\square_{T^\circ,q^\circ}$. The construction of the box $\square_{T^\circ,q^\circ}$ proceeds then by successive insertions of copies of triangles in a tree-like structure according to the following rule (which differs from [4]).

A new copy of the triangle $\triangle_{T^\circ,q'}$ is inserted into the box $\square_{T^\circ,q^\circ}$ in construction if there exists a node $w = (v, T^\circ, q, l)$ in the box in construction and a basic MSC $M \in \Sigma_{T^\circ}$ such that we have $(v, M) \in \mathrm{MISSING}(T^\circ, q, q')$ and no edge labeled by M relates so far w to the initial node of some copy of $\triangle_{T^\circ,q'}$ in the box in construction. In that case an edge labeled by M is added in the box from w to the initial node of the new copy of the triangle $\triangle_{T^\circ,q'}$.

At each step of this procedure we have a morphism from the box in construction to $\mathcal{G}$ which is encoded in the data-structure of nodes. In particular the initial node of each triangle $\triangle_{T^\circ,q}$ maps to node q of $\mathcal{G}$.

By means of Lemma 5.2 the definition of missing edges (Def. 5.3) leads us to the following property.

LEMMA 5.4. *Within a box $\square_{T^\circ,q^\circ}$ associated with a non-connected graph T°, if a word $u \in \Sigma^*$ leads from the initial node of a triangle to the initial node of another triangle then the communication graph of u is precisely T°.*

Recall now that T° is not connected and $\mathcal{G}$ is globally-cooperative. Therefore a branch of the tree-structure of a box in construction cannot involve twice the same triangle, otherwise we get a loop with communication graph T° in $\mathcal{G}$ which contradicts the definition of a globally-cooperative MSG. It follows that this procedure stops and the depth of the resulting tree-structure is at most $|Q|$. As a consequence the size of a box is exponential in the size of the given HMSC.

5.3   Building Boxes with a Connected Graph

We come now to the definition of boxes associated with a connected directed graph. This part is more subtle than the two previous constructions which have a tree-structure: Both do not create new loops in the unfolding. On the contrary the construction of boxes associated with a connected directed graph essentially deals with loops.

Let $T^\circ \subseteq \mathcal{I}^2$ be a connected (non-empty) directed graph. Basically the connected box $\square_{T^\circ,q^\circ}$ collects all triangles $\triangle_{T^\circ,q}$ for all nodes $q \in Q$. Each triangle is replicated a fixed number of times and copies of triangles are connected in some very specific way.

The construction of the box $\square_{T^\circ,q^\circ}$ consists in two steps. First m copies of each triangle $\triangle_{T^\circ,q}$ are inserted in the box. Moreover the first copy of $\triangle_{T^\circ,q^\circ}$ gets rank 1 and the first copy of its initial node becomes the initial node of the box in construction. The actual value of m will be discussed below. For simplicity's sake we require also that copies of the same triangle have consecutive ranks: In particular copies of $\triangle_{T^\circ,q^\circ}$ get ranks 1 to m. In a second step edges are added to connect these triangles to each other. The idea here is to take care of the missing edges in order to get a full morphism: For each triangle $\triangle_{T^\circ,q}$, for each node $q' \in Q$, and for each missing edge $(v, M) \in$ MISSING(T°, q, q') we add an edge labeled by M from each copy of node v to some copy of the initial node of triangle $\triangle_{T^\circ,q'}$.

In this process of connecting triangles we require two key properties:

C_1: *No added edge connects two nodes from the same copy of the same triangle: There is no added edge from node (v, T°, q, l) with rank l to $(\imath_{\triangle,T^\circ,q}, T^\circ, q, l)$.*

C_2: *At most one edge connects one copy of $\triangle_{T^\circ,q}$ to one copy of $\triangle_{T^\circ,q'}$: If we add from a copy of $\triangle_{T^\circ,q}$ of rank l an edge $(v_1, T^\circ, q, l) \xrightarrow{M_1} (\imath_{\triangle,T^\circ,q'}, T^\circ, q', l')$ and an edge $(v_2, T^\circ, q, l) \xrightarrow{M_2} (\imath_{\triangle,T^\circ,q'}, T^\circ, q', l')$ to the same copy of $\triangle_{T^\circ,q'}$ then $v_1 = v_2$ and $M_1 = M_2$.*

Condition C_1 requires simply two copies of each triangle. The number of added edges from a fixed copy of $\triangle_{T^\circ,q}$ to copies of $\triangle_{T^\circ,q'}$ is $|$MISSING$(T^\circ, q, q')|$. It follows that the two conditions above require only

$$m = \max_{q,q' \in Q} |\text{MISSING}(T^\circ, q, q')| + 1$$

copies of each triangle. The construction of the box $\square_{T^\circ,q^\circ}$ starts with the insertion of m copies of each triangle $\triangle_{T^\circ,q}$. Then for a fixed copy of $\triangle_{T^\circ,q}$ and for a fixed node q' we add at most m edges as follows: For each missing edge $(v, M) \in$ MISSING(T°, q, q') the copy of node v is connected to a distinct copy of the initial node of triangle $\triangle_{T^\circ,q'}$. In case $q = q'$ we make sure that v does not get connected along this process to the initial node of the triangle it belongs to.

From the definition of missing edges (Def. 5.3) it follows that the data-structure defines a morphism from the box $\Box_{T^\circ,q^\circ}$ to $\mathcal{G}_{T^\circ,q^\circ}$. Furthermore Lemma 5.2 yields the following useful property.

LEMMA 5.5. *Within a box $\Box_{T^\circ,q^\circ}$ associated with a connected graph T°, if a non-empty word $u \in \Sigma^*$ leads from the initial node of a triangle to the initial node of a triangle then the communication graph of u is precisely T°.*

6 Properties of the Unfolding

6.1 Main Result

The constructions of triangles and boxes yield morphisms to $\mathcal{G}_{T,q}$ that are built inductively on the data-structure. These morphisms are useful in particular to check that the construction of a box with a non-connected directed graph eventually stops because $\mathcal{G}$ is globally-cooperative. We can also check by induction the following useful property.

LEMMA 6.1. *The morphism $\beta_{T,q}$ from a box $\mathcal{B}_{T,q}$ to $\mathcal{G}_{T,q}$ is full.*

Following Definition 5.1 the last box built yields the unfolding MSG $\mathcal{G}_{\mathrm{Unf}}$ together with a full morphism $\beta_{\mathrm{Unf}} : \mathcal{G}_{\mathrm{Unf}} \to \mathcal{G}$ which ensures that $L(\mathcal{G}_{\mathrm{Unf}}) = L(\mathcal{G})$. By Proposition 4.7 we have also $L(\mathcal{G}_{\mathrm{Unf}}) \subseteq L(\widehat{\mathcal{G}_{\mathrm{Unf}}})$. We will prove below that the converse inclusion relation holds (Lemma 6.6) by induction on the structure of boxes and triangles: Thus $L(\mathcal{G}) = L(\widehat{\mathcal{G}_{\mathrm{Unf}}})$. In that way we get our main result.

THEOREM 6.2. *For any globally-cooperative MSG $\mathcal{G}$ the unfolding MSG $\mathcal{G}_{\mathrm{Unf}}$ leads to a netchart $\widehat{\mathcal{G}_{\mathrm{Unf}}}$ such that $L(\mathcal{G}) = L(\widehat{\mathcal{G}_{\mathrm{Unf}}})$.*

Thus our unfolding procedure builds an unfolded globally-cooperative MSG for which the naive construction of a corresponding netchart yields a correct implementation of the specification.

6.2 Properties of Arched Firing Sequences

Let T be a non-empty subgraph of $\mathcal{I}^2$ and $q \in Q$. Let v be a node from the triangle $\mathcal{T}_{T,q}$. By construction of $\mathcal{T}_{T,q}$, v is a quadruple (w, T', q', k') such that w is a node from the box $\Box_{T',q'}$ and $k' \in \mathbb{N}$. Then we say that the *box location* of v is $l^\Box(v) = (T', q', k')$. We define the *sequence of boxes* visited along a path $s = v \xrightarrow{u} v'$ in $\mathcal{T}_{T,q}$ as follows:

- If the length of s is 0 then s corresponds to node v of $\mathcal{T}_{T,q}$ and $\mathbb{L}^\Box(s) = l^\Box(v)$.
- If s is a product $s = s' \cdot t$ where t is the edge $v'' \xrightarrow{a} v'$ then two cases appear:
  - If $l^\Box(v'') = l^\Box(v')$ then $\mathbb{L}^\Box(s) = \mathbb{L}^\Box(s')$;
  - If $l^\Box(v'') \neq l^\Box(v')$ then $\mathbb{L}^\Box(s) = \mathbb{L}^\Box(s').l^\Box(v')$.

Due to the tree-like structure of triangles we have the following obvious property.

PROPOSITION 6.3. *Let $\mathcal{T}_{T,q}$ be a triangle with T a non-empty subgraph of $\mathcal{I}^2$. Let s be an arched firing sequence of the low-level Petri net of $\widehat{\mathcal{T}_{T,q}}$. Then $\mathbb{L}^\Box(s{\downarrow}k) = \mathbb{L}^\Box(s{\downarrow}k')$ for each instance $k, k' \in \mathcal{I}$.*

Similarly to triangles, we define the *triangle location* $l^\triangle(w)$ of a node w in a box $\mathcal{B}_{T,q}$ and the *sequence of triangles* $\mathbb{L}^\triangle(s)$ visited along a path $s = w \xrightarrow{u} w'$ in $\mathcal{B}_{T,q}$. The tree-like structure of *unconnected* boxes yields a property similar to Proposition 6.3. We aim now at establishing a similar property for connected boxes (Prop. 6.5).

Let i, j be two distinct instances. For each firing sequence $s = \mathfrak{m} \, [u\rangle \, \mathfrak{m}'$ of the low-level Petri net of a netchart we define the projection of s on i w.r.t. (i, j) as the sequence of messages $send(s, i, j) = m_1 \dots m_n$ such that the sequence of send actions from i to j in u consists of $i!^{m_1}j, \dots, i!^{m_n}j$. Similarly we define the projection of s on j w.r.t. (i, j) as the sequence of messages $receive(s, i, j) = m_1 \dots m_n$ such that the sequence of receive actions on j from i in u consists of $j?^{m_1}i, \dots, j?^{m_n}i$. It is clear that if a firing sequence $s = \mathfrak{m} \, [u\rangle \, \mathfrak{m}'$ of the low-level Petri net of a netchart corresponds to a FIFO basic MSC then $send(s, i, j) = receive(s, i, j)$ for each pair of distinct instances $i, j \in \mathcal{I}$. This observation leads us to the next result.

LEMMA 6.4. *Let $\mathcal{B}_{T,q}$ be a box with T a non-empty connected subgraph of $\mathcal{I}^2$ and let i, j be two distinct instances such that $(i, j) \in T$. Let s be an arched firing sequence of the low-level Petri net of the netchart $\widehat{\mathcal{B}_{T,q}}$ that corresponds to a FIFO basic MSC. Then $\mathbb{L}^\triangle(s{\downarrow}i) = \mathbb{L}^\triangle(s{\downarrow}j)$.*

Proof. Since s is arched the first (resp. last) triangles coincide in $\mathbb{L}^\triangle(s{\downarrow}i)$ and $\mathbb{L}^\triangle(s{\downarrow}j)$. This result follows now from the three next observations. First, let m be a message in $send(s, i, j)$. Due to the the definition of a low-level Petri net, the message m corresponds to a unique transition $t = i!^m j$ in the low-level Petri net of the netchart $\widehat{\mathcal{B}_{T,q}}$ and moreover $\mathrm{Comp}(t)$ is an edge from $\mathcal{B}_{T,q}$. Thus the sequence of messages $send(s, i, j)$ maps in a natural way to a sequence of edges of the connected box $\mathcal{B}_{T,q}$ and consequently to the sequence of corresponding triangles. Second Lemma 5.5 ensures that at least one send action from i to j occurs when the path $s{\downarrow}i$ goes through a triangle of $\mathcal{B}_{T,q}$. Third, due to Condition $\mathbf{C}_1$ of the construction of connected boxes, when the path $s{\downarrow}i$ goes out of a triangle then it enters into a *distinct* triangle.

These three facts imply that $send(s, i, j)$ is enough to recover the sequence of triangles $\mathbb{L}^\triangle(s{\downarrow}i)$ visited by i along s. A similar observation holds for the process j and $receive(s, i, j)$. We can now conclude easily. If s is an arched firing sequence of the low-level Petri net of $\widehat{\mathcal{B}_{T,q}}$ that corresponds to a FIFO basic MSC, then $send(s, i, j) = receive(s, i, j)$ hence $\mathbb{L}^\triangle(s{\downarrow}i) = \mathbb{L}^\triangle(s{\downarrow}j)$. ∎

PROPOSITION 6.5. *Let $\mathcal{B}_{T,q}$ be a box with T a non-empty connected subgraph of $\mathcal{I}^2$. Let $s = \mathfrak{m} \, [u\rangle \, \mathfrak{m}'$ be an arched firing sequence of the low-level Petri net of $\widehat{\mathcal{B}_{T,q}}$ that corresponds to a FIFO basic MSC M_s. Then there exists an arched firing sequence $s^\dagger = \mathfrak{m} \, [u^\dagger\rangle \, \mathfrak{m}'$ of the low-level Petri net of $\widehat{\mathcal{B}_{T,q}}$ that corresponds to a FIFO basic MSC $M_{s^\dagger}$ such that*

- *$\pi^\circ(M_s) = \pi^\circ(M_{s^\dagger})$,*
- *$\mathbb{L}^\triangle(s^\dagger{\downarrow}k) = \mathbb{L}^\triangle(s^\dagger{\downarrow}k')$ for each pair of instance $k, k' \in \mathcal{I}$.*

Proof. Consider first two *active* instances k and k' of T. Lemma 6.4 ensures that $\mathbb{L}^\triangle(s{\downarrow}k) = \mathbb{L}^\triangle(s{\downarrow}k')$ because T is connected. Now the processes that are not active in T produce in M only ϵ-actions because all transitions that may take place on these processes in the low-level Petri net of $\widehat{\mathcal{B}_{T,q}}$ are ϵ-events. Thus we can force them to behave like a fixed active process k of T. The result is an MSC $M_{s\dagger}$ that differs from M only in ϵ-events located on non-active processes. Consequently, $\pi^\circ(M_s) = \pi^\circ(M_{s\dagger})$ and $\mathbb{L}^\triangle(s^\dagger \downarrow k) = \mathbb{L}^\triangle(s^\dagger \downarrow k')$ for each non-active instance k' of T. $\blacksquare$

6.3   Main Technical Result

LEMMA 6.6. *We have $L(\widehat{\mathcal{G}_{\text{Unf}}}) \subseteq L(\mathcal{G}_{\text{Unf}})$.*

Proof. We proceed by induction. We show for each natural $n \in \{0, 1, 2, ..., |\mathcal{I}^2|\}$ the property $H(n)$ which consists of two similar sub-properties:

1. For each $T \subseteq \mathcal{I}^2$ with $1 \leqslant |T| \leqslant n + 1$ and all nodes $q \in Q$ if $s = \mathfrak{m}_v [u\rangle \mathfrak{m}_{v'}$ is an arched firing sequence of the low-level Petri net of $\widehat{\mathcal{T}_{T,q}}$ that corresponds to a FIFO basic MSC M then $\pi^\circ(M)$ leads in $\mathcal{T}_{T,q}$ from v to v'.
2. For each $T \subseteq \mathcal{I}^2$ with $0 \leqslant |T| \leqslant n$ and all nodes $q \in Q$ if $s = \mathfrak{m}_v [u\rangle \mathfrak{m}_{v'}$ is an arched firing sequence of the low-level Petri net of $\widehat{\mathcal{B}_{T,q}}$ that corresponds to a FIFO basic MSC M then $\pi^\circ(M)$ leads in $\mathcal{B}_{T,q}$ from v to v'.

The proof of Lemma 6.6 follows from $H(n)$ with $n = |\mathcal{I}|^2$. The base case $H(0)$ is obvious because for each $q \in Q$ and each singleton T, the box $\mathcal{B}_{\emptyset,q}$ and the triangle $\mathcal{T}_{T,q}$ consist of a single node.

Induction step of H: We assume now $H(n)$. We show $H(n + 1)$ for connected boxes only, but the cases of triangles and unconnected boxes are similar. We consider some connected subgraph $T \subseteq \mathcal{I}^2$ with $|T| = n + 1$ and some node $q \in Q$. First, we prove by induction that for each natural $d \in \mathbb{N}$ the intermediate property $P(d)$ holds:

$P(d)$: Let L be a sequence of triangles of $\mathcal{B}_{T,q}$ such that $1 \leqslant |L| \leqslant d$. Let $s = \mathfrak{m}_v [u\rangle \mathfrak{m}_{v'}$ be an arched firing sequence of the low-level Petri net of $\widehat{\mathcal{B}_{T,q}}$ that corresponds to a FIFO basic MSC M. If $\mathbb{L}^\triangle(s\downarrow k) = L$ for each process $k \in \mathcal{I}$ then $\pi^\circ(M)$ leads in $\mathcal{B}_{T,q}$ from v to v'.

The base case $P(1)$ follows basically from the induction hypothesis $H(n)$ because in this case s can be viewed as an arched firing sequence of $\widehat{\mathcal{T}_{T,q}}$.

Induction step of P: We assume now that $P(d)$ holds and we prove $P(d+1)$. Let $L.l$ be a sequence of triangles with $|L.l| = d + 1$ and let $s = \mathfrak{m}_v [u\rangle \mathfrak{m}_{v'}$ be an arched firing sequence of the low-level Petri net of $\widehat{\mathcal{B}_{T,q}}$ that corresponds to a FIFO basic MSC M such that $\mathbb{L}^\triangle(s\downarrow k) = L.l$ for each process $k \in \mathcal{I}$. Due to the structure of connected boxes, we claim that we can find an other arched firing sequence $s' = s_1 \cdot s_2 \cdot s_3$ for which $s_1 = \mathfrak{m}_v [u_1\rangle \mathfrak{m}_{v_1}$, $s_2 = \mathfrak{m}_{v_1} [u_2\rangle \mathfrak{m}_{v_2}$ and $s_3 = \mathfrak{m}_{v_2} [u_3\rangle \mathfrak{m}_{v'}$ are three arched firing sequences such that s_2 is non-empty and

S1. $u_1.u_2.u_3$ corresponds to a linear extension of M,
S2. each transition t that appears in u_3 comes from an edge $\text{Comp}(t)$ of $\mathcal{B}_{T,q}$ that occurs within the last triangle l visited along s',
S3. each transition t that appears in u_2 satisfies $\text{Comp}(t) = a$ where a is the unique edge (by Condition C_2 of connected boxes) of $\mathcal{B}_{T,q}$ that relies the two last triangles visited along s'.

In particular, Condition S1 implies that $\mathbb{L}^{\triangle}(s{\downarrow}k) = \mathbb{L}^{\triangle}(s'{\downarrow}k)$ for each process $k \in \mathcal{I}$. Conditions S2 and S3 ensure that $\mathbb{L}^{\triangle}(s_3{\downarrow}k) = l$ and $\mathbb{L}^{\triangle}(s_1{\downarrow}k) = L$. Moreover, Remark 4.8 shows that s_1, s_2 and s_3 correspond respectively to some basic MSCs M_1, M_2 and M_3. Then by Condition S1 we have $M = M_1 \cdot M_2 \cdot M_3$. Therefore these three basic MSCs are FIFO because M is FIFO. Using the induction hypothesis $P(d)$ we deduce that $v \overset{\pi^{\circ}(M_1)}{\longrightarrow} v_1$ and $v_2 \overset{\pi^{\circ}(M_3)}{\longrightarrow} v'$ in $\mathcal{B}_{T,q}$. To conclude, we use Remark 4.8 with S3 and obtain that a is actually the edge $v_1 \overset{\pi^{\circ}(M_2)}{\longrightarrow} v_2$ of $\mathcal{B}_{T,q}$. As a result $v \overset{\pi^{\circ}(M)}{\longrightarrow} v'$ is a path of $\mathcal{B}_{T,q}$. This conclude the proof of $P(d+1)$.

We return now to the proof of $H(n+1)$. Let $s = \mathfrak{m}_v\,[u\rangle\,\mathfrak{m}_{v'}$ be an arched firing sequence of the low-level Petri net of $\widehat{\mathcal{B}_{T,q}}$ that corresponds to some FIFO basic MSC M. By Proposition 6.5, there exists an arched firing sequence $s^{\dagger} = \mathfrak{m}_v\,[u^{\dagger}\rangle\,\mathfrak{m}_{v'}$ of the low-level Petri net of $\widehat{\mathcal{B}_{T,q}}$ that corresponds to a FIFO basic MSC $M^{\dagger}$ such that (*) $\pi^{\circ}(M) = \pi^{\circ}(M^{\dagger})$ and $\mathbb{L}^{\triangle}(s^{\dagger}{\downarrow}i) = \mathbb{L}^{\triangle}(s^{\dagger}{\downarrow}j) = L$ for each pair of instances $i, j \in \mathcal{I}$. Then we can apply $P(|L|)$ together with (*) to get that $v \overset{\pi^{\circ}(M)}{\longrightarrow} v'$. This conclude the proof of $H(n+1)$.   ∎

References

1. Alur R. and Yannakakis M.: *Model Checking of Message Sequence Charts*. CONCUR, LNCS **1664** (1999) 114–129
2. Baudru N. and Morin R.: *Safe Implementability of Regular Message Sequence Charts Specifications*. Proc. of the ACIS 4th Int. Conf. SNDP (2003) 210–217
3. Baudru N. and Morin R.: *The Pros and Cons of Netcharts*. CONCUR, LNCS **3170** (2004) 99–114
4. Baudru N. and Morin R.: *Unfolding Synthesis of Asynchronous Automata*. Accepted to the International Computer Science Symposium in Russia (CSR), LNCS (2006) – 12 pages.
5. Bollig B. and Leucker M.: *Message-Passing Automata are expressively equivalent to EMSO Logic*. CONCUR, LNCS **3170** (2004) 146–160
6. Caillaud B., Darondeau Ph., Hélouët L. and Lesventes G.: *HMSCs as partial specifications... with PNs as completions*. LNCS **2067** (2001) 87–103
7. Diekert V. and Rozenberg G.: *The Book of Traces*. (World Scientific, 1995)
8. Genest B., Muscholl A., Seidl H. and Zeitoun M.: *Infinite-Node High-Level MSCs: Model-Checking and Realizability*. ICALP, LNCS **2380** (2002) 657–668
9. Genest B., Muscholl A. and Kuske D.: *A Kleene Theorem for a Class of Communicating Automata with Effective Algorithms*. DLT, LNCS **3340** (2004) 30–48
10. Henriksen J.G., Mukund M., Narayan Kumar K., Sohoni M. and Thiagarajan P.S.: *A theory of regular MSC languages*. Information and computation **202** (2005) 1–38
11. Holzmann G.J.: *Early Fault Detection*. TACAS, LNCS **1055** (1996) 1–13
12. ITU-TS: *Recommendation Z.120: Message Sequence Charts*. (Geneva, 1996)
13. Lamport L.: *Time, Clocks and the Ordering of Events in a Distributed System*. Communications of the ACM **21,7** (1978) 558–565
14. Métivier Y.: *On Recognizable Subsets of Free Partially Commutative Monoids*. Theor. Comput. Sci. **58** (1988) 201–208
15. Morin R.: *On Regular Message Sequence Chart Languages and Relationships to Mazurkiewicz Trace Theory*. FoSSaCS, LNCS **2030** (2001) 332–346
16. Morin R.: *Recognizable Sets of Message Sequence Charts*. STACS, LNCS **2285** (2002) 523–534

17. Mukund M., Narayan Kumar K. and Thiagarajan P.S: *Netcharts: Bridging the Gap between HMSCs and Executable Specifications*. CONCUR 2003, LNCS **2761** (2003) 296–310
18. Muscholl A. and Peled D.: *Message sequence graphs and decision problems on Mazurkiewicz traces*. MFCS, LNCS **1672** (1999) 81–91
19. Ochmański E.: *Regular behaviour of concurrent systems*. Bulletin of the EATCS **27** (Oct. 1985) 56–67
20. Pratt V.: *Modelling concurrency with partial orders*. International Journal of Parallel Programming **15** (1986) 33–71
21. Thomas W.: *On Logics, Tilings, and Automata*. ICALP, LNCS **510** (1991) 441–454
22. Zielonka W.: *Notes on finite asynchronous automata*. RAIRO, Theoretical Informatics and Applications **21** (Gauthiers-Villars, 1987) 99–135

Non-sequential Behaviour of Dynamic Nets*

Roberto Bruni and Hernán Melgratti

Dipartimento di Informatica, Università di Pisa, Italia
{bruni, melgratt}@di.unipi.it

Abstract. Dynamic nets are an extension of Petri nets where the net topology may change dynamically. This is achieved by allowing (i) tokens to be *coloured* with place names (carried on as data), (ii) transitions to designate places where to spawn new tokens on the basis of the *colours* in the fetched tokens, and (iii) firings to add fresh places and transitions to the net. Dynamic nets have been given step or interleaving semantics but, to the best of our knowledge, their non-sequential truly concurrent semantics has not been addressed in the literature. To fill this gap, we extend the ordinary notions of processes and unfolding to dynamic nets, providing two different constructions: (i) a specific process and unfolding for a particular initial marking, and (ii) processes and unfolding patterns that abstract away from the colours of the token initially available.

1 Introduction

Petri nets, introduced in [13], have became a reference model for studying concurrent systems, mainly due to their simplicity and the intrinsic concurrent nature of their behaviours. In addition to the classical "token game" operational semantics, several alternative approaches have appeared in the literature for characterising the semantics of Petri nets, notably non-sequential processes, unfolding constructions and algebraic models. In particular, non-sequential processes have played a very important rôle when studying the non-interleaved semantics of Petri nets. In essence, non-sequential processes provide a full-fledged account for the causal relations among the steps of a computation, i.e., they provide a full explanation about the causes that led to the firing of a transition.

Recently, the basic Petri net model has been extended to account for mobility, giving birth to the so called *Mobile nets* [1] and *Dynamic nets* [1,5]. The difference between the two is that the structure of dynamic nets is slightly more constrained so to enforce the locality principle: tokens in a place can be consumed only by local transitions (that were spawned together with that place). Dynamic nets are an extension of coloured nets (since tokens carry on information) where token colours are the names of places in the net, transitions may use the information carried on by tokens to designate places where to spawn new tokens, and transitions may add fresh places and transitions when they fire.

---

* Research supported by the EU FET-GC2 IST-2004-16004 Integrated Project SENSORIA.

S. Donatelli and P.S. Thiagarajan (Eds.): ICATPN 2006, LNCS 4024, pp. 105–124, 2006.

The behaviour of dynamic nets has been defined by providing their step or interleaving semantics but, to the best of our knowledge, there is no proposal for their non-sequential truly concurrent semantics. In this work, we pursue this line of research by extending the classical notions of processes and unfolding to dynamic nets. In particular, we provide two different kind of constructions: (i) a specific process and unfolding for a particular initial marking (where the information carried on by tokens of the initial marking is essential), and (ii) a notion of general process and unfolding pattern that do not depend on the information carried on by tokens of the initial marking.

It is worth remarking that the unfolding construction has proved very helpful to define the correct notion of processes for dynamic nets, as otherwise it would have been very difficult to deal with the changes in the net topology due to the introduction of fresh subnets.

The results presented here can find an interesting application in defining net models and causal semantics for distributed mobile calculi and programming languages. More specifically we are thinking of the join calculus [7] and those languages like JoCaml [6] and C-omega [3] whose designs have been strongly influenced by the join paradigm. In fact it has been shown in [5] that join processes may coherently be viewed as dynamic nets (and vice versa).

Structure of the paper. In Section 2 we recall the preliminary definitions from the literature, aiming to keep the paper self-contained. It is worth noting that while we expect the reader to have some confidence with the material in Sections 2.1 and 2.2 (nets, step semantics, causal processes and unfolding), we have chosen to give an extensive introduction to dynamic nets (Sections 2.3), which could be a less familiar subject for many readers. The unfolding construction for marked dynamic nets is carried on in Section 3 accompanied by the definition of deterministic process of dynamic nets. The main result establishes a strong correspondence between the two notions. Section 4 introduces unfolding patterns and process patterns as a framework to give more compact and abstract representations of net behaviour: the same pattern can be instantiated to many different concrete computations. Conclusions and directions for future work are given in Section 5.

2   Background

2.1   P/T Petri Nets

Petri nets are built up from *places* (denoting resources, type of messages), which are repositories of *tokens* (representing instances of resources), and *transitions*, which fetch and produce tokens. In the following we shall consider an infinite set $\mathcal{P}$ of resource names.

Definition 2.1 (Net). *A net N is a 4-tuple $N = (S_N, T_N, \delta_{0N}, \delta_{1N})$ where $S_N \subseteq \mathcal{P}$ is the (nonempty) set of places, $\mathsf{a}, \mathsf{a}', \ldots$, T_N is the set of transitions, $\mathsf{t}, \mathsf{t}', \ldots$ (with $S_N \cap T_N = \emptyset$), and the functions $\delta_{0N}, \delta_{1N} : T_N \to \wp_f(S_N)$ assign finite sets of places, called respectively source and target, to each transition.*

We denote $S_N \cup T_N$ by N, and omit the subscript N if no confusion arises. We abbreviate a transition $\mathsf{t} \in T$ with *preset* $^\bullet\mathsf{t} = \delta_0(\mathsf{t}) = s_1$ and *postset* $\mathsf{t}^\bullet = \delta_1(\mathsf{t}) = s_2$ as $s_1[\rangle s_2$. Similarly for any place $\mathsf{a} \in S$, the preset $^\bullet\mathsf{a} = \{\mathsf{t} | \mathsf{a} \in \mathsf{t}^\bullet\}$ of a is the set of all transitions of which a is target and the postset $\mathsf{a}^\bullet = \{\mathsf{t} | \mathsf{a} \in {}^\bullet\mathsf{t}\}$ of a is the set of all transitions of which a is source. We consider only nets whose transitions have a non-empty preset. If $^\bullet\mathsf{a} \cup \mathsf{a}^\bullet = \emptyset$ the place a is *isolated*. Moreover, we let $^\circ N = \{x \in N | ^\bullet x = \emptyset\}$ and $N^\circ = \{x \in N | x^\bullet = \emptyset\}$ denote the sets of *initial* and *final elements* of N respectively.

While according to Definition 2.1 transitions can consume and produce at most one token in each state, in P/T nets (see Definition 2.2) transitions can fetch and produce several tokens in a particular place, i.e., the pre- and postsets of transitions are multisets instead of sets.

Given a set S, a *multiset* over S is a function $m : S \to \mathbb{N}$ (where $\mathbb{N}$ denotes the set of natural numbers with zero). Let $dom(m) = \{\mathsf{s} \in S \mid m(\mathsf{s}) > 0\}$. The set of all finite multisets (i.e., with finite domain) over S is written $\mathcal{M}_S$. The empty multiset (i.e., with $dom(m) = \emptyset$) is written $\emptyset$. The multiset union $\oplus$ is defined as $(m_1 \oplus m_2)(\mathsf{s}) = m_1(\mathsf{s}) + m_2(\mathsf{s})$ for any $\mathsf{s} \in S$. Given a multiset m, $|m| = \sum_{\mathsf{s} \in S} m(\mathsf{s})$ denotes the *size* of m.

Note that $\oplus$ is associative and commutative, and $\emptyset$ is the identity for $\oplus$. Hence, $\mathcal{M}_S$ is the free commutative monoid $S^\oplus$ over S. We write s for a singleton multiset m such that $dom(\mathsf{s}) = \{\mathsf{s}\}$ and $m(\mathsf{s}) = 1$. Moreover, we write $\{\!|s_1, \ldots, s_n|\!\}$ for $s_1 \oplus \ldots \oplus s_n$. By abusing notation we will apply functions (i.e., f_S) over (multi)sets, meaning the multiset obtained by applying the function element-wise: $f_S(\{\!|a_0, \ldots, a_n|\!\}) = f_S(a_0) \oplus \ldots \oplus f_S(a_n)$. Also, we shall use set operators over multisets to denote the operation over the domain of the multiset, e.g. $s \in m$ and $m_1 \cap m_2$ in place of $s \in dom(m)$ and $dom(m_1) \cap dom(m_2)$.

Definition 2.2 (P/T net). *A marked place / transition Petri net (P/T net) is a tuple $N = (S_N, T_N, \delta_{0N}, \delta_{1N}, m_{0N})$ where $S_N \subseteq \mathcal{P}$ is a set of places, T_N is a set of transitions, the functions $\delta_{0N}, \delta_{1N} : T_N \to \mathcal{M}_{S_N}$ assign respectively, source and target to each transition, and $m_{0N} \in \mathcal{M}_{S_N}$ is the initial marking.*

The notions of pre- and postset, initial and final elements, and isolated places are straightforwardly extended to consider multisets instead of sets. Note that a net can be regarded as a P/T net whose arcs have unary weights.

The operational semantics of P/T nets is given by (the least relation inductively generated by) the inference rules in Figure 1. Given a net N, the proof for $m \to_T m'$ means that a marking m evolves to m' under a *step*, i.e., the concurrent firing of several transitions. We omit the subscript T whenever the set of transitions is clear from the context. Rule (FIRING) describes the evolution of the state of a net (represented by the marking $m \oplus m''$) by applying a transition $m[\rangle m'$, which consumes the tokens m corresponding to its preset and produces the tokens m' corresponding to its postset. The multiset m'' represents idle resources, i.e., the tokens that persist during the evolution. Rule (STEP) stands for the parallel composition of computations. The sequential composition of computations is the reflexive and transitive closure of $\to$, which is written

$$\begin{array}{cc}
\text{(FIRING)} & \text{(STEP)} \\
\dfrac{m \ [\rangle \ m' \in T \quad m'' \in \mathcal{M}_S}{m \oplus m'' \to_T m' \oplus m''} & \dfrac{m_1 \to_T m_1' \quad m_2 \to_T m_2'}{m_1 \oplus m_2 \to_T m_1' \oplus m_2'}
\end{array}$$

Fig. 1. Operational semantics of P/T nets

$\to^*$, i.e., $m \to^* m'$ denotes the evolution of m to m' under a (possibly empty) sequence of steps.

2.2   Unfolding and Process Semantics of P/T Nets

The definition of the processes and unfolding semantics of P/T nets rely on the notions of occurrence nets and deterministic causal nets, which are defined below. (We report here on the presentation given in [12])

Definition 2.3 (Occurrence net). *A net N is an occurrence net if*

- *for all $a \in S_N$, $|{}^\bullet a| \leq 1$*
- *the* causal dependency *relation $\prec$ is irreflexive, where $\prec$ is the transitive closure of the* immediate cause *relation*

$$\prec^1 = \{(a, t) | a \in S_N \wedge t \in a^\bullet\} \cup \{(t, a) | a \in S_N \wedge t \in {}^\bullet a\};$$

moreover, $\forall t \in T_N$, the set $\{t' \in T_N | t' \prec t\}$ is finite and the reflexive closure of $\prec$ is denoted by $\preceq$;
- *the* binary conflict *relation # on $T_N \cup S_N$ is irreflexive, where # is defined in terms of the* binary direct conflict *relation $\#_m$ as below:*

$$\forall t_1, t_2 \in T_N, \quad t_1 \#_m t_2 \Leftrightarrow \delta_{0N}(t_1) \cap \delta_{0N}(t_2) \neq \emptyset \wedge t_1 \neq t_2$$
$$\forall x, y \in S_N \cup T_N, \quad x \# y \Leftrightarrow \exists t_1, t_2 \in T_N : t_1 \#_m t_2 \wedge t_1 \preceq x \wedge t_2 \preceq y.$$

Given $x, y \in T_N \cup S_N$ s.t. $x \neq y$, x and y are *concurrent*, written x *co* y, when $x \not\prec y$, $y \not\prec x$, and $\neg x \# y$. A set $X \in T_N \cup S_N$ is concurrent, written $CO(X)$, if $\forall x, y \in X : x \neq y \Rightarrow x$ *co* y, and $|\{t \in T_N | \exists x \in X, t \preceq x\}|$ is finite.

Definition 2.4 (Causal Net). *A net $K = (S_K, T_K, \delta_{0K}, \delta_{1K})$ is a causal net (also called* deterministic occurrence net*) if it is an occurence net and*

$$\forall a \in S_K, |a^\bullet| \leq 1.$$

It is worth noting that for any causal net K the conflict relation is empty. Consequently, a causal net is an acyclic net where the presets (resp. postsets) of transitions do not share places.

Occurrence nets can represent non-sequential computations: their places represent tokens and their transitions represents events, i.e., firings. The "typing" of tokens and events in the occurence net over places and transitions of the executed net can be expressed as net morphisms, mapping tokens to the places where they have been stored and events to the triggered transitions.

(INI-MK)
$$\frac{m_N(\mathsf{a}) = n}{\{(\emptyset, \mathsf{a})\} \times [n] \subseteq S}$$

(PRE)
$$\frac{B = \{(\epsilon_j, b_j, i_j) | j \in J\} \subseteq S, \; Co(B), \; \mathsf{t} \in T_N, \; \delta_{0N}(\mathsf{t}) = \oplus_{j \in J} b_j}{(B, \mathsf{t}) \in T, \quad \delta_0(B, \mathsf{t}) = B}$$

(POST)
$$\frac{x = (B, \mathsf{t}) \in T}{Q = \{(\{x\}, b, i) \mid 1 \leq i \leq \delta_{1N}(\mathsf{t})(b)\} \subseteq S, \quad \delta_1(x) = Q}$$

Fig. 2. Unfolding rules

Definition 2.5 (Net morphisms). *Let N, N' be nets. A pair $f = (f_S : S_N \to S_{N'}, f_T : T_N \to T_{N'})$ is a net morphism from N to N' (written $f : N \to N'$) if $f_S(\delta_{iN}(\mathsf{t})) = \delta_{iN'}(f_T(\mathsf{t}))$ for any t and $i = 0, 1$. Moreover, N and N' are said to be* isomorphic, *and thus* equivalent, *if f is bijective.*

The above definition extends in the obvious way to the cases in which N and N' are P/T nets.

Definition 2.6 (Deterministic Causal Process). *Let N be P/T net. A deterministic causal process for N is a net morphism P from a causal net K to N such that $P(^\circ K) = m_{0N}$, i.e., P maps the implicit initial marking of K (i.e., the minimal elements $^\circ K$) to the initial marking of N.*

Roughly, a deterministic process represents just a set of causally equivalent computations [8]. Differently, the unfolding of a net N is the least occurrence net that can account for all the possible computations over N, making explicit the causal dependencies, conflicts and concurrency between firings.

Definition 2.7 (Unfolding). *Let N be a P/T net. The occurrence net $\mathcal{U}[N] = (S, T, \delta_0, \delta_1)$ generated inductively by the inference rules in Figure 2 is said the unfolding of N.*

In the case of the unfolding $\mathcal{U}$, it can be readily verified that the mapping over N is just the projection of places and transition names to their second element. In fact tokens are encoded as triples (H, a, i) where H is the set of immediate causes (determining the history of the token), a is the place of N where the token resides and i is a positive integer used to disambiguate tokens in the same place and with the same history, while events are encoded as pairs (H, t), where H is the set of immediate causes and t is the fired transition.

2.3 Dynamic Nets

Different formulations for dynamic nets have been proposed in the literature [1, 5]. The definition we give here is based on [1]. We consider an infinite set of place names $\mathcal{P}$ ranged over by $\mathsf{a}, \mathsf{b}, \ldots$ and an infinite set of variable names $\mathcal{X}$, ranged over by $x, y, \ldots$. We require also variable names to be different from place names, i.e., $\mathcal{X} \cap \mathcal{P} = \emptyset$. We will use $\mathcal{C} = \mathcal{P} \cup \mathcal{X}$, ranged over by $c_1, c_2, \ldots$, to refer both to place and variable names.

Similar to high-level nets, tokens in dynamic nets carry on information (or colours as they are usually known). Although colours can be thought of as data structures of any type, for the sake of simplicity we will assume colours to be sequences of names. Let C be a set of names, C^* stands for the set of all finite (possible empty) sequences of C, i.e., $C^* = \{(c_1, \ldots, c_n) \mid \forall i \text{ s.t. } 0 \leq i \leq n : c_i \in C\}$. The empty sequence, i.e. a token that carries no information, is denoted by $\bullet$ (by analogy with ordinary tokens), and the underlying set of a sequence $(c_1, \ldots, c_n)$ is written:

$$\overline{(c_1, \ldots, c_n)} = \bigcup_i \{c_i\}$$

Then, markings of a dynamic net are just coloured multisets.

Definition 2.8 (Coloured Multiset). *Given two sets S and C, a coloured multiset over S and C is a function $m : S \to C \to \mathbb{N}$. Let $dom(m) = \{(s, c) \in S \times C \mid m(s)(c) > 0\}$. The set of all finite (coloured) multisets over S and C^* is written $\mathcal{M}_{S,C}$. The multiset union is defined as $(m_1 \oplus m_2)(s)(c) = m_1(s)(c) + m_2(s)(c)$.*

We write $s(c)$ for a multiset m such that $dom(m) = \{(s, c)\}$ and $m(s)(c) = 1$. Additionally, $(s, c) \in m$ is a shorthand for $(s, c) \in dom(m)$, while $s \in m$ means $(s, c) \in m$ for some c.

Definition 2.9 (DN). *The set DN is the least set satisfying the following equation (when considering the set of colours $\mathcal{C}$):*

$$\mathcal{N} = \{(S_N, T_N, \delta_{0N}, \delta_{1N}, m_{0N}) \mid$$
$$S_N \subseteq \mathcal{P} \ \wedge \ \delta_{0N} : T_N \to \mathcal{M}_{S_N, \mathcal{C}} \ \wedge \ \delta_{1N} : T_N \to \mathcal{N} \ \wedge \ m_{0N} \in \mathcal{M}_{\mathcal{C}, \mathcal{C}}\}$$

For $(S_N, T_N, \delta_{0N}, \delta_{1N}, m_{0N}) \in \text{DN}$, S_N is the set of places, T_N is the set of the transitions, δ_{0N} and δ_{1N} are the functions assigning the pre- and postset to every transition, and m_{0N} is the initial marking. Note that $\mathcal{N}$ is a domain equation [9] defining the recursive type of dynamic nets. The simplest elements in $\mathcal{N}$ are markings, i.e., the tuples $(\emptyset, \emptyset, \emptyset, \emptyset, m)$ with $m \in \mathcal{M}_{\mathcal{C},\mathcal{C}}$. Then, nets are defined recursively, because the postset of any transition (given by δ_{1N}) is another element of $\mathcal{N}$. The set DN is defined as the least fixed point of the recursive equation above.

As usual, we denote $S_N \cup T_N$ by N, and omit subscript N whenever no confusion arises. Moreover, we abbreviate a transition $\mathbf{t} \in T$ such that $\delta_0(\mathbf{t}) = m$ and $\delta_1(\mathbf{t}) = N$ as $m[\rangle N$, and refer to m as the *preset* of $\mathbf{t}$ (written $^\bullet\mathbf{t}$) and N as the *postset* of $\mathbf{t}$ (written $\mathbf{t}^\bullet$).

Note that the initial marking m_{0N} is not required to be a multiset over the places of the net, i.e., the initial marking can put tokens in places that are not defined by the net. In fact, the initial marking m_{0N} is a multiset in $m_{0N} \in \mathcal{M}_{\mathcal{C},\mathcal{C}}$

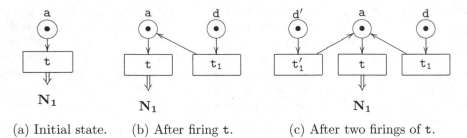

(a) Initial state.   (b) After firing t.   (c) After two firings of t.

Fig. 3. A simple dynamic net

and not over the places of the net S_N (e.g., in $\mathcal{M}_{S_N,C}$). A trivial example is to consider the coloured transition $a(x)[\rangle(\emptyset, \emptyset, \emptyset, \emptyset, b(\bullet))$, where b does not belong to the places of the subnet $(\emptyset, \emptyset, \emptyset, \emptyset, b(\bullet))$. We usually abbreviate transitions as the previous one, where the postset does not define new places, by writing just the initial marking fixed by the postset, i.e., $a(x)[\rangle b(\bullet)$.

Names defined in S_N act as binders on N. Therefore, nets are considered up-to α-conversion on S_N. For instance, the two nets $(\{a\}, \emptyset, \emptyset, \emptyset, a(\bullet))$ and $(\{b\}, \emptyset, \emptyset, \emptyset, b(\bullet))$ are α-equivalent, while $(\emptyset, \emptyset, \emptyset, \emptyset, a(\bullet))$ and $(\emptyset, \emptyset, \emptyset, \emptyset, b(\bullet))$ are not. Analogously, the names of transitions in T_N are binders, and hence we consider nets up-to α-conversion on transition names.

Example 2.1. Consider the net N in Figure 3(a), where circles are places, boxes are transitions, and solid lines connect transitions to their pre and postset, and tokens are represented by their colours. The double-lined arrow indicates the dynamic transition $t = a(\bullet)[\rangle N_1$, which creates an instance of the subnet N_1 when fired. We allow the initial marking of N_1 and the postset of transitions in T_{N_1} to generate tokens in a. Therefore, the following is a valid definition for N_1: $S_{N_1} = \{d\}, T_{N_1} = \{t_1\}, m_{0N_1} = a(\bullet) \oplus d(\bullet)$ and $t_1 = d(\bullet)[\rangle a(\bullet)$. A firing of t will lead to (a net isomorphic to) the net shown in Figure 3(b). (Firings are formally defined in Figure 5.) A fresh place d and a transition t_1 (whose pre- and postset are $d(\bullet)$ and $a(\bullet)$, resp.) have been added to the net. Also two tokens have been produced: one in a and the other in d, accordingly to the initial marking of N_1. This marking enables t, which can be fired again. The new activation of t will create a new subnet containing a new place and a new transition whose names are different from all other names already present in the net (Figure 3(c)).

The colours appearing in the preset of a transition are intended to be the formal parameters of that transition, which are substituted by the actual colours of consumed tokens when a transition is fired. The set of formal parameters (or received names) of a transition is defined below.

Definition 2.10 (Received names of a transition). *The* set of colours of a multiset $m \subseteq \mathcal{M}_{S,C}$ *is defined as* $col(m) = \cup_{(a,c) \in m} \bar{c}$, *the set of constants is* $col_{\mathcal{P}}(m) = col(m) \cap \mathcal{P}$, *and the set of variables or received names of a multiset is* $rn(m) = col_{\mathcal{X}}(m) = col(m) \cap \mathcal{X}$. *Given a transition* $t = m[\rangle N$, *the set of received names of* t *is given by* $rn(t) = col_{\mathcal{X}}(m)$.

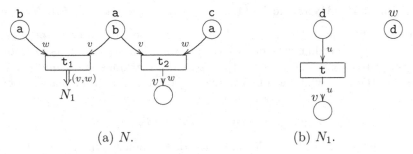

(a) N.   (b) N_1.

Fig. 4. Graphical representation of N and N_1

Note that $col_{\mathcal{P}}(m)$ stands for the set of constant (not variable) colours used as formal parameters of a transition. This feature provides a basic mechanism for pattern matching, since a transition will be enabled only when consumed tokens carry on the same constant colours as those specified by the preset of the transition. Differently, $col_{\mathcal{X}}(m)$ is the set of all variables, which will be instantiated when the transition fires.

Definition 2.11 (Defined and Free names). *The set of* defined names *in a marking* m *is* $dn(m) = \{a | \exists c.(a, c) \in m\}$, *i.e., names appearing in place position. Given* $N = (S_N, T_N, \delta_{0N}, \delta_{1N}, m_{0N}) \in \mathrm{DN}$, *the set of* defined *(dn) and* free *(fn) names of transitions, sets of transitions, and nets are defined as follow:*

$dn(m_1 [\rangle N_1) = dn(m_1)$
$dn(T_N) = \bigcup_{t \in T_N} dn(t)$
$dn(N) = S_N$
$fn(m_1 [\rangle N_1) = dn(m_1) \cup col_{\mathcal{P}}(m_1) \cup (fn(N_1) \setminus rn(m_1))$
$fn(T_N) = \bigcup_{t \in T_N} fn(t) \setminus dn(T_N)$
$fn(N) = (fn(T_N) \cup dn(m_{0N}) \cup col(m_{0N})) \setminus S_N$

Example 2.2. Consider N_1 and N defined as follow and depicted in Figure 4 (Note places v and w in the representation of N_1 are not defined places by N_1):

$$S_N = \{a, b, c\} \qquad T_N = \{t_1, t_2\} \qquad S_{N_1} = \{d\}$$
$$\delta_{0N}(t_1) = a(v) \oplus b(w) \qquad \delta_{1N}(t_1) = N_1 \qquad T_{N_1} = \{t\}$$
$$\delta_{0N}(t_2) = a(v) \oplus c(w) \qquad \delta_{1N}(t_2) = v(w) \qquad \delta_{0N_1}(t) = d(u)$$
$$m_N = a(b) \oplus b(a) \oplus c(a) \qquad\qquad\qquad \delta_{1N_1}(t) = v(u)$$
$$\qquad\qquad\qquad\qquad\qquad\qquad\qquad m_{N_1} = w(d)$$

The sets of defined, received and free names of N and N_1 are as follow

$$rn(t_1) = \{v, w\} \quad rn(t_2) = \{v, w\} \qquad rn(t) = \{u\}$$
$$dn(t_1) = \{a, b\} \quad dn(t_2) = \{a, c\} \qquad dn(t) = \{d\}$$
$$fn(t_1) = \{a, b\} \quad fn(t_2) = \{a, c\} \qquad fn(t) = \{d, v\}$$
$$dn(T_N) = dn(N) = \{a, b, c\} \qquad dn(T_{N_1}) = dn(N_1) = \{d\}$$
$$fn(T_N) = \emptyset \qquad\qquad\qquad\qquad fn(T_{N_1}) = \{v\}$$
$$fn(N) = \emptyset \qquad\qquad\qquad\qquad fn(N_1) = \{v, w\}$$

Definition 2.12 (Dynamic Net). $N \in$ DN *is a dynamic net if* $fn(N) = \emptyset$.

The above definition states that a dynamic net is closed, i.e., it does not generate tokens in places that do not belong to it. The condition $fn(N) = \emptyset$ assures tokens to be generated always in places of the net, since markings are bound to places defined by the net, which are guaranteed to be different from places in other nets.

Remark 2.1. If N is a dynamic net then $m_{0N} \in \mathcal{M}_{S_N, S_N}$.

The net N presented in Example 2.2 is closed, and hence dynamic, even though N_1 is not. In fact, the names v and w are not bound in N_1.

Remark 2.2. As variables in a transition are used to describe parameters, we consider only dynamic nets whose transitions $\mathsf{t} = m[\rangle N$ satisfy the condition $(fn(N) \cap \mathcal{X}) \subseteq rn(\mathsf{t})$. This restriction requires all free variables of the postset of a transition to be bound to some variable in the preset. Note that this is always the case when a net is closed.

Similarly to coloured nets, the firing of a transition t requires the instantiation of the received colours of t, i.e., the formal parameters $rn(\mathsf{t})$ of the transition t have to be replaced by the actual parameters corresponding to the colours of the consumed tokens. Hence, we need suitable notions of substitution and instantiation of nets.

Definition 2.13 (Substitution). *Let* $\sigma : \mathcal{X} \rightharpoonup \mathcal{X} \cup \mathcal{P}$ *be a partial function. The substitution* σ *on a multiset* $m \in \mathcal{M}_{C,C}$ *is given by*

$$(m\sigma)(\mathsf{a}_1)(c_1) = \sum_{a \in \{a' | a'\sigma = a_1\}} \sum_{c \in \{c' | c'\sigma = c_1\}} m(\mathsf{a})(c)$$

Definition 2.14 (Instantiation of transitions and nets). *Let* $\sigma : \mathcal{X} \rightharpoonup \mathcal{P} \cup \mathcal{X}$ *be a substitution. The* instantiation *of a transition* $\mathsf{t} = m_1[\rangle N_1$ *with* σ *is the transition* $\mathsf{t}\sigma = m_1\sigma[\rangle N_1\sigma$. *Given a dynamic net* $N = (S_N, T_N, \delta_{0N}, \delta_{1N}, m_{0N})$, *the* instantiation *of* N *with* σ *s.t.* $range(\sigma) \cap S_N = \emptyset$ *is defined as* $N\sigma = (S_N, T_N, \delta_{0N}, \delta_{1N}\sigma, m_{0N}\sigma)$, *where* $\delta_{1N}\sigma(\mathsf{t}) = (\delta_{1N}(\mathsf{t}))\sigma$ *if* $rn(\mathsf{t}) \cap (dom(\sigma) \cup range(\sigma)) = \emptyset$.

Remark 2.3. (i) The recursive definition given above is well-founded because it is recursive on the structure of a net $N \in$ DN, which is well-founded. (ii) The conditions $range(\sigma) \cap S_N = \emptyset$ and $rn(\mathsf{t}) \cap (dom(\sigma) \cup range(\sigma)) = \emptyset$ avoid name clashes. When such condition is not satisfied, α-conversion either on the places of the net or on the received names of the transition (as defined below) can be applied before the instantiation. (iii) Note that σ is not applied on δ_{0N}, since all variables appearing in the preset of a transition are local.

Definition 2.15 (α-equivalence). *Two transitions* t_1 *and* t_2 *are* α-convertible *if there exists an injective substitution* $\sigma : \mathcal{X} \rightharpoonup \mathcal{X}$, *with* $rn(\mathsf{t}_1) \subseteq dom(\sigma)$, *s.t.* $\mathsf{t}_1\sigma = \mathsf{t}_2$. *As usual,* α-conversion *is a equivalence relation denoted by* $\equiv_\alpha$. *Two nets* N *and* N' *are* α-convertible *if there exist bijective substitutions* $\sigma_S : S_N \rightarrow S_{N'}$ *and* $\sigma_T : T_N \rightarrow T_{N'}$ *s.t.* ${}^\bullet\mathsf{t}\sigma_S[\rangle\mathsf{t}^\bullet\sigma_S \equiv_\alpha {}^\bullet(\mathsf{t}\sigma_T)[\rangle(\mathsf{t}\sigma_T)^\bullet$ *and* $m_{0N}\sigma_S = m_{0N'}$. *We shall always consider transitions and nets up-to* α-equivalence.

(DYN-FIRING)
$$\frac{t = m \; |\rangle \; N_1 \in T \qquad m'' \in \mathcal{M}_{S,C}}{(S,T,m\sigma \oplus m'') \to (S,T,m'') \oslash N_1\sigma} \qquad \begin{array}{l} rn(t) \subseteq dom(\sigma) \text{ and} \\ range(\sigma) \subseteq S \end{array}$$

(DYN-STEP)
$$\frac{(S,T,m_1) \to (S,T,m_1') \oslash N_1 \qquad (S,T,m_2) \to (S,T,m_2') \oslash N_2}{(S,T,m_1 \oplus m_2) \to (S,T,m_1' \oplus m_2') \oslash (N_1 \oplus N_2)}$$

Fig. 5. Operational semantics of dynamic nets

The following definition introduces two different ways for composing nets, which will be used to formalize the operational semantics of dynamic nets.

Definition 2.16 (Composition of nets). *Let N_1, $N_2 \in$ DN s.t. $N_1 \cap N_2 = \emptyset$ (i.e., they share neither places nor transitions) and $fn(N_1) \cap S_{N_2} = \emptyset$ (i.e., N_2 does not define the free names of N_1). Then, the* addition *of N_2 to N_1 (written $N_1 \oslash N_2$) is defined as $N_1 \oslash N_2 = (S_{N_1} \uplus S_{N_2}, T_{N_1} \uplus T_{N_2}, \delta_{0N_1} \uplus \delta_{0N_2}, \delta_{1N_1} \uplus \delta_{1N_2}, m_{0N_1} \oplus m_{0N_2})$. The addition $N_1 \oslash N_2$ is said the* parallel composition *of N_1 and N_2 (written $N_1 \oplus N_2$) if also $fn(N_2) \cap S_{N_1} = \emptyset$.*

Observe that the side conditions required by parallel composition avoid free names of one net to be captured by the transitions defined by the other. Nevertheless, when a subnet N_2 is added to a net N_1 ($N_1 \oslash N_2$) we allow the free names of N_2 to be captured by the definitions in N_1. We remind that we are considering nets up-to α-conversion on the name of places and transitions. Hence, it is always possible to choose N_2' α-equivalent to N_2 s.t. $N_1 \cap N_2' = \emptyset$.

Rules defining the operational semantics of dynamic nets is in Figure 5. Note that the state of a computation considers not only markings but also the structure of the net. For simplicity we write (S,T,m) as a shorthand for $(S,T,\delta_0,\delta_1,m)$. Rule (DYN-FIRING) stands for the firing of t when the marking contains an instance of the preset of t (for a suitable substitution σ). The resulting net consists of the original net, where the consumed tokens have been removed, and a new instance of N_1 (i.e., the postset of t) has been added. The composition $\oslash$ of nets assures the names of the added components to be fresh. The side condition $rn(t) \subseteq dom(\sigma)$ assures all formal parameters of t to be instantiated, while $range(\sigma) \subseteq S$ guarantees all consumed tokens to be coloured with names defined by the net. Rule (DYN-STEP) stands for the parallel composition of computations when the initial marking contains enough tokens for executing them independently. Note that both concurrent steps operate over the same net structure, in fact both start from a net whose places and transitions are S and T. Those steps can add new elements (i.e., N_1 and N_2), which by definition are fresh. Moreover, the new components can be chosen to assure new elements to be disjoint, i.e., such that $(N_1 \oplus N_2)$ is defined.

Remark 2.4 (Subject reduction). When starting from a dynamic net, the application of rules (DYN-FIRING) and (DYN-STEP) generates dynamic nets.

(INI-PL)   (INI-TR)   (INI-MK)

$$\frac{\text{a} \in S_N}{\text{a} \in \mathcal{S}} \qquad \frac{\text{t} \in T_N}{\text{t} \in \mathcal{T}, \quad \xi_0(\text{t}) = \delta_{0N}(\text{t}), \quad \xi_1(\text{t}) = \delta_{1N}(\text{t})} \qquad \frac{m_N(\text{a})(c) = n}{\{(\emptyset, \text{a}(c))\} \times [n] \subseteq S}$$

(PRE)

$$\frac{B = \{(\epsilon_j, b_j, i_j) | j \in J\} \subseteq S, \ Co(B), \ \text{t} \in \mathcal{T}, \ \xi_0(\text{t})\sigma = \oplus_{j \in J} b_j}{(B, \sigma, \text{t}) \in T, \quad \delta_0(B, \sigma, \text{t}) = B}$$

(POST)

$$x = (B, \sigma, \text{t}) \in T, \quad \xi_1(\text{t}) = N_1$$

$$\overline{Q = \{(\{x\}, b(c), i) \mid 0 < i \leq m_{N_1}\rho_x\sigma(\text{b})(c)\} \subseteq S, \quad \delta_1(x) = Q, \quad S_{N_1}\rho_x \subseteq \mathcal{S},}$$
$$T_{N_1}\rho_x \subseteq \mathcal{T}, \quad for \ \text{t} \in T_{N_1} : \xi_0(\text{t}\rho_x) = \delta_{0N_1}(\text{t})\rho_x, \xi_1(\text{t}\rho_x) = \delta_{1N_1}(\text{t})\rho_x\sigma$$

Fig. 6. Definition of $\mathcal{U}[N]$

Example 2.3. Consider the dynamic net presented in Example 2.1 (see Figure 3(a)) but with the initial marking $m' = \text{a}(\bullet) \oplus \text{a}(\bullet)$. In what follows we show a computation that fires concurrently two instances of t. We have

$$\frac{\text{a}(\bullet)[\rangle N_1}{(\{\text{a}\}, \{\text{t}\}, \text{a}(\bullet)) \rightarrow (\{\text{a}\}, \{\text{t}\}, \emptyset) \otimes N_1'} \qquad \frac{\text{a}(\bullet)[\rangle N_1}{(\{\text{a}\}, \{\text{t}\}, \text{a}(\bullet)) \rightarrow (\{\text{a}\}, \{\text{t}\}, \emptyset) \otimes N_1''}$$

$$(\{\text{a}\}, \{\text{t}\}, \text{a}(\bullet) \oplus \text{a}(\bullet)) \rightarrow (\{\text{a}\}, \{\text{t}\}, \emptyset) \otimes (N_1' \oplus N_1'')$$

where $N_1' = (\{\text{d}\}, \{\text{t}_1\}, \text{a}(\bullet) \oplus \text{d}(\bullet))$ and $N_1'' = (\{\text{d}'\}, \{\text{t}_1'\}, \text{a}(\bullet) \oplus \text{d}'(\bullet))$. Note that $N_1' \oplus N_1'' = (\{\text{d}, \text{d}'\}, \{\text{t}_1, \text{t}_1'\}, \text{a}(\bullet) \oplus \text{d}(\bullet) \oplus \text{a}(\bullet) \oplus \text{d}'(\bullet))$. And finally,

$$(\{\text{a}\}, \{\text{t}\}, \emptyset) \otimes (N_1' \oplus N_1'') = (\{\text{a}, \text{d}, \text{d}'\}, \{\text{t}, \text{t}_1, \text{t}_1'\}, \text{a}(\bullet) \oplus \text{d}(\bullet) \oplus \text{a}(\bullet) \oplus \text{d}'(\bullet)).$$

3   Unfolding Dynamic Nets

In this section we characterise the unfolding of dynamic nets. Different from P/T nets, the unfolding of a dynamic net should consider, not only the evolution of markings but also, the changes on the structure of the net. Intuitively, the unfolding of a dynamic net gives two different structures: (i) the places and transitions of the unfolded dynamic net, and (ii) the occurrence net describing the evolution of the states (i.e., markings).

Definition 3.1 (Dynamic net unfolding). *Let $N = (S_N, T_N, \delta_{0N}, \delta_{1N}, m_N)$ be a dynamic net. The unfolding $\mathcal{U}[N] = (S, T, \delta_0, \delta_1, \mathcal{S}, \mathcal{T}, \xi_0, \xi_1)$ is the joint combination of an occurrence net $(S, T, \delta_0, \delta_1)$ and a dynamic net $(\mathcal{S}, \mathcal{T}, \xi_0, \xi_1, \emptyset)$, which are defined inductively by the rules in Figure 6.*

In the above definition $(S, T, \delta_0, \delta_1)$ is the causal net describing the evolution of the marking of the net, while $(\mathcal{S}, \mathcal{T}, \xi_0, \xi_1)$ gives the structure of the unfolded dynamic net. Note that both structures are inductively defined starting from the original dynamic net N: At the beginning (by rules INI-PL and INI-TR) the unfolded structure of the dynamic net $(\mathcal{S}, \mathcal{T}, \xi_0, \xi_1)$ coincides with the definition of N, i.e., with $(S_N, T_N, \delta_{0N}$ and $\delta_{1N})$. Initially, the causal net contains the places corresponding to the initial marking (rule INI-MK), i.e., for any token $a(c)$ in m_{0N}, there is one place in $\mathcal{U}[N]$ named $(\emptyset, a(c), i)$ for some i. Note that the name of a place contains a set (in this case $\emptyset$) that records the history (i.e., the immediate causes) of the particular token associated to that place. In this case $\emptyset$ means that the associated token is part of the initial marking of N (i.e., it has not been generated by a transition). The part $a(c)$ of the name provides the correspondence with the original token (it identifies the corresponding place and colour in N), while the number i univocally identifies one token $a(c)$ when the initial marking contains several copies.

Then, the rules PRE and POST unfold the net by defining the transitions $x = (B, \sigma, t)$ of the causal structure. In fact, rule PRE identifies the set of places B corresponding to concurrent events (i.e., $CO(B)$) in the causal structure that conform an instance (for a substitution σ) of the preset of some transition t in the dynamic structure $\mathcal{T}$. For any such B, a new transition (B, σ, t) is added to the causal net. Note that the preset of (B, σ, t) is B. We assume w.l.o.g. that $dom(\sigma) = rn(t)$, i.e., we consider only the relevant part of substitutions.

Rule POST updates both the dynamic structure and the causal net. Consider transition $x = (B, \sigma, t)$ s.t. the postset of t is N_1 (i.e., $\xi_1(t) = N_1$). Rule POST adds a *fresh* instance of N_1 to the dynamic structure and the events for the initial marking of N_1 to the causal net. Freshness is obtained by applying to every element of S_{N_1} and T_{N_1} the renaming ρ_x, which is defined as follow

$$\forall z \in S_{N_1} \cup T_{N_1} : \rho_x(z) = z_x$$

Note this renaming adds the name of the fired transition x to any element generated by x (i.e., its history). Hence, by applying ρ_x we assure new elements in the generated instance of N_1 to be fresh (and unique).

Then, the causal structure is modified as follows: for any place in the initial marking of N_1 a new place is generated in S. Note the name of such places carries $\{x\}$ as its history (i.e., the name of the transition that generates them), the name of the token $b(c)$ corresponding to the tokens in the initial marking of N_1 suitably renamed by ρ_x (i.e., the names of the fresh instance of N_1) and instantiated by σ (i.e., the colour of the consumed tokens). Moreover, the postset of x is defined as the set Q of all generated places.

Furthermore, rule POST refreshes the structure of the dynamic net by adding the elements of the fresh instance of N_1 (i.e., $S_{N_1}\rho_x$ and $T_{N_1}\rho_x$), and updating the flow relations ξ_i accordingly. In particular, note that σ is also used when defining ξ_1. This is necessary in cases as the following: consider the transition $t = a(x)[\rangle N_1$, s.t. N_1 defines the transition $t_1 = b(y)[\rangle x(y)$. Since x is free in t_1 (and hence in N_1), x is bound to the occurrence of x in the preset of t. Consequently, if t is fired with the substitution σ, then σ has to be applied to the transitions in N_1. In contrast with this, we do not use σ for ξ_0, since the substitution has no effect on the presets of transitions.

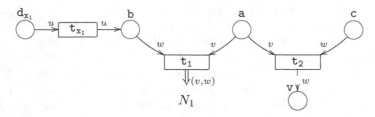

Fig. 7. The dynamic part of the unfolding $\mathcal{U}[N]$: $(\mathcal{S}, \mathcal{T}, \xi_0, \xi_1)$

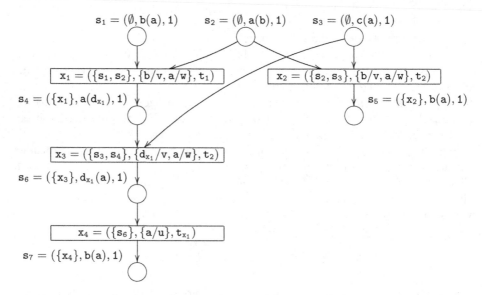

Fig. 8. The causal structure of the unfolding $\mathcal{U}[N]$: $(S, T, \delta_1, \delta_2)$

Remark 3.1. (*i*) The unfolding construction works since transitions generated dynamically can be seen as resources that are concurrent to tokens. When a new transition is added to $\mathcal{T}$ by using rule POST, we are sure that such transition cannot fetch tokens from already existing places. Hence, all elements generated previously are still valid. (*ii*) The unfolding is unique since the result is influenced neither by the order in which productions are applied nor by the multiple applications of a same rule that produce always the same elements.

Example 3.1. Consider the dynamic net N defined in Example 2.2. Its unfolding $\mathcal{U}[N]$ is depicted in Figures 7 and 8.

Proposition 3.1. *Let N be a dynamic net and $\mathcal{U}[N] = (S, T, \delta_0, \delta_1, \mathcal{S}, \mathcal{T}, \xi_0, \xi_1)$ the unfolding of N. Then $(S, T, \delta_0, \delta_1)$ is an occurrence net.*

Proof. By induction on the structure of the proof $x \in T$.

Proposition 3.2. *Let N be a dynamic net and $\mathcal{U}[N] = (S, T, \delta_0, \delta_1, \mathcal{S}, \mathcal{T}, \xi_0, \xi_1)$ the unfolding of N. Then $(\mathcal{S}, \mathcal{T}, \xi_0, \xi_1, \emptyset)$ is a dynamic net.*

Proof. By induction on the structure of the proof that $t \in T$.

The unfolding for the case of an ordinary P/T net (when regarded as a dynamic net) essentially coincides with the ordinary notion of unfolding, since (i) tokens carry on the colour •, which determines a unique possible substitution $\sigma = \emptyset$ in rule PRE, and (ii) rule POST does not modify the dynamic structure.

The dynamic net unfolding $\mathcal{U}[N] = (S, T, \delta_0, \delta_1, \mathcal{S}, \mathcal{T}, \xi_0, \xi_1)$ defines an implicit morphism from the occurrence net $(S, T, \delta_0, \delta_1)$ to $(\mathcal{S}, \mathcal{T}, \xi_0, \xi_1, \emptyset)$, which is given by the projections of places (ϵ, b, i) to the second component b and of transitions (B, σ, t) to the third component t. Exploiting this fact, we can define the notion of deterministic processes of a dynamic net as below.

Definition 3.2 (Process of a dynamic net). *A* deterministic process *for a dynamic net N (written $P : K \rightsquigarrow N$) is a net morphism P from a causal net K to $C = (S, T, \delta_0, \delta_1)$ s.t. $P(^\circ K) = {}^\circ C$, where $\mathcal{U}[N] = (S, T, \delta_0, \delta_1, \mathcal{S}, \mathcal{T}, \xi_0, \xi_1)$.*

We note the set of origins and destinations of $P : K \to N$ respectively by $O(P) = {}^\circ K$ and $D(P) = K^\circ \cap S_K$. Moreover, $pre(P)$ and $post(P)$ stand for the multisets of initial and final markings of P, i.e., $pre(P) = \bigoplus_{(x,b,i) \in P(O(P))} b$ and $post(P) = \bigoplus_{(x,b,i) \in P(D(P))} b$.

The following result relates the operational semantics of dynamic nets with their non-sequential semantics. In particular, we show that, although processes capture more information about the behaviour of a net than reductions, computations in both cases are the same. (We recall that dynamic nets are considered up-to α-conversion).

Theorem 3.1 (Correspondence). *Let N, N' be dynamic nets. Then $N \to^* N'$ iff there exists a process $P : K \rightsquigarrow N$ such that:*

(i) *$pre(P) = m_{0N}$ and $post(P) = m_{0N'}$, and*

(ii) *all places and transitions in N' are either in N or they are added by the computation described by P, i.e., $N \oplus \bigoplus_{x=(B,\sigma,t) \in P(T_K)} t^\bullet(\rho_x, \sigma) = (S_{N'}, T_{N'}, \delta_{0N'}, \delta_{1N'}, m)$, where substitution $t^\bullet(\rho, \sigma)$ for $t^\bullet = N_i$ is defined as $(S_i, T_i, \delta_{i0}, \delta_{i1}, m_{i0})(\rho, \sigma) = (S_i\rho, T_i\rho, \delta_{i0}\rho, \delta_{i0}\rho\sigma, m_{i0}\rho\sigma)$.*

(iii) *Moreover, for the unfolding $\mathcal{U}[N] = (S, T, \delta_0, \delta_1, \mathcal{S}, \mathcal{T}, \xi_0, \xi_1)$ of N, we have $S_{N'} \subseteq \mathcal{S}, T_{N'} \subseteq \mathcal{T}, \delta_{0N'} \subseteq \xi_0, \delta_{1N'} \subseteq \xi_1$.*

Proof. $\Rightarrow$) It follows by induction on the length n of the derivation $N \to^n N'$.

- **Base case ($n = 0$):** Hence $N = N'$. Conditions $(i) - (iii)$ hold by taking $K = (S_K, \emptyset, \emptyset, \emptyset)$ with $S_K = \bigcup_{a(c) \in dom(m_{0N})} (\emptyset, a(c)) \times [m_{0N}(a)(c)]$, and $P : K \to (S, T, \delta_0, \delta_1)$ as the identity on places and transitions.

- **Inductive Step ($n = k + 1$):** Then, $N \to^k N'' \to N'$. By inductive hypothesis on $N \to^k N''$, $\exists P'$ satisfying $(i) - (iii)$. The proof follows by showing (by rule induction on the structure of the proof $N'' \to N'$) that $\exists P$, which is an extension of P', satisfying conditions $(i) - (iii)$.

    * **Rule (DYN-FIRING):** Then, $N'' \to N'$ by firing $t \in T_{N''}$. Note the final elements of P' contains a set M of concurrent elements corresponding to an instance of the preset of t. Since elements in M are

concurrent, the unfolding contains an event e corresponding to the firing of t with preset M. P adds e and its postset to P'.

* **Rule** (DYN-STEP): By inductive hypothesis, there are two processes P_1 and P_2 starting from two different markings m_1 and m_2 of the same net N. Although, P_1 and P_2 are processes of two different unfoldings, they can be seen as being processes of the same unfolding of N with initial marking $m_1 \oplus m_2$. W.l.o.g, we assume K_1 and K_2 only to share initial elements that are not in conflict. Then, $K_1 \cup K_2$ is a causal net. Finally, P is defined as the union of P_2 and P_2.

In both cases, conditions can be verified by straightforward analysis.

$\Leftarrow$) The proof follows by induction on the number of transitions of K.

- **Base Case** ($|T_K| = 0$). It follows immediately by taking $N' = N$.
- **Inductive Step.** $T_K = T_{K'} \cup t$ s.t. $t^\bullet \subseteq K^\circ$ Let K' be the causal net obtained from removing t and $t^\bullet$ from K, and P' the restriction of P to the elements of K'. By defining N'' as
  * $(S_{N''}, T_{N''}, \delta_{0N''}, \delta_{0N''}, m) = N \oplus \bigoplus_{x=(B,\sigma,t)\in P(T_{K'})} t^\bullet(\rho_x, \sigma)$, and
  * $m_{0N''} = post(P') = post((D(P) \cup {}^\bullet t)\backslash t^\bullet)$

and using inductive hypothesis on P', we have $N \to N''$. By condition (iii) and the definition of P, there exists $t \in N''$ whose firing makes N'' to become N'.

4   Unfolding Pattern

The definitions of unfolding and deterministic processes for dynamic nets given in the previous section depend on the colours carried on by tokens. The main disadvantage of such kind of definitions is that they cannot abstract away from colours that are irrelevant for the computation. Consider the net N, where $S_N = \{a, b, c\}$, $T_N = \{t_1, t_2\}$ s.t. $t_1 = a(x,y)[\rangle x(y)$, $t_2 = b(x)[\rangle c(x)$. When constructing the unfoldings for the initial markings $m_1 = a(b, a)$ and $m_2 = a(b, b)$, we obtain isomorphic causal nets as shown in Figure 9 (we omit the representation of the dynamic structure because it stays N in both cases). To quotient out the set of all unfoldings with isomorphic causal nets and the same dynamic structure, we generalise the notions given in the previous section by allowing unfolding and processes to be parametric in (some) colours carried on by tokens of the initial marking. We call such constructions *unfolding* and *process patterns*.

Definition 4.1 (Pattern). *A* (marking) pattern *is a multiset* $p \in \mathcal{M}_{\mathcal{P} \cup \mathcal{X}, \mathcal{P} \cup \mathcal{X}}$. *Moreover, p is a* colour pattern *if* $p \in \mathcal{M}_{\mathcal{P}, \mathcal{P} \cup \mathcal{X}}$, *i.e., variables appear only as colours. A colour pattern p is* linear *if any variable occur at most once in p.*

Definition 4.2 (Unfolding pattern). *Let* $N = (S_N, T_N, \delta_{0N}, \delta_{1N}, m_N) \in$ DN, *s.t. the initial marking m_N is a linear colour pattern. The* unfolding pattern $\mathcal{UP}[N] = (S, T, \delta_0, \delta_1, \mathcal{S}, \mathcal{T}, \xi_0, \xi_1)$ *is defined inductively by the rules in Figure 10.*

The three first rules are quite similar to the definition of unfolding in the previous section. The main difference is that place names (rule INI-MK-PATT) carry on an

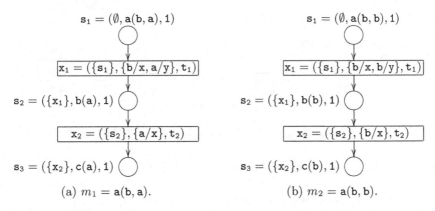

(a) $m_1 = \mathsf{a}(\mathsf{b}, \mathsf{a})$.   (b) $m_2 = \mathsf{a}(\mathsf{b}, \mathsf{b})$.

Fig. 9. Causal nets of the Unfolding of N for different initial markings m_1 and m_2

(INI-PL-PATT)   (INI-TR-PATT)   (INI-MK-PATT)

$$\frac{\mathsf{a} \in S_N}{\mathsf{a} \in S} \qquad \frac{\mathsf{t} \in T_N}{\mathsf{t} \in T, \quad \xi_0(\mathsf{t}) = \delta_{0N}(\mathsf{t}), \quad \xi_1(\mathsf{t}) = \delta_{1N}(\mathsf{t})} \qquad \frac{m_N(\mathsf{a})(c) = n}{\{(\emptyset, \mathsf{a}(c), \emptyset)\} \times [n] \subseteq S}$$

(PRE-PATT)
$$\frac{B = \{(\epsilon_j, b_j, \mu_j, i_j)|j \in J\} \subseteq S, \quad Co(B), \quad \mathsf{t} \in T, \quad \xi_0(\mathsf{t})\sigma = \oplus_{j \in J} b_j \mu_t,}{range(\mu_t) \subseteq S_N, \quad \mu = \mu_t \cup \bigcup_j \mu_j \ well-defined \ substitution}}{(B, \sigma, \mathsf{t}, \mu) \in T, \quad \delta_0(B, \sigma, \mathsf{t}, \mu) = B}$$

(POST-PATT)
$$\frac{x = (B, \sigma, \mathsf{t}, \mu) \in T, \quad \xi_1(\mathsf{t}) = N_1}{Q = \{(\{x\}, \mathsf{b}(c), \mu, i) \mid 0 < i \le m_{N_1} \rho_x \sigma(\mathsf{b})(c)\} \subseteq S, \quad \delta_1(x) = Q, \quad S_{N_1} \rho_x \subseteq S,}{T_{N_1} \rho_x \subseteq T, \quad for \ \mathsf{t} \in T_{N_1} : \xi_0(\mathsf{t}\rho_x) = \delta_{0N_1}(\mathsf{t})\rho_x, \xi_1(\mathsf{t}\rho_x) = \delta_{1N_1}(\mathsf{t})\rho_x \sigma}$$

Fig. 10. Definition of $\mathcal{UP}[N]$

extra element (the third one) that records all substitutions made on variable colours of the initial pattern. Clearly, for tokens corresponding to the initial marking, the third element is set to the empty substitution.

As far as the unfolding is concerned, it works analogously to the previous case. The main difference is that when identifying the preset B of a transition (in rule PRE-PATT) it may be the case that $\oplus_j b_j$ is actually a pattern more general than the preset of the transition we would like to apply, hence its variables may require to be instantiated in order to conform the preset of a transition. Consider the simple case of N containing only the transition $\mathsf{t} = \mathsf{a}(\mathsf{b}, x)[\rangle\mathsf{b}(\mathsf{a})$. When considering the initial pattern $m_{0N} = \mathsf{a}(v, w)$ the causal net of the unfolding contains the place $\mathsf{s}_1 = (\emptyset, \mathsf{a}(v, w), \emptyset, 1)$ (i.e., $\mathsf{s}_1 \in S$). Since $\mathsf{t} \in T$ we would like to fire t with the token s_1. To make $\mathsf{a}(v, w)$ be an instance of ${}^\bullet\mathsf{t}$ we need to substitute the variable v of $\mathsf{a}(v, w)$ by b, and substitute the variable x in ${}^\bullet\mathsf{t}$ by w. Hence, we take $\mu_t = \{\mathsf{b}/v\}$ and $\sigma = \{w/x\}$, so that $\mathsf{a}(v, w)\mu_t = \mathsf{a}(\mathsf{b}, w) = {}^\bullet\mathsf{t}\sigma$.

Moreover, we require μ_t to be *minimal* in order for the unfolding to describe the most general pattern of computation. That is, we will not consider substitutions μ_t

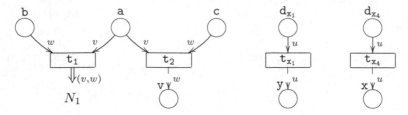

Fig. 11. The dynamic part of the unfolding pattern $\mathcal{UP}[N]$: $(\mathcal{S},\mathcal{T},\xi_0,\xi_1)$

that instantiate more variables than those needed. In the previous example we do not consider the substitutions $\theta_1 = \{b/v, a/w\}$ and $\theta_2 = \{b/v, b/w\}$ as candidate μ_t (although they also satisfy the equation $a(v, w)\theta_i = {}^\bullet t\sigma_i$ with $\sigma_1 = \{a/x\}$ and $\sigma_2 = \{b/x\}$) because they are not minimal instantiations. In particular, μ_t means that there not exists μ' satisfying the instantiation condition and $\mu_t = \mu' \circ \mu''$, with $\mu'' \neq \emptyset$. Substitutions renaming variables are not possible μ_t. In such cases the renaming occurs in σ, as in the previous example.

Since colours in any marking refer to places of the net (i.e., names in S_N) and μ_t is an instantiation of a pattern corresponding to an initial marking, we require $range(\mu_t) \subseteq S_N$. Additionally, this condition assures that if a variable appearing in place position is instantiated, then the chosen transition t belongs to T_N (because transitions generated dynamically cannot fetch messages from already existing places). Consequently, we are also sure that the applied transition is causally independent of the consumed tokens.

Condition $\mu = \mu_t \cup \bigcup_j \mu_j$ *well − defined substitution* assures that all instantiations made by concurrent computations on pattern variables are consistent.

Example 4.1. Consider the dynamic net N presented in Example 3.1. The unfolding pattern for the marking $a(y) \oplus b(x) \oplus c(z)$ is shown in Figures 11 and 12.

While the unfolding is defined for a concrete marking (i.e., all places and colours are constants), an unfolding pattern defines a set of different unfoldings, one for any possible (compatible) instantiation of the pattern.

Definition 4.3 (Instance of an unfolding pattern). *Let $\mathcal{UP}[N]$ the unfolding pattern of $N = (S_N, T_N, \delta_{0N}, \delta_{1N}, p)$, where p is a linear pattern, and a substitution θ s.t. $dom(\theta) \subseteq col_\mathcal{X}(p)$ and $range(\theta) \subseteq S_N$. The instance of $\mathcal{UP}[N]$ by θ, written $\mathcal{UP}[N]\theta = (S\theta, T\theta, \delta\theta_0, \delta\theta_1, S\theta, T\theta, \xi\theta_0, \xi\theta_1)$, is given by the inference rules in Figure 13.*

Note that $\mathcal{UP}[N]\theta$ is obtained by removing all elements of $\mathcal{UP}[N]$ that are incompatible with θ, i.e., elements instantiating the variables of p differently from θ. The following result relates unfoldings and unfolding patterns by showing that the unfolding corresponding to an initial marking $m_0 = p\theta$ can be obtained by instantiating the unfolding pattern for p with θ.

Lemma 4.1. *Let $\mathcal{U}[N] = (S, T, \delta_0, \delta_1, \mathcal{S}, \mathcal{T}, \xi_0, \xi_1)$ be the unfolding of N, and $\mathcal{UP}[N] = (S_P, T_P, \delta_{0P}, \delta_{1P}, \mathcal{S}_P, \mathcal{T}_P, \xi_{0P}, \xi_{1P})$ be the unfolding pattern of N for*

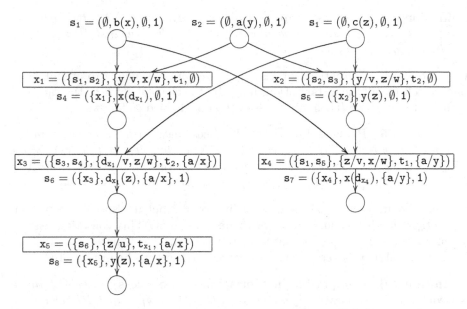

Fig. 12. The causal part of the unfolding pattern $\mathcal{UP}[N]$: (S,T,δ_1,δ_2)

$$\frac{(H,\mathrm{a}(c),\mu,n) \in S,\ (\mu \cup \theta)\ well-defined}{(H,\mathrm{a}(c),\mu,n) \in S\theta} \qquad \frac{\mathrm{a}_x \in S,\ x \in T\theta}{\mathrm{a}_x \in S\theta} \qquad \frac{x \in T\theta}{\delta\theta_i(x) = \delta_i(x)}$$

$$\frac{(B,\sigma,\mathrm{t},\mu) \in T,\ (\mu \cup \theta)\ well-defined}{(B,\sigma,\mathrm{t},\mu) \in T\theta} \qquad \frac{\mathrm{t}_x \in T,\ x \in T\theta}{\mathrm{t}_x \in T\theta} \qquad \frac{\mathrm{t} \in T\theta}{\xi\theta_i(\mathrm{t}) = \xi_i(\mathrm{t})}$$

Fig. 13. Definition of $\mathcal{UP}[N]\theta$, the θ-instance of $\mathcal{UP}[N]$

the linear pattern p s.t. $m_{0N} = p\theta$, and $\mathcal{UP}[N]\theta$ the corresponding instance of $\mathcal{UP}[N]$. Then, there exists a bijective function $f = (f_S : S \to S_P\theta, f_T : T \to T_P\theta, f_S : S \to S_P\theta, f_T : T \to T_P\theta)$ such that:

1. $\forall \mathrm{t} \in T : f_S({}^\bullet\mathrm{t})[\rangle f_S(\mathrm{t}^\bullet) \equiv_\alpha {}^\bullet f_T(\mathrm{t})[\rangle f_T(\mathrm{t})^\bullet$, *i.e., dynamic structures are isomorphic;*
2. $\forall x \in T : f_S({}^\bullet x)[\rangle f_S(x^\bullet) = {}^\bullet f_T(\mathrm{t})[\rangle f_T(\mathrm{t})^\bullet$, *i.e., causal nets are isomorphic;*
3. $\forall x = (B,\sigma,\mathrm{t}) \in T : f_T(x) = (B',\sigma',f_T(\mathrm{t}),\mu)$, *i.e., mapped events correspond to the same transition;*
4. $\forall s = (H,\mathrm{a}(c),i) \in S : f_S(s) = (H',\mathrm{b}(c'),\mu,j)$ *and* $\mathrm{a}(c) = f_S(\mathrm{b}(c')\theta)$, *i.e., mapped places correspond to the same token.*

Proof. The proof follows by rule induction on the definition of $\mathcal{U}[N]$.

Then, starting from the definition of the unfolding pattern, we give the notion of process pattern, which is a process parametric on the colours of the initial tokens. Such definition relies on the following notion of associated instantiation.

Definition 4.4 (Associated instantiation). *Given an* $\mathcal{UP}[N]$, *the instantiation* $inst(s)$ *associated to an element* e *of* $S \cup T$ *is defined as* $inst(H, \sigma, \mu, i) = \mu$ *and* $inst(H, \sigma, \mathsf{t}, \mu) = \mu$. *Given a set* $E \subseteq S \cup T$, $inst(E) = \{inst(e) | e \in E\}$.

The associated instantiations for a set E of elements belonging to an unfolding pattern collects all the instantiations made by the elements of E to the variables of the initial pattern. Then, a process pattern is defined as follows.

Definition 4.5 (Process Pattern). *A* process pattern *for a dynamic net* N *is a net morphism* PP *from a causal net* K *to the net* $(S, T, \delta_0, \delta_1)$, *where* $\mathcal{UP}[N] = (S, T, \delta_0, \delta_1, \mathcal{S}, \mathcal{T}, \xi_0, \xi_1)$ *for a particular pattern* p *as initial marking s.t.* $\forall \mu_1, \mu_2 \in inst(PP(D(PP))) : \mu_1 \cup \mu_2$ *is a well-defined substitution.*

The above definition is analogous to Definition 3.2 but it also requires the instantiations made by final elements to be consistent. This condition assures concurrent events in a process not to instantiate variables differently.

As for unfolding patterns, we allow process patterns to be instantiated.

Definition 4.6 (Compatible instantiation of a process). *A substitution* θ *is a* compatible instantiation *of a process pattern* PP *if* $\forall \mu \in inst(PP(D(PP)))$, $\mu \cup \theta$ *is a well-defined substitution, i.e.,* θ *is compatible with the instantiations of final elements of* P.

Finally, we show that process patterns generalise the definition of processes. We start by showing that an instance of a process pattern is a process of the corresponding instance of the unfolding pattern (Lemma 4.2). Then, we show that the processes of a marked net can be obtained as instantiations of suitable process patterns (Theorem 4.1).

Lemma 4.2. *Let* PP *be a process pattern of* N *for the unfolding pattern* $\mathcal{UP}[N]$, *and* θ *a compatible instantiation. Then,* PP *is a process of* $\mathcal{UP}[N]\theta$.

Proof. The proof follows from the fact that all elements of the unfolding that are image of PP are consistent with θ. Hence, they are also in $\mathcal{UP}[N]\theta$.

Theorem 4.1. *Let* PP *be a process pattern of* N *for the linear pattern* p *s.t.* $m_{0N} = p\theta$. *Then,* PP *describes also a process of* $\mathcal{U}[N]$.

Proof. By Lemma 4.2, PP with θ is a process of $\mathcal{UP}[N]\theta$. By Lemma 4.1, $\mathcal{U}[N]$ and $\mathcal{UP}[N]\theta$ are isomorphic, then there exists P (defined as the composition of PP and the function f of Lemma 4.1), which is a process of $\mathcal{U}[N]$.

5   Concluding Remarks

We have defined the semantic foundations for expressing and analyzing the non-sequential behaviour of dynamic nets. This work extends the line initiated in [4, 11], where suitable notions of process and process pattern have been defined for coloured and reconfigurable nets. Here, we have extended the ordinary notion of unfolding and process to the more expressive setting of dynamic nets.

Since the unfolding semantics have been used for checking properties of, e.g., marked P/T nets and graph grammars [10, 2], we expect the notions given in this paper will allow for the development of techniques for checking properties of dynamic nets. Moreover, unfolding patterns would also enable the study of properties at a more abstract level than unfoldings, for instance, to prove non reachability for a set of initial markings instead of just for a marking.

Since dynamic nets are in one-to-one correspondence with asynchronous join processes, this work provides a process semantics for an asynchronous name passing calculus. Hence, the notions presented in this paper can serve as a starting point for studying the non-sequential behaviour of asynchronous name passing calculus, for instance, for studying process equivalences by considering the true concurrent semantics of processes.

References

1. A. Asperti and N. Busi. Mobile Petri nets. Technical Report UBLCS 96-10, Computer Science Department, University of Bologna, 1996.
2. P. Baldan, A. Corradini, and B. König. A static analysis technique for graph transformation systems. *Proc. CONCUR'01, LNCS* 2154, pp. 381–395. Springer, 2001.
3. N. Benton, L. Cardelli, and C. Fournet. Modern concurrency abstractions for $C^\sharp$. *Proc. ECOOP'02, LNCS* 2374, pp. 415–440. Springer, 2002.
4. R. Bruni, H. Melgratti, and U. Montanari. Extending the zero-safe approach to coloured, reconfigurable and dynamic nets. *Lectures on Concurrency and Petri Nets, Advances in Petri Nets, LNCS* 3098, pp. 291–327. Springer, 2003.
5. M. Buscemi and V. Sassone. High-level Petri nets as type theories in the Join calculus. *Proc. FoSSaCS'01, LNCS* 2030, pp. 104–120. Springer, 2001.
6. S. Conchon and F. Le Fessant. Jocaml: Mobile agents for Objective-Caml. *Proc. ASA'99 / MA'99*, pp. 22–29. IEEE, 1999.
7. C. Fournet and G. Gonthier. The reflexive chemical abstract machine and the Join calculus. *Proc. POPL'96*, pp. 372–385. ACM Press, 1996.
8. U. Goltz and W. Reisig. The non-sequential behaviour of Petri nets. *Inform. and Comput.*, 57:125–147, 1983.
9. C. Gunter and D Scott. Semantic domains. *Handbook of Theoretical Computer Science, Vol. B: Formal Models and Sematics*, pp. 633–674. MIT Press, 1990.
10. K.L. McMillan. *Symbolic Model Checking*. Kluwer, 1993.
11. H. Melgratti. *Models and Languages for Global Computing Transactions*. PhD thesis, Computer Science Department, University of Pisa, 2005.
12. J. Meseguer, U. Montanari, and V. Sassone. Process versus unfolding semantics for place/transition Petri nets. *Theoret. Comput. Sci.*, 153(1-2):171–210, 1996.
13. C.A. Petri. *Kommunikation mit Automaten*. PhD thesis, Institut für Instrumentelle Mathematik, Bonn, 1962.

Complete Finite Prefixes of Symbolic Unfoldings of Safe Time Petri Nets

Thomas Chatain[1] and Claude Jard[2]

[1] IRISA/INRIA,
Campus de Beaulieu, F-35042 Rennes cedex, France
Thomas.Chatain@irisa.fr

[2] IRISA/ENS Cachan-Bretagne,
Campus de Ker Lann, F-35170 Bruz, France
Claude.Jard@bretagne.ens-cachan.fr

Abstract. Time Petri nets have proved their interest in modeling real-time concurrent systems. Their usual semantics is defined in term of firing sequences, which can be coded in a (symbolic and global) state graph, computable from a bounded net. An alternative is to consider a "partial order" semantics given in term of processes, which keep explicit the notions of causality and concurrency without computing arbitrary interleavings. In ordinary place/transition bounded nets, it has been shown for many years that the whole set of processes can be finitely represented by a prefix of what is called the "unfolding". This paper defines such a prefix for safe time Petri nets. It is based on a symbolic unfolding of the net, using a notion of "partial state".

1 Introduction and Related Work

Time Petri nets have proved their interest in modeling real-time concurrent systems. Their usual semantics is defined in term of firing sequences, which can be coded in a (symbolic and global) state graph, computable if the net is bounded. Although efficient for many verification problems, the main drawback of this approach is to mask the concurrent aspects of the model behind the explicit computation of the possible interleavings of actions, leading to the usual state explosion problem. This can be circumvented by the use of heuristics, named as "partial-order reductions". The idea is to explore only the subset of states and transitions that are relevant for properties that are not concerned with commutation of concurrent actions. The main difficulty is the identification of concurrency among the symbolic representation of states. Examples of this approach can be found in [1, 2].

An alternative is to consider a "partial order" semantics given in term of processes, which keep explicit the notions of causality and concurrency without computing arbitrary interleavings. This is our framework. The goal is not only to avoid the cost of having in memory all the interleavings when performing verification, but also to have a graphical representation of the timed processes. This opens new perspectives to other applications like supervision and diagnosis [3].

S. Donatelli and P.S. Thiagarajan (Eds.): ICATPN 2006, LNCS 4024, pp. 125–145, 2006.

The first definition of a (denotational) partial-order semantics for time Petri nets was given in 1996 by Aura and Lilius [4, 5], who formally defined the timed processes of a net. The question of a finite representation of the whole set of timed processes was left open. To do that, two approaches have been explored. The first one is limited to the discrete time framework. It is based on a modeling of clock ticks, which defines an embedding of time Petri nets into ordinary Petri nets. In ordinary place/transition bounded nets, it has been shown for many years that the whole set of processes can be finitely represented by a prefix of what is called the "unfolding" [6, 7]. The drawback is the explosive nature of the transformation. The advantage is to recover the unfolding technology and the existence of a complete finite prefix. This is the historical approach of [8, 9], who defined for the first time the notion of finite prefix for time Petri nets.

We follow a second approach, which definitely uses a symbolic framework to code the time constraints associated with the timed processes. It can deal with dense time and is in the continuation of Berthomieu's works on symbolic representations of the global states of time Petri nets [10]. The first contribution in that direction was made in 1996 by Semenov [11], who has considered a restrictive subclass of safe time Petri nets, named "time independent choice nets". It is a very simple case where the timed aspects do not actually introduce so many constraints. The interest is that the whole set of timed processes can be simply represented as a subset of the unfolding of the underlying ordinary Petri net, by copying the interval constraints. The general case is much more complicated. We have shown in [3] that a new definition of unfoldings is required. These use symbolic constraints on the possible firing dates of transitions, and are based on a notion of "partial state", richer than the usual marking of the input places. This is required by the fact that firing a timed transition cannot be decided locally, and may depend on the dates of arrival of tokens in places feeding transitions that are in conflict (this situation is often called "confusion" in Petri nets). [3] presented this new notion of unfolding and its application for diagnosing timed distributed systems. In this paper, we show that there also exists a notion of complete and finite prefix of this unfolding. It opens new perspectives in the verification of timed systems, based on a time Petri net modeling.

The rest of the paper is organized as follows. Section 2 presents the safe time Petri net model, with their interleaving semantics and their partial order representation given in term of processes. The concurrent operational semantics and the merging of the induced extended processes into a symbolic unfolding is described in Section 3. Section 4 is dedicated to its finite representation using the notion of complete finite prefix, before conclusion.

2   Safe Time Petri Nets

2.1   Definition

Notations. We denote f^{-1} the inverse of a bijection f. We denote $f_{|A}$ the restriction of a mapping f to a set A. The restriction has higher priority than the inverse: $f_{|A}^{-1} = (f_{|A})^{-1}$. We denote $\circ$ the usual composition of functions. Q denotes the set of nonnegative rational numbers.

Time Petri nets were introduced in [12]. A *time Petri net* is a tuple $\langle P, T, pre, post, efd, lfd \rangle$ where P and T are finite sets of *places* and *transitions* respectively, *pre* and *post* map each transition $t \in T$ to its *preset* often denoted ${}^\bullet t \stackrel{\text{def}}{=} pre(t) \subseteq P$ (${}^\bullet t \neq \emptyset$) and its *postset* often denoted $t^\bullet \stackrel{\text{def}}{=} post(t) \subseteq P$; $efd : T \longrightarrow Q$ and $lfd : T \longrightarrow Q \cup \{\infty\}$ associate the *earliest firing delay* $efd(t)$ and *latest firing delay* $lfd(t)$ with each transition t. A time Petri net is represented as a graph with two types of nodes: places (circles) and transitions (rectangles). The closed interval $[efd(t), lfd(t)]$ is written near each transition (see Figure 1).

2.2   Interleaving Semantics

A *state* of a time Petri net is given by a triple $\langle M, dob, \theta \rangle$, where $M \subseteq P$ is a *marking* denoted with tokens (thick dots), $\theta \in Q$ is its date and $dob : M \longrightarrow Q$ associates a *date of birth* $dob(p) \leq \theta$ with each token (marked place) $p \in M$. A transition $t \in T$ is *enabled* in the state $\langle M, dob, \theta \rangle$ if all of its input places are marked: ${}^\bullet t \subseteq M$. Its *date of enabling* $doe(t)$ is the date of birth of the youngest token in its input places: $doe(t) \stackrel{\text{def}}{=} \max_{p \in {}^\bullet t} dob(p)$. All the time Petri nets we consider in this article are *safe*, i.e. in each reachable state $\langle M, dob, \theta \rangle$, if a transition t is enabled in $\langle M, dob, \theta \rangle$, then $t^\bullet \cap (M \setminus {}^\bullet t) = \emptyset$.

A time Petri net starts in an *initial state* $\langle M_0, dob_0, \theta_0 \rangle$, which is given by the *initial marking* M_0 and the initial date θ_0. Initially, all the tokens carry the date θ_0 as date of birth: for all $p \in M_0$, $dob_0(p) \stackrel{\text{def}}{=} \theta_0$.

The transition t can fire at date $\theta' \geq \theta$ from state $\langle M, dob, \theta \rangle$, if:

- t is enabled: ${}^\bullet t \subseteq M$;
- the minimum delay is reached: $\theta' \geq doe(t) + efd(t)$;
- the enabled transitions do not overtake the maximum delays:
  $\forall t' \in T \quad {}^\bullet t' \subseteq M \implies \theta' \leq doe(t') + lfd(t')$.

The firing of t at date θ' leads to the state $\langle (M \setminus {}^\bullet t) \cup t^\bullet, dob', \theta' \rangle$, where $dob'(p) \stackrel{\text{def}}{=} dob(p)$ if $p \in M \setminus {}^\bullet t$ and $dob'(p) \stackrel{\text{def}}{=} \theta'$ if $p \in t^\bullet$.

We call *firing sequence* starting from the initial state S_0 any sequence $((t_1, \theta_1), \ldots, (t_n, \theta_n))$ where there exist states $S_1, \ldots, S_n$ such that for all $i \in \{1, \ldots, n\}$, firing t_i from S_{i-1} at date θ_i is possible and leads to S_i. The empty firing sequence is denoted ϵ.

Finally we assume that time *diverges*: when infinitely many transitions fire, time necessarily diverges to infinity.

In the initial state of the net of Figure 1, p_1 and p_2 are marked and their date of birth is 0. t_1 and t_2 are enabled and their date of enabling is the initial date 0. t_2 can fire in the initial state at any time between 1 and 2. Choose time 1.3. After this firing, p_1 and p_4 are marked, t_1 is the only enabled transition and it has already waited 1.3 time unit. t_1 can fire at any time θ, provided it is greater than 1.3. Let t_1 fire at time 3. p_3 and p_4 are marked in the new state, and transitions t_3 and t_0 are enabled, and their date of enabling is 3 because

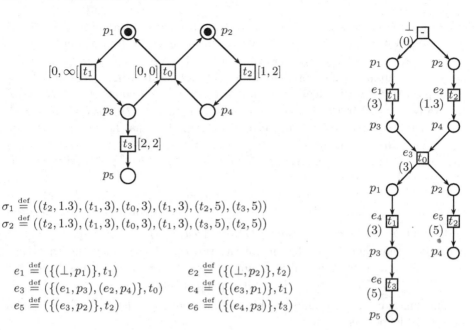

$$\sigma_1 \overset{\text{def}}{=} ((t_2, 1.3), (t_1, 3), (t_0, 3), (t_1, 3), (t_2, 5), (t_3, 5))$$
$$\sigma_2 \overset{\text{def}}{=} ((t_2, 1.3), (t_1, 3), (t_0, 3), (t_1, 3), (t_3, 5), (t_2, 5))$$

$$e_1 \overset{\text{def}}{=} (\{(\bot, p_1)\}, t_1) \qquad e_2 \overset{\text{def}}{=} (\{(\bot, p_2)\}, t_2)$$
$$e_3 \overset{\text{def}}{=} (\{(e_1, p_3), (e_2, p_4)\}, t_0) \qquad e_4 \overset{\text{def}}{=} (\{(e_3, p_1)\}, t_1)$$
$$e_5 \overset{\text{def}}{=} (\{(e_3, p_2)\}, t_2) \qquad e_6 \overset{\text{def}}{=} (\{(e_4, p_3)\}, t_3)$$

Fig. 1. A safe time Petri net and two of its firing sequences σ_1 and σ_2 starting at date 0 from the marking $M_0 \overset{\text{def}}{=} \{p_1, p_2\}$, that lead to the same process, represented on the right, with the date of the events in parentheses

they have just been enabled by the firing of t_1. To fire, t_3 would have to wait 2 time units. But transition t_0 cannot wait at all. So t_0 will necessarily fire (at time 3), and t_3 cannot fire.

Remark. The semantics of time Petri nets are often defined in a slightly different way: the state of the net is given as a pair $\langle M, I \rangle$, where M is the marking, and I maps each enabled transition t to the delay that has elapsed since it was enabled, that is $\theta - doe(t)$ with our notations. It is more convenient for us to attach time information to the tokens of the marking than to the enabled transitions. We have chosen the date of birth of the tokens rather than their age, because we want to make the impact of the firing of transitions as local as possible. And the age of each token in the marking must be updated each time a transition t fires, whereas the date of birth has to be set only for the tokens that are created by t. Furthermore, usual semantics often deal with the delay between the firing of two consecutive transitions. In this paper we use the absolute firing date of the transitions instead. This fits better to our approach in which we are not interested in the total ordering of the events.

2.3   Partial Order Representation of the Runs: Processes

Processes are a way to represent an execution of a Petri net so that the actions (called events) are not totally ordered like in firing sequences: only causality orders the events. We will define the mapping Π from the firing sequences of a safe time Petri net to their partial order representation as processes. These processes are those described in [4]. We use a canonical coding like in [13].

Each process will be a pair $x \stackrel{\text{def}}{=} \langle E, \Theta \rangle$, where E is a set of *events*, and $\Theta : E \longrightarrow Q$ maps each event to its firing date. Θ is sometimes represented as a set of pairs $(e, \Theta(e))$. Each event e is a pair $({}^\bullet e, \tau(e))$ that codes an occurrence of the transition $\tau(e)$ in the process. ${}^\bullet e$ is a set of pairs $b \stackrel{\text{def}}{=} ({}^\bullet b, place(b)) \in E \times P$. Such a pair is called a *condition* and refers to the token that has been created by the event ${}^\bullet b$ in the place $place(b)$. We say that the event $e \stackrel{\text{def}}{=} ({}^\bullet e, \tau(e))$ *consumes* the conditions in ${}^\bullet e$. Symmetrically the set $\{(e, p) \mid p \in \tau(e)^\bullet\}$ of conditions that are *created* by e is denoted $e^\bullet$.

For all set B of conditions, we denote $Place(B) \stackrel{\text{def}}{=} \{place(b) \mid b \in B\}$, and when the restriction of $place$ to B is injective, we denote $place_{|B}^{-1}$ its inverse, and for all $P \subseteq Place(B)$, $Place_{|B}^{-1}(P) \stackrel{\text{def}}{=} \{place_{|B}^{-1}(p) \mid p \in P\}$. The set of conditions that remain at the end of the process $\langle E, \Theta \rangle$ (meaning that they have been created by an event of E, and no event of E has consumed them) is $\uparrow(E) \stackrel{\text{def}}{=} \bigcup_{e \in E} e^\bullet \setminus \bigcup_{e \in E} {}^\bullet e$ (it does not depend on Θ).

The function Π that maps each firing sequence $((t_1, \theta_1), \ldots, (t_n, \theta_n))$ to a process is defined as follows:

- $\Pi(\epsilon) \stackrel{\text{def}}{=} \langle \{\bot\}, \{(\bot, \theta_0)\} \rangle$, where $\bot \stackrel{\text{def}}{=} (\emptyset, \text{-})$ represents the initial event and θ_0 the initial date. Notice that the initial event does not actually represent the firing of a transition, which explains the use of the special value $\text{-} \notin T$. For the same reason, the set of conditions that are created by $\bot$ is defined in a special way: $\bot^\bullet \stackrel{\text{def}}{=} \{(\bot, p) \mid p \in M_0\}$.

- $\Pi(((t_1, \theta_1), \ldots, (t_{n+1}, \theta_{n+1}))) \stackrel{\text{def}}{=} \langle E \cup \{e\}, \Theta \cup \{(e, \theta_{n+1})\} \rangle$, where $\langle E, \Theta \rangle \stackrel{\text{def}}{=} \Pi(((t_1, \theta_1), \ldots, (t_n, \theta_n)))$ and the event $e \stackrel{\text{def}}{=} (Place_{|\uparrow(E)}^{-1}({}^\bullet t_{n+1}), t_{n+1})$ represents the last firing of the sequence.

The set of all the processes obtained as the image by Π of the firing sequences is denoted X.

We define the relation $\rightarrow$ on the events as: $e \rightarrow e'$ iff $e^\bullet \cap {}^\bullet e' \neq \emptyset$. The reflexive transitive closure $\rightarrow^*$ of $\rightarrow$ is called the *causality* relation. Two events of a process that are not causally related are called *concurrent*. For all event e, we denote $\lceil e \rceil \stackrel{\text{def}}{=} \{f \in E \mid f \rightarrow^* e\}$, and for all set E of events, $\lceil E \rceil \stackrel{\text{def}}{=} \bigcup_{e \in E} \lceil e \rceil$.

Figure 1 shows two firing sequences that correspond to the same process. In the representation of the process, the rectangles represent the events, and the circles represent the conditions. An arrow from a condition b to an event e means that $b \in {}^\bullet e$. An arrow from an event e to a condition b means that $b \in e^\bullet$.

3 Symbolic Unfoldings of Safe Time Petri Nets

3.1 Introduction

Symbolic unfoldings have already been addressed in the context of high-level Petri nets [14] to reflect the genericity that appears in the model. In this section we define the symbolic unfolding of time Petri nets, i.e. a compact structure that contains all the processes and exhibits concurrency. When we build unfoldings, we would like to be able to unfold seperately two parts of the system when these two parts do not communicate, like the left part and the right part of the net of Figure 1 when t_1 and t_2 fire. As opposed to untimed Petri nets, in a timed context we accept that this may not yield proper processes but only what we will call *pre-processes*, in which the different parts of the system may not have reached the same date, provided the events that have been built are contained in a real execution. Let us define formally these *pre-process* as prefixes of the processes. Notice that in an untimed context pre-processes would simply be processes.

Definition 1 (pre-processes). *For all process* $\langle E, \Theta \rangle$, *and for all nonempty, causally closed set of events* $E' \subseteq E$ *($\perp \in E'$ and* $\lceil E' \rceil = E'$*),* $\langle E', \Theta_{|E'} \rangle$ *is called a* pre-process. *We sometimes write* $\langle E', \Theta \rangle$ *instead of* $\langle E', \Theta_{|E'} \rangle$ *for short.*

Definition 2 (prefix relation on pre-processes). *We define the* prefix *relation* $\leq$ *on pre-processes as follows:* $\langle E, \Theta \rangle \leq \langle E', \Theta' \rangle$ *iff* $E \subseteq E' \wedge \Theta = \Theta'_{|E}$.

In the case of untimed Petri nets, the unfolding can be defined as the superimposition of all the processes, that is the set of all the events that appear in the processes. This structure is quite compact since an event generally occurs in many different processes. However, it has to be easy to extract a process from the unfolding. Especially, given a causally closed set of events, it has to be easy to tell if there is a process that contains all these events, or containing at least these events. With untimed Petri nets, these two questions are the same and can be solved easily. In some special classes of time Petri nets, especially when the underlying untimed Petri net is extended free choice [15], we could simply define a symbolic unfolding of a time Petri net as in the untimed case. [11] defines such an unfolding for "time independent choice nets".

On the contrary, in the general timed case, a similar definition would not give good results. [4] explains how to know if there is a valid timed execution of a time Petri net that corresponds to a set of events taken from the unfolding of the underlying untimed Petri net. But this does not give a way to build an unfolding of a time Petri net and does not tell if there is a process that contains at least these events.

To be convinced that things are much more complicated than in the untimed case, it may be interesting to remark that in the timed case, the union of two pre-processes $\langle E, \Theta \rangle$ and $\langle E', \Theta' \rangle$ is not necessarily a pre-process, even if $E \cup E'$ is conflict free and $\Theta_{|E \cap E'} = \Theta'_{|E \cap E'}$. In the example of Figure 1, we observe this if $\langle E, \Theta \rangle$ is the process which contains a firing of t_1 at time 0 and a firing of t_2 at time 1, and $\langle E', \Theta' \rangle$ is the pre-process that we obtain by removing the firing of t_2 from the process made of t_1 at time 0, t_2 at time 2 and t_3 at time 2.

These difficulties come from the fact that the condition that allows us to extend a process $x \stackrel{\text{def}}{=} \langle E, \Theta \rangle$ with a new event e concerns all the state reached after the process x, and however the conditions in $\bullet e$ refer only to the tokens in the input places of $\tau(e)$.

3.2   Concurrent Operational Semantics for Safe Time Petri Nets

Although the semantics of time Petri nets requires to check time conditions for all the enabled transitions in the net before firing a transition, there are cases when we know that a transition can fire at a given date θ, even if other transitions will fire before θ in other parts of the net. As an example consider the net of Figure 1 starting at time 0 in the marking $\{p_1, p_2\}$. The semantics forbids to fire t_1 at time 10 from the initial state because t_2 is enabled and must fire before time 2. However we are allowed to run the net until time 10 without firing t_1 (because its latest firing delay is infinite). Then, whatever has occurred until time 10, nothing can prevent t_1 from firing at date 10, because only t_1 can remove the token in place p_1. On the contrary, the firing of t_3 highly depends on the firing date of t_2 because when t_0 is enabled it fires immediately and disables t_3. So if we want to fire t_3 we have to check whether p_2 or p_4 is marked. This intuition leads us to define a **concurrent operational semantics** where it is possible to fire a transition without knowing the entire marking of the net, but only a partial marking made of the consumed tokens plus possibly some tokens which are only read (not consumed) in order to get enough information. Theorems 1 and 2 will validate our concurrent partial order semantics by establishing connections with the processes of Section 2.3.

Assumption. From now on we assume that we know a partition of the set P of places of the net in sets $P_i \subseteq P$ of mutually exclusive places[1]; more precisely we demand that for all reachable marking M, $P_i \cap M$ is a singleton. For all place $p \in P_i$, we denote $\bar{p} \stackrel{\text{def}}{=} P_i \setminus \{p\}$. In the example of Figure 1, we will use the partition $\{p_1, p_3, p_5\}$, $\{p_2, p_4\}$. In fact this partition will be used to test the absence of a token in a marking. For instance if we want to fire t_3, we have to check that t_0 will not fire before t_3 and remove the token in place p_3; if we know that p_2 is marked then we can deduce that p_4 is not, and that t_0 is disabled.

Definition 3 (partial state). *A partial state of a time Petri net is a triple* $\langle L, dob, lrd \rangle$ *where* $L \subseteq P$ *is a partial marking and* $dob, lrd : L \longrightarrow Q$ *associate a date of birth* $dob(p)$ *and a latest reading date* $lrd(p)$ *with each token* $p \in L$.

As opposed to global states, partial states may give only partial information on the state of the net since the partial marking L may not contain one place per set of mutually exclusive places. Notice also that the date θ that appears in global states is replaced by a function that gives the latest reading date of each token of the partial marking, since the global time of the system is not relevant any more in a concurrent semantics.

---

[1] If we do not know any such partition, a solution is to extend the structure of the net with one complementary place for each place of the net and to add these new places in the preset and in the postset of the transitions such that in any reachable marking each place $p \in P$ is marked iff its complementary place is not. This operation does not change the behavior of the time Petri net.

Definition 4 (maximal partial state). *A partial state* $\langle L, dob, lrd \rangle$ *is maximal if L contains one place per set of mutually exclusive places (see the assumption before). From now on the notion of* maximal partial state *or* maximal state *will replace the notion of global state.*

Definition 5 (age of a token in a maximal state). *Let* $S \overset{\text{def}}{=} \langle M, dob, lrd \rangle$ *be a maximal state and let* $p \in M$ *a token (marked place). The date that is reached by the system can be defined as* $\max_{p' \in M} lrd(p')$. *We define the age* $I_S(p)$ *of* p *in the state* S *as the difference:* $I_S(p) \overset{\text{def}}{=} \max_{p' \in M} lrd(p') - dob(p)$.

Definition 6 (temporally consistent maximal state (or consistent state)). *A maximal state* $S \overset{\text{def}}{=} \langle M, dob, lrd \rangle$ *is temporally consistent if for each transition* $t \in T$ *which is enabled in* M *(${}^\bullet t \subseteq M$),* $\min_{p \in {}^\bullet t} I_S(p) \leq lfd(t)$. *A temporally consistent maximal state is also called a* consistent state *for short.*

We will construct a predicate that applies to tuples (L, dob, t, θ), where $L \subseteq P$ is a partial marking, $dob : L \longrightarrow Q$ associates a *date of birth* $dob(p)$ with each token (marked place) $p \in L$, t is a transition and $\theta \geq \max_{p \in L} dob(p)$ is a date.

Such a predicate is called a *local firing condition* and is supposed to tell if knowing that the net is in a state that contains a partial state $\langle L, dob, lrd \rangle$ with $lrd(p) \leq \theta$ for all $p \in L$ is enough to be sure that t can fire at date θ.

Several local firing conditions are possible; some possibilities are discussed in [3]. In this article we will only use a local firing condition LFC that will allow us to build a complete finite prefix of the symbolic unfolding of a time Petri net.

Definition 7 (local firing condition LFC'). *We first define a predicate* LFC' *as follows:* $LFC'(L, dob, t, \theta)$ *holds iff*

- *t is enabled:* ${}^\bullet t \subseteq L$;
- *the minimum delay is reached:* $\theta \geq doe(t) + efd(t)$;
- *the transitions that may consume tokens of L are disabled or do not overtake the maximum delays:*

$$\forall t' \in T \quad {}^\bullet t' \cap L \neq \emptyset \implies \begin{cases} \exists p \in {}^\bullet t' \quad \bar{p} \cap L \neq \emptyset \\ \vee \; \theta \leq \max_{p \in {}^\bullet t' \cap L} dob(p) + lfd(t') \end{cases}$$

The predicate LFC' guarantees that a partial state $\langle L, dob, lrd \rangle$ with $lrd(p) \leq \theta$ for all $p \in L$ is enough to be sure that t can fire at date θ. We define a new local firing condition LFC by demanding that the partial marking L is minimal.

Definition 8 (local firing condition LFC). *LFC is defined as:*
$$LFC(L, dob, t, \theta) \quad \text{iff} \quad \begin{cases} (LFC'(L, dob, t, \theta) \\ \nexists L' \subsetneq L \quad LFC'(L', dob_{|L'}, t, \theta). \end{cases}$$

Semantics of Local Firings. We will now define formally the concurrent operational semantics that we obtain when we allow transitions to fire from a partial state if the local firing condition LFC is satisfied.

The time Petri net starts in an *initial maximal state* $\langle M_0, dob_0, lrd_0 \rangle$, which is given by the *initial marking* M_0 and the initial date θ_0. Initially, all the tokens carry the date θ_0 as date of birth and latest reading date: for all $p \in M_0$, $dob_0(p) \stackrel{\text{def}}{=} lrd_0(p) \stackrel{\text{def}}{=} \theta_0$.

The transition t can fire at date θ using the partial marking $L \subseteq M$, from the maximal state $\langle M, dob, lrd \rangle$ if $(L, dob_{|L}, t, \theta)$ satisfies LFC and for all $p \in L$, $\theta \geq lrd(p)$.

This action leads to the maximal state $\langle (M \setminus {}^\bullet t) \cup t^\bullet, dob', lrd' \rangle$ with

$$dob'(p) \stackrel{\text{def}}{=} \begin{cases} dob(p) & \text{if } p \in M \setminus {}^\bullet t \\ \theta & \text{if } p \in t^\bullet \end{cases} \quad \text{and} \quad lrd'(p) \stackrel{\text{def}}{=} \begin{cases} lrd(p) & \text{if } p \in M \setminus L \\ \theta & \text{if } p \in (L \setminus {}^\bullet t) \cup t^\bullet. \end{cases}$$

We call *sequence of local firings* starting from the initial state S_0 any sequence $((t_1, L_1, \theta_1), \ldots, (t_n, L_n, \theta_n))$ where there exist states $S_1, \ldots, S_n$ such that for all $i \in \{1, \ldots, n\}$, firing t_i from S_{i-1} at date θ_i using the partial marking L_i is possible and leads to S_i. The empty sequence of local firings is denoted ϵ.

Extended Processes. We will define a notion of *extended process*, which is close to the notion of process, but the events are replaced by *extended events* which represent firings from partial states and keep track of all the conditions corresponding to the partial state, not only those that are consumed by the transition: the other conditions will be treated as context of the event. This uses classical techniques of *contextual nets* or nets with *read arcs* (see [16, 17]). It would also be possible to consume and rewrite the conditions in the context of an event, but we feel that the notion of read arc or contextual net is a good way to capture the idea that we develop here.

For all extended event $\dot{e} \stackrel{\text{def}}{=} (B, t)$, denote $\tau(\dot{e}) \stackrel{\text{def}}{=} t$ and $\dot{e}^\bullet \stackrel{\text{def}}{=} \{(\dot{e}, p) \mid p \in t^\bullet\}$. In an extended event, not all the conditions of B are consumed, but only ${}^\bullet\dot{e} \stackrel{\text{def}}{=} Place_{|B}^{-1}({}^\bullet t)$; the conditions in $\underline{\dot{e}} \stackrel{\text{def}}{=} B \setminus {}^\bullet\dot{e}$ are only *read* by $\dot{e}$, which is represented by read arcs. Like for processes, we define the set of conditions that remain at the end of the extended process $\langle \dot{E}, \Theta \rangle$ as $\uparrow(\dot{E}) \stackrel{\text{def}}{=} \bigcup_{\dot{e} \in \dot{E}} \dot{e}^\bullet \setminus \bigcup_{\dot{e} \in \dot{E}} {}^\bullet\dot{e}$.

The function $\dot{\Pi}$ that maps each sequence of local firings $((t_1, L_1, \theta_1), \ldots, (t_n, L_n, \theta_n))$ to an extended process is defined as follows:

- Like for processes, $\dot{\Pi}(\epsilon) \stackrel{\text{def}}{=} \langle \{\bot\}, \{(\bot, \theta_0)\} \rangle$, where $\bot \stackrel{\text{def}}{=} (\emptyset, \text{-})$ represents the initial event. The set of conditions that are created by $\bot$ is defined as: $\bot^\bullet \stackrel{\text{def}}{=} \{(\bot, p) \mid p \in M_0\}$.
- $\dot{\Pi}(((t_1, L_1, \theta_1), \ldots, (t_{n+1}, L_{n+1}, \theta_{n+1}))) \stackrel{\text{def}}{=} \langle \dot{E} \cup \{\dot{e}\}, \Theta \cup \{(\dot{e}, \theta_{n+1})\} \rangle$, where $\langle \dot{E}, \Theta \rangle \stackrel{\text{def}}{=} \dot{\Pi}(((t_1, L_1, \theta_1), \ldots, (t_n, L_n, \theta_n)))$ and the extended event $\dot{e} \stackrel{\text{def}}{=} (Place_{|\uparrow(\dot{E})}^{-1}(L_{n+1}), t_{n+1})$ represents the last local firing of the sequence.

The set of all the extended processes obtained as the image by $\dot{\Pi}$ of the sequences of local firings is denoted $\dot{X}$.

As we use read arcs, the usual causality is not sufficient any more: we have to define an unconditional or strong causality $\to$ and a conditional or weak causality $\nearrow$ between extended events as:

$- \dot{e} \to \dot{f}$ iff $\dot{e}^{\bullet} \cap (^{\bullet}\dot{f} \cup \underline{\dot{f}}) \neq \emptyset$ and
$- \dot{e} \nearrow \dot{f}$ iff $(\dot{e} \to \dot{f}) \vee (\underline{\dot{e}} \cap ^{\bullet}\dot{f} \neq \emptyset)$.

For all extended event $\dot{e}$, we denote $\lceil \dot{e} \rceil \stackrel{\text{def}}{=} \{\dot{f} \in \dot{E} \mid \dot{f} \to^{*} \dot{e}\}$ and for all set $\dot{E}$ of extended events, $\lceil \dot{E} \rceil \stackrel{\text{def}}{=} \bigcup_{\dot{e} \in \dot{E}} \lceil \dot{e} \rceil$.

Figure 2 shows several extended processes. An arrow from a condition b to an extended event $\dot{e}$ means that $b \in {}^{\bullet}\dot{e}$. An arrow from an extended event $\dot{e}$ to a condition b means that $b \in \dot{e}^{\bullet}$. When $b \in \underline{\dot{e}}$, the read arc is represented by a line without arrow between b and $\dot{e}$.

Definition 9 ($RS(\langle \dot{E}, \Theta \rangle)$). *The maximal state that is reached after an extended process $\langle \dot{E}, \Theta \rangle$ is defined as $RS(\langle \dot{E}, \Theta \rangle) \stackrel{\text{def}}{=} \langle Place(\uparrow(\dot{E})), dob, lrd \rangle$ where for all $b = ({}^{\bullet}b, p) \in \uparrow(\dot{E})$, $dob(p) \stackrel{\text{def}}{=} \Theta({}^{\bullet}b)$ and $lrd(p) \stackrel{\text{def}}{=} \max_{\dot{e} \in \dot{E}, \, b \in \dot{e}^{\bullet} \cup \underline{\dot{e}}} \Theta(\dot{e})$.*

Remark that all the sequences of local firings σ such that $\dot{\Pi}(\sigma) = \langle \dot{E}, \Theta \rangle$ lead to $RS(\langle \dot{E}, \Theta \rangle)$.

Definition 10 (temporally complete extended process, $\dot{Y}$). *We say that $\langle \dot{E}, \Theta \rangle$ is temporally complete if $RS(\langle \dot{E}, \Theta \rangle)$ is temporally consistent. The set of all temporally complete extended processes is denoted $\dot{Y}$.*

Correctness and Completeness of LFC. Each extended event $\dot{e}$ can be mapped to the corresponding event

$$h(\dot{e}) \stackrel{\text{def}}{=} \left(\{(h(\dot{f}), p) \mid (\dot{f}, p) \in {}^{\bullet}\dot{e}\}, \tau(\dot{e}) \right).$$

Given an extended process $\langle \dot{E}, \Theta \rangle \in \dot{X}$, $\langle h(\dot{E}), \Theta \circ h_{|\dot{E}}^{-1} \rangle$ is intuitively what we obtain if we remove the read arcs from $\langle \dot{E}, \Theta \rangle$. For example the extended process $\dot{x}$ in Figure 2 would be mapped to the process of Figure 1.

Lemma 1. *For all $\langle \dot{E}, \Theta \rangle \in \dot{X}$, $\langle h(\dot{E}), \Theta \circ h_{|\dot{E}}^{-1} \rangle \in X$ iff $\langle \dot{E}, \Theta \rangle \in \dot{Y}$.*

Proof. Let $\langle \dot{E}, \Theta \rangle \in \dot{X}$ be an extended process and denote $\langle M, dob, lrd \rangle \stackrel{\text{def}}{=} RS(\langle \dot{E}, \Theta \rangle)$ and $\theta \stackrel{\text{def}}{=} \max_{p \in M} lrd(p) = \max_{\dot{e} \in \dot{E}} \Theta(\dot{e})$.

It follows from the definition of the processes that if $\langle h(\dot{E}), \Theta \circ h_{|\dot{E}}^{-1} \rangle \in X$, then $\langle \dot{E}, \Theta \rangle$ is temporally complete.

Conversely, assume that $\langle \dot{E}, \Theta \rangle$ is temporally complete. Choose $\dot{e} \in \dot{E}$ such that $\Theta(\dot{e}) = \theta$ and $\nexists \dot{f} \in \dot{E}$ such that $\dot{e} \nearrow \dot{f}$. Then denote $\langle M', dob', lrd' \rangle \stackrel{\text{def}}{=} RS(\langle \dot{E} \setminus \{\dot{e}\}, \Theta \rangle)$ and $\theta' \stackrel{\text{def}}{=} \max_{p \in M'} lrd'(p)$, and let $t \in T$ such that ${}^{\bullet}t \subseteq M'$. If ${}^{\bullet}t \cap {}^{\bullet}\tau(\dot{e}) = \emptyset$, then $doe'(t) = doe(t) \geq \theta - lfd(t) \geq \theta' - lfd(t)$. Otherwise let $L \stackrel{\text{def}}{=} {}^{\bullet}\dot{e} \cup \underline{\dot{e}}$. As $LFC(L, dob, \tau(\dot{e}), \Theta(\dot{e}))$ holds, then

$$\begin{cases} \exists p \in {}^{\bullet}t \quad \bar{p} \cap L \neq \emptyset \\ \vee \, \theta \leq \max_{p \in {}^{\bullet}t \cap L} dob'(p) + lfd(t) \end{cases}$$

As $^\bullet t \subseteq M'$, then $\nexists p \in {}^\bullet t$ such that $\bar{p} \cap L \neq \emptyset$; thus $\theta \leq \max\limits_{p \in {}^\bullet t \cap L} dob'(p) + lfd(t)$.
Hence $doe'(t) = \max\limits_{p \in {}^\bullet t} dob'(p) \geq \max\limits_{p \in {}^\bullet t \cap L} dob'(p) \geq \theta - lfd(t) \geq \theta' - lfd(t)$. As a
result $\langle \dot{E} \setminus \{\dot{e}\}, \Theta \rangle \in \dot{Y}$.

Assume now that $\langle E, \Theta' \rangle \overset{\text{def}}{=} \langle h(\dot{E} \setminus \{\dot{e}\}), \Theta \circ h_{|\dot{E}}^{-1} \rangle \in X$. It leads to the global
state $\langle M', dob', \theta' \rangle$. As $^\bullet \tau(\dot{e}) \subseteq M'$ and $\theta \geq \theta'$ and $\theta \geq doe'(\tau(\dot{e})) + efd(\tau(\dot{e}))$
and for all $t \in T$, $^\bullet t \subseteq M' \implies \theta \leq doe'(t) + {}^\bullet lfd(t)$, then $\tau(\dot{e})$ can fire at date θ
from $\langle M', dob', \theta' \rangle$, which is coded by the event $(Place_{|\uparrow(E)}^{-1}(\tau(\dot{e})), \tau(\dot{e})) = h(\dot{e})$.
Thus $\langle h(\dot{E}), \Theta \circ h_{|\dot{E}}^{-1} \rangle \in X$.

Theorem 1 (correctness of LFC). *For all extended process $\langle \dot{E}, \Theta \rangle \in \dot{X}$,
$\langle h(\dot{E}), \Theta \circ h_{|\dot{E}}^{-1} \rangle$ is a pre-process (notice that $h_{|\dot{E}}$ is injective). In other terms
there exists a process $\langle E', \Theta' \rangle \in X$ such that $\langle h(\dot{E}), \Theta \circ h_{|\dot{E}}^{-1} \rangle \leq \langle E', \Theta' \rangle$.*

Proof. To prove that LFC is correct, we will prove that for all $\langle \dot{E}, \Theta \rangle \in \dot{X}$,
there exists $\langle \dot{E}', \Theta' \rangle \in \dot{Y}$ such that $\dot{E} \subseteq \dot{E}'$ and $\Theta = \Theta'_{|\dot{E}}$; as a consequence
$\langle h(\dot{E}), \Theta \circ h_{|\dot{E}}^{-1} \rangle \leq \langle h(\dot{E}'), \Theta' \circ h_{|\dot{E}'}^{-1} \rangle \in X$.

Let $\langle \dot{E}, \Theta \rangle \in \dot{X}$. If $\langle \dot{E}, \Theta \rangle$ is temporally complete, then it is sufficient to take
$\langle \dot{E}', \Theta' \rangle \overset{\text{def}}{=} \langle \dot{E}, \Theta \rangle$.

Otherwise, denote $\langle M, dob, lrd \rangle \overset{\text{def}}{=} RS(\langle \dot{E}, \Theta \rangle)$ and $\theta \overset{\text{def}}{=} \max_{p \in M} lrd(p)$,
choose $t \in T$ such that $^\bullet t \subseteq M \wedge \theta > doe(t) + lfd(t)$ and such that t minimizes $\theta_t \overset{\text{def}}{=}$
$doe(t) + lfd(t)$. Let $\dot{F} \overset{\text{def}}{=} \{\dot{f} \in \dot{E} \mid \Theta(\dot{f}) \leq \theta_t\}$. $\langle \dot{F}, \Theta_{|\dot{F}} \rangle$ is a temporally complete
extended process. Denote $\langle M', dob', lrd' \rangle \overset{\text{def}}{=} RS(\langle \dot{F}, \Theta_{|\dot{F}} \rangle)$. $LFC'(M', dob', t, \theta_t)$
holds. Thus there exists $L \subseteq M'$ such that $LFC(L, dob'_{|L}, t, \theta_t)$ holds. Let
$\dot{e} \overset{\text{def}}{=} (Place_{|\uparrow(\dot{F})}^{-1}(L), t)$. We will show that $\langle \dot{E} \cup \{\dot{e}\}, \Theta \cup \{(\dot{e}, \theta_t)\} \rangle \in \dot{X}$. $\Theta \cup \{(\dot{e}, \theta_t)\}$
is compatible with $\nearrow$: if an extended event $\dot{f} \in \dot{E}$ is such that $\underline{\dot{f}} \cap {}^\bullet \dot{e} \neq \emptyset$, then
$\Theta(\dot{f}) \leq \theta_t$ and if $^\bullet \dot{f} \cap \underline{\dot{e}} \neq \emptyset$, then $\Theta(\dot{f}) > \theta_t$. The strict inequality in the second
case also guarantees that $\nearrow$ is acyclic on $\dot{E} \cup \{\dot{e}\}$. As a result, we have built an
extended process $\langle \dot{E} \cup \{\dot{e}\}, \Theta \cup \{(\dot{e}, \theta_t)\} \rangle \in \dot{X}$ by adding the event to $\langle \dot{E}, \Theta \rangle$.

Iterating this until $\langle \dot{E}, \Theta \rangle$ is temporally complete, terminates if we assume
that time diverges: at each step $\langle \dot{F}, \Theta_{|\dot{F}} \rangle$ is temporally complete, so $\langle h(\dot{F}), \Theta \circ$
$h_{|\dot{F}}^{-1} \rangle \in X$; moreover this process has strictly more events at each step and the
dates remain below θ, which does not increase.

Theorem 2 (completeness of LFC). *For all process $\langle E, \Theta \rangle \in X$, there exists
an extended process $\langle \dot{E}, \Theta' \rangle \in \dot{X}$ such that $\langle h(\dot{E}), \Theta' \circ h_{|\dot{E}}^{-1} \rangle = \langle E, \Theta \rangle$.*

Proof. Let $\langle E, \Theta \rangle \in X$ leading to the global state $\langle M, dob, \theta \rangle$, let $t \in T$ be a
transition that can fire at date $\theta' \geq \theta$ from $\langle M, dob, \theta \rangle$, and assume that there
exists an extended process $\langle \dot{E}, \Theta' \rangle \in \dot{X}$ such that $\langle h(\dot{E}), \Theta' \circ h_{|\dot{E}}^{-1} \rangle = \langle E, \Theta \rangle$.
$LFC'(M, dob, t, \theta')$ holds. Thus there exists $L \subseteq M$ such that $LFC(L, dob_{|L}, t, \theta')$

holds. Define $\dot{e} \stackrel{\text{def}}{=} (Place_{|\uparrow \dot{E}}^{-1}(L), t)$. $\langle \dot{E} \cup \{\dot{e}\}, \Theta' \cup \{(\dot{e}, \theta')\} \rangle \in \dot{X}$ and the event $h(\dot{e})$ codes the firing of t at date θ' after $\langle E, \Theta \rangle$.

3.3 Symbolic Unfoldings of Safe Time Petri Nets

We have explained in Section 3.1 that the definition of the unfolding has to rely on a concurrent operational semantics. We will now show how the extended processes obtained from our concurrent operational semantics for time Petri nets can be superimposed to build a symbolic unfolding, and that it is easy to recover the extended processes from their superimposition and to build the unfolding. After the definition, we give two theorems: the first one gives a way to extract extended processes from the unfolding, while the second theorem gives a direct construction of the unfolding.

Definition 11 (symbolic unfolding). *We define the* symbolic unfolding U *of a time Petri net by collecting all the extended events that appear in its extended processes:* $U \stackrel{\text{def}}{=} \bigcup_{\langle \dot{E}, \Theta \rangle \in \dot{X}} \dot{E}$.

For all set B of conditions such that $place_{|B}$ is injective and for all mapping $\Theta : \bigcup_{b \in B} \lceil {}^\bullet b \rceil \longrightarrow Q$, we denote $dob_{B,\Theta}$ the mapping defined as: for all $p \in Place(B)$, $dob_{B,\Theta}(p) \stackrel{\text{def}}{=} \Theta({}^\bullet(place_{|B}^{-1}(p)))$.

Theorem 3. *Let $\dot{E} \subseteq U$ be a nonempty finite set of extended events and $\Theta : \dot{E} \longrightarrow Q$ associate a firing date with each extended event of $\dot{E}$. $\langle \dot{E}, \Theta \rangle$ is an extended process iff:*

$$
\begin{cases}
\lceil \dot{E} \rceil = \dot{E} & (\dot{E} \text{ is causally closed}) \\
\nexists \dot{e}, \dot{e}' \in \dot{E} \quad \dot{e} \neq \dot{e}' \wedge {}^\bullet \dot{e} \cap {}^\bullet \dot{e}' \neq \emptyset & (\dot{E} \text{ is conflict free}) \\
\nexists \dot{e}_0, \dot{e}_1, \ldots, \dot{e}_n \in \dot{E} \quad \dot{e}_0 \nearrow \dot{e}_1 \nearrow \cdots \nearrow \dot{e}_n \nearrow \dot{e}_0 & (\nearrow \text{ is acyclic on } \dot{E}) \\
\forall \dot{e}, \dot{e}' \in \dot{E} \quad \dot{e} \nearrow \dot{e}' \implies \Theta(\dot{e}) \leq \Theta(\dot{e}') & (\Theta \text{ is compatible with } \nearrow) \\
\forall \dot{e} = (B, t) \in \dot{E} \setminus \{\perp\} \quad LFC\,(Place(B), dob_{B,\Theta}, t, \Theta(\dot{e})) & \\
& (\dot{e} \text{ corresponds to a local firing condition})
\end{cases}
$$

Proof. Let $\langle \dot{E}, \Theta \rangle \in \dot{X}$ be an extended process that satisfies the conditions in the curly brace, let $\dot{e} \stackrel{\text{def}}{=} (B, t)$ with $B \subseteq \uparrow(\dot{E})$ and $t \in T$ and $\theta' \geq \max_{\dot{f} \in \dot{E}, \, \dot{f} \nearrow \dot{e}} \Theta(\dot{f})$ such that $LFC(Place(B), dob_{B,\Theta}, t, \theta')$ holds. Then we will show that the extended process $\langle \dot{E}', \Theta' \rangle \stackrel{\text{def}}{=} \langle \dot{E} \cup \{\dot{e}\}, \Theta \cup \{(\dot{e}, \theta')\} \rangle$ also satisfies the conditions in the curly brace. By construction $\dot{E}'$ is causally closed. Moreover for each condition $b \in {}^\bullet \dot{e}$ that is consumed by $\dot{e}$, $b \in \uparrow(\dot{E})$, which implies that b has not been consumed by any event of $\dot{E}$. Thus for all $\dot{f} \in \dot{E}$, ${}^\bullet \dot{e} \cap {}^\bullet \dot{f} = \emptyset$ and $\neg(\dot{e} \nearrow \dot{f})$. So $\dot{E}'$ is conflict free and $\nearrow$ is acyclic on $\dot{E}'$. Θ' is compatible with $\nearrow$ because Θ is compatible with $\nearrow$ and $\Theta'(\dot{e}) = \theta' \geq \max_{\dot{f} \in \dot{E}, \, \dot{f} \nearrow \dot{e}} \Theta(\dot{f})$.

Conversely let $\langle \dot{E}', \Theta' \rangle$ satisfy the conditions in the curly brace. If $\dot{E}' = \{\perp\}$, then $\langle \dot{E}', \Theta' \rangle \in \dot{X}$. Otherwise let $\dot{e} \in \dot{E}'$ be an extended event that has no successor by $\nearrow$ in $\dot{E}'$ (such an extended event exists since $\nearrow$ is acyclic on $\dot{E}'$). $\langle \dot{E}, \Theta \rangle \stackrel{\text{def}}{=} \langle \dot{E}' \setminus \{\dot{e}\}, \Theta'_{|\dot{E}' \setminus \{\dot{e}\}} \rangle$ satisfies the conditions in the curly brace. Assume that $\langle \dot{E}, \Theta \rangle \in \dot{X}$. As $\dot{E}$ is conflict free, $^{\bullet}\dot{e} \subseteq \uparrow(\dot{E})$. And as $\dot{e}$ has no successor by $\nearrow$ in $\dot{E}'$, $\underline{\dot{e}} \subseteq \uparrow(\dot{E})$. Furthermore $\Theta'(\dot{e}) \geq \max_{\dot{f} \in \dot{E},\ \dot{f} \nearrow \dot{e}} \Theta(\dot{f})$ and $LFC\left(Place(^{\bullet}\dot{e} \cup \underline{\dot{e}}), dob_{\bullet_{\dot{e} \cup \underline{\dot{e}}, \Theta}}, \tau(\dot{e}), \Theta'(\dot{e})\right)$ holds. Thus $\langle \dot{E}', \Theta' \rangle = \langle \dot{E} \cup \{\dot{e}\}, \Theta \cup \{(\dot{e}, \Theta'(\dot{e}))\} \rangle \in \dot{X}$.

Theorem 4. *For all* $\dot{e} \overset{\text{def}}{=} (B, t) \in \left(\bigcup_{\dot{f} \in U} \dot{f}^{\bullet}\right) \times T$, $\quad \dot{e} \in U$ *iff*

$$
\begin{cases}
\nexists \dot{f}, \dot{f}' \in \lceil \dot{e} \rceil \quad \dot{f} \neq \dot{f}' \ \wedge \ ^{\bullet}\dot{f} \cap \, ^{\bullet}\dot{f}' \neq \emptyset & (1) \\
\nexists \dot{e}_0, \dot{e}_1, \ldots, \dot{e}_n \in \lceil \dot{e} \rceil \quad \dot{e}_0 \nearrow \dot{e}_1 \nearrow \cdots \nearrow \dot{e}_n \nearrow \dot{e}_0 & (2) \\
\exists \Theta : \lceil \dot{e} \rceil \longrightarrow Q \quad \begin{cases} \forall \dot{f}, \dot{f}' \in \lceil \dot{e} \rceil \quad \dot{f} \nearrow \dot{f}' \implies \Theta(\dot{f}) \leq \Theta(\dot{f}') \\ \forall \dot{f} = (B', t') \in \lceil \dot{e} \rceil \setminus \{\bot\} \\ \quad LFC\left(Place(B'), dob_{B', \Theta}, t, \Theta(\dot{f})\right) \end{cases} & (3)
\end{cases}
$$

This theorem allows us to simply build the unfolding starting from the set $\{\bot\}$ and adding extended events one by one when they satisfy the condition.

Proof. Let $\dot{e} \in U$. There exists $\langle \dot{E}, \Theta' \rangle \in \dot{X}$ such that $\dot{e} \in \dot{E}$. $\langle \dot{E}, \Theta' \rangle$ satisfies the conditions in the curly brace of Theorem 3. As $\lceil \dot{E} \rceil \subseteq \dot{E}$, $\lceil \dot{e} \rceil$ also satisfies them. Then (1) and (2) hold. For (3) a possible Θ is $\Theta'_{|\lceil \dot{e} \rceil}$.

Conversely if $\dot{e} \overset{\text{def}}{=} (B, t)$ satisfies (1), (2) and (3), consider a possible Θ for (3). $\langle \lceil \dot{e} \rceil \setminus \{\dot{e}\}, \Theta \rangle$ satisfies the curly brace of Theorem 3. Then $\langle \lceil \dot{e} \rceil \setminus \{\dot{e}\}, \Theta \rangle \in X$. Moreover (1) implies that $B \subseteq \uparrow(\lceil \dot{e} \rceil \setminus \{\dot{e}\})$. In addition $\Theta(\dot{e}) \geq \max_{\dot{f} \in \lceil \dot{e} \rceil,\ \dot{f} \nearrow \dot{e}} \Theta(\dot{f})$ and $LFC(Place(B), dob_{B,\Theta}, t, \Theta(\dot{e}))$ holds. Thus $\langle \lceil \dot{e} \rceil, \Theta \rangle \in \dot{X}$ and therefore $\dot{e} \in U$.

4   Complete Finite Prefixes

We have defined the symbolic unfolding of a time Petri net. In general this structure is infinite, as well as the unfoldings of untimed Petri nets. However in the untimed case it is possible to define a finite prefix of the unfolding, which contains complete information about the unfolding [6, 7]. To construct this complete finite prefix one remarks that each untimed safe Petri net has finitely many markings, and that if two processes reach the same marking, then they have the same possible futures. With time Petri nets, the same is true with two temporally complete extended processes that reach the same consistent state. But in general there are infinitely many possible maximal states.

This is why we will try to group them as much as possible. The problem of the density of time has already been solved by the use of a symbolic representation of the dates. Another problem is that the time keeps progressing and never loops; this is why the age of the tokens will now be used instead of their date of birth. Recall that the date of birth was first preferred in order to define a concurrent semantics where the system is allowed to reach temporally inconsistent states, that is states where the different parts of the net have not reached the same date; from now we will work with temporally consistent states.

A last problem arises: even the age of a token may progress forever. But we will define a *reduced age* for each token in a marking, which gives enough information to know what actions are possible, and remains bounded.

4.1 Equivalence of Two Maximal States

It was already shown in [9] that the age of the tokens can be reduced to bounded values without losing information about the possible future actions.

Definition 12 (reduced age of a token). *The* reduced age $J_S(p)$ *of the token* $p \in M$ *in the maximal state* $S \stackrel{\text{def}}{=} \langle M, dob, lrd \rangle$ *as:*

$$J_S(p) \stackrel{\text{def}}{=} \min\{I_S(p), \max\{bound(t) \mid t \in T \wedge p \in {}^\bullet t\}\}$$

where $bound(t) \stackrel{\text{def}}{=} \begin{cases} efd(t) & \text{if} \quad lfd(t) = \infty \\ lfd(t) & \text{otherwise.} \end{cases}$

Definition 13 (equivalence of two maximal states). *Two maximal states* $S_1 \stackrel{\text{def}}{=} \langle M_1, dob_1, lrd_1 \rangle$ *and* $S_2 \stackrel{\text{def}}{=} \langle M_2, dob_2, lrd_2 \rangle$ *are* equivalence *(denoted* $S_1 \sim S_2$*) iff* $M_1 = M_2$ *and* $J_{S_1} = J_{S_2}$.

Theorem 5 (firing a transition from two equivalent consistent states). *Let* S_1 *and* S_2 *be two equivalent consistent states. Let* M *be their marking. A transition* t *can fire from* S_1 *at date* $\theta_1 \geq \max_{p \in M} lrd_1(p)$ *using the partial marking* $L \subseteq M$ *iff it can fire from* S_2 *at date* $\theta_1 - \max_{p \in M} lrd_1(p) + \max_{p \in M} lrd_2(p)$ *using the same partial marking* L.

Proof. Denote $\langle M, dob_i, lrd_i \rangle \stackrel{\text{def}}{=} S_i$ and $\theta'_i \stackrel{\text{def}}{=} \max_{p \in M} lrd_i(p)$ for $i \in \{1, 2\}$. Assume that t can fire from S_1 at date $\theta_1 \geq \max_{p \in M} lrd_1(p)$ using the partial marking $L \subseteq M$. To prove that t can fire from S_2 at date $\theta_2 \stackrel{\text{def}}{=} \theta_1 - \theta'_1 + \theta'_2$ using the same partial marking L, we will show that:

1. $\theta_2 \geq doe_2(t) + efd(t)$, with $doe_2(t) \stackrel{\text{def}}{=} \max_{p \in {}^\bullet t} dob_2(p)$,
2. $\forall t' \in T \quad \begin{cases} {}^\bullet t' \cap L \neq \emptyset \\ \nexists p \in {}^\bullet t' \quad \bar{p} \cap L \neq \emptyset \end{cases} \implies \theta_2 \leq \max\limits_{p \in {}^\bullet t' \cap L} dob_2(p) + lfd(t').$

Here are the proofs for these two points:

1. If $\min_{p \in {}^\bullet t} J_{S_1}(p) \geq efd(t)$, then $\min_{p \in {}^\bullet t} I_{S_2}(p) \geq \min_{p \in {}^\bullet t} J_{S_2}(p) = \min_{p \in {}^\bullet t} J_{S_1}(p) \geq efd(t)$ and $\theta_1 \geq \theta'_1$ implies that $\theta_2 \geq \theta'_2$. Furthermore $\theta'_2 = doe_2(t) + \min_{p \in {}^\bullet t} I_{S_2}(p) \geq doe_2(t) + efd(t)$.
   Otherwise, (if $\min_{p \in {}^\bullet t} J_{S_1}(p) < efd(t)$), then $\theta_1 \geq doe_1(t) + efd(t) = \theta'_1 - \min_{p \in {}^\bullet t} I_{S_1}(p) + efd(t)$. And $\min_{p \in {}^\bullet t} I_{S_1}(p) = \min_{p \in {}^\bullet t} I_{S_2}(p)$ because for all p such that $I_{S_1}(p) < efd(t)$, $I_{S_1}(p) = J_{S_1}(p) = J_{S_2}(p) = I_{S_2}(p)$ (the equalities between $I_{S_i}(p)$ and $J_{S_i}(p)$ hold since one of them is strictly smaller than $efd(t)$), and for all p such that $I_{S_1}(p) \geq efd(t)$, $I_{S_2}(p) \geq J_{S_2}(p) = J_{S_1}(p) \geq efd(t)$. Thus $\theta_2 \geq \theta'_2 - \min_{p \in {}^\bullet t} I_{S_2}(p) + efd(t) = doe_2(t) + efd(t)$.

2. Let $t' \in T$ such that ${}^\bullet t' \cap L \neq \emptyset$ and $(\nexists p \in {}^\bullet t' \quad \bar{p} \cap L \neq \emptyset)$. Then $\theta_1 \leq \max_{p \in {}^\bullet t' \cap L} dob_1(p) + lfd(t')$.
If $lfd(t') = \infty$ then $\max_{p \in {}^\bullet t' \cap L} dob_2(p) + lfd(t') = \infty \geq \theta_2$.
Otherwise for $i \in \{1, 2\}$, $\max_{p \in {}^\bullet t' \cap L} dob_i(p) = \theta'_i - \min_{p \in {}^\bullet t' \cap L} I_{S_i}(p)$.
Then $\theta'_1 \leq \theta_1 \leq \theta'_1 - \min_{p \in {}^\bullet t' \cap L} I_{S_1}(p) + lfd(t')$, and consequently $\min_{p \in {}^\bullet t' \cap L} I_{S_1}(p) \leq lfd(t')$. So $\min_{p \in {}^\bullet t' \cap L} I_{S_2}(p) \geq \min_{p \in {}^\bullet t' \cap L} I_{S_1}(p)$ because for all p such that $I_{S_1}(p) \leq lfd(t')$, $I_{S_1}(p) = J_{S_1}(p) = J_{S_2}(p) \leq I_{S_2}(p)$, and for all p such that $I_{S_1}(p) > lfd(t')$, $I_{S_2}(p) \geq J_{S_2}(p) = J_{S_1}(p) \geq lfd(t')$. Thus $\theta_2 \leq \theta'_2 - \min_{p \in {}^\bullet t' \cap L} I_{S_2}(p) + lfd(t') = \max_{p \in {}^\bullet t' \cap L} dob_2(p) + lfd(t')$.

4.2 Substitution of Prefixes in Extended Processes

Knowing that the same actions are possible from equivalent consistent states, if we have two temporally complete extended processes that reach equivalent states, we can translate any continuation of one extended process to the other, providing we also translate the firing dates of the events. This operation is illustrated in Figure 2. It corresponds intuitively to merging the final conditions of the first extended process with the conditions from which the extension starts in the second process.

Definition 14 (substitution of prefixes in extended processes)
Let $\dot{x}_1 \stackrel{\text{def}}{=} \langle \dot{E}_1, \Theta_1 \rangle$ and $\dot{x}_2 \stackrel{\text{def}}{=} \langle \dot{E}_2, \Theta_2 \rangle$ be two extended processes, and $\dot{E}'_2 \subseteq \dot{E}_2$ such that $\dot{x}_1$ and $\langle \dot{E}'_2, \Theta_{2|\dot{E}'_2} \rangle$ are temporally complete extended processes, $RS(\langle \dot{E}'_2, \Theta_{2|\dot{E}'_2} \rangle) \sim RS(\langle \dot{E}_1, \Theta_1 \rangle)$ and for all $\dot{e}' \in \dot{E}'_2$ and $\dot{e} \in \dot{E}_2 \setminus \dot{E}'_2$, $\Theta(\dot{e}'_2) \leq \Theta(\dot{e}_2)$ and $\neg(\dot{e}'_2 \nearrow \dot{e}_2)$. We define the substitution which replaces $\langle \dot{E}'_2, \Theta_{2|\dot{E}'_2} \rangle$ by $\dot{x}_1$ in $\dot{x}_2$ as:

$$subst(\dot{x}_1, \dot{E}'_2, \dot{x}_2) \stackrel{\text{def}}{=} \langle \dot{E}, \Theta \rangle$$

where

$$\dot{E} \stackrel{\text{def}}{=} \dot{E}_1 \cup \phi(\dot{E}_2 \setminus \dot{E}'_2)$$

$$\Theta(\dot{e}) \stackrel{\text{def}}{=} \begin{cases} \Theta_1(\dot{e}) & \text{if } \dot{e} \in \dot{E}_1 \\ \Theta_2(\phi^{-1}(\dot{e})) - \max_{\dot{f} \in \dot{E}'_2} \Theta_2(\dot{f}) + \max_{\dot{f} \in \dot{E}_1} \Theta_1(\dot{f}) & \text{if } \dot{e} \in \phi(\dot{E}_2 \setminus \dot{E}'_2) \end{cases}$$

$$\forall \dot{e} \stackrel{\text{def}}{=} (B, t) \in \dot{E}_2 \setminus \dot{E}'_2 \quad \phi(\dot{e}) \stackrel{\text{def}}{=} (\psi(B), t)$$

$$\forall b \stackrel{\text{def}}{=} (\dot{e}, p) \in \bigcup_{\dot{e} \in \dot{E}_2 \setminus \dot{E}'_2} {}^\bullet \dot{e} \cup \dot{e} \quad \psi(b) \stackrel{\text{def}}{=} \begin{cases} (\phi(\dot{e}), p) & \text{if } \dot{e} \notin \dot{E}'_2 \\ place_{|\uparrow(\dot{E}_1)}^{-1}(p) & \text{if } \dot{e} \in \dot{E}'_2 \end{cases}$$

We generalize this notation to more than two extended processes as:

$$subst(\dot{x}_0, \dot{E}'_1, \dot{x}_1, \ldots, \dot{E}'_n, \dot{x}_n) \stackrel{\text{def}}{=} subst(subst(\dot{x}_0, \dot{E}'_1, \dot{x}_1, \ldots, \dot{E}'_{n-1}, \dot{x}_{n-1}), \dot{E}'_n, \dot{x}_n).$$

Theorem 6 ($\dot{X}$ is closed under substitution of prefixes)
Let $\dot{x}_0, \ldots, \dot{x}_n \in \dot{X}$ and $\dot{E}'_1, \ldots, \dot{E}'_n$ that satisfy the conditions required to define $subst(\dot{x}_0, \dot{E}'_1, \dot{x}_1, \ldots, \dot{E}'_n, \dot{x}_n)$. Then $subst(\dot{x}_0, \dot{E}'_1, \dot{x}_1, \ldots, \dot{E}'_n, \dot{x}_n) \in \dot{X}$.

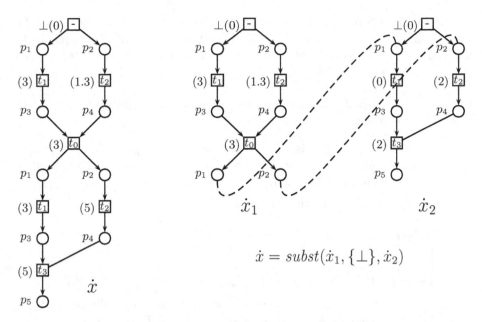

Fig. 2. Substitution of prefixes in extended processes. The dashed curves show how the final conditions of $\dot{x}_1$ are merged with the conditions of $\uparrow(\{\bot\})$ in $\dot{x}_2$.

Proof. We detail the proof for the substitution $subst(\dot{x}_1, \dot{E}'_2, \dot{x}_2)$; the case of more than two extended processes follows immediately.

If $\dot{E}'_2 = \dot{E}_2$ then $subst(\dot{x}_1, \dot{E}'_2, \dot{x}_2) = \dot{x}_1 \in \dot{X}$. Now assume that the theorem is true when $\dot{E}_2 \setminus \dot{E}'_2$ has n elements and consider the case where there are $n+1$ elements. Choose an extended event $\dot{e}_2 \stackrel{\text{def}}{=} (B, t) \in \dot{E}_2 \setminus \dot{E}'_2$ that is minimal in $\dot{E}_2 \setminus \dot{E}'_2$ w.r.t. causality ($\nearrow$) and temporal ordering by Θ_2. Theorem 5 says that t can fire from $RS(\dot{x}_1)$ at date $\theta_1 \stackrel{\text{def}}{=} \Theta_2(\dot{e}_2) - \max_{\dot{e} \in \dot{E}'_2} \Theta_2(\dot{e}) + \max_{\dot{e} \in \dot{E}'_1} \Theta_1(\dot{e})$ using the same partial marking $L \stackrel{\text{def}}{=} Place(B)$. This firing can be coded by the extended event $\dot{e}_1 \stackrel{\text{def}}{=} (Place^{-1}_{|\uparrow(\dot{E}_1)}(L), t)$, and $\dot{x}'_1 \stackrel{\text{def}}{=} \langle \dot{E}_1 \cup \{\dot{e}_1\}, \Theta_1 \cup \{(\dot{e}_1, \theta_1)\}\rangle \in \dot{Y}$. Moreover $subst(\dot{x}_1, \dot{E}'_2, \dot{x}_2)$ equals $subst(\dot{x}'_1, \dot{E}'_2 \cup \{\dot{e}_2\}, \dot{x}_2)$, which belongs to $\dot{X}$ since $\dot{E}_2 \setminus (\dot{E}'_2 \cup \{\dot{e}_2\})$ has n elements.

4.3   Study of the Form of the Constraints

Now we have to deal with the fact that the unfolding we have defined is symbolic, and thus each event represents an action that may occur at several dates. We will show how to check that all the actions that are possible after a symbolic extended process are possible after another one. For this we have to take into account all the possible values for the date of the events of the symbolic extended processes. As well as Berthomieu defined a finite graph of symbolic state classes in [10] using the interleaving semantics, we show that the set of possible reduced ages after a symbolic extended process is taken in a finite set, which allows us to define a complete finite prefix of the symbolic unfolding of a time Petri net.

Definition 15 (constraints $predJ(\dot{E})$). *Let $\dot{E}$ be a nonempty, causally closed, finite set of extended events and $M \stackrel{\text{def}}{=} Place(\uparrow(\dot{E}))$. We define the predicate $predJ(\dot{E})$ as follows: for all $J : M \longrightarrow Q$, $predJ(\dot{E})(J)$ holds iff there exists $\Theta : \dot{E} \to Q$ such that $\langle \dot{E}, \Theta \rangle \in \dot{Y}$ and $J = J_{RS(\langle \dot{E}, \Theta \rangle)}$.*

Theorem 7. *For each maximal marking M, the set of the $predJ(\dot{E})$ with $M \stackrel{\text{def}}{=} Place(\uparrow(\dot{E}))$ is finite.*

Proof. Recall that $predJ(\dot{E})$ is a predicate on the $J(p)$, $p \in M$. If we denote $\dot{e}_1, \ldots, \dot{e}_n$ the events of $\dot{E}$ and introduce a variable θ to represent $\max_{\dot{e} \in \dot{E}} \Theta(\dot{e})$, which plays a role in the definition of $J_{RS(\langle \dot{E}, \Theta \rangle)}$ and also in checking that $\langle \dot{E}, \Theta \rangle$ is temporally complete, we can write $predJ(\dot{E})(J)$ as:

$$\exists \Theta(\dot{e}_1), \ldots, \Theta(\dot{e}_n), \theta \in Q \text{ such that}$$
$$\begin{cases} \langle \dot{E}, \Theta \rangle \in \dot{X} \\ \theta = \max_{\dot{e} \in \dot{E}} \Theta(\dot{e}) \\ \forall t \in T \quad {}^{\bullet}t \subseteq M \implies \theta \leq \max_{p \in {}^{\bullet}t} \Theta({}^{\bullet}(place^{-1}_{|\uparrow(\dot{E})}(p))) + lfd(t) \\ \qquad\qquad \text{(to check that } \langle \dot{E}, \Theta \rangle \text{ is temporally complete)} \\ \forall p \in M \quad J(p) = \min\{I(p), \max\{bound(t) \mid t \in T \wedge p \in {}^{\bullet}t\}\} \\ \qquad\qquad \text{(with } I(p) \stackrel{\text{def}}{=} \theta - \Theta({}^{\bullet}(place^{-1}_{|\uparrow(\dot{E})}(p)))) \end{cases}$$

Consider the system in the curly brace and rewrite all the quantifiers that concern information about the structure of the time Petri net or the structure of $\dot{E}$ (including those coming from ($\langle \dot{E}, \Theta \rangle \in \dot{X}$)) as disjunctions or conjunctions. For instance ($\exists p \in {}^{\bullet}t_0 \quad f(p)$) becomes ($f(p_3) \vee f(p_4)$). The result is a system of inequalities, some of which containing one "min" or one "max". These inequalities can be rewritten without "min" and "max", so that we obtain a boolean combination of inequalities of the following types:

$$\Theta(\dot{e}) \# \Theta(\dot{e}') + c \qquad\qquad \Theta(\dot{e}) \# \theta + c \qquad\qquad \Theta(\dot{e}) \# \theta - J(p)$$

where $\dot{e}$ and $\dot{e}'$ are events of $\dot{E}$, p is a place, c is a constant taken among the $efd(t)$ and $lfd(t)$ and $\#$ is an operator in $\{<, \leq, \geq, >\}$ (= is not necessary).

Rewrite now this boolean combination of inequalities in normal disjunctive form. The quantifiers ($\exists \Theta(\dot{e}_1), \ldots, \Theta(\dot{e}_n), \theta \in Q$) can be distributed in each term of the disjunction. $predJ(\dot{E})(J)$ becomes a disjunction of quantified conjunctions of inequalities of the types we have described before. We will now eliminate one by one the quantifiers $\exists \Theta(\dot{e}_i)$ in one quantified conjunction of inequalities: we show that there remains a quantified conjunction of inequalities of a slightly more general type than before:

$$\begin{aligned} &\Theta(\dot{e}) \# \Theta(\dot{e}') + c \\ &\Theta(\dot{e}) \# \theta + c \\ &\Theta(\dot{e}) \# \theta - J(p) + c \end{aligned} \qquad\qquad \begin{aligned} &J(p) \# c \\ &J(p) \# J(p') + c \end{aligned}$$

where the constants c may now be linear combinations of the $efd(t)$ and $lfd(t)$. To eliminate $\Theta(\dot{e}_i)$, we isolate it in each inequality when it appears, which leads to a conjunction C of $a < \Theta(\dot{e}_i)$, $b \leq \Theta(\dot{e}_i)$, $c \geq \Theta(\dot{e}_i)$ and $d > \Theta(\dot{e}_i)$, where a, b, c and d are terms of the form $(\Theta(\dot{e}) + c)$, $(\theta + c)$ or $(\theta - J(p) + c)$. Then $(\exists \Theta(\dot{e}_i)\ C)$ is equivalent to the conjunction of all the inequalities $(a < c)$, $(a < d)$, $(b \leq c)$ and $(b < d)$, which all have one of the expected forms.

Once all the $\Theta(\dot{e}_i)$ have been eliminated, the remaining inequalities can be only of the form $J(p)\ \#\ c$ or $J(p)\ \#\ J(p') + c$. That is, θ does not appear any more. So the quantifier $\exists \theta$ can be removed.

Notice now that by definition $0 \leq J(p) \leq \max_p \overset{\text{def}}{=} \max\{bound(t) \mid t \in T \wedge p \in {}^\bullet t\}$ for all $p \in M$. Therefore all the inequalities of type $(J(p)\ \#\ c)$ with $|c| > \max_p$ can be immediately evaluated to **true** or **false**. The same happens for the inequalities of the form $(J(p)\ \#\ J(p') + c)$ with $|c| > \max\{\max_p, \max_{p'}\}$. The constants c that remain are bounded. Recall also that they are linear combinations of the $efd(t)$ and $lfd(t)$. Since the $efd(t)$ and $lfd(t)$ are rationals, there are finitely many acceptable values for the constants c.

As a consequence, there are finitely many interesting inequalities, and then finitely many conjunctions of such inequalities, and then finitely many disjunctions of such conjunctions. Finally there is a finite number of $predJ(\dot{E})$.

Let us take the example of $predJ(\{\bot, \dot{e}_1\})(J)$ using the extended events that are represented in Figure 3. It can be written, after some simplifications, as:

$$\exists \Theta(\bot), \Theta(\dot{e}_1), \theta \quad \begin{cases} \Theta(\dot{e}_1) \geq \Theta(\bot) \\ \theta = \max\{\Theta(\bot), \Theta(\dot{e}_1)\} \\ \theta - \Theta(\dot{e}_1) \leq 2 \\ \theta - \Theta(\bot) \leq 2 \\ J(p_2) = \min\{\theta - \Theta(\bot), 2\} \\ J(p_3) = \min\{\theta - \Theta(\dot{e}_1), 2\} \end{cases}$$

These inequalities can be rewritten without "min" and "max", so that we obtain a boolean combination of inequalities, which can be written in normal disjunctive form. Then the quantified variables can be eliminated one by one, and we obtain: $J(p_3) = 0\ \wedge\ 0 \leq J(p_2) \leq 2$.

4.4   Complete Finite Prefix

Now we can define the complete finite prefix of the symbolic unfolding of a time Petri net, by keeping only a finite number of extended events, that contain all the information about the unfolding. More precisely, we show that every temporally complete extended process can be obtained by substitution of prefixes in extended processes that belong to the prefix. Figure 2 shows an example of such a decomposition. The complete finite prefix is represented in Figure 3. The idea behind the construction of the prefix is that a temporally complete extended process $\langle \dot{E}, \Theta \rangle$ will not be continued if the predicate $predJ(\dot{E})$ is equal to a $predJ(\dot{E}')$ with $|\dot{E}'| < |\dot{E}|$. As a possible improvement, the idea of adequate order, introduced by Esparza [7] seems to be usable in our framework.

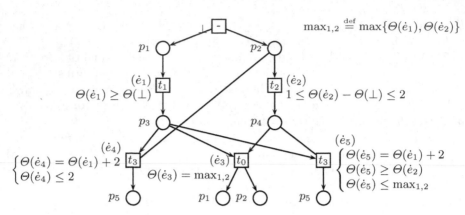

Fig. 3. The complete finite prefix of the symbolic unfolding of the time Petri net of Figure 1. The predicate $LFC(Place(B), dob_{B,\Theta}, t, \Theta(\dot{e}))$ is written near each extended event $\dot{e} \overset{\text{def}}{=} (B, t)$.

Definition 16 ($\bar{X}$ and complete finite prefix $\bar{U}$). *We define the subset $\bar{X}$ of $\dot{X}$ as:*

$$\langle \dot{E}, \Theta \rangle \in \bar{X} \quad \textit{iff} \quad \exists \sigma \quad \begin{cases} \dot{\Pi}(\sigma) = \langle \dot{E}, \Theta \rangle \\ \nexists n', \sigma'' \quad |\sigma''| < n' < |\sigma| \; \wedge \; predJ(\dot{E}') = predJ(\dot{E}'') \end{cases}$$

where $\sigma \overset{\text{def}}{=} ((t_1, L_1, \theta_1), \ldots, (t_{|\sigma|}, L_{|\sigma|}, \theta_{|\sigma|}))$ and σ'' are sequences of local firings and $\dot{E}'$ (respectively $\dot{E}''$) is the set of events that appear in $\dot{\Pi}((t_1, L_1, \theta_1), \ldots, (t_{n'}, L_{n'}, \theta_{n'}))$ (respectively $\dot{\Pi}(\sigma'')$).

We define the finite complete prefix $\bar{U}$ of the symbolic unfolding U of a time Petri net by collecting all the extended events that appear in the extended processes of $\bar{X}$: $\bar{U} \overset{\text{def}}{=} \bigcup_{\langle \dot{E}, \Theta \rangle \in \bar{X}} \dot{E}$.

Denote N the cardinality of $\{predJ(\dot{E}) \mid \langle \dot{E}, \Theta \rangle \in \bar{X}\}$, which was proved finite in Section 4.3. The extended process $\dot{\Pi}(\sigma)$ may belong to $\bar{X}$ only if $|\sigma| < N$. Since there is a finite number of sequences of local firings that are shorter than N, there is a finite number of extended processes in $\bar{X}$ and each of them has less than N events (without $\perp$). Thus $\bar{U}$ is finite.

Theorem 8 (decomposition of an extended process in $\bar{U}$). *For all temporally complete extended process $\langle \dot{E}, \Theta \rangle \in \dot{Y}$, there exist extended processes $\dot{x}_0, \ldots, \dot{x}_n \in \bar{X}$ and $\dot{E}'_1, \ldots, \dot{E}'_m$ with $\dot{x}_i \overset{\text{def}}{=} \langle \dot{E}_i, \Theta_i \rangle$ for all $i \in \{0, \ldots, m\}$ and $\dot{E}'_i \subsetneq \dot{E}_i$ for all $i \in \{1, \ldots, m\}$, such that $\langle \dot{E}, \Theta \rangle = subst(\dot{x}_0, \dot{E}'_1, \dot{x}_1, \ldots, \dot{E}'_m, \dot{x}_m)$.*

Proof. To build the substitution we take the extended events in a total order that respects causality and temporal ordering. Each time we add an event, we try to add it at the end of the last extended process ($\dot{x}_m$) if it remains in $\bar{X}$. Otherwise we add an extended process $\dot{x}_{m+1}$ in the substitution. More formally, let $\langle \dot{E}, \Theta \rangle \in \dot{Y}$, $\dot{e} \overset{\text{def}}{=} (B, t) \in \dot{E}$ a maximal extended event in $\dot{E}$ w.r.t. causality ($\nearrow$) and temporal ordering by Θ, and assume that there exist $\dot{x}_0, \ldots, \dot{x}_n \in \bar{X}$ and $\dot{E}'_1, \ldots, \dot{E}'_m$

such that $\langle \dot{E} \setminus \{\dot{e}\}, \Theta \rangle = subst(\dot{x}_0, \dot{E}'_1, \dot{x}_1, \ldots, \dot{E}'_m, \dot{x}_m)$. Let $\langle \dot{E}_m, \Theta_m \rangle \stackrel{\text{def}}{=} \dot{x}_m$ and $\dot{x}'_m \stackrel{\text{def}}{=} \langle \dot{E}_m \cup \{\dot{e}'\}, \Theta'_m \rangle \stackrel{\text{def}}{=} subst(\dot{x}_m, \dot{E} \setminus \{\dot{e}\}, \langle \dot{E}, \Theta \rangle)$, where $\dot{e}' \stackrel{\text{def}}{=} \phi(\dot{e})$ in the substitution. If $\dot{x}'_m \in \bar{X}$ then $subst(\dot{x}_0, \dot{E}'_1, \dot{x}_1, \ldots, \dot{E}'_m, \dot{x}'_m)$ fits. Otherwise let $\dot{F} \stackrel{\text{def}}{=} \dot{E}_m \cup \{\dot{e}'\}$ and find a sequence of local firings $\sigma_n \stackrel{\text{def}}{=} ((t_1, L_1, \theta_1), \ldots, (t_n, L_n, \theta_n))$ such that $\dot{\Pi}(\sigma_n) = \dot{x}_m$. Let $t_{n+1} \stackrel{\text{def}}{=} t$, $L_{n+1} \stackrel{\text{def}}{=} B$, and $\theta_{n+1} \stackrel{\text{def}}{=} \Theta'_m(\dot{e}')$ so that $\dot{x}'_m = \dot{\Pi}((t_1, L_1, \theta_1), \ldots, (t_{n+1}, L_{n+1}, \theta_{n+1}))$. There exist σ'' and $n' < n+1$ such that $|\sigma''| < n'$ and $predJ(\dot{F}') = predJ(\dot{F}'')$, where $\dot{F}'$ (respectively $\dot{F}''$) is the set of events that appear in $\dot{\Pi}((t_1, L_1, \theta_1), \ldots, (t_{n'}, L_{n'}, \theta_{n'}))$ (respectively $\dot{\Pi}(\sigma'')$). Choose σ'' of minimal length so that $\dot{x}_{m+1} \stackrel{\text{def}}{=} \dot{\Pi}(\sigma'' \cdot (t_{n+1}, L_{n+1}, \theta_{n+1})) \in \bar{X}$. It holds that $n' = n$ since $\dot{x}_m \in \bar{X}$. There exists $\Theta'_{m+1} : \dot{F}'' \to Q$ such that $\langle \dot{F}'', \Theta_{m+1} \rangle$ is temporally complete and $RS(\langle \dot{F}'', \Theta_{m+1} \rangle) \sim RS(\langle \dot{F}', \Theta_m \rangle)$. And $\langle \dot{E}, \Theta \rangle = subst(\dot{x}_0, \dot{E}'_1, \dot{x}_1, \ldots, \dot{E}'_m, \dot{x}_m, \dot{F}'', \dot{x}_{m+1})$.

5   Conclusion

We have presented a new notion of complete finite prefix for safe time Petri nets. It is based on the construction of a symbolic unfolding, and on the study of the form of constraints associated with the transitions. This required to define a notion of partial state in order to equip time Petri nets with a concurrent operational semantics. A prototype has been implemented (a few thousands lines of Lisp code, and the help of a Simplex subroutine). Several improvements could be done. In particular, the idea of adequate order, introduced by Esparza [7] seems to be usable in our framework. Our next work on the subject will be to make some experiments to try to experimentally prove (or disprove) the interest of our technique in the context of model-checking or others.

References

1. Kitai, T., Oguro, Y., Yoneda, T., Mercer, E., Myers, C.: Partial order reduction for timed circuit verification based on a level oriented model. IEICE Trans. **E86-D**(12) (2001) 2601–2611
2. Penczek, W., Pólrola, A.: Abstractions and partial order reductions for checking branching properties of time Petri nets. In: ICATPN. Volume 2075 of LNCS. (2001) 323–342
3. Chatain, T., Jard, C.: Time supervision of concurrent systems using symbolic unfoldings of time Petri nets. In: FORMATS. Volume 3829 of LNCS. (2005) 193–207 Extended version available in INRIA Research Report RR-5706.
4. Aura, T., Lilius, J.: Time processes for time Petri nets. In: ICATPN. Volume 1248 of LNCS. (1997) 136–155
5. Lilius, J.: Efficient state space search for time Petri nets. In: MFCS Workshop on Concurrency '98. Volume 18 of ENTCS., Elsevier (1999)
6. McMillan, K.L.: A technique of state space search based on unfolding. Formal Methods in System Design **6**(1) (1995) 45–65
7. Esparza, J., Römer, S., Vogler, W.: An improvement of McMillan's unfolding algorithm. Formal Methods in System Design **20**(3) (2002) 285–310

8. Bieber, B., Fleischhack, H.: Model checking of time Petri nets based on partial order semantics. In: CONCUR. Volume 1664 of LNCS. (1999) 210–225
9. Fleischhack, H., Stehno, C.: Computing a finite prefix of a time Petri net. In: ICATPN. (2002) 163–181
10. Berthomieu, B., Diaz, M.: Modeling and verification of time dependent systems using time Petri nets. IEEE Trans. Software Eng. **17**(3) (1991) 259–273
11. Semenov, A.L., Yakovlev, A.: Verification of asynchronous circuits using time Petri net unfolding. In: DAC, ACM Press (1996) 59–62
12. Merlin, P., Farber, D.: Recoverability of communication protocols – implications of a theorical study. IEEE Transactions on Communications **24** (1976)
13. Engelfriet, J.: Branching processes of Petri nets. Acta Inf. **28**(6) (1991) 575–591
14. Chatain, T., Jard, C.: Symbolic diagnosis of partially observable concurrent systems. In: FORTE. Volume 3235 of LNCS. (2004) 326–342
15. Best, E.: Structure theory of Petri nets: the free choice hiatus. In: Proceedings of an Advanced Course on Petri Nets: Central Models and Their Properties, Advances in Petri Nets 1986-Part I, London, UK, Springer-Verlag (1987) 168–205
16. Baldan, P., Corradini, A., Montanari, U.: Contextual Petri nets, asymmetric event structures, and processes. Inf. Comput. **171**(1) (2001) 1–49
17. Vogler, W., Semenov, A.L., Yakovlev, A.: Unfolding and finite prefix for nets with read arcs. In: CONCUR. Volume 1466 of LNCS. (1998) 501–516

On the Computation of Stubborn Sets
of Colored Petri Nets

Sami Evangelista and Jean-François Pradat-Peyre

CEDRIC - CNAM Paris
292, rue St Martin, 75003 Paris
{evangeli, peyre}@cnam.fr

Abstract. Valmari's Stubborn Sets method is a member of the so-called *partial order methods*. These techniques are usually based on a selective search algorithm: at each state processed during the search, a stubborn set is calculated and only the enabled transitions of this set are used to generate the successors of the state. The computation of stubborn sets requires to detect dependencies between transitions in terms of conflict and causality. In colored Petri nets these dependencies are difficult to detect because of the color mappings present on the arcs: conflicts and causality connections depend on the structure of the net but also on these mappings. Thus, tools that implement this technique usually unfold the net before exploring the state space, an operation that is often untractable in practice. We present in this work an alternative method which avoids the cost of unfolding the net. To allow this, we use a syntactically restricted class of colored nets. Note that this class still enables wide modeling facilities since it is the one used in our model checker Helena which has been designed to support the verification of software specifications. The algorithm presented has been implemented and several experiments which show the benefits of our approach are reported. For several models we obtain a reduction close or even equal to the one obtained after an unfolding of the net. We were also able to efficiently reduce the state spaces of several models obtained by an automatic translation of concurrent software.

1   Introduction

State space analysis is a powerful formal method for proving that finite systems match their specification. It consists of enumerating all the possible configurations (or states) of the system to track the erroneous ones. A major obstacle to the application of this technique to industrial systems is the famous state explosion problem: the number of reachable states can be far too large to fit in memory or even on disk.

The state explosion has been the subject of many researches in the last two decades and techniques that alleviate this problem have been introduced. This includes the family of partial order methods which tackle one of the main sources of the combinatorial explosion: the concurrent execution of several components. These are based on the following observation: due to the interleaving semantic

S. Donatelli and P.S. Thiagarajan (Eds.): ICATPN 2006, LNCS 4024, pp. 146–165, 2006.

of concurrent systems, a set of different executions can have exactly the same effect on the system and be only a permutation of the same sequence. Thus, an efficient way to reduce the state explosion would be to explore only a single or some representative executions and ignore all the others permutations that are equivalent to the chosen ones. This is why the term of *model checking using representatives* [14] seems more appropriate than the one of *partial order model checking*.

Valmari's stubborn sets method [17, 16, 18] is a member of this family. It is based on a selective search algorithm: at each state processed, a stubborn set of transitions is computed and only the enabled transitions of this set are used to generate the successors of the state. The elimination of some transitions from the stubborn set can cause some states not to be explored and can thus greatly reduce the number of visited states.

The problem of deciding if a set of transitions is stubborn at a state is at least as hard as the reachability problem [5]. Thus, selective search algorithms exploit the structure of the modeled system to build sets for which the stubbornness conditions are guaranteed to hold. Construction algorithms are thus tightly linked to the formalism of the model, e.g., Promela [10], Petri nets, but, whatever the formalism, they always rely on the notion of dependency between transitions. When adding a transition to the stubborn set, one has to find the transitions that could disable or enable the considered transition. For ordinary PT-nets, these dependencies can be directly deduced from the structure of the net. For this reason, stubborn sets have been widely studied in the field of ordinary Petri nets and many algorithms were introduced to solve the problem. Examples include the *deletion algorithm* [17, 22], and the *incremental algorithm* [17, 22, 16].

The computation of stubborn sets for colored Petri nets is more problematic. Indeed, the detection of dependencies between transitions can not solely rely on the structure of the net but must also consider the color mappings that label arcs between places and transitions. A brute force approach consists then in unfolding the net, i.e., building the equivalent ordinary Petri net, in order to apply traditional stubborn sets algorithm for PT-nets. However, this unfolding step is not always possible because of large color domains. A possible way to avoid this unfolding is to work at the "colored" level, and to detect the dependencies between transitions symbolically. As a counterpart, we can not expect to obtain as good results as if the net was unfolded since the analysis of dependencies can not be as precise.

We investigate in this paper a solution for the computation of stubborn sets of colored Petri nets that does not rely on the unfolding of the net nor works at the transition binding level when constructing a stubborn set. Our algorithm detects dependencies directly on the colored Petri net. A keypoint is to use a slightly restricted class of colored Petri net that enables a symbolic detection of the dependencies. However, this class still allows wide modeling facilities since it is the one used in Helena [7] an explicit state model checker based on high level Petri nets and aimed at software verification.

Related works. Stubborn sets of colored Petri nets were initially introduced by Valmari in [19, 20]. His algorithm performs an implicit unfolding of the net: the net is not explicitly unfolded but the algorithm systematically enumerates transition bindings. The possibility to ignore the colors of the net, i.e., treat each colored transition as if it was an agglomeration of all its unfolded transitions, was also mentioned but not recommended, since it usually leads to unnecessary large stubborn sets.

Brgan and Poitrenaud proposed in [1] an optimized version of Valmari's algorithm for well formed Petri nets which exploits the good structuring of the color domains and mappings of this class. The idea of their algorithm is to translate some structural relations into equivalent constraints systems before the search. These systems express dependencies between transition bindings and are thus repeatedly solved during the search to compute stubborn sets. Though their method can efficiently speed up the detection of dependencies, their construction of stubborn sets still works at the transition binding level. Consequently, their approach is not much different from unfolding the net.

To our best knowledge, the only algorithm proposed so far, that does not work at the binding level is [13]. In this paper, the authors suggest that a possible way to obtain "good" stubborn sets for a colored Petri net without unfolding it is to add some extra informations on top of the structure of the net. The model designer must thus supply to the model checking tool some additional inputs, such as the type of the places (e.g., communication buffer, shared resource), that give crucial hints on the structure and the dynamic of the unfolded net. These informations are then used during the search in the stubborn sets construction. Their method seems to provide good reductions for a very low cost as acknowledged by the experimental results. The author argue that it is reasonable to assume that the type of information needed can be provided by the user. An analogy is made to the fact of typing variables in a program. However, there is no other possibility for the tool for validating these informations than unfolding the net or generating the complete state space, which is, by hypothesis, infeasible. Thus, in case of "typing error", the tool could produce wrong results. Another interesting contribution of [13] is a theoretical result about the complexity of the problem: the size of the unfolded PT-net is the worst-case time complexity of any algorithm that computes non-trivial stubborn sets. A trivial stubborn set include all the transition bindings of the net.

This paper is organized as follows. The next section recalls some basic definitions and notations on colored Petri nets and stubborn sets. Our algorithm is presented in Section 3. Section 4 presents a set of experimental results. Lastly we conclude in Section 5.

2   Formal Background

We recall in this section the basic definitions and notions on colored Petri nets [11] and stubborn sets that are needed for the comprehension of this paper. We assume that the reader is familiar with PT-nets and their dynamic behavior.

$\mathbb{N}$ will denote the set of positive integers, $\mathbb{N}^+$ the set $\mathbb{N} \setminus \{0\}$ and $\mathbb{B} = \{false, true\}$ the set of booleans.

The definition of colored Petri nets is based on multi-sets. A multi-set over a set S is a mapping from S to $\mathbb{N}$. The set of multi-set over a set S is denoted by $Bag(S)$. The addition, substraction, and comparison of multi-sets are defined in the usual way. $\emptyset$ denotes the empty multi-set. If m is a multi-set then $e \in m \Leftrightarrow m(e) > 0$.

If S is a set, S^* is the set of finite words over S and S^∞ is the set of infinite words over S. The "." operator shall be used to denote the concatenation of two sequences. ϵ shall denote the empty sequence.

We shall denote by $\mathcal{P}(S)$ the powerset of a set S, i.e., the set of its sub-sets.

2.1   Colored Petri Nets

Definition 1 (Colored Petri net). *A colored Petri net (or CPN) is a tuple* $\langle P, T, C, W^-, W^+, \phi, m_0 \rangle$ *where P is a finite set of* **places***; T is a finite set of* **transitions** *such that $P \cap T = \emptyset$; C a* **color function** *is a mapping from $P \cup T$ to Σ, a set of finite and non empty sets; W^- and W^+ the* **forward and backward incidence matrixes** *associate to each pair (p, t) of $P \times T$ a mapping from $C(t)$ to $Bag(C(p))$; ϕ a* **guard function** *associates to each $t \in T$ a mapping from $C(t)$ to $\mathbb{B}$; and m_0 an* **initial marking** *is an element of $\mathbb{M}$ the set of mappings which associate to each $p \in P$ an element of $Bag(C(p))$.*

From now on a CPN N will implicitly define the tuple $\langle P, T, C, W^-, W^+, \phi, m_0 \rangle$. Given a node n of $P \cup T$, $C(n)$ will be called the color domain of n. The set of inputs (resp. outputs) of a place p is the set ${}^\bullet p$ (resp. $p^\bullet$) defined by: ${}^\bullet p = \{t | W^+(p, t) \neq 0\}$ [1] (resp. $p^\bullet = \{t | W^-(p, t) \neq 0\}$). The same sets can be defined for a transition t : ${}^\bullet t = \{p | W^-(p, t) \neq 0\}$ and $t^\bullet = \{p | W^+(p, t) \neq 0\}$. These notations are extended to set of nodes as usual.

The firing rule defines the dynamic of the net.

Definition 2 (Firing rule). *Let N be a CPN and $m \in \mathbb{M}$. The instance c of transition t is firable (or enabled) at m (denoted by $m[(t, c)\rangle$) if and only if $\phi(t)(c) \wedge m(p) \geq W^-(p, t)(c)$. The firing of the instance (t, c) at m leads to a marking m' (denoted by $m[(t, c)\rangle m'$) defined by: $\forall p \in P, m'(p) = m(p) - W^-(p, t)(c) + W^+(p, t)(c)$.*

The state space of a CPN is the set of all the markings of the net which can be reached from the initial marking by a sequence of firings.

Definition 3 (State space). *Let N be a CPN. The state space (or reachability set) of N is the minimal set $S \subseteq \mathbb{M}$ such that $m_0 \in S$ and if $\exists m \in S, t \in T, c \in C(t)$ such that $m[(t, c)\rangle m'$ then $m' \in S$.*

In the remainder we will often refer to the unfolded net of a CPN. Such a net is an ordinary PT-net obtained from the colored one by creating a node for each place or transition instance of the net. The flow relation of the unfolded net can

---

[1] 0 denotes here the empty mapping from $C(t)$ to $Bag(C(p))$, i.e., $\forall c \in C(t), 0(c) = \emptyset$.

then be derived by a direct application of the color mappings which label the arcs of the CPN.

2.2   Stubborn Sets

The stubborn sets method can be used to build a reduced state space of the system. The construction of this reduced state space can be done by introducing a little modification into a standard search algorithm: at each state processed, a stubborn set of transitions is computed and only the enabled transitions of this set are used to generate the immediate successors of the processed state. The set is said to be stubborn because the transitions outside it can not affect its behavior: it remains stubborn after the firing of any sequence of non stubborn transitions. Consequently, some transitions may never be executed, reducing the number of explored states. In the best case the reduction is exponential. Such a modified algorithm is usually called a selective search algorithm.

The computation of the stubborn set must respect certain rules in order to preserve the desired property in the reduced state space. This has led to the introduction of several versions of the stubborn sets method. However, the notion of *dynamic stubbornness* is a common basis of many of these versions.

Definition 4 (Dynamic Stubbornness). *Let N be a Petri net, m be a marking of N and $S \subseteq T$. The set S is dynamically stubborn at m if conditions $\mathbf{D1}$ and $\mathbf{D2}$ hold where:*

D1 $\forall t \in S, \sigma \in (T \setminus S)^{*}, m[\sigma.t\rangle \Rightarrow m[t.\sigma\rangle$
D2 $(\exists t \in T \mid m[t\rangle) \Rightarrow \exists t \in S \mid \forall \sigma \in (T \setminus S)^{*}, m[\sigma\rangle \Rightarrow m[\sigma.t\rangle$

Condition D1 states that the firing of a sequence of transitions that are outside the stubborn set can not enable the firing of a disabled transition of S while D2 ensures that if the marking is not dead, then there is an enabled transition in the stubborn set (called a *key transition* in the stubborn set terminology) which remains firable after the firing of any sequence of non stubborn transitions.

It is well known that a selective search algorithm which computes stubborn sets having the dynamic stubbornness property builds a reduced state space which contains all the dead markings of the initial state space and at least one infinite sequence if such a sequence exists. Because of the *ignoring phenomenon* [19], that's basically all that this definition preserves: a firable transition may be infinitely forgotten in the stubborn set computation. In order to verify more elaborated properties, such as liveness properties, it is crucial to prevent this ignoring phenomenon and to add additional constraints in the computation of the stubborn set, possibly leading to less reductions. For instance it is usually needed that along each cycle of the reduced reachability graph, a transition enabled at a marking of the cycle belongs to at least one of the stubborn sets computed for the markings of the cycle. However, this subject is beyond the scope of this work and we will focus our attention on the dynamic stubbornness from definition 4. Thus in the remainder of the paper, we will call a stubborn set a set of transitions which respect the dynamic stubbornness property.

3   Symbolic Computation of Stubborn Sets

We present in this section an algorithm to compute stubborn sets of colored Petri nets. The proposed method does not rely on an explicit or implicit unfolding of the net but rather on the two following ideas:

- We treat instances classes instead of explicitly enumerating instances. We thus obtain an algorithm which complexity is independent from the size of the unfolded net.
- We define a class of colored Petri nets allowing an efficient detection of the dependencies between the transitions of the net.

Our method is therefore inspired by two works [1, 13] on the same subject which have been mentioned in the previous section.

In the remainder of the section we will proceed as follows. Firstly we give a static stubbornness definition for PT-nets on which our algorithm is based. Then we make an informal presentation of our algorithm with the help of an example which illustrates several difficulties that arise when directly handling colored nets instead of unfolding. A syntactically restricted class of CPN is introduced in the third part of the section along with additional definitions needed to handle bindings classes. The last part of the section is more technical and deals with a concrete implementation of the algorithm for our CPN class.

3.1   Static Stubbornness for PT-Nets

The dynamic stubbornness definition is based on the semantic of the net, i.e., its reachability graph, and it is thus difficult to obtain an "implementation" of this property. Moreover, it has been proved that the problem of deciding if a set is stubborn is at least as hard as checking reachability for the full reachability graph [5] which is exactly what we want to avoid. Consequently, rather than checking the dynamic stubbornness property for an arbitrary set of transitions, algorithms usually exploit the structure of the modeled system to produce sets of transitions for which the dynamic stubbornness is guaranteed to hold. Such algorithms have been proposed for various description languages, e.g., Promela [10], Variable/Transition systems [19], and Petri nets [17, 16, 22].

We now introduce a stubbornness definition for Petri nets. It is far from being optimal and could be refined, e.g., see [22], but we will use it for its simplicity.

Definition 5 (Stubbornness). *Let N be a Petri net, m be a marking and $S \subseteq T$. S is stubborn at m if the three following conditions are satisfied:*

1. $(\exists t \in T \mid m[t\rangle) \Rightarrow \exists t \in S \mid m[t\rangle$
2. **if** $t \in S$ **and** $m[t\rangle$ **then** $({}^\bullet t)^\bullet \subseteq S$
3. **if** $t \in S$ **and** $\neg m[t\rangle$ **then** $\exists p \in {}^\bullet t$ **such that** $m(p) < W^-(p,t)$ **and** ${}^\bullet p \subseteq S$

Item 1 prevents from picking an empty set if there are enabled transitions. Item 2 ensures that a firable transition t remains firable after the firing of a sequence which only includes transitions outside the stubborn set. Indeed, all the transitions which could prevent the firing of t, i.e., the set $({}^\bullet t)^\bullet$, are also in the stubborn set. Let us note that all the enabled transitions selected in the stubborn sets are thus key transitions since they remain firable. At last item 3 ensures that a disabled transition t remains disabled after the firing of a sequence of non stubborn transitions since place p prevents the firing of t and transitions that could put tokens in it are also stubborn. The place p of this item is usually called a *scapegoat* in the stubborn sets terminology.

These three items are thus sufficient to prove the following proposition.

Proposition 1. *Let N be a Petri net, m be a marking and $S \subseteq T$. If S is stubborn at m then it is dynamically stubborn at m.*

The previous definition can be used to define a stubborn sets computation algorithm for PT-nets. This algorithm initiates the construction with an enabled transition and successively applies item 2 and 3 until reaching a fix point. If the marking does not enable any transition then it directly returns an empty set.

The goal of our work if to present a "colored" version of this algorithm. We will see that difficulty arise with the introduction of color mappings in the net and how to deal with this additional complexity.

3.2   Informal Presentation of the Algorithm

At first we informally present our algorithm with the help of the net of figure 1. This net models the part of a distributed system in which different client and server process interact. Clients compete for acquiring access to a pool of shared objects. To each object of the pool is an associated lock guaranteeing an exclusive access to the object. An idle client C can evolve in two different ways. First he can try to grab the lock of object O (transition $takeLock$). Alternatively, if a server S has acknowledged his request (transition $sendAck$) he can choose to receive and treat this acknowledgment (transition $receiveAck$). We denote $\{c_1, \ldots, c_n\}$ the set of clients, $\{s_1, \ldots, s_m\}$ the set of servers and $\{o_1, \ldots, o_l\}$ the set of shared objects.

We propose to detail a few steps of the construction of the stubborn set $\mathcal{S}$ at the marking m depicted.Binding $tb_1 = (takeLock, \langle C = c_1, O = o_1 \rangle)$ is enabled at m and we choose it to initialize the stubborn set $\mathcal{S}$.

*Binding $tb_1 = (takeLock, \langle C = c_1, O = o_1 \rangle$ Since tb_1 is enabled we must apply the second item of definition 5 and identify the instances in conflict with it to

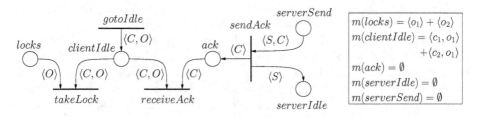

Fig. 1. An illustrative CPN example

include them in $\mathcal{S}$. It appears that binding $tb_2 = (receiveAck, \langle C = c_1, O = o_1 \rangle)$ is clearly in conflict with tb_1. Indeed, they both withdraw the token $\langle c_1, o_1 \rangle$ from place $clientIdle$. We therefore include tb_2 in $\mathcal{S}$. What seems less clear from a PT-net point of view is that tb_1 is also in conflict with other bindings of transition $takeLock$. An analysis of the unfolded net would indeed reveal that bindings of set $\cup_{c \in \{c_2, \ldots, c_n\}}\{(takeLock, \langle C = c, O = o_1 \rangle)\}$ also withdraw token $\langle o_1 \rangle$ from place $locks$ and must thus be included in $\mathcal{S}$. Algorithms of [20] and [1] would directly insert these $n - 1$ bindings. The size of the stubborn set would then quickly increase as color domains grow. This may even fail with infinite color domains. Our approach is different since we choose to not explicitly enumerate these $n - 1$ bindings but rather to treat this set as a single unit. For this purpose we introduce the $\star$ symbol which will be used to denote "any item" of a color. Thus we will say that tb_1 is in conflict with binding $tb_3 = (takeLock, \langle C = \star, O = o_1 \rangle)$, i.e., all the clients who wishes to acquire the same lock. Since tb_3 is a compact representation of a set of bindings we will rather use the term bindings class or more simply class.

Binding $tb_2 = (receiveAck, \langle C = c_1, O = o_1 \rangle)$ To be enabled tb_2 needs a token $\langle c_1 \rangle$ in place ack. The instance $(ack, \langle c_1 \rangle)$ is therefore chosen as a scapegoat. The set $\cup_{s \in \{s_1, \ldots, s_m\}}\{(sendAck, \langle S = s, C = c_1 \rangle)\}$ includes all the bindings which can produce this token. We thus add class $tb_4 = (sendAck, \langle S = \star, C = c_1 \rangle)$ to $\mathcal{S}$. In other words, the only event that can cause the reception by c_1 of an acknowledgment is the sending of this acknowledgment whatever the sending server.

Binding $tb_3 = (takeLock, \langle C = \star, O = o_1 \rangle)$ The introduction of the concept of bindings class raises here a difficulty. Indeed, since we want to treat all the bindings of this class as a unit without differentiating them we have to assign the same status (disabled / enabled) to all these bindings in order to apply the second or third item of definition 5. In practice, the number of enabled bindings of a CPN always remains relatively low even when the color domains are large. We thus choose to proceed as follows. We treat each class as if all the bindings within it were disabled and we extract from it the enabled bindings which are inserted into the stubborn and treated separately. Thus an enabled binding added by this way to the stubborn set will be both considered as disabled and enabled. Though this will produce larger stubborn sets, this does not affect the validity of our algorithm. For instance, the two enabled bindings $tb_1 = (takeLock, \langle C = c_1, O = o_1 \rangle)$ and $tb_5 = (takeLock, \langle C = c_2, O = o_1 \rangle)$ both belong to the class tb_3 and must therefore be included in $\mathcal{S}$.

Let us come back to tb_3. Each disabled binding $b = (takeLock, \langle C = c_i, O = o_1 \rangle)$ of this class needs a token $\langle o_1 \rangle$ in $locks$ and a token $\langle c_i, o_1 \rangle$ in $clientIdle$. Since $m(locks)(\langle o_1 \rangle) = 1$, the absence of token $\langle c_i, o_1 \rangle$ is the reason why b is disabled. Once again we avoid an explicit enumeration of these scapegoats by using the $\star$ symbol: the scapegoat of class tb_3 is $(clientIdle, \langle \star, o_1 \rangle)$. Such an approximation is valid since any disabled binding of tb_3 has a scapegoat place in this class. Item 3 requires that we insert in $\mathcal{S}$ the bindings which put in $clientIdle$ tokens belonging to class $\langle \star, o_1 \rangle$. The single input of $clientIdle$ is $gotoIdle$. By looking at the color mapping which labels the arc from $gotoIdle$ to $clientIdle$ we notice that a binding $(gotoIdle, \langle C = c, O = o \rangle)$ can produce a token of class $\langle \star, o_1 \rangle$ if and only if $o = o_1$. Consequently, we must add $(gotoIdle, \langle C = \star, O = o_1 \rangle)$ to $\mathcal{S}$.

We mentioned in the introduction of this section that the time complexity of this algorithm is not related to the size of the unfolded net. As a matter of fact, the worst time complexity is indeed related to this size since each enabled binding can be processed and the number of enabled bindings is bounded by the number of transitions in the unfolded net. However, in practice, we observe that, even when the types of places and transitions are very large, the number of enabled bindings remains reasonable. So the worst time complexity of our algorithm is rarely reached.

This example highlighted the necessity to have CPNs with a well structured syntax. Indeed, each time we processed a binding we exploited the structuring of the arc functions to detect dependencies between transitions. This approach would typically fail with CPNs having arbitrarily complex arc functions or color domains. We introduce in the next sub-section a class of CPNs inspired by Well Formed colored nets [3] which, on one hand, allows us a symbolic detection of dependencies and, on the other hand, still enables large modeling facilities.

3.3   A Class of Colored Petri Nets

Color domains of our colored nets are Cartesian products of finite and non empty sets called basic types.

Definition 6 (Color domain). *The set of basic types Δ is a finite set of finite and non empty sets. A color domain C is a product $C_1 \times \cdots \times C_{s(C)}$ where $C_i \in \Delta$ and $s(C) \in \mathbb{N}$ is the size of C. The set of color domains is noted $\mathcal{C}$.*

A color domain item will be noted as a tuple, e.g., $\langle 2, true \rangle$. In addition, for each transition t, we will assume a bijective mapping associating each element of its color domain to a variable which will appear in the tuple, e.g., $\langle X = 2, B = true \rangle$.

Color mappings of the net are built with the help of elementary expressions. Two types of elementary expressions are allowed: the projection (or variable) and the functional expression. The first one is used to choose a specific element in an item of a color domain. Functional expressions are provided to enable complex operations on basic types. All the valid expressions from a color domain to a basic type can be put in this family. Projections are thus particular cases of functional expressions.

Definition 7 (Elementary expression). *Let $C \in \mathcal{C}$ and $\delta \in \Delta$. $\mathcal{E}_{C,\delta}$, the set of elementary expressions from C to δ is the set $\mathcal{E}_{C,\delta} = \mathcal{V}_{C,\delta} \cup \mathcal{F}_{C,\delta}$ where*

- $\mathcal{V}_{C,\delta} = \{V_i \mid i \in [1..s(C)] \wedge C_i = \delta\}$
  is the set of **projections** *from C to δ. V_i is defined by:*
  $\forall c = \langle c_1, \ldots, c_{s(C)} \rangle \in C, V_i(c) = c_i.$
- $\mathcal{F}_{C,\delta} = \{(f, (e_1, \ldots, e_n)) \mid f \in \delta_1 \times \cdots \times \delta_n \to \delta \wedge \forall i \in [1..n], e_i \in \mathcal{E}_{C,\delta_i}\}$
  is the set of **functional expressions** *from C to δ. $(f, (e_1, \ldots, e_n))$ is defined by:* $\forall c \in C, (f, (e_1, \ldots, e_n))(c) = f(e_1(c), \ldots, e_n(c))$

In our examples, instead of using the formal notation V_i we will prefer to use the variable of the transition at position i. For example V_2 will be directly noted B if the second element of the transition domain is associated to this variable.

Expressions tuples are basic components of color mappings. They can be guarded by a boolean expression which condition the tokens production. The syntax of guards will appear later in this section.

Definition 8 (Elementary expressions tuple). *Let $C, C' \in \mathcal{C}$. The set of elementary expressions tuples (or tuples) from C to $Bag(C')$, is noted $Tup_{C,C'}$ and is the set of triplets (γ, α, E) such that: $\gamma \in \mathcal{G}_C$ is the guard of the tuple; $\alpha \in \mathbb{N}^+$ is the factor of the tuple; $E = \langle e_1, \ldots, e_{s(C')} \rangle$, with $\forall i \in [1..s(C')], e_i \in \mathcal{E}_{C,C'_i}$, is the expressions list of the tuple. $tup = (\gamma, \alpha, \langle e_1, \ldots, e_n \rangle)$ is defined by: $\forall c \in C, tup(c) = $ if $\gamma(c)$ then $\alpha \cdot \langle e_1(c), \ldots, e_n(c) \rangle$ else $\emptyset$.*

A tuple $(\gamma, \alpha, \langle e_1, \ldots, e_n \rangle)$ will also be noted $[\gamma] \, \alpha \cdot \langle e_1, \ldots, e_n \rangle$. For instance $[X > Y] \, 2 \cdot \langle X, 0, f(Y) \rangle$ is a valid example tuple. Given an instantiation of the variables X and Y of the transition it produces 2 items of type $\langle X, 0, f(Y) \rangle$ if $X > Y$. Otherwise, it produces the empty multi-set.

At last, color mappings that label the arcs of the net are sums of elementary expressions tuples.

Definition 9 (Color mapping). *Let $C, C' \in \mathcal{C}$. A color mapping f from C to $Bag(C')$ is a sum $f = \sum_{i=1}^{k} tup_i$ (with $\forall i \in [1..k], tup_i \in Tup_{C,C'}$) defined naturally. The set of color mappings from C to $Bag(C')$ is noted $Map_{C,C'}$.*

A color mapping will sometimes be considered as the union of the tuples which constitute it, i.e., $tup \in map \Leftrightarrow map = tup + map'$.

Expressions tuples and transitions can be guarded by a boolean expression which states under which conditions the tuple produces items or the transition if firable. We do not impose special constraints on these guards. Any boolean expression on the color domain of the transition is a valid guard.

Definition 10 (Guard). *Let $C \in \mathcal{C}$. $\mathcal{G}_C = \mathcal{E}_{C,\mathbb{B}}$ is the set of guards over C.*

Dealing with instances classes. We first "extend" basic types by including to them the $\star$ symbol. The same extension is done for color domains. The definition assumes that this symbol does not belong to any basic type.

Definition 11 (Extended color domain). *Let $\delta \in \Delta$. The extended type $\delta^\star$ is the set $\delta \cup \{\star\}$. The set of extended types $\Delta^\star$ is the set $\{\delta^\star | \delta \in \Delta\}$. Let $C \in \mathcal{C}$. The extended color domain $C^\star$ is the Cartesian product $C_1^{\star} \times \cdots \times C_{s(C)}^{\star}$. The set of extended color domains is noted $\mathcal{C}^\star$.*

We must also modify the semantics of the elementary expressions to take into account this extension. The value of a projection is unchanged. If a sub-expression of a functional expression e is evaluated to $\star$ so is e. Otherwise its value does not change.

Definition 12 (Extended elementary expression). Let $C \in \mathcal{C}$, $\delta \in \Delta$, and $e \in \mathcal{E}_{C,\delta}$. The extended expression $e^\star$ from $C^\star$ to $\delta^\star$ is defined by: $\forall c \in C^\star$ such that $c = \langle c_1, \ldots, c_n \rangle$:

$$e^\star(c) = \begin{cases} \text{if } e = V_i \text{ then } c_i \\ \text{if } e = (f, \langle e_1, \ldots, e_m \rangle) \text{ then } \begin{cases} \text{if } \forall i \in [1..m], e_i^\star(c) \neq \star \text{ then } e(c) \\ \text{else } \star \end{cases} \end{cases}$$

We will often use in this section the unfolding mapping defined below which is used to enumerate all the items within a class, e.g., $Unf_{\mathbb{B} \times \mathbb{B}}(\langle \star, true \rangle) = \{\langle false, true \rangle, \langle true, true \rangle\}$.

Definition 13 (Class unfolding). Let $\delta \in \Delta$ and $C \in \mathcal{C}$. The mappings Unf_δ from $\delta^\star$ to $\mathcal{P}(\delta)$ and Unf_C from $C^\star$ to $\mathcal{P}(C)$ are defined by:

- $Unf_\delta(e) \qquad = \text{if } e = \star \text{ then } \delta \text{ else } \{e\}$
- $Unf_C(\langle c_1, \ldots, c_n \rangle) = Unf_{C_1}(c_1) \times \cdots \times Unf_{C_n}(c_n)$

Lastly we introduce the inclusion relation $\succeq_C$ defined for every basic type or color domain C. We have $c \succeq_C c'$ if each item c_i of c is either the $\star$ symbol either c_i', the item at the same position in c'. For instance, it holds that $\langle true, \star \rangle \succeq_{\mathbb{B} \times \mathbb{B}} \langle true, false \rangle$, but $\langle true, \star \rangle \succeq_{\mathbb{B} \times \mathbb{B}} \langle false, \star \rangle$ does not. Trivially, if $c \succeq c'$ then $Unf_C(c') \subseteq Unf_C(c)$. We will then say that c' is a subclass of c.

Definition 14 (Inclusion relation). Let $C \in \mathcal{C}$. The relation $\succeq_C$ over $C^\star \times C^\star$ is defined by: $\langle c_1, \ldots, c_n \rangle \succeq_C \langle c_1', \ldots, c_n' \rangle \Leftrightarrow \forall i \in [1..n], c_i = \star \lor c_i = c_i'$.

We will omit in the sequel the subscript and superscript of Unf_C, $\succeq_C$ or $e^\star$ when there will be no ambiguity.

3.4   The Algorithm

We now introduce the general algorithm (figure 2) to compute a stubborn set of transitions of a CPN. Its input is a marking m of the CPN and it returns a stubborn set at m. Three main data structures are used. $\mathcal{S}$ is the stubborn set computed. $\mathcal{U}$ is the set of bindings classes which have not been treated yet. $\mathcal{N}$ is the set of binding classes which must be included in the stubborn set.

   An enabled binding is randomly chosen to initialize the stubborn set and the set of unprocessed classes (line 1). If there is no enabled binding at m, the empty set is directly returned. At each iteration, a binding class (t, c_t) is removed from $\mathcal{U}$ and treated by the algorithm (lines 3-4). If this class is composed of a single binding, i.e. the $\star$ symbol does not appear in it, enabled at m we apply item 2 of the stubbornness definition and we compute the set of bindings classes in conflict with (t, c_t) (line 6). Otherwise we consider it as a class of disabled bindings. We first check (line 8) if c_t is not a subclass of a previously treated class c_t', i.e., all the bindings of c_t are in c_t'. In this case the stubborn set is unchanged. Else the algorithm applies item 3 and computes the classes of bindings which produce tokens needed by the disabled bindings of (t, c_t). Additionally, we must include in $\mathcal{N}$ all the enabled bindings which belong to the bindings class (t, c_t) (line 11).

STUBBORN (m)
1    $S \leftarrow$ **if** $\exists(t,c_t)$ **such that** $m[(t,c_t)\rangle$ **then** $\{(t,c_t)\}$ **else** $\emptyset$; $\mathcal{U} \leftarrow S$
2   **while** $\mathcal{U} \neq \emptyset$ **do**
3   **let** $(t,c_t) \in \mathcal{U}$ **with** $c_t = \langle c_{t,1}, \ldots, c_{t,n} \rangle$
4        $\mathcal{U} \leftarrow \mathcal{U} \setminus \{(t,c_t)\}$
5   **if** $\forall i \in [1..n], c_{t,i} \neq \star$ **and** $m[(t,c_t)\rangle$ **then**
6            $\mathcal{N} \leftarrow disablingClasses(t,c_t)$ (* apply item 2 of definition 5 *)
7   **else**
8   **if** $\exists(t,c_t') \in S$ **such that** $c_t' \succeq c_t$ **then** $\mathcal{N} \leftarrow \emptyset$
9   **else**
10                $\mathcal{N} \leftarrow enablingClasses(t,c_t,m)$ (* apply item 3 of definition 5 *)
11                $\mathcal{N} \leftarrow \mathcal{N} \cup \{(t,c_t')$ **such that** $m[(t,c_t')\rangle$ **and** $c_t \succeq c_t'\}$
12   **end if**
13   **end if**
14        $\mathcal{U} \leftarrow \mathcal{U} \cup (\mathcal{N} \setminus S)$; $S \leftarrow S \cup \mathcal{N}$
15   **end while**
16   **return** S

Fig. 2. A stubborn sets computation algorithm for colored Petri nets

3.5   Implementing Algorithm's Operations

The algorithm of figure 2 is a generic one in the sense that it is not related to our CPN class and could theoretically be implemented for any colored net. We have seen in our introductory example that an efficient implementation of the mappings *enablingClasses* and *disablingClasses* seems to require colored nets well structured enough to enable symbolic computations or alternatively some user supplied informations as it is done in [13].

We detail now an implementation of these operations for our CPN class.

Reversing color mappings. A frequently used operation in our algorithm is the reverse operation. This one consists of finding for a given color c and a color mapping f the set of colors c' such that $f(c')(c) > 0$. Different methods have been proposed in the literature to address this problem in an efficient way (e.g., [6], [1], [2], [8]) for Well Formed nets or similar classes. Even for this formalism, the problem is quite hard, and the solutions proposed either extend the formalism [2] or add some extra restrictions on arc functions [8].

For our colored nets, the reverse operation is, in general, impossible to apply without an explicit enumeration of colors. This is due to the possibility to insert in tuples some functional expressions which can obviously not be reversed. However, we can still approximate the result of this operation by exploiting the elementary expressions that are "well formed", i.e., the projections, and which are easily reversible.

We now introduce the mapping $reverseMap_{t,p}$ defined for every couple (p,t) of $P \times T$. Given a color mapping map from $C(t)$ to $Bag(C(p))$ and a class $c_p \in C(p)^\star$, $reverseMap_{t,p}$ computes a set of classes of transition t which is such that any binding of t which image by map contains a token of class c_p belongs to one of the classes computed. More formally, the following proposition must hold for all $c \in Unf(c_p)$ and $c' \in C(t)$:

$$map(c')(c) > 0 \Rightarrow \exists c_t \in reverseMap_{t,p}(c_p, map) \mid c' \in Unf(c_t)$$

Using the fact that a color mapping is a sum of tuples we reduce the problem to the definition of the tuple reversal mapping $reverseTup_{t,p}$ (definition 15).

We can clearly identify two steps in this procedure. In the first one we identify the class of bindings (t, c_t) such that $tup(c_t)(c) > 0$ for some c of class c_p. This is done by initializing the resulting class to $\langle \star, \ldots, \star \rangle$ which covers all the bindings of t and by looking for projections in tup. When a projection V_j is found at position i in the tuple we replace in r the $\star$ at position j by the item at position i in c_p. For instance $reverseTup_{t,p}(\langle 0, 1, 2 \rangle, \langle X, Z, f(Y) \rangle) = \{\langle X = 0, Y = \star, Z = 1 \rangle\}$. The class computed is clearly an over-approximation of the bindings which are really needed. This is mainly due to the fact that functional expressions may appear in the tuple and that such expressions can not be handled without being "unfolded". Since this is typically what we want to avoid, we have no other choice than ignoring these expressions in this first step. A possible optimization would be to identify expressions which can be reversed, and to exploit this reversibility, e.g., $reverseTup_{t,p}(\langle 1 \rangle, \langle X + 1 \rangle) = \{\langle X = 0 \rangle\}$.

The second step of the reverse operation consists of checking if the computed class is inconsistent with respect to the tuple or the transition. A first inconsistency may appear if there is an expression e_i in the tuple which can be evaluated with binding r, i.e., $e_i(r) \neq \star$, and which value is different from c_i, the item at the same position in class c_p (if $c_i \neq \star$). We can then directly return the empty set since there is obviously no binding of t which can produce by tup a token of class c. For instance, $reverseTup_{t,p}(\langle 2, 2 \rangle, \langle X, X + 1 \rangle) = \emptyset$. A second inconsistency is detected if either the guard of the tuple or the guard of the transition can be evaluated and does not hold. For instance, $reverseTup(\langle 0, \star \rangle, [X > 0] \langle X, Y \rangle) = \emptyset$. Let us recall that these guards or the expressions in the tuple may not be evaluable since the variables which appear in these can have an undefined value.

Definition 15. *Let* $t \in T$ *and* $p \in P$. *The mappings* $reverseMap_{t,p}$ *from* $C(p)^* \times Map_{C(t),C(p)}$ *to* $\mathcal{P}(C(t)^*)$ *and* $reverseTup_{t,p}$ *from* $C(p)^* \times Tup_{C(t),C(p)}$ *to* $\mathcal{P}(C(t)^*)$ *are defined by:*

$$reverseMap_{t,p}(c_p, map) = \bigcup_{tup \in map} reverseTup_{t,p}(c_p, tup)$$

$reverseTup_{t,p}(c_p, tup)$
1   **let** $c_p = \langle c_1, \ldots, c_n \rangle$
2   **let** $tup = [\gamma] \, \alpha \cdot \langle e_1, \ldots, e_n \rangle$
3     $r \leftarrow \langle \star, \ldots, \star \rangle$
4   **for** $i \in [1..n]$ **do if** $e_i = V_j$ **and** $r_j = \star$ **then** $r_j \leftarrow c_i$
5   (* *check inconsistencies* *)
6   **if** $\exists i \in [1..n]$ **such that** $e_i(r) \neq \star$ **and** $c_i \neq \star$ **and** $e_i(r) \neq c_i$ **then return** $\emptyset$
7   **if** $\gamma(r) = false$ **or** $\phi(t)(r) = false$ **then return** $\emptyset$
8   **return** $\{r\}$

Computing scapegoats. The treatment of a disabled bindings class c involves to identify the bindings firing of which can enable the elements of c. This detection is based, in our static stubbornness definition, on the ability to compute a *scapegoat*. A scapegoat of a low-level transition t at a marking m is a place which disables t, i.e. $W^-(p,t) > m(p)$. For a high-level transition binding (t, c_t) it is simply a couple (p, c_p) such that $p \in P, c_p \in C(p)$ and $W^-(p,t)(c_t)(c_p) > m(p)(c_p)$. When directly handling bindings classes instead of explicit bindings a difficulty appears that was not highlighted by our introductory example. Indeed, we must find a scapegoat for all the bindings within the class. Thus, in some cases, we will have to choose several scapegoats. We illustrate this problem with the help of the following net.

Let us consider the class $c_t = \langle X = 2, Y = \star \rangle$ of transition t for which we have two possible scapegoats: $(q, \langle \star, false \rangle)$ and $(p, \langle 2, \star \rangle)$. For instance, at the two markings m and m' our algorithm will proceed as follows:

- for m defined by $m(p) = \langle 2, 3 \rangle, m(q) = \langle 4, true \rangle$
  Class $(q, \langle \star, false \rangle)$ is a valid scapegoat for all the bindings of c_t since no token in q has $false$ as its second component.
- for m' defined by $m'(p) = \langle 2, 3 \rangle, m'(q) = \langle 4, false \rangle$
  Class $(q, \langle \star, false \rangle)$ can not be chosen since $m'(q) \geq W^-(t, q)(c)$ for $c = \langle X = 2, Y = 4 \rangle$ which belongs to the class c_t. For the same reason, $(p, \langle 2, \star \rangle)$ can not be chosen since $m'(p) \geq W^-(p, t)(\langle X = 2, Y = 3 \rangle)$. Consequently, both classes, i.e., $(q, \langle \star, false \rangle)$ and $(p, \langle 2, \star \rangle)$, must be chosen to ensure that each binding of c_t has a scapegoat in one of the two classes.

Once again, this example shows that working at the high-level has a cost insofar as we compute a set of scapegoats which, from a PT-net point of view, is clearly unnecessarily large.

The purpose of function *scapegoat* is to find, given a bindings class (t, c_t), and a marking m, a set of scapegoat classes for (t, c_t) at m: each disabled binding of (t, c_t) has a scapegoat in a class of $scapegoat(t, c_t, m)$. More formally, the following must hold for all $c \in Unf(c_t)$ such that $\neg m[(t, c)\rangle$:

$$\exists (p, c_p) \in scapegoat(t, c_t, m), c' \in Unf(c_p) \mid W^-(p, t)(c)(c') > m(p)(c')$$

The function proceeds in two steps. It first tries to find a unique class which is an acceptable scapegoat for all the bindings of c_t. A sufficient condition is that there is a tuple $[\gamma]\alpha \cdot \langle e_1, \ldots, e_n \rangle$ appearing in the input arc from p to t such that all the tokens present in p at m fulfill one of these two conditions:

- The multiplicity of the token is strictly less than the factor of the tuple α.
- There is an expression at position i in the tuple which produces with c_t a value different from $\star$ and different from the item at the same position in the token.

In addition, the guard of the tuple must evaluate to *true* with c_t. Otherwise, i.e., $\gamma(c_t) = false$ or $\gamma(c_t) = \star$, there could be instances in c_t for which the tuple does not produce any item. For these instances, the token consumed by the tuple is obviously not a valid scapegoat. If there exists such a tuple tup, then it is straightforward to see that $\forall c \in Unf(c_t), tup(c) > m(p)$. The token produced by the tuple can therefore be chosen as a scapegoat.

If we fail to find such a tuple then we pick all the tokens consumed by the bindings of c_t. This set is obviously a correct scapegoat for c_t. However, all the couples (p, c') for which it necessarily holds that $m(p)(c') \geq W^-(p,t)(c)(c')$ for any c of c_t can be safely withdrawn from this set. A sufficient condition for this to hold is that there exists a tuple tup in $W^-(p,t)$ such that (1) c_p, the image of c_t by tup does not contain the $\star$ symbol, (2) the multiplicity of c_p at m is greater than the maximal multiplicity of any item produced by $W^-(p,t)$, i.e., $\Gamma(W^-(p,t))$.

Definition 16. *The mapping scapegoat from $T \times C^\star \times \mathbb{M}$ to $\mathcal{P}(P \times C^\star)$ is defined by: scapegoat$(t, c_t, m) =$*

$$
\left\{
\begin{array}{ll}
\textbf{if} & \exists p \in {}^\bullet t, (\gamma, \alpha, \langle e_1, \ldots, e_n \rangle) \in W^-(p,t) \text{ such that} \\
& \quad \gamma(c_t) = true \\
& \textbf{and} \ \ \forall c_p = \langle c_{p,1}, \ldots, c_{p,n} \rangle \in m(p), \ m(p)(c_p) < \alpha \\
& \hspace{4.5cm} \textbf{or } \exists i \in [1..n] \mid e_i(c_t) \notin \{\star, c_{p,i}\} \\
\textbf{then} & \{(p, \langle e_1(c_t), \ldots, e_n(c_t) \rangle)\} \\
\\
\textbf{else} & \{(p, c_p) \mid p \in {}^\bullet t \text{ and } \exists (\gamma, \alpha, \langle e_1, \ldots, e_n \rangle) \in W^-(p,t) \text{ such that} \\
& \quad c_p = \langle e_1(c_t), \ldots, e_n(c_t) \rangle \ \text{ and } \ \gamma(c_t) \neq false \\
& \textbf{and} \ \neg(\forall i \in [1..n], e_i(c_t) \neq \star \ \text{ and } \ m(p)(c_p) \geq \Gamma(W^-(p,t)))\}
\end{array}
\right.
$$

where $\Gamma(map) = \sum_{i=1}^k \alpha_i$ *with* $map = \sum_{i=1}^k (\gamma_i, \alpha_i, E_i)$.

The size of the reduced reachability graph depends to a large extent on the stubborn sets computed. Though always choosing the stubborn set with the lowest number of enabled bindings does not necessarily yields the best reduction it seems however to be the best and simplest heuristic. The choice of the scapegoats is a nondeterministic factor which can affect the number of enabled bindings in the resulting stubborn set. For PT-nets different strategies have been proposed in [21]. Our implementation of mapping *scapegoat* sorts "scapegoat candidates" according to three criteria and chooses the first candidate according to this sorting. These three criteria are (we note C the set of bindings classes which are directly inserted into the stubborn set if the scapegoat if chosen): (1) the number of enabled bindings in C, (2) the number of $\star$ which appear in the classes of C, and (3) the number of classes in C. We thus try to limit the number of enabled bindings and the number of transitions of the unfolded net inserted to the stubborn set right after the choice of the scapegoat. Indeed, the number of low-level transitions covered by a class directly depends on the number of stars which appear in the class.

After intensive experiments we observed that this strategy competes favorably against a pseudo-random strategy.

Mappings *disablingClasses* and *enablingClasses*. Concluding, we define the mappings *disablingClasses* and *enablingClasses*.

The definition of mapping *disablingClasses* is based on this simple observation: a binding (t, c_t) remains firable as long as any binding which could withdraw tokens needed by (t, c_t) are not fired. It is thus sufficient to inspect the tokens consumed by (t, c_t) and to identify the bindings which consume these tokens by an application of the reverse operation. Therefore we closely follow the stubbornness definition for PT-nets (definition 5).

Definition 17. *The mapping disablingClasses from $T \times C^*$ to $\mathcal{P}(T \times C^*)$ is defined by: disablingClasses$(t, c_t) =$*

$$\bigcup_{p \in {}^\bullet t, t' \in p^\bullet, c_p \in W^-(p,t)(c_t)} reverseMap_{t',p}(c_p, W^-(p, t'))$$

To identify the bindings which can enable the instances of a class c_t of transition t we enumerate the scapegoats of (t, c_t) and, for each scapegoat (p, c_p), we look at each input transition t' of p. An application of the reverse operation gives us the bindings of t' which put tokens of class c_p in p. Once again this definition respects the static stubbornness definition for PT-nets.

Definition 18. *The mapping enablingClasses from $T \times C^* \times \mathbb{M}$ to $\mathcal{P}(T \times C^*)$ is defined by: enablingClasses$(t, c_t, m) =$*

$$\bigcup_{(p,c_p) \in scapegoat(t,c_t,m), t' \in {}^\bullet p} reverseMap_{t',p}(c_p, W^+(p, t')) \cdot$$

4   Experiments

The algorithm described in this work has been implemented in our model checker Helena [7]. This section reports the results of two series of experiments that have been carried out. In the first series considered we analyzed some models obtained from concurrent programs by an automatic translation using the Quasar [9] tool. In the second one we considered several academic models included in the Helena distribution (available at `http://helena.cnam.fr`) of which some are recurrent examples of the CPN literature.

All measures were obtained on a Pentium IV with a 2.4 Ghz processor.

Models extracted from programs (table 1). Four real concurrent programs were first translated using the tool Quasar: an implementation of Chang and Roberts election protocol, two different implementations of the dining philosophers and a client-server protocol with dynamic creation of servers to handle client requests. Helena could not unfold these CPNs due to the huge color domains of the nodes. Indeed, some places of the net model variables having high-level data types, e.g., records or arrays. Each program is scalable by a parameter and we considered several values of this parameter (the value is given in the first column) and ran two tests: one without the stubborn method enabled (column

Table 1. Data collected for some models extracted from concurrent software

	Complete Graph			Reduced Graph		
	$|\mathcal{N}|$	$|\mathcal{A}|$	$\mathcal{T}$	$|\mathcal{N}|$	$|\mathcal{A}|$	$\mathcal{T}$
	The leader election protocol					
7	198 039	1 041 750	61	45 780	93 361	30
8	1 037 209	6 175 069	644	201 943	430 120	184
9	5 961 241	40 179 197	10 035	824 362	2 061 193	1 038
	The dining philosophers (first implementation)					
7	1 398 615	5 050 508	138	29 412	44 166	4
8	5 416 243	21 585 453	1 038	83 670	127 191	14
9	24 842 432	112 433 417	13 581	221 865	319 833	40
	The dining philosophers (second implementation)					
4	26 539	55 245	1	6 322	7 667	1
5	219 304	505 765	11	37 139	46 911	7
6	1 789 459	4 582 322	145	214 853	359 901	43
	The client / server program					
2	4 141	14 461	1	108	131	0
3	130 221	593 583	14	1 131	1 434	0
4	5 445 681	30 593 745	2 508	13 921	19 232	1

Complete Graph), the second with it (column *Reduced Graph*). The columns $|\mathcal{N}|$, $|\mathcal{A}|$, and $\mathcal{T}$ indicate for each run the number of nodes and arcs of the graph, and the exploration time of the graph in seconds.

We observe that, despite the complexity of the CPNs obtained from an automatic translation of programs, our algorithm gives a significant reduction for the four programs considered. The reductions factor goes from 7 in the worst case (the leader election program) to almost 400 in the best case (the client / server program) and makes realistic the automatic verification of concurrent software.

To further enhance the reduction we plan to combine our method with [13]. Concurrent programs can indeed easily be mapped to process-partitioned CP nets of [13] with a simple static analysis of the program. For instance, places corresponding to variables local to a process can be typed as local and places corresponding to variables accessed by several processes can be typed as shared.

Academic models (table 2). We then considered several academic models which size allows (except for one) an unfolding. For each model we therefore ran an additional test with Prod and its stubborn set algorithm activated. Let us recall that Prod unfolds the net in order to apply the stubborn set method. We used the deletion algorithm of Prod (option -d) which is, to our best knowledge, the most advanced algorithm for PT-nets. The reduction observed with Prod must be seen as a lower bound which is hard to reach without unfolding the net. Comparing the size of the graph reduced by Helena to the size of the complete graph tells us how good the reduction is while comparing it to the graph reduced by Prod tells us how good the reduction could be.

The examples studied can be classified in four categories.

Table 2. Data collected for some academic models

| Helena | | | | | | Prod | | |
| Complete Graph | | | Reduced Graph | | | Reduced Graph | | |
$|\mathcal{N}|$	$|\mathcal{A}|$	$\mathcal{T}$	$|\mathcal{N}|$	$|\mathcal{A}|$	$\mathcal{T}$	$|\mathcal{N}|$	$|\mathcal{A}|$	$\mathcal{T}$
The resource allocation system								
2 550 759	11 435 684	49	72 637	100 925	2	72 637	100 925	7
The distributed database system (from [11])								
649 540	4 330 282	19	67 585	112 662	1	232	242	0
The dining philosophers								
1 153 351	10 416 483	34	602 493	2 131 338	18	265 143	616 555	41
Eisenberg and McGuire's mutual exclusion algorithm								
624 790	2 490 418	9	414 555	1 345 417	129	223 482	428 297	491
The distributed Sieve of Eratosthenes								
2 028 969	9 947 808	59	273	272	0	Net not unfoldable		
Lamport's fast mutual exclusion algorithm (from [12]								
1 672 728	7 944 684	41	959 494	3 176 750	188	197 554	338 504	58
Chang and Roberts leader election protocol								
218 931	1 836 299	7	156 254	799 871	26	123 979	212 531	51
The load balancing system								
9 324 768	54 723 965	295	275 090	499 615	11	252 458	477 487	867
The multiprocessor system (from [4])								
7 322 076	85 522 635	356	138 239	283 646	76	138 239	283 646	572
Peterson's mutual exclusion algorithm								
1 242 528	4 970 112	19	186 037	386 272	46	80 193	152 565	38
The slotted ring protocol (from [15])								
439 296	2 897 664	11	287 508	97 8514	26	20 613	37 806	2

In a first category we can put the models for which the graph is weakly reduced by Helena but efficiently reduced by Prod. The two models which belong to this category are the slotted ring protocol and to a lesser extent Lamport's algorithm. We are currently not able to explain the bad results obtained for these two models but plan to investigate this.

The dining philosophers, Eisenberg and McGuire's algorithm, and the election protocol constitute a second category. Their characteristic is that their graph is weakly reduced both by Helena and Prod. A surface analysis may let us think that our algorithm performs a bad reduction for these models. However the poor results obtained with Prod show that we can not expect much better. Indeed, there are some problems for which partial order methods are inefficient. Models such as the dining philosophers or the Eisenberg and McGuire's algorithm which make heavy use of shared resources such as global variables usually fall into this category. These shared resources are a major source of conflicts which lead to compute large stubborn sets yielding a small reduction.

A third category is made up of the distributed database system and Peterson's algorithm. For these two models Helena builds a very reduced graph but Prod can do even better by unfolding the net. This particularly holds for the database system.

Its initial graph is in $\mathcal{O}(N \cdot 3^N)$. Helena builds a reduced graph in $\mathcal{O}(N^2 \cdot 2^N)$. Prod produces a reduced graph in $\mathcal{O}(N^2)$. An analysis of these nets reveals a certain complexity in arc mappings which makes it hard to efficiently detect dependencies without unfolding the net. Let us note that our reduced graph has exactly the same size as in [13] where they use a semi-automatic algorithm.

At last, for the four other models Helena performs a reduction close or equal to the reduction performed by Prod but outperforms significantly Prod with respect to time. The best results are observed for Eratosthene's algorithm: Helena builds a reduced reachability graph which size is linear with respect to the parameter of the system despite the high complexity of the arc functions. In addition, we notice that the net could not be unfolded because of the size of the color domains. For the multiprocessor and the resource allocation systems we obtain a reduced graph which is exactly the same as the one obtained by Prod. Lastly, for the load balancing system, the gain obtained by unfolding the net is completely marginal with respect to the brutal increase of the exploration time.

5   Conclusion

The contribution of this paper is a stubborn sets computation algorithm for colored Petri nets which avoids the unfolding by mixing two approaches. First we do not directly handle explicit bindings but rather bindings classes. We therefore stay at the high-level and never explicitly enumerate the transitions of the unfolded net. Second we define a syntactically restricted class of colored Petri nets for which it is possible to detect dependencies in a symbolic manner while preserving high expressiveness. As a counterpart, the detection of the dependencies between transitions can not be as fine as on the unfolded net and some approximations are done which lead to larger stubborn sets.

A set of experimental results have shown the benefits of our approach. For many academic models we achieved a reduction close or equal to the one obtained after an unfolding of the net. We were also able to significantly reduce the state spaces of several concurrent programs automatically translated to colored nets by the Quasar tool. The unfolding approach fails for these colored nets having huge unfolded nets.

In future works we will combine our algorithm with the method of [13]. Our algorithm exploits the structuring of our CPN class whereas their method is based on user supplied informations. Both should therefore be fully compatible and lead to better reductions.

References

1. R. Brgan and D. Poitrenaud. An efficient algorithm for the computation of stubborn sets of well formed petri nets. In *Application and Theory of Petri Nets*, volume 935 of *LNCS*, pages 121–140. Springer, 1995.
2. L. Capra, G. Franceschinis, and M. De Pierro. A high level language for structural relations in well-formed nets. In *Applications and Theory of Petri Nets*, volume 3536 of *LNCS*, pages 168–187. Springer, 2005.

3. G. Chiola, C. Dutheillet, G. Franceschinis, and S. Haddad. On well-formed coloured nets and their symbolic reachability graph. In *Application and Theory of Petri Nets*, pages 373–396. Springer, 1990.
4. G. Chiola, G. Franceschinis, and R. Gaeta. A symbolic simulation mechanism for well-formed coloured petri nets. In *Simulation Symposium*, 1992.
5. E.M. Clarke, O. Grumberg, and D. Peled. *Model Checking*. The MIT Press, 1999.
6. C. Dutheillet and S. Haddad. Structural analysis of coloured nets. application to the detection of confusion. Technical Report 16, IBP/MASI, 1992.
7. S. Evangelista. High level petri nets analysis with Helena. In *Applications and Theory of Petri Nets*, volume 3536 of *LNCS*. Springer, 2005.
8. S. Evangelista, S. Haddad, and J.-F. Pradat-Peyre. Syntactical colored petri nets reductions. In *Automated Technology for Verification and Analysis*, volume 3707 of *LNCS*. Springer, 2005.
9. S. Evangelista, C. Kaiser, C. Pajault, J.-F. Pradat-Peyre, and P. Rousseau. Dynamic tasks verification with Quasar. In *Reliable Software Technologies*, volume 3555 of *LNCS*. Springer, 2005.
10. G.J. Holzmann and D. Peled. An improvement in formal verification. In *Formal Description Techniques*, pages 197–211, 1994.
11. K. Jensen. Coloured petri nets: A high level language for system design and analysis. In *Advances in Petri Nets*, volume 483 of *LNCS*. Springer, 1991.
12. J.B. Jørgensen and L.M. Kristensen. Computer aided verification of lamport's fast mutual exclusion algorithm using colored petri nets and occurrence graphs with symmetries. *IEEE Transactions on Parallel and Distributed Systems*, 10(7), 1999.
13. L.M. Kristensen and A. Valmari. Finding stubborn sets of coloured petri nets without unfolding. In *Application and Theory of Petri Nets*, volume 1420 of *LNCS*, pages 104–123. Springer, 1998.
14. D. Peled. All from one, one for all: on model checking using representatives. In *Computer Aided Verification*, volume 697 of *LNCS*, pages 409–423. Springer, 1993.
15. D. Poitrenaud and J.-F. Pradat-Peyre. Pre- and post-agglomerations for LTL model checking. In *Application and Theory of Petri Nets*, volume 1825 of *LNCS*, pages 387–408. Springer, 2000.
16. A. Valmari. Error detection by reduced reachability graph generation. In *Application and Theory of Petri Nets*, volume 424 of *LNCS*. Springer, 1988.
17. A. Valmari. *State Space Generation : Efficiency and Practicality*. PhD thesis, Tampere University of Technology, 1988.
18. A. Valmari. Eliminating redundant interleavings during concurrent program verification. In *Parallel Architectures and Languages*, volume 366 of *LNCS*, pages 89–103. Springer, 1989.
19. A. Valmari. Stubborn sets for reduced state space generation. In *Advances in Petri Nets*, volume 483 of *LNCS*, pages 491–515. Springer, 1991.
20. A. Valmari. Stubborn sets of coloured petri nets. In *Application and Theory of Petri Nets*, pages 102–121, 1991.
21. K. Varpaaniemi. On choosing a scapegoat in the stubborn set method. In *Workshop on Concurrency, Specification & Programming*, pages 163–171, 1993.
22. K. Varpaaniemi. *On the Stubborn Set Method in Reduced State Space Generation*. PhD thesis, Helsinki University of Technology, 1998.

On the Construction of Pullbacks
for Safe Petri Nets

Eric Fabre

IRISA/INRIA
Campus de Beaulieu
35042 Rennes cedex, France
Eric.Fabre@irisa.fr

Abstract. The product of safe Petri nets is a well known operation : it generalizes to concurrent systems the usual synchronous product of automata. In this paper, we consider a more general way of combining nets, called a pullback. The pullback operation generalizes the product to nets which interact both by synchronized transitions and/or by a shared sub-net (*i.e.* shared places and transitions). To obtain all pullbacks, we actually show that all equalizers can be defined in the category of safe nets. Combined to the known existence of products in this category, this gives more than what we need : we actually obtain that all small limits exist, *i.e.* that safe nets form a complete category.

1 Introduction

We consider the category *Nets* of safe Petri nets (PN) as defined by Winskel in [2]. Safe Petri nets provide a natural and widespread model for concurrent systems. A product × was defined in [2] for safe PNs, that can be considered as a generalization of the usual synchronous product of automata. In practice, this product is essentially interesting when specialized to labeled nets : roughly speaking, it would then synchronize transitions of two nets as soon as they carry the same label. It therefore offers a very natural way to build large concurrent systems from elementary components. As a nice property, × is the categorical product in *Nets*. Pushing forward this idea, it can be interesting to derive a notion of pullback for PNs. While the product assumes that nets interact through common events, the pullback goes further and also allows interactions by shared places and transitions. Pullbacks can be used, for example, to combine two concurrent systems that synchronize through common events and at the same time share some resources (*e.g.* locks to access data).

The notion of pullback has been extensively explored for other models of concurrency (transition graphs, graph grammars, etc.) [7], or for other categories of Petri nets [3] (proposition 11). But the choice of net morphisms plays a crucial role, and apparently the construction of pullbacks in the category *Nets* of [2] is still missing. This category remains of great interest however, because it allows foldings (and consequently unfoldings!), and already has a product. Unfoldings have become an important tool for the verification of concurrent systems [10, 11, 12, 13, 14, 15]. They have also been advocated for the monitoring

S. Donatelli and P.S. Thiagarajan (Eds.): ICATPN 2006, LNCS 4024, pp. 166–180, 2006.

of concurrent systems [16]. In particular, this second application domain relies intensively on factorization properties of unfoldings: the fact that the unfolding of a product system can be expressed as a product of unfoldings of its components [17]. This property is actually the key to distributed or modular monitoring algorithms (surprisingly, this approach has not been explored in model checking applications, to the knowledge of the author). The derivation of the factorization property on unfoldings (or on other structures like trellises [18, 19]) relies on categorical arguments, and in particular on the fact that the unfolding operation preserves limits, like the product for example. In order to obtain a similar property for other ways of combining components, it is therefore crucial to characterize them as categorical limits. This is the main motivation of the present work.

Let us mention some contributions to the topic. B. Koenig provides in [9] a definition for specific pullback diagrams. M. Bednarczyk *et al.* prove in [8] that *Nets* is finitely complete, so all pullbacks exist. But the result is obtained in a much more general setting, and is hard to specialize to the case of safe nets. Finally, let us stress that [8] mentions in its introduction (p.3) that the existence of a pullback construction for safe Petri nets has been reported... although the authors have not been able to locate any reference! It is therefore useful to provide a simple and direct definition for this construction.

We proceed in several steps. We first consider unlabeled nets. It is a well known fact that the labeling is essentially a decoration that can be reincorporated at no cost in net operations (see [5]), which we do at the end of the paper (section 4). Secondly, we recall (section 2) that a pullback operation can be derived from a product and an equalizer (see [1], chap. V-2, thm. 1, and [7], sec. 5). Since all products exist in *Nets*, we simplify the construction (and proofs) by building equalizers, which is the heart of the contribution (section 3). We finally gather all pieces to give a comprehensive definition of the pullback of labeled Petri nets (section 4), first in the general case, then in the specific case where morphisms are partial functions. The conclusion underlines some important consequences of this construction.

2   Notations

Net. We denote Petri nets by $\mathcal{N} = (P, T, \rightarrow, P^0)$, representing respectively places, transitions, initially marked places and the flow relation. For each place $p \in P$, we assume $|p^\bullet \cup {}^\bullet p| \geq 1$, and for each transition $t \in T$, $|t^\bullet| \geq 1$ and $|{}^\bullet t| \geq 1$. For labeled nets, we take $\mathcal{N} = (P, T, \rightarrow, P^0, \lambda, \Lambda)$ where $\lambda : T \rightarrow \Lambda$ is the labeling function.

Morphism. A morphism [2] $\phi : \mathcal{N}_1 \rightarrow \mathcal{N}_2$ between nets $\mathcal{N}_i = (P_i, T_i, \rightarrow_i, P_i^0)$ is a pair (ϕ_P, ϕ_T) where

C1. $\phi_T : T_1 \rightarrow T_2$ is a partial function, and ϕ_P a relation between P_1 and P_2,

C2. $P_2^0 = \phi_P(P_1^0)$ and $\forall p_2 \in P_2^0, \exists$ a unique $p_1 \in P_1^0 : p_1 \xleftrightarrow{\phi_P} p_2$,

C3. if $p_1 \xleftarrow{\phi_P} p_2$ then the restrictions $\phi_T : {}^\bullet p_1 \to {}^\bullet p_2$ and $\phi_T : p_1{}^\bullet \to p_2{}^\bullet$ are total functions,

C4. if $t_2 = \phi_T(t_1)$ then the restrictions $\phi_P^{op} : {}^\bullet t_2 \to {}^\bullet t_1$ and $\phi_P^{op} : t_2{}^\bullet \to t_1{}^\bullet$ are total functions.

where ϕ_P^{op} denotes the opposite relation to ϕ_P. Observe that condition C3 implies that if ϕ_P is defined at $p_1 \in P_1$, then ϕ_T is defined at all transitions $t_1 \in T_1$ connected to p_1. In the sequel, we will simply write ϕ for ϕ_P or ϕ_T, and $\phi(X)$ to denote places in relation with at least one place in X. By $Dom(\phi)$, we represent the elements of $\mathcal{N}_1$ (places or transitions) where ϕ is defined, $i.e.$ $\phi^{op}(P_2 \cup T_2)$.

Notice that condition C3 entails that the pair (ϕ_P, ϕ_T) preserves the flow relation (on its domain of definition). Together with C4 and C2, this guarantees that a run of $\mathcal{N}_1$ is mapped into a run of $\mathcal{N}_2$ by ϕ_T (see [2]), which is the least one should require from net morphisms. Simpler definitions of net morphisms would ensure this property, but C1-C4 are actually necessary to provide extra categorical properties, as we shall see in the sequel.

Remark. Notice that condition C2 becomes a consequence of C3 and C4 when one assumes the existence of a fake initial transition $t_{i,0}$ in each $\mathcal{N}_i$, fed with a fake initial place $p_{i,0} \to_i t_{i,0}$, such that $t_{i,0}{}^\bullet = P_i^0$ and $t_{2,0} = \phi(t_{1,0})$, $p_{2,0} \xleftarrow{\phi} p_{1,0}$. We shall use this trick in the sequel to simplify proofs (focusing on C3, C4 and omitting to check C2).

Safe Petri nets with the above definition of morphisms define the category $Nets$ [2, 4]. For labeled nets, we naturally consider label-preserving morphisms to define the category $\lambda Nets$. Section 4 will detail the definition of this category.

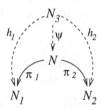

Fig. 1. Commutative diagram of the product $\mathcal{N} = \mathcal{N}_1 \times \mathcal{N}_2$

Product. Let $\mathcal{N}_1, \mathcal{N}_2$ be nets, their categorical product $\mathcal{N}_1 \times \mathcal{N}_2$ in $Nets$ is a net $\mathcal{N}$ associated to morphisms $\pi_i : \mathcal{N} \to \mathcal{N}_i$, $i = 1, 2$, satisfying the so-called universal property of the product (fig. 1): for every other candidate triple $(\mathcal{N}_3, h_1, h_2)$ with $h_i : \mathcal{N}_3 \to \mathcal{N}_i$, there exists a unique morphism $\psi : \mathcal{N}_3 \to \mathcal{N}$ such that $h_i = \pi_i \circ \psi$. This net $\mathcal{N} = (P, T, \to, P^0)$ and the π_i are given by [4, 6]

1. $P = \{(p_1, \star) : p_1 \in P_1\} \cup \{(\star, p_2) : p_2 \in P_2\}$: disjoint union of places, $\pi_i(p_1, p_2) = p_i$ if $p_i \neq \star$ and is undefined otherwise,
2. $P^0 = \pi_1^{-1}(P_1^0) \cup \pi_2^{-1}(P_2^0)$,
3. $T = (T_1 \times \{\star\}) \cup (\{\star\} \times T_2) \cup (T_1 \times T_2)$, $\pi_i(t_1, t_2) = t_i$ if $t_i \neq \star$ and is undefined otherwise,

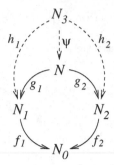

Fig. 2. Commutative diagram of the pullback $\mathcal{N} = \mathcal{N}_1 \wedge \mathcal{N}_2$

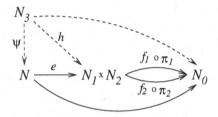

Fig. 3. Equalizing $f_1 \circ \pi_1$ and $f_2 \circ \pi_2$

4. the flow $\to$ is defined as follows: for $t \in T$, $^{\bullet}t = \pi_1^{-1}(^{\bullet}\pi_1(t)) \cup \pi_2^{-1}(^{\bullet}\pi_2(t))$ and symm. for $t^{\bullet}$, assuming $^{\bullet}\pi_i(t) = \pi_i(t)^{\bullet} = \emptyset$ if π_i is undefined at t.

At first sight, this categorical product may look useless since every transition is free to fire alone or jointly with any transition of the other net. Again, the interest of this construction appears when it is applied to labeled nets, in association with a synchronization algebra [4]. Its practical interest then becomes obvious to build large systems starting from elementary components. Since labels bring no technical difficulty other than notational, we put them aside until section 4.

Decomposition of the pullback. Let $\mathcal{N}_0, \mathcal{N}_1, \mathcal{N}_2$ be nets, and $f_i : \mathcal{N}_i \to \mathcal{N}_0$, $i = 1, 2$ be net morphisms, so $\mathcal{N}_0$ forms a kind of interface between $\mathcal{N}_1$ and $\mathcal{N}_2$. We look for a terminal net $\mathcal{N} = (P, T, \to, P^0)$, associated to morphisms $g_i : \mathcal{N} \to \mathcal{N}_i$, $i = 1, 2$, such that (fig. 2):

$$f_1 \circ g_1 = f_2 \circ g_2 \tag{1}$$

By "terminal," we mean the universal property of the pullback: whenever there exists another triple $(\mathcal{N}_3, h_1, h_2)$ satisfying the same commutative diagram, there exists a unique mediating morphism $\psi : \mathcal{N}_3 \to \mathcal{N}$ such that $h_i = g_i \circ \psi$. We denote the pullback by $\mathcal{N}_1 \wedge^{\mathcal{N}_0} \mathcal{N}_2$, or by $\mathcal{N}_1 \wedge \mathcal{N}_2$ for short.

It is well known that the pullback operation can be decomposed into a product, followed by an equalization. Consider the product net $\mathcal{N}_1 \times \mathcal{N}_2$, and the associated canonical projections $\pi_i : \mathcal{N}_1 \times \mathcal{N}_2 \to \mathcal{N}_i$, $i = 1, 2$. In general, $\mathcal{N}_1 \times \mathcal{N}_2$ and the π_i do not satisfy the pullback condition, *i.e.* $f_1 \circ \pi_1 \neq f_2 \circ \pi_2$. However, by equalizing them, one gets the desired result. $(\mathcal{N}, e)$ equalizes $f_1 \circ \pi_1$ and $f_2 \circ \pi_2$ iff $(f_1 \circ \pi_1) \circ e = (f_2 \circ \pi_2) \circ e$, and for any other candidate $(\mathcal{N}_3, h)$ there exists a unique $\psi : \mathcal{N}_3 \to \mathcal{N}$ such that $h = e \circ \psi$ (fig. 3). It is straightforward to check that $(\mathcal{N}, \pi_1 \circ e, \pi_2 \circ e)$ then yields the desired pullback. For details, we refer the reader to [1], chap. V-2, thm. 1, or to [7], sec. 5 where this construction is also used.

3   Equalizer in *Nets*

Consider two nets $\mathcal{N}_i = (P_i, T_i, \to_i, P_i^0), i = 1, 2$ related by two morphisms $f, g : \mathcal{N}_1 \to \mathcal{N}_2$. We want to build the equalizer $(\mathcal{N}, e)$ of f and g, *i.e.* a net $\mathcal{N}$ and a morphism $e : \mathcal{N} \to \mathcal{N}_1$ satisfying $f \circ e = g \circ e$, and such that for any other candidate pair $(\mathcal{N}_3, h)$ there exists a unique morphism $\psi : \mathcal{N}_3 \to \mathcal{N}$ satisfying $h = e \circ \psi$ (fig. 4).

3.1   Equalizer and Coequalizer in *Sets*

We recall here two classical results that will be instrumental in the sequel.

Equalizer. We consider the category of sets with *partial* functions as morphisms (or equivalently pointed sets with total functions). Let T_1, T_2 be two sets related by partial functions $f, g : T_1 \to T_2$. The equalizer of f and g is the pair (T, e) where

$$T = \{t_1 \in T_1 : f(t_1) = g(t_1) \text{ or both } f \text{ and } g \text{ are undefined at } t_1\} \quad (2)$$

and e is the canonical injection of T into T_1 (we'll use the shorthand $t_1 \in T$ instead of $t \in T, t_1 = e(t)$). In the setting of pointed sets, where functions point to the special value ϵ of a set to mean "undefined," (2) takes the simplest form $f(t_1) = g(t_1)$.

Given another candidate pair (T_3, h), the unique morphism (partial function) $\psi : T_3 \to T$ is obtained by $\psi = e^{-1} \circ h$ (it is easy to check that $Im(h) \subseteq T$).

Coequalizer. We now consider the category of sets with *total* functions. The coequalizer diagram corresponds to fig. 4 with all arrows reversed. Let S_2, S_1 be two sets related by total functions $F, G : S_2 \to S_1$, and denote by (S, E) the coequalizer of F and G. The construction is a bit more complex.

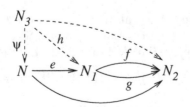

Fig. 4. A pair $(\mathcal{N}, e)$ equalizing f and g

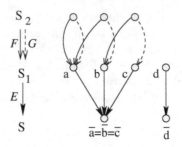

Fig. 5. Coequalizing the total functions F and G

Define the relation R on elements of S_1 by

$$p_1 \, R \, p_1' \Leftrightarrow \exists p_2 \in S_2, \ \{p_1, p_1'\} = \{F(p_2), G(p_2)\} \tag{3}$$

and consider the equivalence relation $\equiv$ generated by R. We denote by $[p_1]$ the class of p_1 for $\equiv$. Then

$$S = \{[p_1] : p_1 \in S_1\} \tag{4}$$

and the function $E : S_1 \to S$ is simply the quotient operation, *i.e.* $E(p_1) = [p_1]$. See fig. 5 for an example.

Given another candidate pair (S_3, H), the unique morphism (total function) $\Psi : S_3 \to S$ is obtained by $\Psi = H \circ E^{-1}$, or in other words by $\forall [p_1] \in S, \Psi([p_1]) = H(p_1)$. Indeed, it is easy to check that H is necessarily class invariant.

3.2 Candidate Equalizer in *Nets*

Let $(\mathcal{N}, e)$ denote the desired equalizer, with $\mathcal{N} = (P, T, \to, P^0)$ and $e : \mathcal{N} \to \mathcal{N}_1$.

Transitions. On transition sets, $f, g : T_1 \to T_2$ are partial function, so we adopt definition (2) for T and e on T.

Places. On place sets, the definition is a bit more complex. The morphism definition in *Nets* actually states in C4 that $\phi^{op} : {}^\bullet t_2 \to {}^\bullet t_1$ and $\phi^{op} : t_2{}^\bullet \to t_1{}^\bullet$ are total functions, for $t_2 = \phi(t_1)$, which orients us to co-equalizers in *Sets*. So let t be a transition of T, with $t_1 = e(t) \in T_1$.

Assume first that f, g are defined at t_1, and $f(t_1) = g(t_1) = t_2 \in T_2$. We take for e^{op} in ${}^\bullet t_1$ the coequalizer of $f^{op}, g^{op} : {}^\bullet t_2 \to {}^\bullet t_1$. Eq. (3) thus defines $R^{\bullet t_1}$, the equivalence relation $\equiv^{\bullet t_1}$ and place classes $[p_1]^{\bullet t_1}$. And similarly in the post-set of t_1.

When f, g are both undefined at t_1, we take for e^{op} in ${}^\bullet t_1$ (or $t_1{}^\bullet$) the co-equalizer of functions f^{op}, g^{op} from the empty set. So e^{op} is simply the identity.

In summary, the place set P of $\mathcal{N}$ is a subset of 2^{P_1} given by

$$P = \{[p_1]^{\bullet t_1} : t_1 \in T, \ p_1 \in {}^\bullet t_1\} \cup \{[p_1]^{t_1 \bullet} : t_1 \in T, \ p_1 \in t_1{}^\bullet\} \tag{5}$$

and the relation e on places is simply given by $p \xleftarrow{\;e\;} p_1$ iff $p_1 \in p$. Observe that a place $p_1 \in P_1$ not connected to a transition of T has no counterpart in P.

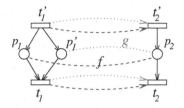

Fig. 6. Identity of equivalence classes

Lemma 1. *Let $t_1, t'_1 \in T$. Assume $p_1, p'_1 \subseteq t'^\bullet_1 \cap {}^\bullet t_1$, then*

$$p_1 \equiv^{t'^\bullet_1} p'_1 \Longleftrightarrow p_1 \equiv^{\bullet t_1} p'_1 \qquad (6)$$

Proof. Assume $p_1 \neq p'_1$ and $p_1 R^{t'^\bullet_1} p'_1$. This means f, g are defined at t'_1, $f(t'_1) = t'_2 = g(t'_1)$, and for example[1] $\exists p_2 \in t'^\bullet_2 : p_1 \xleftrightarrow{f} p_2 \xleftrightarrow{g} p'_1$. Let $t_2 = f(t_1) = g(t_1)$, by C3 on f or g, one has $p_2 \in {}^\bullet t_2$, whence $p_1 R^{\bullet t_1} p'_1$. This proves $[p_1]^{t'^\bullet_1} \subseteq [p_1]^{\bullet t_1}$. One can show in the same way the reverse inclusion, which proves the lemma. $\qquad \square$

Naturally, the lemma holds also for the other arrow orientations, *i.e.* for $p_1, p'_1 \subseteq t'^\bullet_1 \cap t^\bullet_1$ and for $p_1, p'_1 \subseteq {}^\bullet t'_1 \cap {}^\bullet t_1$.

Initial places. In eq. (5), we assume the existence of (fake) transitions $t_{i,0}$ with $t_{i,0}{}^\bullet = P_i^0$ and $f(t_{1,0}) = g(t_{1,0}) = t_{2,0}$. So initial places in P are given by

$$P^0 = \{[p_1]^{t_{1,0}{}^\bullet} : p_1 \in P_1^0\} \qquad (7)$$

For $p_1 \in P_1$ and $t_1 \in T_1$, notice that the equivalence class $[p_1]^{\bullet t_1}$ (or equivalently $[p_1]^{t_1{}^\bullet}$) may both contain marked places of P_1^0 and unmarked places of $P_1 \setminus P_1^0$. Such a class is not taken as an initial place of $\mathcal{N}$. See the example of p' in fig. 7.

Conversely, assume an equivalence class $[p_1]^{\bullet t_1}$ (for ex.) satisfies $[p_1]^{\bullet t_1} \subseteq P_1^0$. By lemma 1, $[p_1]^{\bullet t_1} = [p_1]^{t_1^0{}^\bullet}$ which corresponds to an initial place of $\mathcal{N}$. We could thus take as an alternate definition:

$$P^0 = \{p \in P : e(p) \subseteq P_1^0\} \qquad (8)$$

Flow relation. It is obviously defined by $p \to t$ when $e(t) = t_1$ and $p = [p_1]^{\bullet t_1}$ for some $p_1 \in {}^\bullet t_1$. But, using lemma 1, we can derive the simpler criterion:

$$p \to t \Longleftrightarrow e(p) \subseteq {}^\bullet e(t) \text{ in } \mathcal{N}_1 \qquad (9)$$

We proceed symmetrically for $t \to p$.

Example. Fig. 7 illustrates this construction. Observe that $p_1 R^{t'^\bullet_1} p'_1$ and $p_1 R^{\bullet t_1} p''_1$, which results in two classes/places in $\mathcal{N}$, both related to p_1 by e. These places must indeed be distinguished: by merging places p' and p in $\mathcal{N}$, *i.e.* by aggregating classes sharing one or more places of P_1, the resulting e wouldn't be a morphism (C3 violated).

---

[1] The other possibility is $p_1 \xleftrightarrow{g} p_2 \xleftrightarrow{f} p'_1$, but this doesn't affect the proof.

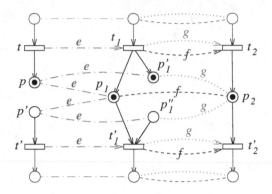

Fig. 7. The equalizer $(\mathcal{N}, e)$ (left) for nets $\mathcal{N}_1$ (center) and $\mathcal{N}_2$ (right) related by two morphisms f, g. Notice that t', t_1', t_2' could be the "fake" initial transitions.

3.3 Coherence of the Definition

$e : \mathcal{N} \to \mathcal{N}_1$ *is a net morphism.* C1 holds by definition, and with the trick of fake initial transitions, C2 is a consequence of C3 and C4, which we only need to examine.

C4 obviously holds by construction of places of P: let $t_1 = e(t)$, then $e^{op} : {}^\bullet t_1 \to {}^\bullet t$ defined by $e^{op}(p_1) = [p_1]^{\bullet t_1}$ is a total function. And similarly for $e^{op} : t_1{}^\bullet \to t^\bullet$.

For C3, consider $p \to t$ in $\mathcal{N}$, such that $p \xleftrightarrow{e} p_1$ and $e(t) = t_1$. We want to check that $p_1 \to_1 t_1$ in $\mathcal{N}_1$. By definition of the flow in $\mathcal{N}$, one has $p \to t$ iff $e(p) \subseteq {}^\bullet e(t) = {}^\bullet t_1$, and $p \xleftrightarrow{e} p_1$ iff $p_1 \in p$, so $p_1 \to_1 t_1$ holds. The same reasoning proves that $e : {}^\bullet p \to {}^\bullet p_1$ is also a total function.

$\mathcal{N}$ *is a safe net.* By a standard argument [2]: since $e : \mathcal{N} \to \mathcal{N}_1$ is a net morphism, it maps runs of $\mathcal{N}$ to runs of $\mathcal{N}_1$. So if $\mathcal{N}$ is not safe, one of its run fills some place with more than one token, which reveals by e a non safe run in $\mathcal{N}_1$, because e is a total function on T.

$(\mathcal{N}, e)$ *satisfies the commutative diagram.* This is true by construction for the partial functions on transitions. It also holds locally for relations on places, *i.e.* around triples of transitions (t, t_1, t_2) with $t_1 = e(t), t_2 = f(t_1) = g(t_1)$. This allows to reach completely the place relations e, f, g.

3.4 Universal Property

Assume the pair $(\mathcal{N}_3, h)$ satisfies $f \circ h = g \circ h$, with $N_3 = (P_3, T_3, \to_3, P_3^0)$ and $h : \mathcal{N}_3 \to \mathcal{N}_1$. We look for a (unique) $\psi : \mathcal{N}_3 \to \mathcal{N}$ satisfying $h = e \circ \psi$ (see fig. 4).

Definition of ψ. On transitions, ψ is uniquely given by $\psi = e^{-1} \circ h$, as it was seen in section 3.1.

For places, consider a triple $(t_3, t, t_1) \in T_3 \times T \times T_1$ of related transitions: $\psi(t_3) = t$ and $h(t_3) = t_1 = e(t)$. We say that such a triple (t_3, t, t_1) *forms a triangle.* From the construction of co-equalizers in section 3.1, we know that $\psi^{op} : {}^{\bullet}t \to {}^{\bullet}t_3$ is uniquely defined from $h^{op} : {}^{\bullet}t_1 \to {}^{\bullet}t_3$ by

$$\forall p_1 \in {}^{\bullet}t_1, \quad \psi^{op}([p1]^{{}^{\bullet}t_1}) = h^{op}(p_1) \cap {}^{\bullet}t_3 \tag{10}$$

Specifically, $h^{op}(p_1) \cap {}^{\bullet}t_3$ exists and is formed by a single place p_3 because h is a net morphism and thus satisfies C4. Moreover, this value p_3 doesn't depend on the choice of p_1 in $[p1]^{{}^{\bullet}t_1}$ because, as a co-equalizer h^{op} is necessarily class invariant on ${}^{\bullet}t_1$ (see 3.1). We proceed similarly to define $\psi^{op} : t^{\bullet} \to t_3^{\bullet}$.

ψ satisfies the commutative diagram. By construction of ψ, $h = e \circ \psi$ is obvious on transitions, and locally on places (*i.e.* around triangles of transitions). To show that the relation holds globally on places, consider $p_3 \in P_3$. By assumption, p_3 is connected to at least one transition t_3 in $\mathcal{N}_3$. If h is defined at p_3 and $p_3 \overset{h}{\longleftrightarrow} p_1$, then h is also defined at t_3 (by C3), $h(t_3) = t_1 \in T$ and p_1 is connected to t_1. We then use $h = e \circ \psi$ around the triangle (t_3, t, t_1), where $t = \psi(t_3)$.

ψ is a net morphism. It obviously satisfies C1, and C4 is imposed by the construction of ψ on places. So only C3 has to be checked, which is the difficult part of the proof.

For C3, consider a pair of places $(p_3, p) \in P_3 \times P$ related by ψ (*i.e.* $p_3 \overset{\psi}{\longleftrightarrow} p$) and assume $p_3 \to t_3$ in $\mathcal{N}_3$. We want to show that ψ is defined at t_3, and $\psi(t_3) \in p^{\bullet}$ in $\mathcal{N}$. By definition of ψ on places, there exists a triangle $(t_3', t', t_1') \in T_3 \times T \times T_1$ such that for example[2] $t_3' \to_3 p_3$, $t_1' \to_1 p_1$, $t' \to p$ and $p = [p_1]^{t_1'{}^{\bullet}}$ (see Fig. 8).

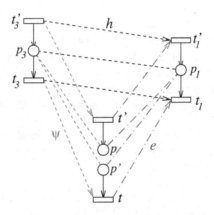

Fig. 8. Proof that ψ satisfies C3

h is defined at p_3, thus also at t_3 by C3. Since $f \circ h = g \circ h$, one has $t_1 = h(t_3) \in T$. So there exists $t \in T$ with $e(t) = t_1$ and thus we already know that ψ is defined at $t_3 : \psi(t_3) = t$. In other words, $(t_3, t, t_1) \in T_3 \times T \times T_1$ forms another

---

[2] Equivalently, we could have assumed that the related places are in the presets (instead of post-sets) of a transition triangle.

triangle. Since e is a morphism, let p' be the image of p_1 by $e^{op} : {}^\bullet t_1 \to {}^\bullet t$, so $p' = [p_1]^{\bullet t_1}$. By definition of ψ in the presets of the triangle (t_3, t, t_1), see (10), one has $p_3 \overset{\psi}{\longleftrightarrow} p'$. To conclude the proof, we thus have to show that $p = p'$. We essentially use the fact that h is a morphism satisfying $f \circ h = g \circ h$.

Let p_1' be a place of $t_1'^{\bullet}$ such that $p_1 \equiv^{t_1'^{\bullet}} p_1'$. We know that $p_3 \overset{h}{\longleftrightarrow} p_1'$, because $h^{op} : t_1'^{\bullet} \to t_3'^{\bullet}$ is class invariant (a consequence of $f \circ h = g \circ h$). From $p_3 \to_3 t_3$ in $\mathcal{N}_3$ and $p_3 \overset{h}{\longleftrightarrow} p_1'$, we derive by C3 that $p_1' \to_1 t_1 = h(t')$. We are now exactly in the situation of lemma 1, so $p_1 \equiv^{\bullet t_1} p_1'$. We have thus proved that $[p_1]^{t_1'^{\bullet}}$ and $[p_1]^{\bullet t_1}$ are identical, or in other words $p = p'$.

4 Application to Pullbacks of Labeled Nets

We now reassemble all elements to provide a definition for pullbacks of safe labeled nets. The first task is to define the category $\lambda Nets$. Consider labeled nets $\mathcal{N}_i = (P_i, T_i, \to_i, P_i^0, \lambda_i, \Lambda_i)$, $\phi : \mathcal{N}_1 \to \mathcal{N}_2$ is a morphism in $\lambda Nets$ iff ϕ is a net morphism (as defined in section 2 by C1-C4), with the extra requirements:

C5. ϕ_T preserves labels,
C6. $\Lambda_1 \supseteq \Lambda_2$,
C7. $Dom(\phi_T) = \lambda_1^{-1}(\Lambda_2)$.

The next section recalls the definition of the product in this category, that we combine to the equalizer to obtain the pullback.

4.1 Product

Let $\mathcal{N}_i = (P_i, T_i, \to_i, P_i^0, \Lambda_i, \lambda_i), i = 1, 2$ be two labeled nets. To build net products, we assume a simple synchronization algebra [5]: two transitions carrying the same label have to synchronize, while transitions carrying a private label remain private. Private labels are those in $(\Lambda_1 \setminus \Lambda_2) \cup (\Lambda_2 \setminus \Lambda_1)$. The product $\bar{\mathcal{N}} = \mathcal{N}_1 \times \mathcal{N}_2$ and the associated projections $\pi_i : \bar{\mathcal{N}} \to \mathcal{N}_i$ are defined as follows[3]:

1. $\bar{P} = \{(p_1, \star) : p_1 \in P_1\} \cup \{(\star, p_2) : p_2 \in P_2\}$: disjoint union of places, $\pi_i(p_1, p_2) = p_i$ if $p_i \neq \star$ and is undefined otherwise,
2. $\bar{P}^0 = \pi_1^{-1}(P_1^0) \cup \pi_2^{-1}(P_2^0)$,
3. the transition set $\bar{T}$ is given by

$$\bar{T} = \{(t_1, \star) : t_1 \in T_1, \lambda_1(t_1) \in \Lambda_1 \setminus \Lambda_2\}$$
$$\cup \{(\star, t_2) : t_2 \in T_2, \lambda_2(t_2) \in \Lambda_2 \setminus \Lambda_1\}$$
$$\cup \{(t_1, t_2) \in T_1 \times T_2 : \lambda_1(t_1) = \lambda_2(t_2) \in \Lambda_1 \cap \Lambda_2\}$$

$\pi_i(t_1, t_2) = t_i$ if $t_i \neq \star$ and is undefined otherwise,

---

[3] Remark: if ones wishes to use the trick of fake initial transitions t_i^0 to define initial markings P_i^0 by $P_i^0 = t_i^{0\bullet}$, one has to assume that each Λ_i contains a special label ϵ^0 reserved to the transition t_i^0.

4. the flow $\rightarrow$ is defined by ${}^{\bullet}t = \pi_1^{-1}({}^{\bullet}\pi_1(t)) \cup \pi_2^{-1}({}^{\bullet}\pi_2(t))$ and symm. for $t^{\bullet}$, assuming ${}^{\bullet}\pi_i(t) = \pi_i(t){}^{\bullet} = \emptyset$ if π_i is undefined at t,
5. $\bar{\Lambda} = \Lambda_1 \cup \Lambda_2$ and $\bar{\lambda}$ is the unique labeling preserved by the π_i.

Let us recall that the product of labeled nets can also be obtained by taking the product of non-labeled nets, and then discarding transition pairs that violate the rules of the synchronization algebra.

For our choice of morphisms, it is straightforward to check that the above definition actually yields the categorical product in $\lambda Nets$: The π_i are net morphisms that obviously satisfy C5-C7. And for the universal property, with notations of fig. 1, the ψ computed in $Nets$ (ignoring labels) is defined by[4] $\forall t_3 \in T_3$, $\psi(t_3) = (h_1(t_3), h_2(t_3))$, so it clearly satisfies C5-C7 when h_1, h_2 do.

4.2 Equalizer

Similarly, the construction of equalizers derived in $Nets$ naturally extends to equalizers of labeled nets. With notations of fig. 4, we take $\Lambda = \Lambda_1$ for the label set of $\mathcal{N}$, and define the labeling function by $\lambda = \lambda_1 \circ e$. The morphism $e : \mathcal{N} \rightarrow \mathcal{N}_1$ then clearly satisfies C5-C7. For the universal property, the morphism $\psi :$ $\mathcal{N}_3 \rightarrow \mathcal{N}$ is defined on transitions by $\psi_T = e_T^{-1} \circ h_T$. So $Dom(\psi_T) = Dom(h_T)$, and ψ clearly satisfies C5-C7.

4.3 Pullback

Assume the $f_i : \mathcal{N}_i \rightarrow \mathcal{N}_0$ are morphisms of labeled nets. The pullback $\mathcal{N} = \mathcal{N}_1 \wedge \mathcal{N}_2$ is defined as follows, by combining the definitions of product and equalizer (section 2).

Transitions. We distinguish "shared" transitions in $\mathcal{N}_1$ and $\mathcal{N}_2$, *i.e.* those having an image in $\mathcal{N}_0$, from "private" ones, the others. For private transitions, the definition of the pullback mimics the definition of the product. For shared transitions, only pairs that match through the f_i are preserved.

$$T_s = \{(t_1, t_2) \in T_1 \times T_2 : t_i \in Dom(f_i), f_1(t_1) = f_2(t_2)\} \quad (11)$$

$$T_p = \{(t_1, t_2) \in T_1 \times T_2 : t_i \notin Dom(f_i), \lambda_1(t_1) = \lambda_2(t_2)\}$$
$$\cup \{(t_1, \star) : t_1 \in T_1, t_1 \notin Dom(f_1), \lambda_1(t_1) \in \Lambda_1 \setminus \Lambda_2\}$$
$$\cup \{(\star, t_2) : t_2 \in T_2, t_2 \notin Dom(f_2), \lambda_2(t_2) \in \Lambda_2 \setminus \Lambda_1\} \quad (12)$$

$$T = T_s \cup T_p \quad (13)$$

Notice that the label condition doesn't appear in (11) : it comes as a consequence of $f_1(t_1) = f_2(t_2)$, since morphisms preserve labels.

Places. Places are obtained by inspecting transitions selected in T.

Consider first a private transition $(t_1, t_2) \in T_p$, where one (at most) of the t_i can be $\star$. Assume $p_i \rightarrow_i t_i$ (or equivalently $t_i \rightarrow_i p_i$) in $\mathcal{N}_i$, with $t_i \neq \star$. Observe that necessarily $p_i \notin Dom(f_i)$, otherwise f_i would be defined at t_i. Such a place

---

[4] With the convention that $\psi(t_3) = (\star, \star)$ means "undefined."

p_i induces a singleton equivalence class in P, either $(p_1, \star)$, or $(\star, p_2)$. We denote by P_p all such "private" places.

Consider now a pair of shared transitions $(t_1, t_2) \in T_s$, where $f_1(t_1) = t_0 = f_2(t_2)$. Consider for example a place $p_1 \in {}^\bullet t_1$ (or equivalently $p_1 \in t_1{}^\bullet$, and symmetrically for a place $p_2 \in {}^\bullet t_2{}^\bullet$).

a. If $p_1 \notin Dom(f_1)$, then $[(p_1, \star)]^{{}^\bullet(t_1, t_2)}$ is reduced to $(p_1, \star)$, which yields another private place in P_p.

b. If $p_1 \in Dom(f_1)$, let $p_0 \in P_0 \cap {}^\bullet t_0$ satisfy $p_1 \xrightarrow{f_1} p_0$. By C4 applied to f_2, there exists $p_2 \in {}^\bullet t_2$ such that $p_2 \xrightarrow{f_2} p_0$, so $(p_1, \star)\, R^{{}^\bullet(t_1, t_2)}\, (\star, p_2)$ in the product $\mathcal{N}_1 \times \mathcal{N}_2$. The resulting equivalence class $[(p_1, \star)]^{{}^\bullet(t_1, t_2)}$, takes the form (Q_1, Q_2), with $\emptyset \neq Q_i \subseteq P_i$, and yields a "shared" place in the pullback.

In summary:

$$P_p = \{ (p_1, \star) \; : \; p_1 \in P_1, \; p_1 \notin Dom(f_1), \; \exists(t_1, \cdot) \in T, \; p_1 \in {}^\bullet t_1{}^\bullet \}$$
$$\cup \{ (\star, p_2) \; : \; p_2 \in P_2, \; p_2 \notin Dom(f_2), \; \exists(\cdot, t_2) \in T, \; p_2 \in {}^\bullet t_2{}^\bullet \} \qquad (14)$$
$$P_s = \{ (Q_1, Q_2) \; : \; Q_i \subseteq P_i, \; Q_i \subseteq Dom(f_i), \; \exists(t_1, t_2) \in T_s,$$
$$Q_1 \uplus Q_2 \text{ equiv. class of } \equiv^{{}^\bullet(t_1, t_2)} \text{ or of } \equiv^{(t_1, t_2)^\bullet} \} \qquad (15)$$
$$P = P_p \cup P_s \qquad (16)$$

In (14), the dot in $(t_1, \cdot)$ stands for either t_2 or $\star$, and symmetrically for the second line.

Initial places. By abuse of notation, let us identify a private place like $(p_1, \star)$ to $(Q_1, Q_2) = (\{p_1\}, \emptyset)$, and $(\star, p_2)$ to $(Q_1, Q_2) = (\emptyset, \{p_2\})$. So (Q_1, Q_2) denotes a general place in P.

$$P^0 = \{(Q_1, Q_2) \in P : Q_1 \subseteq P_1^0, \; Q_2 \subseteq P_2^0\} \qquad (17)$$

Flow. Let $(Q_1, Q_2) \in P$ and $(t_1, t_2) \in T$ (where one of the t_i can be $\star$). Then

$$(Q_1, Q_2) \to (t_1, t_2) \iff Q_1 \subseteq {}^\bullet t_1 \text{ in } \mathcal{N}_1, \; Q_2 \subseteq {}^\bullet t_2 \text{ in } \mathcal{N}_2 \qquad (18)$$
$$(t_1, t_2) \to (Q_1, Q_2) \iff Q_1 \subseteq t_1{}^\bullet \text{ in } \mathcal{N}_1, \; Q_2 \subseteq t_2{}^\bullet \text{ in } \mathcal{N}_2 \qquad (19)$$

with the convention that $\emptyset \subseteq {}^\bullet \star$ and $\emptyset \subseteq \star^\bullet$ hold.

Morphisms g_i. Let (t_1, t_2) be a transition of T, one has $g_i(t_1, t_2) = t_i$ if $t_i \neq \star$, and is undefined otherwise. Let (Q_1, Q_2) be a general place in P, one has $(Q_1, Q_2) \xrightarrow{g_i} p_i$ iff $p_i \in Q_i$.

4.4   Special Case

We examine here the special case where morphisms $f_i : \mathcal{N}_i \to \mathcal{N}_0$ are partial functions not only on transitions, but also on places (instead of being relations on

places). The definition changes only for P_s in (15): when place duplications are
forbidden, equivalence classes of shared places are reduced to two elements only.

$$P_s = \{ (p_1, p_2) : p_i \in P_i \cap Dom(f_i),\ f_1(p_1) = f_2(p_2) = p_0,$$
$$\exists (t_1, t_2) \in T_s,\ f_1(t_1) = f_2(t_2) = t_0,\ p_0 \in {}^\bullet t_0{}^\bullet \} \tag{20}$$

This definition coincides with the proposition of [9] (and also to an early version
of the present notes), apart from the extra condition that places created in (14)
and (20) be connected to at least one transition of the pullback. An example is
given in fig. 9.

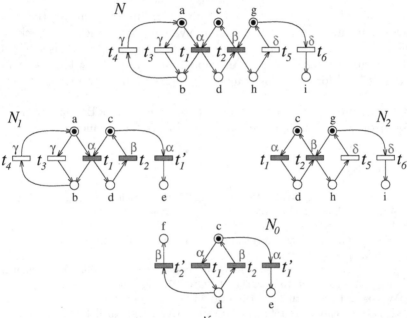

Fig. 9. Example of a pullback: $\mathcal{N} = \mathcal{N}_1 \overset{\mathcal{N}_0}{\wedge} \mathcal{N}_2$, in the simple case of injective morphisms. Morphisms are represented by common names on transitions and places. Tran-
sition labels are indicated by Greek letters. Observe that transition t_1' of $\mathcal{N}_1$ disappears
in $\mathcal{N}$ since it finds no partner in $\mathcal{N}_2$ with the same image in the interface net $\mathcal{N}_0$. This
example doesn't reflect the full generality of the pullback construction since outside the
domains of f_1 and f_2, transitions of $\mathcal{N}_1$ and $\mathcal{N}_2$ don't synchronize: $(\Lambda_1 \cap \Lambda_2) \setminus \Lambda_0 = \emptyset$.

5   Conclusion

The original motivation for this work was the derivation of a simple construction
for pullbacks of safe nets, thus providing a way to express in a categorical framework the combination of nets that interact by sharing places and transitions.
We actually obtained more: we proved the existence of all equalizers in *Nets*,
which, in conjunction with the existence of all products, proves the existence of
all (small) limits in *Nets*.

Expressing the combination of nets as a categorical limit has some advantages. Consider for example the unfolding operation [4], that associates the unfolding $\mathcal{U}(\mathcal{N})$ to a safe net $\mathcal{N}$. $\mathcal{U}$ is actually a functor from $Nets$ to the subcategory Occ of occurrence nets, and we know that $\mathcal{U} : Nets \rightarrow Occ$ has a left adjoint, and so preserves limits. As a consequence, when $\mathcal{N} = \mathcal{N}_1 \wedge^{\mathcal{N}_0} \mathcal{N}_2$, one immediately obtains $\mathcal{U}(\mathcal{N}) = \mathcal{U}(\mathcal{N}_1) \wedge_O^{\mathcal{U}(\mathcal{N}_0)} \mathcal{U}(\mathcal{N}_2)$ where $\wedge_O$ denotes the pullback in Occ. This result expresses that the factorized form of a net immediately gives rise to a factorized form on runs of this net. Moreover, one obtains for free the existence of pullbacks in Occ, with a formal expression for $\wedge_O$: let $\mathcal{O}_0, \mathcal{O}_1, \mathcal{O}_2$ be occurrence nets, one has $\mathcal{O}_1 \wedge_O^{\mathcal{O}_0} \mathcal{O}_2 \equiv \mathcal{U}(\mathcal{O}_1 \wedge^{\mathcal{O}_0} \mathcal{O}_2)$, where the last pullback is computed in $Nets$, and where $\equiv$ means "isomorphic to."

The results above naturally extend to general limits : whatever the way one combines elementary nets to build a larger system (by products, pullbacks, etc.), a similar decomposition holds on the unfolding (or on the trellis [19]) of the global system. We believe this is an important key to study large systems by parts (see [17, 18] for examples of modular diagnosis based on these ideas).

Acknowledgment. The author would like to thank Marek Bednarczyk for fruitful discussions, and Philippe Darondeau for his useful comments.

References

1. S. Mac Lane : Categories for the Working Mathematician, Springer-Verlag, 1971.
2. G. Winskel : A new Definition of Morphism on Petri Nets, LNCS 166, pp. 140-149, 1984.
3. M. Nielsen, G. Winskel : Petri Nets and Bisimulation, BRICS report no. RS-95-4, Jan. 1995.
4. G. Winskel : Categories of models for concurrency, Seminar on Concurrency, Carnegie-Mellon Univ. (July 1984), LNCS 197, pp. 246-267, 1985.
5. G. Winskel : Event structure semantics of CCS and related languages , LNCS 140, 1982, also as report PB-159, Aarhus Univ., Denmark, April 1983.
6. G. Winskel : Petri Nets, Algebras, Morphisms, and Compositionality , Information and Computation, vol. 72, pp. 197-238, 1987.
7. M.A. Bednarczyk, L. Bernardinello, B. Caillaud, W. Pawlowski, L. Pomello : Modular System Development with Pullbacks, ICATPN 2003, LNCS 2679, pp. 140-160, 2003.
8. M.A. Bednarczyk, A. Borzyszkowski, R. Somla : Finite Completeness of Categories of Petri Nets, Fundamenta Informaticae, vol. 43, no. 1-4, pp. 21-48, 2000.
9. B. Koenig : Parallel Composition and Unfolding of Petri Nets (Including Some Examples), private communication, 2005.
10. K.L. McMillan, Using unfoldings to avoid the state explosion problem in the verification of asynchronous circuits, in Proc. 4th Workshop of Computer Aided Verification, Montreal, 1992, pp. 164-174.
11. J. Esparza, Model checking using net unfoldings, Science of Computer Programming 23, pp. 151-195, 1994.
12. J. Esparza, C. Schröter, Reachability Analysis Using Net Unfoldings, Workshop of Concurrency, Specification and Programming, volume II of Informatik-Bericht 140, pp. 255-270, Humboldt-Universität zu Berlin, 2000.

13. S. Melzer, S. Römer, Deadlock checking using net unfoldings, CAV'97, LNCS 1254, pp. 352-363, 1997.
14. V. Khomenko, M. Koutny, A. Yakovlev, Detecting State Encoding Conflicts in STG Unfoldings Using SAT, Fundamenta Informaticae, Volume 62, Issue 2, pp. 221-241 IOS Press, 2004 (Special Issue on Best Papers from ACSD 2003).
15. V. Khomenko, M. Koutny, A. Yakovlev, Logic Synthesis for Asynchronous Circuits Based on STG Unfoldings and Incremental SAT, Fundamenta Informaticae, Volume 70, Issue 1-2, pp. 49-73, IOS Press, 2006.
16. A. Benveniste, E. Fabre, S. Haar, C. Jard, Diagnosis of asynchronous discrete event systems, a net unfolding approach, IEEE Trans. on Automatic Control, vol. 48, no. 5, pp. 714-727, May 2003.
17. E. Fabre, A, Benveniste, S. Haar, C. Jard: Distributed Monitoring of Concurrent and Asynchronous Systems, Journal of Discrete Event Systems, special issue, pp. 33-84, May 2005.
18. E. Fabre: Distributed Diagnosis based on Trellis Processes, 44th Conf. on Decision and Control (CDC), Seville, Spain, 12-15 Dec. 2005.
19. E. Fabre, Trellis processes: a compact representation for runs of concurrent systems, INRIA research report, no. RR-5554, March 2005.

From Petri Nets to Differential Equations - An Integrative Approach for Biochemical Network Analysis

David Gilbert[1] and Monika Heiner[2]

[1] Bioinformatics Research Centre, University of Glasgow
Glasgow G12 8QQ, Scotland, UK
drg@brc.dcs.gla.ac.uk
[2] Department of Computer Science, Brandenburg University of Technology
Postbox 10 13 44, 03013 Cottbus, Germany
monika.heiner@informatik.tu-cottbus.de

Abstract. We report on the results of an investigation into the integration of Petri nets and ordinary differential equations (ODEs) for the modelling and analysis of biochemical networks, and the application of our approach to the model of the influence of the Raf Kinase Inhibitor Protein (RKIP) on the Extracellular signal Regulated Kinase (ERK) signalling pathway. We show that analysis based on a discrete Petri net model of the system can be used to derive the sets of initial concentrations required by the corresponding continuous ordinary differential equation model, and no other initial concentrations produce meaningful steady states. Altogether, this paper represents a tutorial in step-wise modelling and analysis of larger models as well as in structured design of ODEs.

1   Motivation

Classical, i.e. time-less discrete Petri nets combine an intuitive modelling style with well-founded analysis techniques. It is for this reason that they are widely used in various application areas, where they have been proven to be useful for a qualitative verification of technical as well as "natural" systems, i.e. biochemical networks like metabolic networks, signal transduction networks, or gene regulatory networks.

However, any real system behaviour happens in time. Thus the next step following on from a qualitative analysis typically consists in quantitative analyses taking into account timing information. In the case of biochemical systems, all atomic actions take place continuously. Moreover, the rates of all the atomic actions typically depend on the continuous concentrations of the involved substances. Hence systems of ordinary differential equations (ODEs) appear to be a natural choice for quantitative modelling of biochemical networks.

In this paper we bridge the gap between these two worlds, i.e. the (time-less) discrete and the (timed) continuous one, and demonstrate by means of

S. Donatelli and P.S. Thiagarajan (Eds.): ICATPN 2006, LNCS 4024, pp. 181–200, 2006.

one of the standard examples used in the systems biology community – the core model of the influence of the Raf-1 Kinase Inhibitor Protein (RKIP) on the ERK signalling pathway – how both sides can play together by providing different, but complementary viewpoints on the same subject.

This paper can be considered as a tutorial in the step-wise modelling and analysis of larger models as well as in the structured design of ODEs. The discrete model is introduced as a supplementary intermediate step, at least from the viewpoint of the biochemist accustomed to ODE modelling only, and serves mainly for model validation since this cannot be performed on the continuous level. Having successfully validated the discrete model, the continuous model is derived from the discrete one by assigning rate equations to all of the atomic actions in the network. Thus the continuous model preserves the structure of the discrete one, and the continuous Petri net is nothing else than a structured description of ODEs.

The approach is presented by a small example, which is however sophisticated enough to highlight the main ideas — it is common sense to practice new techniques on small examples at first, before attempting larger ones, where the outcome to be expected tends to be less well-defined.

Moreover we demonstrate how the discrete model can be used to drive the continuous model by automatically generating sets of biochemically plausible values for the initial concentrations of protein species.

This paper is organized as follows. The next section provides an overview on the biochemical context on hand and introduces the running example. Afterwards, we demonstrate the step-wise modelling and analysis, where section 3 deals with the contributions by the discrete viewpoint, while section 4 is devoted to the continuous viewpoint. Having presented our own approach, we discuss some related work in section 5. We conclude with a summary and outlook on intended further research directions.

2   Biochemical Context

There are many networks of interacting components known to exist as part of the machinery of living organisms. Biochemical networks can be metabolic, regulatory or signal transduction networks. The role of metabolic networks is to synthesize essential biochemical compounds from basic components, or to degrade compounds. Regulatory networks are used to control the ways in which genes are expressed as RNAs or proteins, whereas signal transduction networks transmit biochemical signals between or within cells.

The two terms "pathway" and "network" tend to be used interchangeably in the literature, with "pathway" being (implicitly) taken to be a part of a more general network. In this paper we follow the generally accepted use of the term "pathway" to refer to the core of a biochemical network, comprising a sequence of activities, for example a kinase cascade. Thus, for example, we will describe the ERK pathway as being embedded in a more general signal transduction network, and that the ERK pathway is a member of a large family of MAP Kinase pathways.

In this paper we focus on signal transduction, which is the mechanism which enables a cell to sense changes in its environment and to make appropriate responses. The basis of this mechanism is the conversion of one kind of signal into another. Extracellular signaling molecules are detected at the cell membrane by being bound to specific trans-membrane receptors that face outwards from the membrane and trigger intracellular events, which may eventually effect transcriptional activities in the nucleus. The eventual outcome is an alteration in cellular activity including changes in the gene expression profiles of the responding cells. These events, and the molecules that they involve, are referred to as (intracellular) "signalling pathways"; they contribute to the control of processes such as proliferation, cell growth, movement, apoptosis, and inter-cellular communication. Many signal transduction processes are "signalling cascades" which comprise a series of enzymatic reactions in which the product of one reaction acts as the catalytic enzyme for the next. The effect can be amplification of the original signal, although in some cases, for example the MAP kinase cascade, the signal gain is modest [1], suggesting that a main purpose is regulation [2] which may be achieved by positive and negative feedback loops.

The main factor which distinguishes signal transduction pathways from metabolic networks is that in the former the product of an enzymatic reaction becomes the enzyme for the next step in the pathway, whereas in the latter the product of one reaction becomes the substrate for the next, see Fig 1. In general, it is *dynamic* behaviour which is of interest in a signalling pathway, as opposed to the steady state in a metabolic network. In gene regulatory networks, on the other hand, the inputs are proteins such as transcription factors (produced from signal transduction or metabolic activity), which then influence the expression of genes – enzymatic activity plays no direct role here. However, the products of gene regulatory networks can play a part in the transcription of other proteins, or can act as enzymes in signalling or metabolic pathways.

The ERK pathway (also called Ras/Raf, or Raf-1/MEK/ERK pathway) is a ubiquitous pathway that conveys cell division and differentiation signals from the cell membrane to the nucleus. Ras is activated by an external stimulus, via one of many growth factor receptors; it then binds to and activates Raf-1 to become Raf-1*, or activated Raf, which in turn activates MAPK/ERK Kinase (MEK) which in turn activates Extracellular signal Regulated Kinase (ERK).

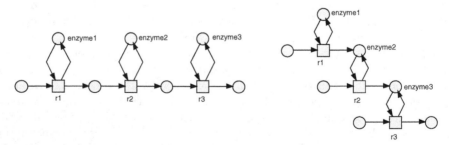

Fig. 1. The essential structural difference between metabolic networks (left) and signal transduction networks (right)

This cascade (Raf-1 → Raf-1* → MEK → ERK) of protein interaction controls cell differentiation, the effect being dependent upon the activity of ERK. An important area of experimental scientific investigation is the role that the Raf-1 Kinase Inhibitor Protein (RKIP) plays in the behaviour of this pathway. The hypothesis is that RKIP can inhibit activation of Raf-1 by binding to it, disrupting the interaction between Raf-1 and MEK, thus playing a part in regulating the activity of the ERK pathway.

3   The Discrete Approach

In this section we apply place/transition Petri nets to model the pathway of interest, and interpret them in the standard way. The reader is assumed to be familiar with the basic terms and typical analysis techniques; for an introduction see e.g. [3]. The software tools which have been used in this section are: for modelling – Snoopy [4], and for analysis – the Integrated Net Analyser (INA) [5], and the Model Checking Kit [6].

3.1   Qualitative Modelling

We apply the well-established modelling principles to represent biochemical networks by (various versions of) Petri nets, outlined e.g. in [7], [8].

Accordingly, we create a place/transition Petri net, see Figure 2, of the RKIP pathway, given in [9] in the style of a bichromatic graph. Circles (places) stand for the states of a protein or protein complex and are labelled with the corresponding name; complexes are indicated by an underscore "_" between the protein names. For example, Raf-1* and RKIP are proteins, and Raf-1*_RKIP is a protein complex formed from Raf-1* and RKIP. A suffix -P or -PP denotes a single or double phosphorylated protein, for example RKIP-P and ERK-PP. In the pathway under consideration there are 11 proteins or complexes; a discrete concentration $m1, m2, \ldots$ is associated with each protein or complex. In the case of the qualitative model, these concentrations can be thought of as being 'high' or 'low' (present or absent).

Rectangles (transitions) stand for reactions, with reversible reactions being indicated by a pair of two complementary transitions. In this pathway, reactions comprise protein complexation and decomplexation events, often accompanied by phosphorylation or dephosphorylation. For example, Raf-1* and RKIP combine in a forwards reaction to form Raf-1*_RKIP which can disassociate in a backwards reaction into Raf-1* and RKIP, or combine with ERK-PP to form the complex Raf1*_RKIP_ERK-PP. In this qualitative model, $k1, k2, \ldots$ stand for reaction labels.

3.2   Qualitative Analysis

The Petri net enjoys all the pleasant general properties a Petri net insider could dream of: boundedness, liveness, reversibility, which are three orthogonal basic behavioural net properties [3]. The decision about the first two properties can be

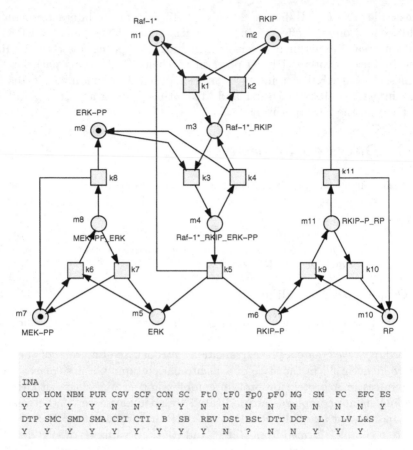

INA

ORD	HOM	NBM	PUR	CSV	SCF	CON	SC	Ft0	tF0	Fp0	pF0	MG	SM	FC	EFC	ES
Y	Y	Y	Y	N	N	Y	Y	N	N	N	N	N	N	N	N	Y

DTP	SMC	SMD	SMA	CPI	CTI	B	SB	REV	DSt	BSt	DTr	DCF	L	LV	L&S
Y	Y	Y	Y	Y	Y	Y	Y	Y	N	?	N	N	Y	Y	Y

Fig. 2. The Petri net for the core model of the RKIP pathway. Places stand for the states of the concentration of a protein; complexes are indicated by an underscore "_" between the protein names. Pairs of two complementary transitions stand for reversible reactions (there are four of them). The layout follows the suggestions by the graphical notation used in [9]. At the bottom the two-lines result vector as produced by the Integrated Net Analyser [5] is provided. The properties of interest in the given context of biochemical network analysis are explained in the text. The initial marking is constructed systematically using standard Petri net analysis techniques, compare step (3) in the qualitative analysis section.

made for our example in a static way, while the last property requires dynamic analysis techniques. The essential steps of the systematic analysis procedure for the example are given in more detail as follows.

(1) Structural properties. The following three structural properties reflect the modelling approach and can be read as preliminary consistency checks.

The net is *ordinary*, i.e. all arc weights equal to 1. This includes homogeneity, i.e. the outgoing arcs of each place have the same multiplicity, which is a necessary prerequisite for the Deadlock Trap Property (DTP), see step (4) below.

The net is *pure*, i.e. there are no side-conditions. So, the net structure is fully represented by the incidence matrix, which is used for the calculation of the P- and T-invariants, see next step.

The net is *strongly connected*, which involves the absence of boundary nodes. So, the net is self-contained, i.e. a closed system. Therefore, in order to make the net live, we have to construct an initial marking, compare step (3).

Moreover, the net belongs to the structural class "extended simple". Hence, we know that the net has the ability to be live independent of time, i.e. if it is live, then it remains live under any duration timing [10].

(2) Static decision of marking-independent behavioural properties.

Model validation should include a check of all minimal P/T-invariants [11] for their biological plausibility [8].

A P-invariant stands for a set of places, over which the weighted sum of tokens is constant, independently of any firing. So, P-invariants represent token-preserving sets of places. In the context of metabolic networks, P-invariants reflect substrate conservations, while in signal transduction networks P-invariants often correspond to the several states of a given species (protein or protein complex). A place belonging to a P-invariant is obviously bounded.

In the net under consideration there are five minimal P-invariants covering the net (CPI), consequently the net is bounded. All the P-invariants x_i contain only entries of 0 and 1, which allows a short-hand specification by just giving the names of the places involved.

$$x_1 = (\text{Raf-1*}, \text{Raf-1*\_RKIP}, \text{Raf-1*\_RKIP\_ERK-PP}),$$
$$x_2 = (\text{MEK-PP}, \text{MEK-PP\_ERK}),$$
$$x_3 = (\text{RP}, \text{RKIP-P\_RP}),$$
$$x_4 = (\text{ERK}, \text{ERK-PP}, \text{MEK-PP\_ERK}, \text{Raf-1*\_RKIP\_ERK-PP}),$$
$$x_5 = (\text{RKIP}, \text{Raf-1*\_RKIP}, \text{Raf-1*\_RKIP\_ERK-PP}, \text{RKIP-P\_RP}, \text{RKIP-P}).$$

Each P-invariant x_i stands obviously for a reasonable conservation rule. The first name given indicates the species preserved within each P-invariant. Due to the chosen naming convention, this particular name also appears in all the other place names of the same P-invariant.

A T-invariant has two interpretations in the given biochemical context. The entries of a T-invariant represent a multiset of transitions which by their partially ordered firing reproduce a given marking, i.e. they occur basically one after the other. The partial order sequence of the firing events of the T-invariant's transitions may contribute to a deeper understanding of the net behaviour.

The entries of a T-invariant may also be read as the relative transition firing rates of transitions, all of them occurring permanently and concurrently. This activity level corresponds to the steady state behaviour [12]. Independently of the chosen interpretation, the net representation of minimal T-invariants (the T-invariant's transitions plus their pre- and post-places and all arcs in between) characterize typically minimal self-contained subnetworks with an enclosed biological meaning.

The net under consideration is covered by T-Invariants (CTI), which is a necessary condition for bounded nets to be live. Besides the expected four trivial T-invariants for the four reversible reactions, there is only one non-trivial minimal T-invariant $y = (k1, k3, k5, k6, k8, k9, k11)$. The net representation of this T-invariant describes the essential partial order model of our system, given in text style: (k1; k3; k5; (k6; k8), (k9; k11)), where ";" stands for "sequentiality" and "," for "concurrency". The automatic identification of non-trivial minimal T-invariants is in general useful as a method to highlight important parts of a network, and hence aid its comprehension by biochemists, especially when the entire network is too complex to easily comprehend.

All the properties above relate only to the structure, i.e. they are valid independently of the initial marking. In order to proceed we first need to generate an initial marking.

(3) Initial marking construction. For a systematic construction of the initial marking, the following criteria have to be taken into consideration.

- Each P-invariant needs at least one token.
- All (non-trivial) T-invariants should be realizable, meaning, the transitions, making up the T-invariant's multi-set can be fired in an appropriate order.
- Additionally, it is common sense to look for a minimal marking (as few tokens as possible), which guarantees the required behaviour.
- Within a P-invariant, choose the species with the most *inactive* (e.g. non-phosphorylated) or the *monomeric* (i.e. non-complexed) state.

Taking all these criteria together, the initial marking on hand is: Raf-1*, RKIP, ERK, MEK-PP, RP get each one token, while all remaining places are empty. With this initial marking, the net is covered by 1-P-invariants (exactly one token in each P-invariant), therefore the net is 1-bounded (also called safe). That is in perfect accordance with the understanding that in signal transduction networks a P-invariant comprises all the different states of one species. Obviously, each species can be only in one state at any time.

In the following, however, we will use an initial marking derived from the initial concentrations used by Cho et al. [9] as part of their method to estimate rate parameters required for their ODE model of the RKIP pathway (see Table 1 in Section 4.2). This marking is represented in Figure 2. We use their initial marking because we focus on the Cho et al. model throughout this paper for illustrative purposes. We will later in this paper demonstrate that the Cho et al. initial marking is equivalent to the initial marking which we have constructed, and that in fact both markings are members of a larger equivalence class of markings.

With the chosen marking we can check the non-trivial minimal T-invariant (see step (2)) for realizability, which then involves the realizability of all the trivial T-invariants. We obtain an infinite run, the beginning of which is given as labelled condition/event net in Figure 3, and characterize this in a short-hand notation by the following set of partially ordered words out of the alphabet of all transition labels: (k1; k3; k5; [((k9; k11; k1), (k6; k8)); k3; k5]*). This partial

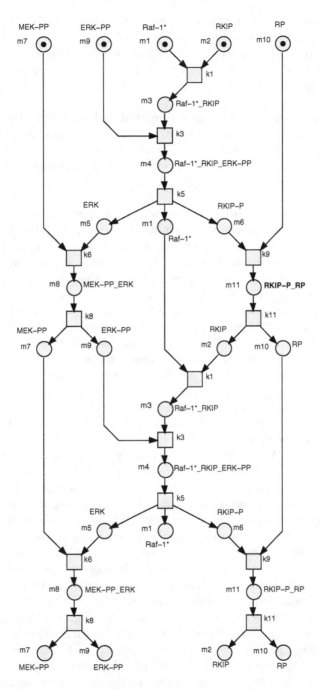

Fig. 3. The beginning of the infinite partial order run of the non-trivial minimal T-invariant of the place/transition Petri net given in Figure 2. Here, transitions represent events, labelled by the name of the reaction taking place, while places stand for binary conditions, labelled by the name of the species, set or reset by the event, respectively.

order run gives further insight into the dynamic behaviour of the network which may not be apparent from the standard net representation, e.g. it becomes now clear that there is no requirement for k6 and k8 to occur before k1.

Having established and justified our initial marking we proceed to the next steps of the analysis.

(4) Static decision of marking-dependent behavioural properties. The net belongs to the structural class "extended simple" and the Deadlock-Trap Property (DTP) holds (any structural deadlock contains a marked trap), therefore the net is live, see e.g. [3], [5]. However, most biochemical networks (as well as non-trivial technical networks) do not fulfill the DTP.

(5) Dynamic decision of behavioural properties. In order to decide reversibility we have to calculate the reachability graph. The nodes of a reachability graph represent all possible states (markings) of the net. The arcs in between are labelled by single transitions, the firing of which causes the related state change. Altogether, the reachability graph gives a finite automaton representation of all possible single step firing sequences. Consequently, concurrent behaviour is described by enumerating all interleaving firing sequences (interleaving semantics).

Because we already know that the net is bounded, we do also know that the reachability graph has to be finite. Here, the reachability graph has 13 states (out of $2048 = 2^{11}$ theoretically possible ones), forming one strongly connected component. Therefore, the Petri net is reversible, i.e. the initial system state is always reachable again, or in other words - the system has the capability of self-reinitialization. Further, the liveness of the net has already been decided structurally, so we know that each transition (reaction) appears at least once in this strongly connected component.

Moreover, from the viewpoint of the discrete model, all these 13 states are equivalent, i.e. any of those 13 states could be taken as initial state resulting in exactly the same total (discrete) system behaviour. That is in perfect accordance with the observations gained during quantitative analyses, see Section 4.2.

For reasons of completeness we explored all other possible sensible initial states. Following our understanding that P-invariants in signal transduction networks reflect different states of a given species, the net should be covered by 1-P-invariants. Therefore we had to consider only those initial markings which do not contradict this assumption. None of these potential initial markings results in a net whose behaviour is reversible and live, and only a few of them produce a terminal strongly connected component in the reachability graph (meaning that at least this part, consisting of mutually reversible reactions, is live).

This concludes the analysis of *general* behavioural net properties, i.e. of properties we can speak about in syntactic terms only, without any semantic knowledge. The next step consists in a closer look at *special* behavioural net properties, reflecting the expected special functionality of the network.

(6) Model checking of special behavioural properties. Special properties are best expressed using temporal logics, e.g. Computational Tree Logic

(CTL) – a branching time logic, which we are going to interpret in interleaving semantics. For an introduction into the specification of biologically relevant properties of biochemical networks using CTL see [13].

Because we are in the fortunate position of having a bounded model, these temporal-logic formulae can be checked using standard model checking techniques. Furthermore, the model under consideration is 1-bounded. Therefore, we can rely on a particularly rich choice of model checkers to solve this task [6]. In the case of our rather simple example the variety of model checkers is not important, and the properties could even be checked manually. However, the state space of more complex networks exceeds typically several millions.

We instantiate some of the generic property patterns provided in [13] and get the following samples of meaningful statements for our running example, whose truth can be determined via model checking:

- **property 1:** There are reachable states where ERK is phosphorylated and at the same time RKIP is not phosphorylated.

$$\mathbf{EF} \, [\, (\text{ERK-PP} \lor \text{Raf-1*\_RKIP\_ERK-PP}) \, \land \, \text{RKIP} \,]$$

- **property 2:** The phosphorylation of ERK (to ERK-PP) does not dependent on a phosphorylated state of RKIP.

$$\mathbf{EG} \, [\, \text{ERK} \rightarrow \mathbf{E} \, (\, \neg(\text{RKIP-P} \lor \text{RKIP-P\_RP}) \, \mathbf{U} \, \text{ERK-PP}) \,]$$

- **property 3:** A cyclic behaviour w.r.t. the presence/absence of RKIP is possible forever.

$$\mathbf{AG} \, [\, (\, \text{RKIP} \rightarrow \mathbf{EF} \, (\neg\text{RKIP}) \,) \land (\neg\text{RKIP} \rightarrow \mathbf{EF} \, (\text{RKIP}) \,) \,]$$

3.3   Summary

To summarize the preceding validation steps, the model has passed the following validation criteria.

validation criterion 0

- All expected structural properties hold.
- All expected general behavioural properties hold.

validation criterion 1

- CPI.
- No minimal P-invariant without biological interpretation.

validation criterion 2

- CTI.
- No minimal T-invariant without biological interpretation.
- No known biological behaviour without corresponding, not necessarily minimal T-invariant.

validation criterion 3

- All expected special behavioural properties, expressed as temporal-logic formulae, hold.

It is worth noting that not all of the validation criteria outlined above are always feasible. E.g. it only makes sense to ask for CPI as well as CTI for self-contained (closed) systems, i.e. without boundary nodes. In the case of signal transduction networks it depends on the modelling style whether the essential system behaviour can be explained by the discussion of only T-invariants. Finally, validation criterion 3 relies on temporal logics as a flexible language to describe special properties. Thus it requires seasoned understanding of the network under investigation combined with the skill to correctly express the expected correct behaviour in temporal logics.

Therefore, the set of meaningful validation criteria has to be adjusted to the case study on hand, but it should become common practice to do some model validation and to make the criteria applied explicit.

Now we are ready for a more sophisticated quantitative analysis of our model.

4   The Continuous Approach

In this section we transform our validated time-less discrete model, given as place/transition Petri net, into a timed continuous one, specified as continuous Petri net. For an introduction into continuous Petri nets see e.g. [14], [15].

The software tools, which have been used in this section, are: an extended version of Snoopy [16], supporting modelling as well as analysis by some standard numerical integration algorithms. Additionally, we use Gepasi [17] and Matlab [18] for more detailed analyses.

4.1   Quantitative Modelling

In a continuous Petri net the marking of a place is no longer an integer, but a positive real number, called token value (which we are going to interpret as the concentration of a given species). The instantaneous firing of a transition is carried out like a continuous flow, whereby the current firing rate depends generally on the current marking.

To be precise we need the following notations: The preset of a node $x \in P \cup T$ is defined as $^\bullet x := \{y \in P \cup T | f(y, x) \neq 0\}$, and its postset as $x^\bullet := \{y \in P \cup T | f(x, y) \neq 0\}$.

Definition 1 (Continuous Petri net). *A continuous Petri net is a quintuple* $\mathcal{CON} = \langle P, T, f, v, m_0 \rangle$, *where*

- *P and T are finite, non empty, and disjoint sets. P is the set of continuous places, T is the set of continuous transitions.*
- *$f : (P \times T) \cup (T \times P) \to \mathbb{R}_0^+$ defines the set of directed arcs, weighted by non-negative real values.*

– $v : T \to H$ *assigns to each transition a* firing rate function, *whereby*
$H := \bigcup_{t \in T} \left\{ h | h : \mathbb{R}^{|\bullet t|} \to \mathbb{R} \right\}$ *is the set of all firing rate functions, and*
$\mathrm{dom}\,(v\,(t)) = {}^{\bullet}t.$
– $m_0 : P \to \mathbb{R}_0^+$ *gives the* initial marking.

A continuous transition t is enabled by m, iff $\forall p \in {}^{\bullet}t : m(p) > 0$. Due to the influence of time, a continuous transition is forced to fire as soon as possible. The firing rate of an atomic (re-) action depends typically on the current concentrations of the substances involved, i.e. of the token values of the transition's preplaces. So we get marking-dependent, i.e. variable firing rates. Please note, a firing rate may also be negative, in which case the reaction takes place in the reverse direction. This feature is commonly used (but not in this paper) to model reversible reactions by just one transition, where positive firing rates correspond to the forward direction, and negative ones to the backward direction.

Altogether, the semantics of a continuous Petri net is defined by a system of ordinary differential equations (ODEs), where one equation describes the continuous changes over time on the token value of a given place by the continuous increase of its pretransitions' flow and the continuous decrease of its posttransitions' flow:

$$\frac{m\,(p)}{dt} = \sum_{t \in {}^{\bullet}p} f\,(t, p)\,v\,(t) - \sum_{t \in p^{\bullet}} f\,(p, t)\,v\,(t).$$

Each equation corresponds basically to a line in the incidence matrix, whereby now the matrix elements consist of the rate functions multiplied by the arc weight, if any. Moreover, as soon as there are transitions with more than one preplace, we get a non-linear system, which calls for a numerical treatment of the system on hand.

With other words, the continuous Petri net becomes the structured description of the corresponding ODEs. Due to the explicit structure we expect to get descriptions which are less error prone compared to those ones created manually from the scratch. In fact, writing down a system of ODEs by designing a continuous Petri net instead of just using a text editor might be compared to high-level instead of assembler programming. In order to simulate the continuous Petri net, exactly the same algorithms are employed as for numerical differential equation solvers, see e.g. [16].

To transform our qualitative model, see Figure 2, into a continuous one, we interpret the reaction labels $k1, k2, \ldots (k_i)$ as rate constants, which define – multiplied by the preplaces – the reaction *rates* (mass action equation pattern). Then, our continuous Petri net generates the following system of ordinary differential equations given below in a structured notation (generated by our continuous Petri net tool). We use the $m1, m2, \ldots (m_i)$ as synonyms for the lengthy species (place) names. The initial concentrations reappear in Table 1.

$$\frac{dm_1}{dt} = r_2 + r_5 - r_1 \qquad \frac{dm_5}{dt} = r_5 + r_7 - r_6 \qquad \frac{dm_9}{dt} = r_4 + r_8 - r_3$$

$$\frac{dm_2}{dt} = r_2 + r_{11} - r_1 \qquad \frac{dm_6}{dt} = r_5 + r_{10} - r_9 \qquad \frac{dm_{10}}{dt} = r_{10} + r_{11} - r_9$$

$$\frac{dm_3}{dt} = r_1 + r_4 - r_2 - r_3 \qquad \frac{dm_7}{dt} = r_7 + r_8 - r_6 \qquad \frac{dm_{11}}{dt} = r_9 - r_{10} - r_{11}$$

$$\frac{dm_4}{dt} = r_3 - r_4 - r_5 \qquad \frac{dm_8}{dt} = r_6 - r_7 - r_8$$

$$r_1 = k_1 * m_1 * m_2 \qquad r_5 = k_5 * m_4 \qquad r_9 = k_9 * m_6 * m_{10}$$
$$r_2 = k_2 * m_3 \qquad r_6 = k_6 * m_5 * m_7 \qquad r_{10} = k_{10} * m_{11}$$
$$r_3 = k_3 * m_3 * m_9 \qquad r_7 = k_7 * m_8 \qquad r_{11} = k_{11} * m_{11}$$
$$r_4 = k_4 * m_4 \qquad r_8 = k_8 * m_8$$

$$k_1 = 0.53 \qquad k_5 = 0.0315 \qquad k_9 = 0.92$$
$$k_2 = 0.0072 \qquad k_6 = 0.6 \qquad k_{10} = 0.00122$$
$$k_3 = 0.625 \qquad k_7 = 0.0075 \qquad k_{11} = 0.87$$
$$k_4 = 0.00245 \qquad k_8 = 0.071$$

4.2 Quantitative Analysis

In general, biochemists will wish to use ODE models of biochemical systems to explore in a general manner possible observable behaviours, for example the concentration change of a component over time, or the steady-state properties of the system including oscillatory behaviour. Specifically in the case of signalling pathways the system components are proteins in both complexed and uncomplexed forms and in phosphorylated and unphosphorylated states. The kinds of experimental observations that can be made often result in very coarse data points — for example immuno-blotting will give quite inexact data on the *relative* concentrations of species at a few time-points, the data varying quite a lot between repeated experiments. In addition, experiments are often conducted *in-vivo* in cells, and immuno-blotting applied to the entire cell contents (after lysing, or breaking down the cell wall) — hence there is very little exactness possible in terms of concentrations since in reality these may vary through the cell, but local concentrations may not be measurable by this technique. Moreover it is often not possible to distinguish between the complexed and non-complexed form of proteins – thus for example the relative concentration of phosphorylated ERK (ERK-PP) will be given as a combination of ERK-PP alone ($m8$) plus the ERK-PP component of the Raf-1*_RKIP_ERK-PP complex ($m4$).

Given these inexactitudes, biochemists will want to know the answers to general questions, such as "Will the concentration of the phosphorylated form of protein X-PP rise for the first 10 minutes after a particular stimulus is given to the cell, and then remain constant?", and in the same experiment "Will the concentration of the unphosphorylated form of protein Y rise from the start of the experiment, peaking at 20 minutes at a concentration higher than that of X-PP, and then fall off during the remainder of the time, eventually becoming less than the concentration of X-PP?".

The ODE solvers which are normally used to interpret ODE models of biochemical networks rely on exact values of rate constants and initial concentrations in order for the computations to be performed. Thus the results produced by simulations of ODE models of networks may be over-exact with respect to the characteristics of the real data. For this reason, biochemists will often interpret the results of ODE-based simulations as indicators of the behaviour of the components of the network, rather than being concerned with the exact value of the concentration of a particular species at a particular point in time.

We have performed a quantitative analysis of the results of simulating the behaviour of the network using a system of ODEs. The aim of this analysis was to determine whether the 13 'good' initial states suggested by the qualitative Petri net analysis were indeed in some way equivalent (they all result in the same steady state), and that no other possible initial states can be used to give the same results.

The differential equation model of the pathway, taken from Cho et al. [9] and reproduced in Section 4 above, was coded in MatLab. Although these authors do not explicitly state the initial concentrations of the 11 species when computing the simulation of the network, we have deduced by inspection of Fig. 5 in their paper which presents their simulation results that they are as given in the μM_{Cho} column of Table 1. For the purposes of our computations we have mapped any non-zero concentrations to 1, as in column μM_{PN} of that table, hence our initial concentrations correspond to the marking in the Petri net in Figure 2.

Table 1. Initial concentrations

Species	μM_{Cho}	μM_{PN}
Raf1*	2.5	1
RKIP	2.5	1
Raf1_RKIP	0	0
RAF_RKIP_ERK	0	0
ERK	0	0
RKIP-P	0	0
MEK-PP	2.5	1
MEK-PP_ERK	0	0
ERK-PP	2.5	1
RP	2.5	1
RKIP-P_RP	0	0

Since there are eleven species, there are $2048 = 2^{11}$ possible initial states, including that given in the original paper. Of these, 13 were identified by the reachability graph analysis (Section 3.2) to form one strongly connected component, making the net live and reversible, and thus to be 'good' initial states (see Table 2).

These are 'sensible' initial states from the point of view of biochemistry, in that in all these 13 cases, and in none of the other 2035 states, each protein species is in a high initial concentration in only one of the following states: uncomplexed, complexed, unphosphorylated or phosphorylated. These conditions relate exactly to the 1-P-invariant interpretation given in our initial marking construction procedure in Section 3.2.

We then computed the final steady state of the set of species for each possible initial state, using the MatLab ODE solver ode45, which is based on an explicit Runge-Kutta formula, the Dormand-Prince pair [19], with 100 time steps.

We found that all of the 13 'good' initial states result in the same final state, within the bounds of computational error of the ODE solver. These results are summarized in Table 3 which reports the mean steady state concentration and standard deviation for each of the 11 species.

Table 2. Initial 13 'good' state configurations

Species	S1	S2	S3	S4	S5	S6	S7	S8	S9	S10	S11	S12	S13
Raf-1*	1	0	0	1	1	1	1	1	0	0	1	1	1
RKIP	1	0	0	0	0	0	0	1	0	0	1	0	0
Raf-1*_RKIP	0	1	0	0	0	0	0	0	1	1	0	0	0
Raf-1*_RKIP_ERK-PP	0	0	1	0	0	0	0	0	0	0	0	0	0
ERK	0	0	0	1	0	0	1	1	1	0	0	0	0
RKIP-P	0	0	0	1	1	0	0	0	0	0	0	0	1
MEK-PP	1	1	1	1	0	0	1	1	1	0	0	1	1
MEK-PP_ERK	0	0	0	0	1	1	0	0	0	1	1	0	0
ERK-PP	1	1	0	0	0	0	0	0	0	0	0	1	1
RP	1	1	1	1	1	0	0	1	1	1	1	0	1
RKIP-P_RP	0	0	0	0	0	1	1	0	0	0	0	1	0

In Figure 4 we reproduce two simulations of the model: State 1 corresponding to the initial marking suggested by Cho et al [9] where the initial concentration of ERK-PP is high and ERK is low, and State 8 corresponding to the initial marking, suggested by our approach described above in Section 3.2, with ERK-PP low and ERK high. State 8 has been confirmed by an expert signal transduction researcher as the most sensible starting state [20]. The equivalence of the final states, compared with the difference in some intermediate states is clearly illustrated in these figures. For example, the concentration of Raf-1*_RKIP behaves overall in a similar manner in both State 1 and State 8, peaking before 10 minutes although the peak is greater when ERK is not phosphorylated at the start of the experiment. In Figure 5 we reproduce the computed behaviour of ERK-PP for all 13 good initial states, showing that despite differences in the concentrations at early time-points, the steady state concentration is the same in all 13 states.

Table 3. Mean values for steady states for the 13 'good' initial states

Species	Mean steady state concentration	Standard Deviation
Raf-1*	0.2133	0.1225 * 1.0e-04
RKIP	0.1727	0.0854 * 1.0e-04
Raf-1*_RKIP	0.2163	0.5546 * 1.0e-04
Raf-1*_RKIP_ERK-PP	0.5704	0.4346 * 1.0e-04
ERK	0.0332	0.0135 * 1.0e-04
RKIP-P	0.0200	0.0169 * 1.0e-04
MEK-PP	0.7469	0.6020 * 1.0e-04
MEK-PP_ERK	0.2531	0.6020 * 1.0e-04
ERK-PP	0.1433	0.1846 * 1.0e-04
RP	0.9793	0.0471 * 1.0e-04
RKIP-P_RP	0.0207	0.0473 * 1.0e-04

Fig. 4. Dynamic behaviour for state 1 (left) and state 8 (right)

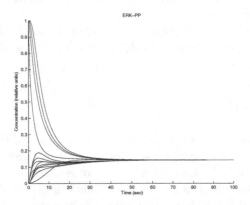

Fig. 5. Dynamic behaviour of ERK-PP for all 13 'good' states

We computed the Euclidean distances between the vector of mean values of the final steady states of the 13 states in the reachability graph and each of the final steady states for the states not identified by the reachability graph. These distances ranged from 0.7736 to 6.0889, and we summarize these results

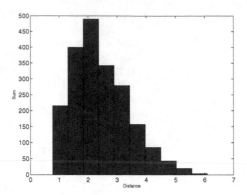

Fig. 6. Distribution of 'bad' steady states as Euclidean distances from the 'good' final steady state

in Figure 6. None of the initial states which is not identified by the reachability analysis resulted in a final steady state which was near that of the set of the 13 'good' states.

4.3 Summary

In summary, our results show that

1. All of the 13 states identified by the reachability graph of the validated discrete Petri net result in the same set of steady state values for the 11 species in the pathway.
2. None of the remaining 2035 possible initial states of the discrete Petri net results in a final steady state close to that generated by the 13 markings in the reachability graph.
3. The transient behaviour — the crucial point of interest in signal transduction networks — of the continuous model is sensible for the 13 states identified by our method.

5 Related Work

There are several research groups, applying various kinds of Petri nets to model and analyse biochemical networks. However, most of them are devoted to hybrid Petri nets, see e.g. [7], [21]; for a bibliography of related papers see [22]. Hybrid Petri nets comprise the discrete as well as the continuous case. Thus, they have to be treated by dedicated simulation techniques, instead by standard ODE solvers.

An approach combining qualitative and quantitative analysis techniques is proposed in [12]. In this paper time Petri nets are used to describe the steady state behaviour of a given biochemical network, whereby the time intervals are derived from the time-less model by help of the T-invariants, which are interpreted as firing count vectors. Interval-timed Petri nets provide a continuous time scale, but keep the discrete firing behaviour. Therefore they can still be treated in a discrete way, but they do not help in investigating continuous firing behaviour, e.g. in the transient state of a given network.

An approach describing the automatic derivation of ODEs from stochastic process algebra models of signalling pathways is presented in [23]. Consequently, the authors employ different analysis methods, which might be complementary to our ones, but they do not generate initial good markings (configurations).

Investigations on the relation between the properties of discrete and continuous Petri nets are fairly recent, as mentioned in [24]. In contrast to our approach, these authors focus on technical applications, which are inherently discrete. Then, they use the continuous model as a relaxation/approximation of the discrete one to get a better efficiency of analysis. In any case, they find themselves confronted with exactly the same questions we face: how do the properties of the one model relate to the properties in the other one. In particular the paper mentions some open questions of great interest, the solution of which coincide perfectly with our requirements.

6   Summary

We have created a discrete Petri net model of the influence of the Raf Kinase Inhibitor Protein (RKIP) on the Extracellular signal Regulated Kinase (ERK) signalling pathway, based on the ODE model presented by Cho et al [9]. We have then analysed the discrete model using a set of Petri net based tools and shown that the model enjoys several nice properties, among them boundedness, liveness, and reversibility. Moreover, the net is covered by P-invariants and T-invariants, all of them having sensible biological interpretation, and it fulfills several special functional properties, which have been expressed in temporal logic. Reachability graph analysis identifies 13 strongly connected states out of 2048 theoretically possible ones, which permit self-reinitialization of the Petri net. From the viewpoint of the discrete model, all these 13 states are equivalent and could be taken as an initial state resulting in exactly the same total (discrete) system behaviour.

We have then transformed the discrete Petri net into a continuous Petri net, defining ODEs. We have shown empirically that in the ODE model the 13 initial states, derived from the validated discrete model, result in the same (continuous) steady state. This analysis was performed by numerically solving the system of ordinary differential equations. Moreover, none of the other 2035 possible initial states result in a steady state close to that derived using those identified by reachability graph analysis.

Altogether we advocate a two-step technology for the modelling and analysis of biochemical networks in a systematic manner: (1) qualitative, i.e. (time-less) discrete modelling and analysis, esp. for the beneficial effect of confidence-increasing model validation, and (2) quantitative, i.e. (timed) continuous modelling and analysis, esp. with the hope of the reliable prediction of behaviour. For both steps we favour the deployment of both discrete as well as continuous Petri nets, sharing the same net structures for a given case. The quantitative model is derived from the qualitative one only by the addition of the quantitative parameters. Hence both models are likely to share some behavioural properties. However, the meticulous rules of this approach are the subject of further ongoing investigations by the authors.

Acknowledgements

We would like to thank Rainer Breitling for the constructive discussions as well as Alex Tovchigrechko and Simon Rogers for their support in the computational experiments. This work has been supported by the DTI Bioscience Beacons Projects programme.

References

1. Schoeberl, B., Eichler-Jonsson, C., Gilles, E., Muller, G.: Computational modeling of the dynamics of the MAP kinase cascade activated by surface and internalized EGF receptors. Nature Biotechnology **20** (2002) 370–375
2. Kolch, W., Calder, M., Gilbert, D.: When kinases meet mathematics: the systems biology of MAPK signalling. FEBS Letters **579** (2005) 1891–5
3. Murata, T.: Petri nets: Properties, analysis and applications. Proc.of the IEEE 77 **4** (1989) 541–580
4. Fieber, M.: Design und implementation of a generic and adaptive tool for graph manipulation, (in German). Master thesis, BTU Cottbus, Dep. of CS (2004)
5. Starke, P., Roch, S.: INA - The Intergrated Net Analyzer, Humboldt University Berlin, http://www.informatik.hu-berlin.de/~starke/ina.html. (1999)
6. Schröter, C., Schwoon, S., Esparza, J.: The Model Checking Kit. In: Proc. ICATPN, LNCS 2697, Springer (2004) 463–472
7. Matsuno, H., Fujita, S., Doi, A., Nagasaki, M., Miyano, S.: Towards pathway modelling and simulation. In: Proc. 24th ICATPN, LNCS 2679. (2003) 3–22
8. Heiner, M., Koch, I.: Petri net based model validation in systems biology. In: Proc. 25th ICATPN 2004, LNCS 3099, Springer (2004) 216–237
9. Cho, K.H., Shin, S.Y., Kim, H.W., Wolkenhauer, O., McFerran, B., Kolch, W.: Mathematical modeling of the influence of RKIP on the ERK signaling pathway. Lecture Notes in Computer Science **2602** (2003) 127–141
10. Starke, P.H.: Some properties of timed nets under the earliest firing rule. Lecture Notes in Computer Science; Advances in Petri Nets **424** (1989) 418–432
11. Lautenbach, K.: Exact liveness conditions of a Petri net class (in German). Technical report, GMD Report 82, Bonn (1973)
12. Popova-Zeugmann, L., Heiner, M., Koch, I.: Time Petri nets for modelling and analysis of biochemical networks. Fundamenta Informaticae **67** (2005) 149–162
13. Chabrier-Rivier, N., Chiaverini, M., Vincent Danos, F.F., Schächter, V.: Modeling and querying biomolecular interaction networks. Theoretical Computer Science **325** (2004) 25–44
14. David, R., Alla, H.: Discrete, Continuous, and Hybrid Petri Nets. Springer (2005)
15. Lonitz, K.: 'Hybrid Systems Modelling in Engineering and Life Sciences'. Master thesis, Universität Koblenz-Landau (2005)
16. Scheibler, D.: A software tool for design and simulation of continuous Petri nets, (in German). Master thesis, BTU Cottbus, Dep. of CS (2006)

17. Mendes, P.: GEPASI: A software package for modelling the dynamics, steady states and control of biochemical and other systems. Comput. Applic. Biosci. **9** (1993) 563–571
18. Shampine, L.F., Reichelt, M.W.: The MATLAB ODE Suite. SIAM Journal on Scientific Computing **18** (1997) 1–22
19. Dormand, J.R., Prince, P.J.: A family of embedded runge-kutta formulae. J. Comp. Appl. Math. **6** (1980) 1–22
20. Kolch, W. personal communication (2005)
21. Chen, M., Hofestädt, R.: A medical bioinformatics approach for metabolic disorders: Biomedical data prediction, modeling, and systematic analysis. J Biomedical Informatics **39(2)** (2006) 147–59
22. Will, J., Heiner, M.: Petri nets in biology, chemistry, and medicine - bibliography. Technical Report 04/2002, BTU Cottbus, Computer Science (2002)
23. Calder, M., Gilmore, S., Hillston, J.: Automatically deriving ODEs from process algebra models of signalling pathways. In: Proc. Computational Methods in Systems Biology (CSMB 2005), LFCS, University of Edinburgh (2005) 204–215
24. Silva, M., Recalde, L.: Continuization of timed Petri nets: From performance evaluation to observation and control. In: Proc. 26th ICATPN 2005, LNCS 3536, Springer (2005) 26–47

How Expressive Are Petri Net Schemata?

Andreas Glausch and Wolfgang Reisig

Humboldt-Universität zu Berlin
Institut für Informatik
{glausch, reisig}@informatik.hu-berlin.de

Abstract. *Petri net schemata* are an intuitive and expressive approach
to describe high-level Petri nets. A Petri net schema is a Petri net with
edges and transitions inscribed by terms and Boolean expressions, respectively. A concrete high-level net is gained by interpreting the symbols in
the inscriptions by a *structure*. The semantics of this high-level net can
be described in terms of a *transition system*. Therefore, the semantics of
a Petri net schema can be conceived as a family of transition systems
indexed by structures.

In this paper we characterize the expressive power of a general version
of Petri net schemata. For that purpose we examine families of transition
systems in general and characterize the families as generated by Petri net
schemata. It turns out that these families of transition systems can be
characterized by simple and intuitive requirements.

1   Introduction

Petri net schemata are an expressive formalism to represent algorithms in a
highly abstract manner. Figure 1 shows an example of a Petri net schema: The
edges and the transitions of the underlying Petri net are inscribed by *terms*
and by *Boolean expressions*, respectively. Terms and Boolean expressions are
constructed from *function symbols* (reachable, true, triple, first, second, third)
and *variable symbols* (p, x, y, t). Furthermore, places are usually inscribed by
multiset symbols (P, F, Q).

We will refer to the Petri net schema in Fig. 1 by PHIL. An *interpretation*
A of PHIL provides a concrete function f_A for each function symbol f and a
concrete multiset M_A for each multiset symbol M. This fixes a high-level Petri
net. Markings and transition occurrences of this net are defined as usual, yielding
a transition system, PHIL_A.

As an example, let A be an interpretation where triple_A is the function of
arity 3, creating a triple from its three arguments. The functions first_A, second_A
and third_A, applied to a triple (u_1, u_2, u_3), return u_1, u_2, and u_3, respectively. As
usual, true is interpreted as the truth value *true*. The multiset symbols P, T and
Q are interpreted by $P_A = [a, b, c]$, $F_A = [f, f]$, and $Q_A = [\,]$, with $[\,]$ denoting
the empty multiset.

Intuitively, this interpretation assumes three philosophers a, b, c, and two
indistinguishable forks. The predicate symbol reachable is interpreted by

$$\text{reachable}_A(a, f) = \text{reachable}_A(b, f) = \text{reachable}_A(c, f) = true.$$

S. Donatelli and P.S. Thiagarajan (Eds.): ICATPN 2006, LNCS 4024, pp. 201–220, 2006.

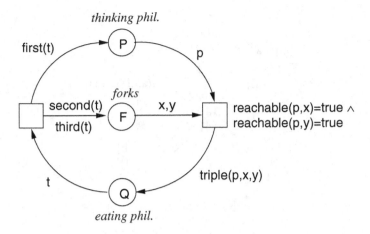

Fig. 1. The Petri net schema PHIL of dining philosophers

Intuitively formulated, each philosopher reaches each fork.

Figure 2 represents the transition system PHIL$_A$ generated by interpretation A. Each marking is represented as a three entry column, representing (from up to down) the token load of *thinking phil.*, *forks*, and *eating phil.*, respectively. The initial marking is indicated by an incoming dashed arrow.

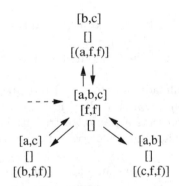

Fig. 2. Transition system PHIL$_A$

A different interpretation of the multiset symbols P, F, Q yields a different transition system. For example, let A' be an interpretation of PHIL similar to A, except for F interpreted by $[f, f, f, f]$. Therefore, four forks are available in A'. Figure 3 shows the corresponding transition system.

An entirely different interpretation, B, assumes four different forks 1, 2, 3, 4. The truth value of reachable(*phil, fork*) depends on both arguments according to the table

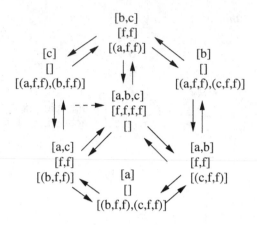

Fig. 3. Transition system PHIL$_{A'}$

reachable	a	b	c
1	*true*	*false*	*false*
2	*true*	*true*	*false*
3	*false*	*true*	*true*
4	*false*	*false*	*true.*

Hence, philosopher a may reach forks 1 and 2, b may reach 2 and 3, and c may reach 3 and 4.

With $P_B = [a, b, c]$, $F_B = [1, 2, 3, 4]$ and $Q_B = [\,]$, B creates a totally different transition system PHIL$_B$ given in Fig. 4. For the sake of simplicity, states that differ only in the order of the forks in the triples on place *eating phil.* are identified.

These small examples already hint to the rich expressive power of Petri net schemata: For each interpretation A, a Petri net schema N creates a transition system N_A. Hence, N represents a *family* of transition systems. This leads to the following question: Which families of transition systems can be represented by Petri net schemata?

Nielsen, Rozenberg, and Thiagarajan address this problem in [3] for the class of *elementary Petri nets*, i.e. nets where each place either is empty or holds a single, unspecified token. The semantics of an elementary Petri net is described as a transition system. They characterize the class of *elementary transition systems* and show that this class exactly comprises the transition systems as generated by elementary Petri nets. In [6] this idea has been extended to a basic class of Petri net schemata without variables, where the semantics of a Petri net schema is described as a transition system. Subsequently, the class of *algorithmic transition system* is characterized, and is proven to comprise all transition systems as generated by basic Petri net schemata.

In this paper we expand this work for a general class of Petri net schemata with variables, which is closely related to the class of predicate/transition nets [1]. In contrast to [3] and [6], we allow places to carry multisets of tokens, and

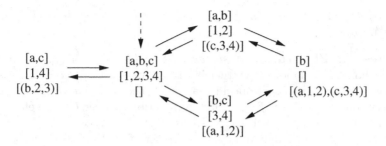

Fig. 4. Transition system PHIL$_B$

describe the semantics of a Petri net schema as a *family* of transition systems. Subsequently, we characterize a class of families of transition systems, and prove this class to comprise all families as generated by Petri net schemata.

The rest of this paper is structured as follows: The next section introduces syntax and semantics of Petri net schemata formally. Section 3 presents a characterization of the families of transition systems as generated by Petri net schemata. Finally, in Sec. 4 the correctness of this characterization is proven.

2   Petri Net Schemata

We start with the formal background used in the sequel, and proceed with the syntax of Petri net schemata. Subsequently, the semantics of Petri net schemata is defined in terms of transition systems.

2.1   Some Basic Notions from Algebra

In this section we recall some basic notions, including multisets, the well-known formalism of signatures and structures, and transition systems. In addition, we introduce the notations used in the rest of this paper.

Multisets. A *multiset* is a collection of elements where an element may occur more than once. Formally, for a set V, a *multiset over V* is a function $M : V \to \mathbb{N}$ where $M(v) > 0$ holds only for finitely many $v \in V$. The set of all multisets over V is denoted by bags(V). $v \in V$ *occurs* in M (v is an *element* of M) if $M(v) > 0$. $E(M)$ denotes the set of all elements occurring in M. The size of M is

$$|M| =_{\text{def}} \sum_{x \in V} M(x).$$

A multiset is frequently represented as a list in square brackets, e.g. $M = [a, b, b, c, c, c]$. As a special case, $[\,]$ denotes the empty multiset.

The *addition* $M_1 + M_2$ of two multisets M_1 and M_2 is defined element-wise by $(M_1 + M_2)(v) =_{\text{def}} M_1(v) + M_2(v)$. Analogously, the *scaling* $n \cdot M$ for $n \in \mathbb{N}$ and a multiset M is defined element-wise by $(n \cdot M)(v) =_{\text{def}} n \cdot M(v)$. The partial order $\leq$ over multisets is defined by $M_1 \leq M_2$ iff $M_1(v) \leq M_2(v)$ for all $v \in V$.

In case $M_1 \leq M_2$, the *difference* $M_2 - M_1$ is defined by $(M_2 - M_1)(v) =_{\text{def}}$ $M_2(v) - M_1(v)$.

Signatures and Structures. A *signature* $\Sigma = (f_1, ..., f_k, n_1, ..., n_k)$ consists of a set of function symbols f_i and their respective arities n_i $(i = 1, ..., k)$. A Σ-*structure* $A = (U, g_1, ..., g_k)$ specifies a set U, the *universe* of A, and interprets every function symbol f_i by a n_i-ary function g_i over U. To refer to the components of a Σ-structure A, the universe of A is denoted by $U(A)$, and the interpretation of f_i in A is denoted by f_{iA}. The set of all Σ-structures is denoted by $\text{Str}(\Sigma)$.

For two Σ-structures A and B, an *isomorphism* from A to B is a bijective function $\phi : U(A) \to U(B)$ where

$$\phi(f_A(u_1, \ldots, u_n)) = f_B(\phi(u_1), \ldots, \phi(u_n))$$

for all n-ary function symbols f from Σ and $u_1, \ldots, u_n \in U(A)$. If there is an isomorphism from A to B, A and B are *isomorphic*.

Terms. Terms are constructed from function symbols and variable symbols. Given a signature Σ and a set X of variable symbols, the set of Σ-X-*terms* is constructed inductively: Every variable symbol from X and every 0-ary function symbol from Σ is a Σ-X-term. If $t_1, \ldots, t_n$ are Σ-X-terms and f is an n-ary function symbol in Σ, $f(t_1, \ldots, t_n)$ is a Σ-X-term. The set of all Σ-X-terms is denoted by $T_\Sigma(X)$. The set of variable symbols occurring in a term t is denoted by $\text{var}(t)$.

Assignments, Evaluation, and Σ-X-modes. For a set of variable symbols X and a set of values V, a function $\alpha : X \to V$ is an *assignment of X over V*. Given a Σ-structure A and an assignment of X over $U(A)$, every Σ-X-term t can be *evaluated* to a unique value $t_{A,\alpha}$: If t is a variable symbol from X, $t_{A,\alpha} =_{\text{def}} \alpha(t)$. If t is a 0-ary function symbol from Σ, $t_{A,\alpha} =_{\text{def}} t_A$. In case $t = f(t_1, \ldots, t_n)$, $t_{A,\alpha} =_{\text{def}} f_A(t_{1A,\alpha}, \ldots, t_{nA,\alpha})$. The pair $m = (A, \alpha)$ is a Σ-X-*mode*. Hence, the evaluation of t is written t_m. By $U(m) =_{\text{def}} U(A)$ we denote the *universe of m*.

Boolean Expressions. For two Σ-X-terms t_1 and t_2, $t_1 = t_2$ is a Σ-X-*equation*. $t_1 = t_2$ is *satisfied* by a Σ-X-mode m iff $t_{1m} = t_{2m}$. Σ-X-equations can be combined to *Boolean Σ-X-expressions* by the usual Boolean operators: Every Σ-X-equation is a Boolean Σ-X-expression. If e_1 and e_2 are Boolean Σ-X-expressions, $e_1 \wedge e_2$ and $\neg e_1$ are Boolean Σ-X-expressions. The set of all Σ-X-expressions is denoted by $E_\Sigma(X)$. $e_1 \wedge e_2$ is *satisfied* in a Σ-X-mode m iff e_1 *and* e_2 are satisfied in m, and $\neg e_1$ is *satisfied* in m iff e_1 is *not* satisfied in m. For an arbitrary Boolean Σ-X-expression e, $m \models e$ indicates that e is satisfied in m. The set of variable symbols occurring in an expression e is denoted by $\text{var}(e)$.

Multiterms. A multiset of terms is a *multiterm*. Hence, for a signature Σ and a set of variable symbols X, the set of all Σ-X-multiterms is

$$MT_\Sigma(X) =_{\text{def}} \text{bags}(T_\Sigma(X)).$$

In analogy to terms, a multiterm u is evaluated by a Σ-X-mode m by replacing each t in u by its evaluation t_m:

$$u_m \quad =_{\text{def}} \quad \sum_{t \in T_\Sigma(X)} u(t) \cdot [t_m].$$

Hence, u_m is a multiset over $U(m)$. The set of all variable symbols occurring in the terms in u is denoted by $\text{var}(u)$.

Transition Systems. Let Ω be a set and let $\to \,\subseteq\, \Omega \times \Omega$. Then $\mathfrak{T} = (\Omega, \to)$ is a *transition system*. Each $s \in \Omega$ is a *state* of $\mathfrak{T}$ and $\to$ is the *step relation* of $\mathfrak{T}$. Usually, transition systems are equipped with distinguished *initial states*. We assume each state as initial, i.e. skip this notion entirely. Consequently, a *run* of $\mathfrak{T}$ may start in any state: A run $\rho = (s_0, s_1, s_2, \dots)$ is a (finite or infinite) sequence of states from Ω such that $s_{i-1} \to s_i$ for all indices i.

2.2   Syntax of Petri Net Schemata

A Petri net schema is an inscribed *Petri net*. As usual, a Petri net is a triple $(\mathbf{P}, \mathbf{T}, \mathbf{F})$, where $\mathbf{P}$ is the set of *places*, $\mathbf{T}$ the set of *transitions*, and $\mathbf{F} \subseteq (\mathbf{P} \times \mathbf{T}) \cup (\mathbf{T} \times \mathbf{P})$ the set of *edges* of N. For $x \in \mathbf{P} \cup \mathbf{T}$, ${}^\bullet x =_{\text{def}} \{y | (y, x) \in \mathbf{F}\}$ denotes the *pre-set* of x, and $x^\bullet =_{\text{def}} \{y | (x, y) \in \mathbf{F}\}$ denotes the *post-set* of x.

A *Petri net schema* is a finite Petri net where each edge is inscribed by a multiterm, and each transition is inscribed by a Boolean expression:

Definition 1 (Petri net schema). *Let Σ be a signature and let X be a set of variable symbols. Let $(\mathbf{P}, \mathbf{T}, \mathbf{F})$ be a finite Petri net and let $\psi : \mathbf{T} \to E_\Sigma(X)$ and $\omega : \mathbf{F} \to MT_\Sigma(X)$ be functions. Then $N = (\mathbf{P}, \mathbf{T}, \mathbf{F}, \Sigma, X, \psi, \omega)$ is a Petri net schema.*

For technical convenience, we write $\omega(x, y)$ instead of $\omega((x, y))$ for $(x, y) \in \mathbf{F}$. Furthermore, we extend ω to $(\mathbf{P} \times \mathbf{T}) \cup (\mathbf{T} \times \mathbf{P})$ by $\omega(x, y) =_{\text{def}} [\,]$ in case $(x, y) \notin \mathbf{F}$.

To give an example, we reconsider the dining philosophers from the introduction. With $\Sigma = (\mathsf{reachable}, \mathsf{true}, \mathsf{triple}, \mathsf{first}, \mathsf{second}, \mathsf{third}, 2, 0, 3, 1, 1, 1)$ and $X = \{\mathsf{p}, \mathsf{x}, \mathsf{y}, \mathsf{t}\}$, Fig. 1 shows a Petri net schema where each place additionally is inscribed by a symbol. For the sake of simplicity, we refrain from place inscriptions in the above definition, as in Fig. 5.

2.3   Semantics of Petri Net Schemata

A Petri net schema N over a signature Σ yields to each Σ-structure A a transition system N_A. The semantics of N can then be conceived as the family of transition systems $\{N_A\}_{A \in \text{Str}(\Sigma)}$.

To define the transition system N_A formally, we have to specify its states and its step relation. A state of N_A is represented by a *marking* of the places in N:

Definition 2 (Marking). *Let $\mathbf{P}$ and V be sets. Then a function $\mu : \mathbf{P} \to \text{bags}(V)$ is a marking of $\mathbf{P}$ over V.*

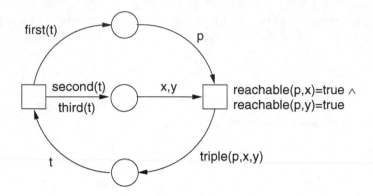

Fig. 5. The Petri net schema PHILS without place inscriptions

We abbreviate "marking of the places of N" to "marking of N". A *state* of N_A is a marking of N over $U(A)$:

Definition 3 (State space Ω_A^N). *Let $N = (\mathbf{P}, \mathbf{T}, \mathbf{F}, \Sigma, X, \psi, \omega)$ be a Petri net schema and let A be a Σ-structure. Then Ω_A^N denotes the set of all markings of N over $U(A)$.*

Thus, Ω_A^N constitutes the state space of N_A.

The multiset operations $+, -, \cdot$ and the multiset relation $\leq$ can be extended to markings by applying them component-wise for each $p \in \mathbf{P}$. For example, $\mu_1 + \mu_2$ is the marking with

$$(\mu_1 + \mu_2)(p) \quad =_{\text{def}} \quad \mu_1(p) + \mu_2(p)$$

for all $p \in \mathbf{P}$. For a marking μ, the *elements of μ* are the elements occurring in the multisets of μ:

$$E(\mu) \quad =_{\text{def}} \quad \bigcup_{p \in \mathbf{P}} E(\mu(p)).$$

A marking of N can be updated by removing some elements from and adding some new elements to the places of N. The elements to be removed and to be added are specified by the transitions and edges of N and their inscriptions.

Definition 4 (t_m^-, t_m^+). *Let $N = (\mathbf{P}, \mathbf{T}, \mathbf{F}, \Sigma, X, \psi, \omega)$ be a Petri net schema, let $t \in T$, and let m be a Σ-X-mode. Then t_m^- and t_m^+ are markings of N defined for $p \in \mathbf{P}$ by*

$$t_m^-(p) \quad =_{\text{def}} \quad \omega(p, t)_m,$$
$$t_m^+(p) \quad =_{\text{def}} \quad \omega(t, p)_m.$$

Then a step of N_A is obtained by

- choosing an assignment α of X such that (A, α) satisfies $\psi(t)$ for some transition t,

- removing the elements $t_{A,\alpha}^-$ from the marking, and
- adding $t_{A,\alpha}^+$ to the marking.

Hence, the step relation is defined as follows:

Definition 5 (Step relation $\to_A^N$). *Let $N = (\mathbf{P}, \mathbf{T}, \mathbf{F}, \Sigma, X, \psi, \omega)$ be a Petri net schema and let A be a Σ-structure. Then $\to_A^N \subseteq \Omega_A^N \times \Omega_A^N$ is defined as follows: $\mu \to_A^N \mu'$ iff there is a transition $t \in \mathbf{T}$ and an assignment α of X over $U(A)$ such that*

1. *(A, α) satisfies $\psi(t)$,*
2. *$t_{A,\alpha}^- \leq \mu$,*
3. *$\mu' = (\mu - t_{A,\alpha}^-) + t_{A,\alpha}^+$.*

By Ω_A^N and $\to_A^N$, we defined both components of the transition system N_A, i.e.

$$N_A =_{\text{def}} (\Omega_A^N, \to_A^N).$$

According to this definition, for a fixed Σ-structure A, *every* marking over the universe of A is a state of N_A. Hence, we do not distinguish initial states, and accept every marking as an initial marking of N. The transition system N_A comprises all transition systems generated by specific initial states. For example, the transition systems of Fig. 2 and Fig. 3 are just components of the transition system PHILS$_A$, with PHILS as in Fig. 5.

3   The Expressive Power of Petri Net Schemata

In this section we firstly identify a class of Petri net schemata which we call *well-formed*. Secondly, we characterize the expressive power of the semantics of well-formed Petri net schemata by five requirements.

3.1   Well-Formed Petri Net Schemata

Informally, we call a Petri net schema *well-formed*, if every variable occurring at a transition t also occurs as a term at some edge of t. Hence, in case transition t performs a step, the value of each variable is bound to a token consumed or produced. This restriction is rather natural: In a Petri net schema describing a distributed algorithm, variables are intended to symbolize tokens residing on the places. Consequently, Petri net schemata describing distributed algorithms are always well-formed. [4] presents a large collection of distributed algorithms specified by Petri net schemata.

As an example, consider the two Petri net schemata in Fig. 6(a). In both schemata the variable x occurs in the inscriptions, but no edge is inscribed by x. Hence, in a step the value of x is not bound to a token produced or consumed. Figure 6(b) shows a possible "repair" of the schemata in Fig. 6(a): In both cases an additional place is supplied such that the value of x is always bound to a token consumed or produced at the new place.

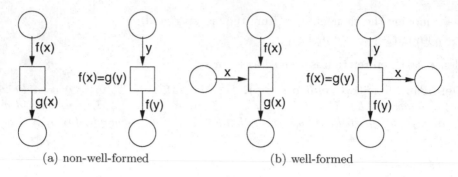

(a) non-well-formed   (b) well-formed

Fig. 6. Some examples of well-formed and non-well-formed Petri net schemata

To define well-formedness formally, we denote, for a transition t, the set of all variables occurring in the inscription of t and in the inscriptions of the edges at t by $\mathrm{var}(t)$:

Definition 6 (Variables at a transition). *Let* $N = (\mathbf{P}, \mathbf{T}, \mathbf{F}, \Sigma, X, \psi, \omega)$ *be a Petri net schema and let* $t \in \mathbf{T}$. *Then*

$$\mathrm{var}(t) \quad =_{\mathrm{def}} \quad \mathrm{var}(\psi(t)) \cup \bigcup_{p \in {}^{\bullet}t} \mathrm{var}(\omega(p, t)) \cup \bigcup_{p \in t^{\bullet}} \mathrm{var}(\omega(t, p))$$

denotes the set of all variables at t.

A Petri net schema is *well-formed* if, for every transition t and for every variable $x \in \mathrm{var}(t)$, at least one edge at t is inscribed by x:

Definition 7 (Well-formed Petri net schema). *Let* $N = (\mathbf{P}, \mathbf{T}, \mathbf{F}, \Sigma, X, \psi, \omega)$ *be a Petri net schema such that for every transition* $t \in \mathbf{T}$ *and every* $x \in \mathrm{var}(t)$ *holds: There is a* $p \in {}^{\bullet}t$ *with* $\omega(p, t)(x) > 0$ *or there is a* $p \in t^{\bullet}$ *with* $\omega(t, p)(x) > 0$. *Then* $N = (\mathbf{P}, \mathbf{T}, \mathbf{F}, \Sigma, X, \psi, \omega)$ *is a well-formed Petri net schema.*

For the rest of this paper, we restrict ourself to well-formed Petri net schemata, and assume well-formedness even if not explicitly stated.

3.2   The Expressive Power of Well-Formed Petri Net Schemata

According to Sec. 2, the semantics of a Petri net schema N is a family of transition systems $\{N_A\}_{A \in \mathrm{Str}(\Sigma)}$. As declared already in the introduction, our aim is to answer the following question:

Which families of transition systems can be described by Petri net schemata?

The rest of this paper answers this question. To this end, we fix a signature Σ and an arbitrary family of transition systems $\mathfrak{T} = \{\mathfrak{T}_A\}_{A \in \mathrm{Str}(\Sigma)}$. Then we can reformulate the above question more precisely:

Is there a Petri net schema N such that $\mathfrak{T}_A = N_A$ for each Σ-structure A?

We answer this question by formulating five requirements to $\mathfrak{T}$. These requirements are inspired by Gurevich's characterization of the expressive power of Abstract State Machines in [2], which is critically re-examined in [5]. At the end of this section we present a theorem stating that the above question is answered positively if and only if $\mathfrak{T}$ meets all five requirements.

The first requirement is merely technical, demanding the state spaces of $\mathfrak{T}$ be compatible with the state spaces of the interpretations of a Petri net schema. For each $A \in \mathrm{Str}(\Sigma)$, let $\Omega_A^{\mathfrak{T}}$ denote the state space of $\mathfrak{T}_A$. Then the first requirement is:

> *There is a finite set $\mathbf{P}$ such that for all $A \in \mathrm{Str}(\Sigma)$, $\Omega_A^{\mathfrak{T}}$ is the set of markings of $\mathbf{P}$ over $U(A)$.*   (R1)

The second requirement demands $\mathfrak{T}$ to respect isomorphism. To formalize this, we extend isomorphisms to multisets and markings: Let $A, B \in \mathrm{Str}(\Sigma)$, let ϕ be an isomorphism from A to B, and let M be a multiset over $U(A)$. Then

$$\phi(M) \quad =_{\mathrm{def}} \sum_{u \in U(A)} M(u) \cdot [\phi(u)].$$

Intuitively, every element u in M is replaced by its isomorphic element $\phi(u)$. ϕ extends to markings μ of $\mathbf{P}$ over $U(A)$: $\phi(\mu)$ is the marking of $\mathbf{P}$ over $U(B)$ with

$$(\phi(\mu))(p) \quad =_{\mathrm{def}} \quad \phi(\mu(p))$$

for all $p \in \mathbf{P}$. Hence, all elements in μ are replaced according to the isomorphism ϕ. For every Σ-structure A, let $\to_A^{\mathfrak{T}}$ denote the step relation of $\mathfrak{T}_A$. Then the second requirement is:

> *For two isomorphic Σ-structures A, B with an isomorphism ϕ from A to B holds $\mu \to_A^{\mathfrak{T}} \mu'$ iff $\phi(\mu) \to_B^{\mathfrak{T}} \phi(\mu')$.*   (R2)

The third requirement demands for each $A \in \mathrm{Str}(\Sigma)$ the state transition relation $\to_A^{\mathfrak{T}}$ to be *monotonous*: A step $\mu \to_A^{\mathfrak{T}} \mu'$ remains executable in case μ and μ' are extended by the same marking:

> *For all $A \in \mathrm{Str}(\Sigma)$ and all markings ν of $\mathbf{P}$ over $U(A)$ holds: If $\mu \to_A^{\mathfrak{T}} \mu'$, then $(\mu + \nu) \to_A^{\mathfrak{T}} (\mu' + \nu)$.*   (R3)

Hence, infinitely many steps of $\mathfrak{T}_A$ can be derived from a single step $\mu \to_A^{\mathfrak{T}} \mu'$ by extending μ and μ'. Nevertheless, there are some steps that cannot be derived in this way. We call such steps *minimal*:

Definition 8 (Minimal step). *Let $A \in \mathrm{Str}(\Sigma)$. A step $\mu \to_A^{\mathfrak{T}} \mu'$ is minimal if there is no nonempty marking ν with $\nu \leq \mu$ and $\nu \leq \mu'$ such that $(\mu - \nu) \to_A^{\mathfrak{T}} (\mu' - \nu)$.*

According to (R3), for every step $\mu \to_A^{\mathfrak{T}} \mu'$ there exist a minimal step $\nu \to_A^{\mathfrak{T}} \nu'$ and a marking ξ such that $\mu = \nu + \xi$ and $\mu' = \nu' + \xi$. The minimal step $\nu \to_A^{\mathfrak{T}} \nu'$ specifies the *actual change* of the step, and ξ specifies the *context* of the step. As all steps of $\mathfrak{T}_A$ can be derived from the minimal steps of $\mathfrak{T}_A$ by adding some context, the last two requirements will deal with minimal steps only. These requirements are the most decisive ones: Informally, they require the actual change of all steps to be bounded, and the set of evaluated terms to be bounded in all steps.

More precisely, the fourth requirement demands the size of all minimal steps to be bounded. To this end, we define the *size* of a marking μ as

$$|\mu| =_{\text{def}} \sum_{p \in \mathbf{P}} |\mu(p)|.$$

The fourth requirement reads:

> *There is a constant $k \in \mathbb{N}$ such that for each $A \in \text{Str}(\Sigma)$ and for each minimal step $\mu \to_A^{\mathfrak{T}} \mu'$ holds $|\mu| + |\mu'| \leq k$.* (R4)

The fifth requirement adopts the *bounded exploration* principle from [2]: There is a *finite* set T of terms such that for each Σ-structure A the step relation $\to_A^{\mathfrak{T}}$ is characterized by evaluations of the terms in T. More precisely, there exists a finite set X of variable symbols and a finite set of Σ-X-terms T such that for all $A \in \text{Str}(\Sigma)$ and all $\mu, \mu' \in \Omega_A^{\mathfrak{T}}$ holds: Whether or not $\mu \to_A^{\mathfrak{T}} \mu'$ is a minimal step, depends only on the evaluation of the terms in T by A and by assignments of X over the elements in μ and μ'.

To formalize this requirement, we introduce *indistinguishability* of Σ-structures wrt a set of terms T and a set of elements E. As an example, consider the signature $\Sigma = (\mathsf{R}, \mathsf{true}, 2, 0)$ and two Σ-structures ORD and DIV, both over the universe $\mathbb{N} \cup \{\mathit{true}, \mathit{false}\}$ such that

- $\mathsf{true}_{\text{ORD}} = \mathsf{true}_{\text{DIV}} = \mathit{true}$,
- For $i, j \in \mathbb{N}$ holds $\mathsf{R}_{\text{ORD}}(i, j) = \mathit{true}$ iff $i \leq j$,
- For $i, j \in \mathbb{N}$ holds $\mathsf{R}_{\text{DIV}}(i, j) = \mathit{true}$ iff $i|j$.

Hence, R_{ORD} is the usual ordering relation and R_{DIV} is the usual divisor relation of natural numbers. $T = \{\mathsf{R}(\mathsf{x}, \mathsf{y}), \mathsf{true}\}$ is a set of Σ-X-terms (where $X = \{\mathsf{x}, \mathsf{y}\}$) and $E = \{2, 4\}$ is a set of elements from the universe of ORD and DIV. Now evaluation of *every* term in T with *every* assignment of X over E yields:

$$\mathsf{true}_{\text{ORD}} = \mathit{true} = \mathsf{true}_{\text{DIV}}$$
$$\mathsf{R}_{\text{ORD}}(2, 2) = \mathit{true} = \mathsf{R}_{\text{DIV}}(2, 2)$$
$$\mathsf{R}_{\text{ORD}}(2, 4) = \mathit{true} = \mathsf{R}_{\text{DIV}}(2, 4)$$
$$\mathsf{R}_{\text{ORD}}(4, 2) = \mathit{false} = \mathsf{R}_{\text{DIV}}(4, 2)$$
$$\mathsf{R}_{\text{ORD}}(4, 4) = \mathit{true} = \mathsf{R}_{\text{DIV}}(4, 4).$$

Hence, though ORD and DIV are completely different structures, they cannot be distinguished by evaluating the terms in T with variable assignments over E.

The sets T and E only provide a local view to ORD and DIV, and both ORD and DIV are equal on those views. Formally, we define indistinguishability as follows:

Definition 9 (Indistinguishable structures). *Let X be a set of variable symbols and let $T \subseteq T_\Sigma(X)$. Let $A, B \in \text{Str}(\Sigma)$ and let $E \subseteq U(A) \cap U(B)$ such that for all $t \in T$ and for all assignments α of X over E holds $t_{A,\alpha} = t_{B,\alpha}$. Then A and B are* indistinguishable *by T and E.*

Finally, we formulate the fifth and last requirement:

> *There is a finite set X of variable symbols and a finite set $T \subseteq T_\Sigma(X)$ such that for all $A, B \in \text{Str}(\Sigma)$ holds: If $\mu \to_A^{\mathfrak{T}} \mu'$ is a minimal step, and if A and B are indistinguishable by T and $E(\mu) \cup E(\mu')$, then $\mu \to_B^{\mathfrak{T}} \mu'$.*   (R5)

A set T of terms fulfilling the properties in (R5) is called *characteristic* for $\mathfrak{T}$.

The following lemma states that the requirements (R1), ..., (R5) are fulfilled for the semantics of Petri net schemata:

Lemma 1. *Let N be a Petri net schema over Σ. Then $\{N_A\}_{A \in \text{Str}(\Sigma)}$ fulfills (R1), ..., (R5).*

The proof of this lemma is rather simple. For example, a bound k for (R4) is the number of inscriptions at the edges of N, and a characteristic set T of terms for (R5) is the set of all terms occurring in the inscriptions at the edges and transitions of N.

Surprisingly, the reverse of Lemma 1 holds, too. A family of transition systems fulfilling (R1), ..., (R5) can always be represented by a Petri net schema:

Theorem 1. *Let $\mathfrak{T} = \{\mathfrak{T}_A\}_{A \in \text{Str}(\Sigma)}$ be a family of transition systems fulfilling (R1), ..., (R5). Then there is a Petri net schema N such that $N_A = \mathfrak{T}_A$ for all Σ-structures A.*

The proof of this theorem is considerably harder than the proof of Lemma 1, and will be given in the next section.

4   Proof of the Theorem

In this section we prove Theorem 1. We start by introducing the basic notions *mode isomorphism* and *T-equivalence*. Next, for an arbitrary Σ-structure A, we show how a Petri net schema can be derived from a step of $\mathfrak{T}_A$. After this, we introduce *isomorphisms* between Petri net schemata and *composition* of Petri net schemata. Finally, we give the proof of Theorem 1.

4.1   Some Basic Tools

We first extend the notion of isomorphism from structures to modes: Two modes (A, α) and (B, β) are *isomorphic*, if A and B are isomorphic and the assignments α and β respect the isomorphism.

Definition 10 (Mode isomorphism). *Let Σ be a signature and let X be a set of variable symbols. Let $a = (A, \alpha)$ and $b = (B, \beta)$ be Σ-X-modes such that ϕ is an isomorphism from A to B and $\beta(x) = \phi(\alpha(x))$ for all $x \in X$. Then ϕ is a mode isomorphism from a to b, and a and b are isomorphic.*

It is easy to prove that for two isomorphic Σ-X-modes a, b holds: If t, t' are Σ-X-terms then $t_a = t'_a$ iff $t_b = t'_b$. Hence, this is a necessary requirement for a and b to be isomorphic.

In case this requirement does not hold for all Σ-X-terms t, t' but only for all t, t' from a subset T of Σ-X-terms, we call a and b *T-equivalent*:

Definition 11 (T-equivalence). *Let Σ be a signature, let X be a set of variable symbols and let $T \subseteq T_\Sigma(X)$. Let a, b be Σ-X-modes such that $t_a = t'_a$ iff $t_b = t'_b$ for all $t, t' \in T$. Then a and b are T-equivalent.*

The following lemma provides a simple characterization of T-equivalence:

Lemma 2. *Let Σ be a signature, let X be a set of variable symbols and let $T \subseteq T_\Sigma(X)$. Then two Σ-X-modes a and b are T-equivalent iff there exists a Σ-X-mode c such that a and c are isomorphic and $t_c = t_b$ for all $t \in T$.*

Proof. ($\Rightarrow$) Create c from a by replacing for all $t \in T$ the element t_a by t_b and by replacing every other element from $U(a)$ by a new element not contained in $U(a)$. As a and b are T-equivalent, this construction is well-defined. By construction, a and c are isomorphic, and $t_c = t_b$ for all $t \in T$.

($\Leftarrow$) For all $t, t' \in T$ holds

$$t_a = t'_a \Leftrightarrow t_c = t'_c \quad \text{(as a and c are isomorphic)}$$
$$ \Leftrightarrow t_b = t'_b \quad \text{(as $t_c = t_b$ for all $t \in T$).}$$

Hence, a and b are T-equivalent.   □

We finish this section by presenting two simple properties of a Petri net schema N: A transition s evaluated by isomorphic modes a and b yields isomorphic markings s_a^-, s_b^- and s_a^+, s_b^+. Furthermore, if all terms in the inscriptions of N are evaluated equally in a and b, the markings s_a^-, s_b^- and s_a^+, s_b^+ are equal, respectively.

Lemma 3. *Let $N = (\mathbf{P}, \mathbf{T}, \mathbf{F}, \Sigma, X, \psi, \omega)$ be a Petri net schema, let $s \in \mathbf{T}$, and let a, b be two Σ-X-modes.*

(i) *If there is an mode isomorphism ϕ from a to b, then $\phi(s_a^-) = s_b^-$ and $\phi(s_a^+) = s_b^+$.*

(ii) *Let T be the set of all terms occurring in ψ and ω. If $t_a = t_b$ for all $t \in T$, then $s_a^- = s_b^-$ and $s_a^+ = s_b^+$.*

Proof. Follows from Definition 4.   □

4.2   Construction of Component Schemata

We prove Theorem 1 in a constructive manner, i.e. we construct from $\mathfrak{T}$ a Petri net schema N. The foundations of this construction are laid in this section: For a Σ-structure A and a minimal step $\mu \to_A^{\mathfrak{T}} \mu'$, we construct a Petri net schema C such that

1. $\mu \to_A^C \mu'$,
2. for every Σ-structure B, $\to_B^C \subseteq \to_B^{\mathfrak{T}}$.

Hence, C is able to execute some behaviour of $\mathfrak{T}$ (1.) and does not introduce any behaviour impossible in $\mathfrak{T}$ (2.). Later we compose the desired schema N from schemata constructed in this way.

We introduce the construction of component schemata in detail now. For the rest of this paper, let $T \subseteq T_\Sigma(X)$ be a characteristic set of terms for $\mathfrak{T}$ (see (R5)), let k be the size bound of minimal steps (see (R4)), and let Y be a set of variable symbols such that $|Y| = k$. Furthermore, let A be an arbitrary Σ-structure and let $\mu \to_A^{\mathfrak{T}} \mu'$ be an arbitrary minimal step.

First, choose a subset $\hat{Y} \subseteq Y$ such that $|\hat{Y}| = |E(\mu) \cup E(\mu')|$. This is always possible, as $|E(\mu) \cup E(\mu')| \leq k = |Y|$ according to (R4). Let $\hat{\alpha} : \hat{Y} \to E(\mu) \cup E(\mu')$ be bijective and set $\hat{a} := (A, \hat{\alpha})$ (hence, $\hat{a}$ is a Σ-$\hat{Y}$-mode).

We will now transform the set of Σ-X-terms T to a set of Σ-$\hat{Y}$-terms $\hat{T}$ by replacing the variable symbols X by the variable symbols $\hat{Y}$. For this, we need the notion of *variable substitution*:

Definition 12 (Variable substitution). *Let Σ be a signature and let X, Y be sets of variable symbols. Then a function $\sigma : X \to Y$ is a variable substitution. For $t \in T_\Sigma(X)$, the application of σ to t is a Σ-Y-term, defined inductively as*

$$\sigma(t) = \begin{cases} \sigma(t) & , if \ t \in X \\ f(\sigma(t_1), \ldots, \sigma(t_n)) & , if \ t \notin X \ and \ t = f(t_1, \ldots, t_n). \end{cases}$$

Lemma 4. *Let Σ be a signature and let X, Y be sets of variable symbols and let $\sigma : X \to Y$. Let t be a Σ-X-term and let (A, α) be a Σ-Y-mode. Then*

$$\sigma(t)_{A,\alpha} = t_{A,\alpha \circ \sigma}.$$

Proof. Follows from the definition of evaluation of terms and from Def. 12.   □

We now introduce a construction whose purpose will become clear in a succeeding lemma. We apply *all* variable substitution from X to $\hat{Y}$ to *all* terms in T, add the variable symbols $\hat{Y}$, and denote the result by $\hat{T}$:

$$\hat{T} =_{\text{def}} \{\sigma(t) | t \in T \text{ and } \sigma : X \to \hat{Y}\} \cup \hat{Y}.$$

Obviously,

$$\begin{aligned} |\hat{T}| &\leq |T| \cdot |\hat{Y}|^{|X|} + |\hat{Y}| \\ &\leq |T| \cdot |Y|^{|X|} + |Y|, \end{aligned} \tag{1}$$

i.e. the size of $\hat{T}$ is bounded.

As an example, consider $\Sigma = (\mathsf{R}, \text{true}, 2, 0)$ and $T = \{\mathsf{R}(\mathsf{x},\mathsf{y}), \text{true}\}$. With $\hat{Y} = \{\mathsf{a}, \mathsf{b}, \mathsf{c}\}$ this construction yields:

$$\hat{T} = \{\text{true}, \mathsf{R}(\mathsf{a},\mathsf{a}), \mathsf{R}(\mathsf{a},\mathsf{b}), \mathsf{R}(\mathsf{a},\mathsf{c}),$$
$$\mathsf{R}(\mathsf{b},\mathsf{a}), \mathsf{R}(\mathsf{b},\mathsf{b}), \mathsf{R}(\mathsf{b},\mathsf{c}),$$
$$\mathsf{R}(\mathsf{c},\mathsf{a}), \mathsf{R}(\mathsf{c},\mathsf{b}), \mathsf{R}(\mathsf{c},\mathsf{c}), \mathsf{a}, \mathsf{b}, \mathsf{c}\}.$$

The following lemma reveals the decisive aspect of this construction (remember that we fixed a Σ-$\hat{Y}$-mode $\hat{a} = (A, \hat{\alpha})$):

Lemma 5. *Let $b = (B, \beta)$ be a Σ-$\hat{Y}$-mode such that $t_{\hat{a}} = t_b$ for all $t \in \hat{T}$. Then A and B are indistinguishable by T and $E(\mu) \cup E(\mu')$.*

Hence, though indistinguishability of A and B is decided by evaluating the terms in T by *multiple* assignments (see Def. 9), we can verify indistinguishability of A and B by evaluating the terms in $\hat{T}$ by the *single* assignments $\hat{\alpha}$ and β, respectively.

Proof. As $\hat{Y} \subseteq \hat{T}$, $v_{\hat{a}} = v_b$ for all $v \in \hat{Y}$. Hence, $\hat{\alpha} = \beta$. Let α be an arbitrary assignment of X over $E(\mu) \cup E(\mu')$. As $\hat{\alpha}$ is a bijection from $\hat{Y}$ to $E(\mu) \cup E(\mu')$, there exists a variable substitution σ from X to $\hat{Y}$ such that

$$\alpha = \hat{\alpha} \circ \sigma. \tag{2}$$

Therefore, for all terms $t \in T$ holds:

$$\begin{aligned}
t_{A,\alpha} &= t_{A,\hat{\alpha}\circ\sigma} &&\text{(by (2))}\\
&= \sigma(t)_{A,\hat{\alpha}} &&\text{(by Lemma 4)}\\
&= \sigma(t)_{B,\hat{\alpha}} &&\text{(as } \sigma(t) \in \hat{T} \text{ and } \hat{\alpha} = \beta)\\
&= t_{B,\hat{\alpha}\circ\sigma} &&\text{(by Lemma 4)}\\
&= t_{B,\alpha} &&\text{(by (2))}.
\end{aligned}$$

As α was chosen arbitrarily, A and B are indistinguishable by T and $E(\mu)\cup E(\mu')$. $\square$

From the terms in $\hat{T}$ we construct an expression e:

$$e =_{\text{def}} \bigwedge_{t,t'\in\hat{T}} \begin{cases} t = t' & \text{, if } t_{\hat{a}} = t'_{\hat{a}}\\ \neg(t = t') & \text{, otherwise.} \end{cases} \tag{3}$$

According to (1), the length of e (denoted by $|e|$) is bounded: For a constant c depending on the length of the longest term in T holds

$$|e| \leq c \cdot |\hat{T}|^2 \leq c \cdot (|T| \cdot |Y|^{|X|} + |Y|)^2. \tag{4}$$

Finally, we construct *component schemata*: Let $C = (\mathbf{P}, \mathbf{T}, \mathbf{F}, \Sigma, Y, \psi, \omega)$ be a Petri net schema with

- **T** contains only a single transition s, i.e. $T = \{s\}$,
- $\mathbf{F} = \{(p,s)|\mu(p) \neq \emptyset\} \cup \{(s,p)|\mu'(p) \neq \emptyset\}$,
- $\psi(s) = e$,
- for all $p \in \mathbf{P}$,

$$\omega(p,s) = \sum_{v \in \hat{Y}} \mu(p)(\hat{\alpha}(v)) \cdot [v],$$

$$\omega(s,p) = \sum_{v \in \hat{Y}} \mu'(p)(\hat{\alpha}(v)) \cdot [v]. \qquad (5)$$

Then C is a *component schema to step* $\mu \to_A^{\mathfrak{T}} \mu'$. Notice that in the construction only s may be chosen freely. Therefore, different component schemata to step $\mu \to_A^{\mathfrak{T}} \mu'$ differ in the transition element s only. The following lemma verifies the properties demanded for C at the beginning of this section:

Lemma 6. *Let C be component schema to step $\mu \to_A^{\mathfrak{T}} \mu'$. Then*

(i) $\mu \to_A^C \mu'$,
(ii) *for every Σ-structure B, $\to_B^C \subseteq \to_B^{\mathfrak{T}}$.*

Proof. (i) For every $p \in \mathbf{P}$,

$$s_{\hat{a}}^-(p) = \omega(p,s)_{\hat{a}} \qquad \text{(Def. 4)}$$

$$= \left(\sum_{v \in \hat{Y}} \mu(p)(\hat{\alpha}(v)) \cdot [v] \right)_{\hat{a}} \qquad \text{(by (5))}$$

$$= \sum_{v \in \hat{Y}} \mu(p)(\hat{\alpha}(v)) \cdot [v_{\hat{a}}]$$

$$= \sum_{v \in \hat{Y}} \mu(p)(\hat{\alpha}(v)) \cdot [\hat{\alpha}(v)]$$

$$= \sum_{u \in E(\mu) \cup E(\mu')} \mu(p)(u) \cdot [u] = \mu(p) \quad \text{(as $\hat{\alpha}$ is bijective).}$$

Analogously, for every $p \in \mathbf{P}$,

$$s_{\hat{a}}^+(p) = \omega(s,p)_{\hat{a}} = \mu'(p).$$

Hence, $s_{\hat{a}}^- = \mu$ and $s_{\hat{a}}^+ = \mu'$. By construction of e in (3), $\psi(s)$ is satisfied in $\hat{a}$. Therefore, according to Def. 5, $\mu \to_A^C \mu'$.

(ii) We have to show that $\nu \to_B^C \nu'$ implies $\nu \to_B^{\mathfrak{T}} \nu'$. So, let $\nu \to_B^C \nu'$. According to Def. 5, there is an assignment β of $\hat{Y}$ over $U(B)$ such that e is satisfied in mode $b := (B, \beta)$ and $\nu' = (\nu - s_b^-) + s_b^+$.

By construction of e in (3), and as e is satisfied in b and $\hat{a}$, for all $t, t' \in \hat{T}$ holds

$$t_b = t'_b \quad \Leftrightarrow \quad t_{\hat{a}} = t'_{\hat{a}}.$$

Hence, $\hat{a}$ and b are $\hat{T}$-equivalent. According to Lemma 2, there exists a mode $d = (D, \delta)$ such that

$$t_{\hat{a}} = t_d \text{ for all } t \in \hat{T}, \tag{6}$$

and d and b are isomorphic with an isomorphism ϕ.

According to the proof of (i), $s_{\hat{a}}^- \to_A^{\mathfrak{T}} s_{\hat{a}}^+$. According to (6) and Lemma 5, A and D are indistinguishable by T and $E(\mu) \cup E(\mu')$. Hence, (R5) implies $s_{\hat{a}}^- \to_D^{\mathfrak{T}} s_{\hat{a}}^+$. As ϕ is a mode isomorphism from d to b, ϕ is an isomorphism from D to B. Then (R2) implies

$$\phi(s_{\hat{a}}^-) \to_B^{\mathfrak{T}} \phi(s_{\hat{a}}^+). \tag{7}$$

As d and b are isomorphic with ϕ, by Lemma 3(i) holds $s_b^- = \phi(s_d^-)$ and $s_b^+ = \phi(s_d^+)$. As $\hat{T}$ contains all terms occurring in the inscriptions of schema C, (6) and Lemma 3(ii) imply $s_d^- = s_{\hat{a}}^-$ and $s_d^+ = s_{\hat{a}}^+$. Together this yields:

$$\begin{aligned}
s_b^- &= \phi(s_d^-) = \phi(s_{\hat{a}}^-), \\
s_b^+ &= \phi(s_d^+) = \phi(s_{\hat{a}}^+).
\end{aligned} \tag{8}$$

(7) and (8) together imply $s_b^- \to_B^{\mathfrak{T}} s_b^+$. Let $\xi := \nu - s_b^-$. According to (R3), $(s_b^- + \xi) \to_B^{\mathfrak{T}} (s_b^+ + \xi)$. Furthermore,

$$\begin{aligned}
s_b^- + \xi &= s_b^- + (\nu - s_b^-) = \nu, \\
s_b^+ + \xi &= s_b^+ + (\nu - s_b^-) = \nu'.
\end{aligned}$$

Hence, $\nu \to_B^{\mathfrak{T}} \nu'$. □

4.3 Schema Isomorphisms and Composition of Schemata

In this section we introduce two basic notions regarding schemata. First, a *schema isomorphism* identifies two schemata with equal inscriptions. Second, *schema composition* allows construction of a new schema from two given schemata.

Definition 13 (Schema isomorphism). *Let* $N_1 = (\mathbf{P}, \mathbf{T}_1, \mathbf{F}_1, \Sigma, X, \psi_1, \omega_1)$ *and* $N_2 = (\mathbf{P}, \mathbf{T}_2, \mathbf{F}_2, \Sigma, X, \psi_2, \omega_2)$ *be Petri net schemata. Let* $\phi : \mathbf{T}_1 \to \mathbf{T}_2$ *be a bijective function with*

- *$\psi_2(\phi(x)) = \psi_1(x)$ for all $x \in \mathbf{T}$,*
- *$(\phi(x), \phi(y)) \in \mathbf{F}_2$ iff $(x, y) \in \mathbf{F}_1$,*
- *$\omega_2(\phi(x), \phi(y)) = \omega_1(x, y)$ for all $(x, y) \in \mathbf{F}_1$.*

Then ϕ is a schema isomorphism *between N_1 and N_2. If there exists a schema isomorphism between N_1 and N_2, N_1 and N_2 are* isomorphic.

Classically, the places of two isomorphic schemata can be different, too. In our case, the place sets of considered schemata will always be equal. Therefore, our notion of schema isomorphism is sufficient and will simplify some technical details.

The following lemma shows that isomorphic schemata are equivalent in a strong sense:

Lemma 7. *Let N_1 and N_2 be isomorphic Petri net schemata. Then $\rightarrow_A^{N_1} = \rightarrow_A^{N_2}$ for all Σ-structures A.*

Proof. Follows from Def. 4 and 5.   □

Two schemata are *composed* by uniting their transitions, edges, and inscriptions. Again, we consider only schemata with equal place sets.

Definition 14 (Schema composition). *Let $N_1 = (\mathbf{P}, \mathbf{T}_1, \mathbf{F}_1, \Sigma, X, \psi_1, \omega_1)$ and $N_2 = (\mathbf{P}, \mathbf{T}_2, \mathbf{F}_2, \Sigma, X, \psi_2, \omega_2)$ be be Petri net schemata such that $\mathbf{T}_1$ and $\mathbf{T}_2$ are disjoint. Then the union of N_1 and N_2 is defined as*

$$N_1 \cup N_2 =_{\text{def}} (\mathbf{P}, \mathbf{T}_1 \cup \mathbf{T}_2, \mathbf{F}_1 \cup \mathbf{F}_2, \Sigma, X, \psi_1 \cup \psi_2, \omega_1 \cup \omega_2).$$

Lemma 8. *Let N_1 and N_2 be Petri net schemata as in Def. 14. Then*

$$\rightarrow_A^{N_1 \cup N_2} = \rightarrow_A^{N_1} \cup \rightarrow_A^{N_2}.$$

Proof. Follows from Def. 5.   □

4.4   Main Proof

Proof (of Theorem 1). For a Σ-structure A, let $\mathfrak{C}(A)$ be the set of *all* component schemata of *all* minimal steps of $\mathfrak{T}_A$. Let

$$\mathfrak{C} =_{\text{def}} \bigcup_{A \in \text{Str}(\Sigma)} \mathfrak{C}(A). \tag{9}$$

Hence, $\mathfrak{C}$ contains all component schemata that can be constructed from $\mathfrak{T}$. By construction of component schemata (see Sec. 4.2), for every $C \in \mathfrak{C}$ holds:

- $\mathbf{P}$ is the place set of C,
- C contains only one transition s,
- Σ is the signature of C, and Y is the set of variable symbols of C,
- the number of terms at each edge of C is bounded by k,
- the length of every term at each edge of C bounded by 1,
- the length of the Boolean expression at s is bounded by $c \cdot (|T| \cdot |Y|^{|X|} + |Y|)^2$ for a constant c.

Hence, the place set, the signature and the variable symbols are fixed, the number of transitions and inscriptions is bounded, and the size of the inscriptions is bounded. Therefore, up to isomorphism, $\mathfrak{C}$ contains only finitely many different

Petri net schemata, i.e. $\mathfrak{C}$ decomposes into finitely many isomorphism classes $\mathfrak{C}_1$, ..., $\mathfrak{C}_n$.

For each class $\mathfrak{C}_i$, choose a $C_i \in \mathfrak{C}_i$ such that C_i and C_j have different transitions for $i \neq j$. Set $N = C_1 \cup \cdots \cup C_n$. By Lemma 8, for all Σ-structures A holds

$$\rightarrow_A^N \; = \; \rightarrow_A^{C_1} \cup \cdots \cup \rightarrow_A^{C_n} \; . \tag{10}$$

According to Lemma 7, for all $C \in \mathfrak{C}_i$ $(i = 1, \ldots, n)$,

$$\rightarrow_A^{C_i} \; = \; \rightarrow_A^{C} \; . \tag{11}$$

(10) and (11) yield

$$\rightarrow_A^N \; = \; \bigcup_{C \in \mathfrak{C}} \rightarrow_A^C \; . \tag{12}$$

Finally, we prove that $\rightarrow_A^{\mathfrak{T}} = \rightarrow_A^N$ for all Σ-structures A:

($\subseteq$) Let $\mu \rightarrow_A^{\mathfrak{T}} \mu'$. In case this step is minimal, let C be a component schema of $\mu \rightarrow_A^{\mathfrak{T}} \mu'$. By Lemma 6(i), $\mu \rightarrow_A^C \mu'$. By (9), $C \in \mathfrak{C}$. By (12), $\mu \rightarrow_A^N \mu'$.
Now assume that $\mu \rightarrow_A^{\mathfrak{T}} \mu'$ is not minimal. Then there is a minimal step $\nu \rightarrow_A^{\mathfrak{T}} \nu'$ such that $\mu = \nu + \xi$ and $\mu' = \nu' + \xi$ for a marking ξ. According to the first case, $\nu \rightarrow_A^N \nu'$. According to (R3), $(\nu + \xi) \rightarrow_A^N (\nu' + \xi)$, therewith $\mu \rightarrow_A^N \mu'$.

($\supseteq$) Let $\mu \rightarrow_A^N \mu'$. By (12), there is a component schema $C \in \mathfrak{C}$ with $\mu \rightarrow_A^C \mu'$. By Lemma 6(ii), $\mu \rightarrow_A^{\mathfrak{T}} \mu'$. $\qquad\square$

5 Conclusion

We introduced a simple, yet expressive version of Petri net schemata, and described the semantics of each schema as a family of transition systems. We identified the subclass of well-formed Petri net schemata, and characterized their expressive power by five basic requirements. Two requirements of this characterization are decisive and were inspired by [2]: The amount of change in a step is bound (R4), and the steps can be characterized by a finite set of terms (R5). Therefore, we successfully transferred the principles of *bounded change* and *bounded exploration* from Gurevich's Abstract State Machines to Petri net schemata.

These principles seem to be fundamental for a wide range of system models. Our future research will concentrate on the extension of these principles to distributed system models, in particular to distributed semantics of Petri net schemata and Abstract State Machines.

References

1. Hartmann J. Genrich. Predicate/Transition Nets. In Wilfried Brauer, Wolfgang Reisig, and Grzegorz Rozenberg, editors, *Advances in Petri Nets*, volume 254 of *Lecture Notes in Computer Science*, pages 207–247. Springer, 1986.

2. Y. Gurevich. Sequential Abstract State Machines Capture Sequential Algorithms. *ACM Transactions on Computational Logic*, 1(1):77–111, Juli 2000.
3. M. Nielsen, G. Rozenberg, and P. S. Thiagarajan. Elementary transition systems. *Theor. Comput. Sci.*, 96(1):3–33, 1992.
4. W. Reisig. *Elements of Distributed Algorithms: Modeling and Analysis with Petri Nets*. Springer-Verlag, 1998.
5. Wolfgang Reisig. On Gurevich's Theorem on Sequential Algorithms. *Acta Informatica*, 39(5):273–305, 2003.
6. Wolfgang Reisig. On the Expressive Power of Petri Net Schemata. In Gianfranco Ciardo and Philippe Darondeau, editors, *ICATPN*, volume 3536 of *Lecture Notes in Computer Science*, pages 349–364. Springer, 2005.

A New Approach to the Evaluation of Non Markovian Stochastic Petri Nets

Serge Haddad[1], Lynda Mokdad[1], and Patrice Moreaux[2]

[1] LAMSADE, UMR CNRS 7024, Universit Paris Dauphine
Place du Marchal de Lattre de Tassigny
75775 PARIS Cedex 16, France
{haddad, mokdad}@lamsade.dauphine.fr
[2] LISTIC, ESIA-Universit de Savoie
Domaine universitaire d'Annecy le Vieux
BP 806, 74016 ANNECY Cedex, France
patrice.moreaux@univ-savoie.fr

Abstract. In this work, we address the problem of transient and steady-state analysis of a stochastic Petri net which includes non Markovian distributions with a finite support but without any additional constraint. Rather than computing an approximate distribution of the model (as done in previous methods), we develop an exact analysis of an approximate model. The design of this method leads to a uniform handling of the computation of the transient and steady state behaviour of the model. This method is an adaptation of a former one developed by the same authors for general stochastic processes (which was shown to be more robust than alternative techniques). Using Petri nets as the modelling formalism enables us to express the behaviour of the approximate process by tensorial expressions. Such a representation yields significant savings w.r.t. time and space complexity.

1 Introduction

Non Markovian process analysis. The transient and steady-state analysis of Markovian discrete event systems is now well-established with numerous tools at the disposal of the modellers. The main open issue is the reduction of the space complexity induced by this analysis. However in a realistic system, the distribution of the occurrence (or the duration) of some events cannot be described by an exponential law (e.g., the triggering of a time-out). Theoretically any "reasonable" distribution is approximated by a phase-type distribution enabling again a Markovian analysis [1]. Unfortunately the continuous time Markov chain (CTMC) associated with this approximation is so huge that it forbids its analysis (indeed even its construction). Such a phenomenon often occurs when the non exponential distribution has a finite support i.e., when the whole probability mass is included in a finite subset of $\mathbb{R}^+$ (non null Dirac, uniform, etc.); then a good phase-type approximation requires too much stages for a close approximation.

Hence the research has focused on alternative methods. In the case of a single realization of a non Markovian distribution at any time, successful methods have

S. Donatelli and P.S. Thiagarajan (Eds.): ICATPN 2006, LNCS 4024, pp. 221–240, 2006.

been proposed [2] both for the transient and steady state analysis, especially in the Stochastic Petri Net (SPN) modelling framework. Let us cite, for instance, the method of supplementary variables [3, 4] or the method of the subordinated Markov chains [5].

The general case (i.e., simultaneous multiple realizations of such distributions) is more intricate. The method of supplementary variables is still theoretically applicable but the required space and the computation time limit its use to very small examples. An alternative approach is described for non null Dirac distributions (i.e., "deterministic" durations) in [6]. The stochastic process, which is a General State space Markov Process (GSMP) is observed at periodic moments of time ($\{h\Delta \mid h \in \mathbb{N}\}$) and this new process is expressed by a system of integro-differential equations and solved numerically. The steady-state distributions of these processes are identical and, with another computation, one obtains the transient distribution of the original process from some transient distribution of the transformed process. This method has been implemented in the DSPNexpress tool [7] (but currently for only two concurrent "deterministic" events with same duration). By imposing conditions on the simultaneous occurrences of concurrent activities, other authors have also designed efficient algorithms [8, 9, 10, 11, 12] (see section 6 for more details).

Our previous contribution. In [13] we have proposed a different approach to deal with *multiple concurrent* events with finite support distributions. Moreover, in contrast with other works, our solution does not require specific synchronization between these events such as non overlapped or nested events. The main idea is to define an *approximate* model on which we perform an *exact* analysis. To this end, given a time interval (say Δ) we describe the behaviour of the stochastic model by two components: a CTMC and a discrete time Markov chain (DTMC). During an interval $(h\Delta, (h + 1)\Delta)$ the behaviour is driven by the CTMC which corresponds to Markovian events occurring in $(h\Delta, (h + 1)\Delta)$. Non Markovian activities are taken into account at $h\Delta$ instants only: the corresponding untimed probabilistic changes of state are processed according to a DTMC.

In the approximate process, the Markovian events are in fact exactly modelled since the set $\{h\Delta \mid h \in \mathbb{N}\}$ has a null measure. The approximation comes from non Markovian events: the distribution of a non Markovian event is approximated by a discrete random variable expressing the number of points $h\Delta$ that must be reached before its occurrence. Thus the residual number of points to be met is included in the state of the approximate process. At any moment $h\Delta$, the current residual numbers are decreased and the corresponding events occur when their residues are null.

The approximate process may be analysed either in transient mode or in steady-state. The transient analysis is done by successively computing the state distribution at the instants $\Delta, 2\Delta, \ldots, h\Delta, \ldots$ applying a transient analysis of the CTMC during an interval Δ (via the uniformization technique [14]) followed by a "step" of the DTMC. In order to smooth the effect of the discretization, we average the distribution upon the last interval with a variant of uniformization. Since the asymptotic behaviour of the process depends on the relative position w.r.t. the points $h\Delta$, the approximate process does not admit a steady-state distribution but it is asymptotically periodic. Hence, for the steady-state analysis, one computes the steady-state distribution at the instants $h\Delta$ and then starting from this distribution, one again averages the steady-state distribution upon an interval.

Our current contribution. Standard benchmarks (like $M/D/S/K$ queue) have shown that the implementation of our method is at least as efficient as tools like DSPNexpress-NG [15]. Furthermore it is robust, i.e., it still provides good approximations under extreme situations like loaded heavy queues unlike other tools. However, the space complexity is the main limitation of our method since a state includes its logical part and the delays of each enabled non Markovian event. In order to tackle this problem, we start with a high-level description given by a stochastic Petri net with general distributions. Then we define its semantics by specifying the associated stochastic process. Afterwards, we introduce an approximate stochastic process for the stochastic Petri net whose behaviour is structured as described above.

Thus we give tensorial expressions for the infinitesimal generator of the continuous part of our process and for the probability transition matrix of its discrete part. More precisely, these matrices are decomposed into blocks and each block is associated with a tensorial expression. Such a structure is a consequence of the representation of the state space. Indeed rather than (over-)approximating the state space by a cartesian product (as done in the original tensorial methods), we approximate it by a finite union of such products yielding a significant space reduction. A similar approach has been successfully experimented with in [16] for SPNs with phase-type distributions.

Here, we face an additional difficulty. Inside an interval, the delay of each enabled general transitions is non null whereas it can be null at the bounds of the interval. Thus the state space of the discrete part of the process is an extension of the one of the continuous part. Furthermore, the block decomposition must be refined. Indeed inside the interval, states are grouped w.r.t the general enabled transitions. At the bounds, states are grouped w.r.t both the general enabled transitions and the fireable ones (i.e., with null delay). Thus the alternation between the continuous part and the discrete part requires a tricky expansion of the probability vector. Hopefully, the complexity of this operation is negligible w.r.t. the other parts of the computation.

The experimentations are still in progress and their results will be provided in a forthcoming LISTIC technical report. Nevertheless we will detail here some key features of the implementation.

Organization of the paper. In section 2 we recall the principle of our approach for a general stochastic process. Then we present the approximate process of a stochastic Petri net in section 3. Afterwards we develop in section 4 the tensorial expressions of the matrices defining the behaviour of this approximate process. We give information about our implementation in section 5. In section 6, we show that all the previous methods handle particular cases of the systems that we are able to analyse. Finally we conclude and we give indications on future developments of our work.

2   The Approximate Method

2.1   Principle

Semantics of the approximate process As mentioned in the introduction, we define an *approximate* model (say $Y^{(\Delta)}$) of the initial model (say X), on which we perform an *exact* analysis. The main idea is to choose a time interval Δ and to restrict in $Y^{(\Delta)}$ the non Markovian events to only occur at times $t_h = h\Delta$. We then study in an exact way the evolution of the stochastic process $Y^{(\Delta)}$ in each interval (t_h, t_{h+1}) and during the state changes at time t_h . We stress that the starting times of the active non Markovian events are *in no way related*.

We obtain such a model $Y^{(\Delta)}$ from a general model with non Markovian finite support distributions as follows. The distribution of every non Markovian event is approximated by a discrete time distribution lying on points $h\Delta$. Let us note that although Δ seems to be the approximation parameter, the appropriate parameter is the maximum number of points used to express the distribution. Moreover this indicator is the key factor for the complexity of our analysis.

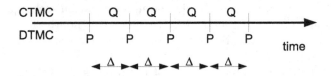

Fig. 1. Time decomposition

In the approximate process, the Markovian events occur during the intervals $(h\Delta, (h + 1)\Delta)$. Non Markovian events always occur in $\{h\Delta \mid h \in \mathbb{N}\}$. Let us describe how they are scheduled. When a non Markovian event is enabled in an interval $(h\Delta, (h + 1)\Delta)$ due to the occurrence of a Markovian event, then its approximate distribution is interpreted as the number of points $k\Delta$ that must be met before its occurrence. Here we can choose whether we count the next point (i.e., an under-evaluation of the approximated distribution) or not (i.e., an overestimation of the approximated distribution). The impact of this choice will be discussed later. Thus the residual number of points to be met is included in the state of $Y^{(\Delta)}$. At any moment $h\Delta$, the current residual numbers correspond-ing to non Markovian events are decreased. If some residues are null then the corresponding (non Markovian) events occur with possibly some probabilistic choice in case of conflicts. The occurrence of these events may enable new non Markovian events. Such events are handled similarly except that the next point is always counted since now it corresponds to a complete interval. If we denote by t_h^- (t_h^+) the "time" before (after) the state change in t_h, the process $Y^{(\Delta)}$ is defined by two components:

- the subordinated process in (t_h, t_{h+1}) associated with states at t_h^+ records only exponential events. It is then a CTMC defined by its generator $\mathbf{Q}$;
- the state changes at t_h are defined by a stochastic matrix
  $\mathbf{P}[i, j] = \Pr(Y^{(\Delta)}(h\Delta^+) = j \mid Y^{(\Delta)}(h\Delta^-) = i)$.

Thus the Markov Regenerative Process (MRGP) $Y^{(\Delta)}$ is fully defined by its initial probability vector $\boldsymbol{\pi}(0)$ and the matrices $\mathbf{P}, \mathbf{Q}$ (figure 1). These three components depend on Δ since the state space includes the residual number of instants per activated event. We stress however that, even if non Markovian events occur at $h\Delta$, *all kinds of concurrency are allowed between the activities of the system, contrary to previous methods*.

Furthermore, it is important to note that in the approximate process, the Markovian events are in fact *exactly* modelled since the set $\{h\Delta \mid h \in \mathbb{N}\}$ has a null measure. The only approximation comes from non Markovian events: their approximate distribution is interpreted as the number of points $k\Delta$ that must be met before their occurrence.

This approximate process may be analysed either in transient mode or in steady-state. The proposed analysis is an adaptation of the classical Markovian renewal theory methods.

Transient analysis. The transient analysis is performed by successively computing the state distribution at the instants $\Delta, 2\Delta, \ldots$ applying a transient analysis of the CTMC during an interval Δ (via the uniformization technique [14]) followed by a "step" of one of the DTMC. In order to smooth the effect of the discretization, we average the distribution on the last interval (with a variant of the uniformization).

Let $\boldsymbol{\pi}(h\Delta^+)$ be the probability vector of the process $Y^{(\Delta)}$ at time $h\Delta$ after the discrete time change, and $\boldsymbol{\pi}_X$ the probability vector of the initial model. We have:

$$\boldsymbol{\pi}(h\Delta^+) = \boldsymbol{\pi}((h-1)\Delta^+)e^{\mathbf{Q}\Delta}\mathbf{P} = \boldsymbol{\pi}(0)(e^{\mathbf{Q}\Delta}\mathbf{P})^h$$

Since we want to smooth the discretization effect, we define the approximate value $\boldsymbol{\pi}^{(a)}(h\Delta)$ of $\boldsymbol{\pi}_X(h\Delta)$ as the averaged value of the probabilities of the states of $Y^{(\Delta)}$ in $[t_h, t_{h+1}]$:

$$\boldsymbol{\pi}^{(a)}(h\Delta) = \frac{1}{\Delta} \int_{h\Delta}^{(h+1)\Delta} \boldsymbol{\pi}(\tau)d\tau = \frac{1}{\Delta}\boldsymbol{\pi}(0)(e^{\mathbf{Q}\Delta}\mathbf{P})^h \int_0^{\Delta} e^{\mathbf{Q}\tau}d\tau \tag{1}$$

Finally, we are in general interested by performance measures defined on the states *of the system*, and not on the states of the stochastic process $Y^{(\Delta)}$. Hence, all components of $\boldsymbol{\pi}^{(a)}(t)$ corresponding to a given state of the original system (i.e., when neglecting the residual numbers) are summed up to compute performance measures.

Steady-state analysis. Since the distribution depends on the relative position w.r.t. the points $h\Delta$, the approximate process does not admit a steady-state distribution but it is asymptotically periodic, with period Δ. We first compute the (approximate) steady-state distribution at times $h\Delta$: $\boldsymbol{\pi}^{(\Delta)} \stackrel{def}{=} \lim_{h\to\infty} \boldsymbol{\pi}(h\Delta^+)$.

This steady-state distribution is computed by a transient analysis stopped when the distribution is stabilised. Since $Y^{(\Delta)}$ is asymptotically periodic with Δ as period, we average the steady-state distributions on an interval $[0, \Delta]$. Then the approximate steady-state distribution is given by:

$$\pi^{(a)} = \frac{1}{\Delta} \pi^{\Delta} \int_0^{\Delta} e^{\mathbf{Q}\tau} d\tau \tag{2}$$

As in the transient case, all components of $\pi^{(a)}$ corresponding to a given state of the system are summed up to compute performance indices.

2.2   Numerical Considerations

Formulae (1) and (2) for transient and steady-state probabilities involve vector-matrix products with possibly very large matrices, either $e^{\mathbf{Q}\Delta}$ or $I^{(\Delta)} = \int_0^{\Delta} e^{\mathbf{Q}\tau} d\tau$. Moreover, it is well-known (and reported as the "fill in" phenomenon) that, although $\mathbf{Q}$ is generally very sparse, $e^{\mathbf{Q}\tau}$ is not sparse at all. Since these matrices are only required through vector-matrix products, the usual approach [17] is to never compute these matrices explicitly but to compute directly the vector-matrix products avoiding the fill in phenomenon. The products of a vector by an exponential matrix are based on the series expansion of the exponential matrix (uniformization) and numerical summation until a required precision level is reached. This is the method that we have implemented.

When we need $e^{\mathbf{Q}\tau}$ we follow the uniformization approach [18]. If $\mathbf{A}_u = \mathbf{I} + \frac{1}{u}\mathbf{Q}$ is the uniformised matrix of $\mathbf{Q}$ with rate $u > \max_i\{|q_{ii}|\}$, we have

$$e^{\mathbf{Q}\tau} = \sum_{k \geq 0} e^{-u\tau} \frac{(u\tau)^k}{k!} (\mathbf{A}_u)^k \tag{3}$$

For the transient solution (1), $\pi(0)\,(e^{\mathbf{Q}\Delta}\mathbf{P})^h$ is computed iteratively. During the algorithm only one current vector $\mathbf{V}$ indexed on the state space is required (two for the intermediate computations) and for each step we apply the vector-matrix product method to $\mathbf{V} \cdot e^{\mathbf{Q}\Delta}\mathbf{P}$. The computation of $I^{(\Delta)} = \int_0^{\Delta} e^{\mathbf{Q}\tau} d\tau$ is based on (3). By definition,

$$I^{(\Delta)} = \sum_{k \geq 0} \left[\int_0^{\Delta} e^{-ut} \frac{(ut)^k}{k!} dt \right] (\mathbf{A}_u)^k$$

An elementary derivation with integration by parts and summation gives:

$$I^{(\Delta)} = \frac{1}{u} \sum_{k \geq 0} \left[1 - e^{-u\Delta} \sum_{h=0}^{h=k} \frac{(u\Delta)^h}{h!} \right] (\mathbf{A}_u)^k \tag{4}$$

As for $e^{\mathbf{Q}\Delta}$, we only need $I^{(\Delta)}$ through products $\frac{1}{\Delta} \cdot V \cdot I^{(\Delta)}$. We compute these products iteratively to avoid the fill in. An analogous approach was used in [19] for steady-state solution of Deterministic Stochastic Petri Nets (DSPN) but restricted to one deterministic event at any given time.

Steady-state solution is obtained in a similar way, steps $\mathbf{V}^{(m+1)} = \mathbf{V}^{(m)} e^{\mathbf{Q}\Delta}\mathbf{P}$ being computed until convergence.

Algorithm 2.1. Computing the approximate probability distribution (time horizon τ)

// ϵ is the required precision
// n_0 is the initial value of n, the subdivision factor
// inc is the additive term applied to n at each step of the iteration
begin
$\quad n \leftarrow n_0$
$\quad$ compute $\boldsymbol{\pi}_n^{(L)}(\tau)$ and $\boldsymbol{\pi}_n^{(H)}(\tau)$
$\quad \mathbf{V} \leftarrow (\boldsymbol{\pi}_n^{(L)}(\tau) + \boldsymbol{\pi}_n^{(H)}(\tau))/2$
$\quad$ **repeat**
$\quad\quad n \leftarrow n + inc$
$\quad\quad old\mathbf{V} \leftarrow \mathbf{V}$
$\quad\quad old\boldsymbol{\pi}_n^{(L)}(\tau) \leftarrow \boldsymbol{\pi}_n^{(L)}(\tau)$
$\quad\quad old\boldsymbol{\pi}_n^{(H)}(\tau) \leftarrow \boldsymbol{\pi}_n^{(H)}(\tau)$
$\quad\quad$ compute $\boldsymbol{\pi}_n^{(L)}(\tau)$ and $\boldsymbol{\pi}_n^{(H)}(\tau)$
$\quad\quad d_n^{(L)} \leftarrow 1/\parallel \boldsymbol{\pi}_n^{(L)}(\tau) - old\boldsymbol{\pi}_n^{(L)}(\tau) \parallel$
$\quad\quad d_n^{(H)} \leftarrow 1/\parallel \boldsymbol{\pi}_n^{(H)}(\tau) - old\boldsymbol{\pi}_n^{(H)}(\tau) \parallel$
$\quad\quad \mathbf{V} \leftarrow \left(d_n^{(L)} \boldsymbol{\pi}_n^{(L)}(\tau) + d_n^{(H)} \boldsymbol{\pi}_n^{(H)}(\tau) \right) / \left(d_n^{(L)} + d_n^{(H)} \right)$
$\quad\quad d \leftarrow \parallel \mathbf{V} - old\mathbf{V} \parallel$
$\quad$ **until** $d \leq \epsilon$
$\quad$ // $\mathbf{V}$ is the approximation
end

Choosing an approximate probability vector. Recall that our goal is to give an approximate probability vector $\boldsymbol{\pi}^{(a)}$ either at time τ or in steady-state for models with finite support distributions. The parameter of the approximation is given by n leading to the interval length $\Delta = 1/n$.

The computation of the approximate $\boldsymbol{\pi}^{(a)}(\tau)$ is given in Algorithm 2.1 for the transient case with a given time horizon τ. The main idea is to compute successive approximation vectors until a given level ϵ of precision is reached. At each step we increase the precision of the approximation by decreasing the size Δ of the elementary interval. In the algorithm, we use the L_1 norm $\parallel \boldsymbol{\pi}_1 - \boldsymbol{\pi}_2 \parallel = \sum_i |\boldsymbol{\pi}_1[i] - \boldsymbol{\pi}_2[i]|$ to compare two probability distributions $\boldsymbol{\pi}_1$ and $\boldsymbol{\pi}_2$ and the precision of the approximation is given by the distance between successive vectors.

The special feature of the algorithm lies in the definition of our approximate vector ($\mathbf{V}$). Recall (see section 2.1), that for a given n (hence Δ) we can choose between two approximations depending whether we count or not the next $k\Delta$ to be met in the value returned by the discrete random variable corresponding the distribution of a non Markovian event. This gives us two approximate vectors at time $h\Delta$ denoted by $\boldsymbol{\pi}_n^{(L)}(\tau)$ and $\boldsymbol{\pi}_n^{(H)}(\tau)$. We observed during our experiments that the sequences $(\boldsymbol{\pi}_n^{(L)}(\tau))_{n \in \{n_0 + k \cdot inc\}}$ and $(\boldsymbol{\pi}_n^{(H)}(\tau))_{n \in \{n_0 + k \cdot inc\}}$ are both convergent but that $\parallel \boldsymbol{\pi}_n^{(L)}(\tau) - \boldsymbol{\pi}_n^{(H)}(\tau) \parallel_{n \in \{n_0 + k \cdot inc\}}$ does not necessarily converge to 0. Moreover, several comparisons have shown that depending on the parameters, one of the two sequences converges faster than the other and that the corresponding limit is closer to the exact distribution (when available) than the other one. These behaviours have led us to define the approximate distribution for n as a weighed sum of $\pi_n^{(L)}(\tau)$ and $\pi_n^{(H)}(\tau)$ based on their respective convergence rate as given in the algorithm. Note that, as usual with efficient iterative methods, we are not able to estimate analytically the convergence rate.

The steady-state approximate distribution algorithm is defined similarly except that the successive approximations are computed with the method explained in the steady-state analysis paragraph.

Note that we compute iteratively the sums (1) and (2) so that we only store two probability vectors during computation and no (full) exponential matrix.

3   Application to Stochastic Petri Nets

3.1   Presentation of Stochastic Petri Nets

Syntax. A stochastic Petri net is a Petri net enhanced by distributions associated with transitions. In the following definition, we distinguish two kinds of transitions depending on whether their distribution is *exponential* or *general*.

Definition 1. *A (marked) stochastic Petri net (SPN) $N = (P, T, Pre, Post, \mu,$ $\Phi, w, m_0)$ is defined by:*

- *P, a finite set of places*
- *$T = T_X \uplus T_G$ with $P \cap T = \emptyset$, a finite set of transitions, disjoint union of exponential transitions T_X and general transitions T_G sets*
- *Pre (resp. Post), the backward (resp. forward) incidence matrix from $P \times T$ to $\mathbb{N}$*
- *μ, a function from T_X to $\mathbb{R}^{+*}$, the strictly positive rate of exponential transitions*
- *Φ, a function from T_G to the set of finite support distributions defining the distributions of the general transitions*
- *w, a function from T_G to $\mathbb{R}^{+*}$, the weight of the general transitions*
- *m_0, a integer vector of $\mathbb{N}^P$ the initial marking*

Notations

- **dmax**$(t) = \lceil Inf(x \mid \Phi(t)(x) = 1) \rceil$ denotes the integer least upper bound of the support of $\Phi(t)$ (i.e., the range of a random variable with distribution $\Phi(t)$).
- Let m be a marking, $En(m) = \{t \mid m \geq Pre(t)\}$ denotes the set of enabled transitions. $En^X(m) = En(m) \cap T_X$ (resp. $En^G(m) = En(m) \cap T_G$) denotes the set of enabled exponential (resp. general) transitions.
- $T_G = \{t_1, \ldots, t_{n_g}\}$.

We assume that $\forall t \in T_G$, $\Phi(t)(0) = 0$, meaning that a general transition cannot be immediately fired. This also excludes the possibility of immediate transitions. This restriction is introduced only for readability purposes. In a forthcoming technical report, we will indicate how we handle this case which requires more complicated computation also encountered with discrete time SPN [20, 21, 11].

Semantics. We briefly sketch the dynamic behaviour of the stochastic process associated with an SPN. In fact, we give two *equivalent* descriptions. The former is a standard one whereas the latter takes into account the properties of the exponential distribution.

- *First description.* At some time τ, a *tangible state* of the stochastic process is given by a marking m and a vector $\mathbf{d}$ of (residual) non null delays over $En(m)$. The process lets time elapse until $\tau + dmin$ where $dmin = Inf(\mathbf{d}(t))$ decrementing the delays. Let $Fired$ be the subset of transitions such that the corresponding delay is now null. The process performs a probabilistic choice between these transitions whose distribution is defined according to their weights (for this semantics, we associate also weights with exponential transitions). Let t be the selected transition, an intermediate state is reached with marking $m' = m - Pre(t) + Post(t)$. If $(Fired \setminus \{t\}) \cap En(m') \neq \emptyset$, the process performs again a choice between these transitions and fires the selected transition. This iterative step ends when all the transitions in $Fired$ have been selected or disabled at least once. Given m' the reached marking, a new tangible state is now obtained by choosing a delay for every $t \in En(m')$ such that either t was not enabled in one of the previous markings or t was fired. This probabilistic choice is done according to $\Phi(t)$ or to the exponential distribution of parameter $\mu(t)$ (depending on the type of the transition). Note that delaying these choices after $Fired$ has been exhausted does not modify the semantics due to our assumption about $\Phi(t)(0)$.

- *Second description.* At some time τ, a *tangible state* of the stochastic process is given by a marking m and a vector $\mathbf{d}$ of (residual) non null delays over $En^G(m)$. The process "computes" probabilities of some events related an *hypothetical* delay (say also $\mathbf{d}(t)$) for every $t \in En^X(m)$ chosen according to the exponential distribution of parameter $\mu(t)$ and an *hypothetical* induced $dmin = Inf(\mathbf{d}(t) \mid t \in En(m))$. These (mutually exclusive) events are: $dmin$ is associated with a *single* exponential transition t or $dmin$ is associated with a set of general transitions, $Fired$. The other cases have a null probability to occur. We note $dmin^G = Inf(\mathbf{d}(t) \mid t \in En^G(m))$ Thus the process randomly selects one of these cases according to these probabilities and acts as follows:

  - The hypothetical $dmin$ is associated with a *single* exponential transition t. The process selects a delay $dcur$ for firing t according to a conditional distribution obtained from the exponential one by requiring that $dcur < dmin^G$. Then the process lets time elapse until $\tau + dcur$ and the delays of general transitions are decremented. Afterwards t is fired and for every transition $t' \in En^G(m') \setminus En^G(m)$ a new delay is chosen according to $\Phi(t')$.

- $dmin = dmin^G$ is associated with a set of general transitions denoted $Fired$. The process lets time elapse until $\tau + dmin^G$ decrementing the delays. Then the process performs a probabilistic choice between the transitions of $Fired$ w.r.t. their weights. Let t be the selected transition, an intermediate state is reached with marking $m' = m - Pre(t) + Post(t)$. If $(Fired \setminus \{t\}) \cap En(m') \neq \emptyset$, the process performs again a choice between these transitions and fires the selected transition. This iterative step ends when all the transitions in $Fired$ have been selected or disabled at least once. Given m' the reached marking, a new tangible state is now obtained by choosing a delay for every $t \in En^G(m')$ such that either t was disabled in one of the previous markings or t was fired. This probabilistic choice is done according to $\Phi(t)$.

Discussion. When defining semantics for SPNs one must fix three policies: the service policy, the choice policy and the memory policy. Here in both cases we have chosen the simplest policies. However most of the other choices do not yield significant additional difficulties w.r.t. the application of our generic method. Let us detail our policies. First, we have chosen the *single server* policy meaning that whatever is the enabling degree of an enabled transition t we consider a single instance of firing for t. Second, we have chosen the *race* policy meaning that the selection of the next transition to be fired is performed according to shortest residual delay: this is the standard assumption. In order to select transitions with equal delays, we perform a probabilistic choice defined by weights. Last, we have chosen the *enabling* memory meaning that the delay associated with a transition is kept until the firing or the disabling of the transition. Finally, the firing of a transition t is considered as *atomic* meaning that we do not look at the intermediate marking obtained after consuming the tokens specified by $Pre(t)$ in order to determine which transitions are still enabled.

3.2   An Approximate Stochastic Process for SPNs

The approximate process we propose behaves as the generic process of section 2. It is parameterized by n and $\Delta = 1/n$ where the greater is n the better is the approximation.

First we compute for $t \in T_G$, $\Phi^n(t)$ an approximate discrete distribution of $\Phi(t)$.

Definition 2. *Let $\Phi(t)$ be a finite support distribution with integer l.u.b. $\mathbf{dmax}(t)$ such that $\Phi(t)(0) = 0$. Then the distribution $\Phi^n(t)$ is defined by the random variable X:*

- *whose range is defined by $range^n(t) = \{1, 2, \ldots, n \cdot \mathbf{dmax}(t) - 1, n \cdot \mathbf{dmax}(t)\}$,*
- *and $Prob(X = i) = \Phi(t)(i/n) - \Phi(t)((i-1)/n)$.*

The semantics of the approximate process is close to the second description of the semantics of the SPN. However, here the time is divided in intervals of length Δ and the firings of a general transition may only occur at some $h\Delta$. At a time $\tau \in [h\Delta, (h+1)\Delta)$, a tangible state of the approximate process is defined by a marking m and and a vector $\mathbf{d}$ of (residual) non null delays over $En^G(m)$. The residual delay $\mathbf{d}(t) \in \{1, 2, \ldots, n \cdot \mathbf{dmax}(t) - 1, n \cdot \mathbf{dmax}(t)\}$ and is now interpreted as the number of intervals to elapse before the firing of t.

The process "computes" probabilities of some events related to an *hypothetical* delay (say also $\mathbf{d}(t)$) for every $t \in En^X(m)$ chosen according to the exponential distribution of parameter $\mu(t)$ and an *hypothetical* induced $dmin = Inf(\mathbf{d}(t) \mid t \in En^X(m))$. These (mutually exclusive) events are: $dmin < (k+1)\Delta - \tau$ and $dmin$ is associated with a *single* exponential transition t or $dmin > (k+1)\Delta - \tau$. The other cases have a null probability to occur. Thus the process randomly selects one of these cases according to these probabilities and acts as follows:

- The hypothetical $dmin < (k+1)\Delta - \tau$ is associated with a *single* exponential transition t. The process selects a delay $dcur$ for firing t according to a conditional distribution obtained from the exponential one by requiring that $dcur < (k+1)\Delta - \tau$. Then the process lets time elapse until $\tau + dcur$ and the delays of general transitions are *unchanged* as the process lies in the same interval. Afterwards t is fired and for every transition $t' \in En^G(m') \backslash En^G(m)$ a new delay is chosen according to $\Phi^n(t')$.
- $dmin > (k+1)\Delta - \tau$. The process lets time elapse until $(k+1)\Delta$ decrementing by one unit the delays. Let $Fired$ be the subset of transitions of $En^G(m)$ with a null delay. Then the process performs an probabilistic choice between the transitions of $Fired$ defined by their weight. Let t be the selected transition, an intermediate state is reached with marking $m' = m - Pre(t) + Post(t)$. If $(Fired \backslash \{t\}) \cap En(m') \neq \emptyset$, the process performs again a choice between these transitions and fires the selected transition. This iterative step ends when all the transitions in $Fired$ have been selected or disabled at least once. Given m' the reached marking, a new tangible state is now obtained by choosing a delay for every $t \in En^G(m')$ such that either t was disabled in one of the previous markings or t was fired. This probabilistic choice is done according to $\Phi^n(t)$.

At this stage, it should be clear that this approximate process is a special case of the process we have described in section 2. During an interval $(h\Delta, (h+1)\Delta)$, only the Markovian transitions fire whereas at time points $h\Delta$, only the general transitions fire. It remains to define the state space of this stochastic process and the associated matrices $\mathbf{P}$ and $\mathbf{Q}$.

4   From a SPN to the Tensorial Expressions of Its Approximate Process

4.1   The State Space Associated with the Matrix Q

As usual with the tensorial based methods, we build an over-approximation of the state space. However here, this over-approximation is reduced since we represent the state space as a finite union of cartesian products of sets instead of a single cartesian product.

First, one builds the reachability graph of the untimed version of the SPN. It is well-known that all the reachable markings of the SPN are also reachable in this setting. Let us denote the set of reachable markings M, the second step consists in partitioning M w.r.t. the equivalence relation $m \equiv m'$ iff $En^G(m) = En^G(m')$. Thus $M = M(TE_1) \uplus \ldots \uplus M(TE_e)$ with $TE_i \subseteq T_G$ and $M(TE_i) = \{m \in M \mid En^G(m) = TE_i\}$.

According to section 3.2, MS the set of tangible states may be decomposed as follows:

$$MS = \biguplus_{i=1..e} M(TE_i) \times \left(\prod_{g=1}^{n_g} range^n(t_g, TE_i) \right)$$

where $range^n(t, TE_i)$ is:

- $\{0\}$ if $t \notin TE_i$,
- $range^n(t)$ otherwise

In this expression, we have associated an artificial delay (0) with each disabled general transition. This does not change the size of the state space and makes easier the design of the tensorial expressions.

4.2   Tensorial Expression of Q

Let us recall that the matrix $\mathbf{Q}$ expresses the behaviour of the SPN inside an interval $(h\Delta, (h+1)\Delta)$. Thus a state change is only due to the firing of an exponential transition. As usual with tensorial methods for continuous time Markov chains, we represent $\mathbf{Q} = \mathbf{R} - \mathbf{Diag}(\mathbf{R} \cdot \mathbf{1})$. The matrix $\mathbf{R}$ includes only the rates of state changes whereas the matrix $\mathbf{Diag}(\mathbf{R} \cdot \mathbf{1})$ accumulates the rates of each row and s it on the diagonal coefficient. Generally the latter matrix is computed by a matrix-vector multiplication with the tensorial expression of $\mathbf{R}$ and then stored as a (diagonal) vector. Thus we focus on the tensorial expression of $\mathbf{R}$. More precisely, due to the representation of the state space, $\mathbf{R}$ is a block matrix where each block has a tensorial expression. We denote $\mathbf{R}_{i,j}$ the block corresponding to the state changes from $M(TE_i) \times (\prod_{g=1}^{n_g} range^n(t_g, TE_i))$ to $M(TE_j) \times (\prod_{g=1}^{n_g} range^n(t_g, TE_j))$

We first give its expression and then we detail each component.

$$\mathbf{R}_{i,j} = \left[\sum_{t \in T_X} \mu(t)\mathbf{R}'_{i,j}(t) \right] \otimes \left(\bigotimes_{g=1}^{n_g} \mathbf{R}'_{i,j}(t_g) \right)$$

For $t \in T_X$, the matrix $\mathbf{R}'_{i,j}(t)$ is a binary matrix (i.e., with 0's and 1's) only depending on the reachability relation between markings: the presence of a 1 for row m and column m' witnesses the fact that $m \xrightarrow{t} m'$.

For $t_g \in T_G$, the matrix $\mathbf{R}'_{i,j}(t_g)$ expresses what happens to the delay of t_g when an exponential transition is fired. An important observation is that this matrix does not depend on the fired transition. Indeed it only depends on the enabling of t_g in $M(TE_i)$ and $M(TE_j)$, i.e., whether $t_g \in TE_i$ and $t_g \in TE_j$. If t_g belongs to both the subsets, the delay is unchanged (yielding the identity matrix $\mathbf{I}$). If t_g does not belong to TE_j, the delay is reset to 0. Finally, if t_g belongs to TE_j and does not belong to TE_i the delay is randomly selected w.r.t. the distribution $\Phi^n(t_g)$ (see table 1).

Table 1. Structure of matrices $\mathbf{R}'_{i,j}(t_g)$

	$t_g \in TE_j$	$t_g \notin TE_j$
$t_g \in TE_i$	$\mathbf{I}$	$\begin{bmatrix} 1 \\ 1 \\ \vdots \\ 1 \end{bmatrix}$
$t_g \notin TE_i$	$\Phi^n(t_g)$	$[1]$

Due to the previous observation, rather than computing the matrices the matrix $\mathbf{R}'_{i,j}(t)$ for every $t \in T_X$, it is more efficient to compute their weighted sum:

$$\mathbf{RX}_{i,j} = \sum_{t \in T_X} \mu(t)\mathbf{R}'_{i,j}(t)$$

Furthermore, this computation can be performed on the fly when building the reachability graph of the net.

4.3 The State Space Associated with the Matrix P

The matrix $\mathbf{P}$ expresses the instantaneous changes at instants $h\Delta$. Such a change consists in decrementing the delays of the enabled general transitions followed by successive firings of general transitions with null delays (see section 3.2).

First, observe that the delay of a general enabled transition t during the intermediate stages may be either 0 (when ready to fire) or any value in $range^n(t)$ (when newly enabled). Thus the state space needs to be expanded to include null delays for enabled transitions. Furthermore, the matrix decomposition into blocks needs to be refined in order to obtain a tensorial expression. Indeed the decomposition associated with $\mathbf{Q}$ was based on the set of the enabled general transitions. Here a block will correspond to a pair (TE_i, TF) where $TF \subseteq TE_i$ represents the enabled transitions with null delay. We call such a transition a *fireable* transition.

Thus MS' the set of vanishing states may be decomposed as follows:

$$MS' = \biguplus_{i=1..e} \biguplus_{TF \subseteq TE_i} M(TE_i) \times \left(\prod_{g=1}^{n_g} range^n(t_g, TE_i, TF) \right)$$

where $range^n(t, TE_i, TF)$ is:

- $\{0\}$ if $t \notin TE_i$ or $t \in TF$,
- $range^n(t)$ otherwise

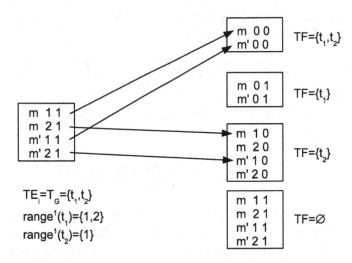

Fig. 2. Expanding the state vector while decrementing the delay

During the computations, we will expand the current probability vector then we will multiply it by $\mathbf{P}$ and afterwards contract it. Figure 2 illustrates the expansion of a block $M(TE_i) \times \left(\prod_{g=1}^{n_g} range^n(t_g, TE_i)\right)$ into blocks $\biguplus_{TF \subseteq TE_i} M(TE_i) \times \left(\prod_{g=1}^{n_g} range^n(t_g, TE_i, TF)\right)$. We have represented inside the boxes the indices of the vectors (and not their contents). In fact, we perform simultaneously the expansion and the decrementation of the delay. The copy of contents are represented by the arrows. The contraction is not represented as it is straightforward to design. It consists in copying the contents of the block associated with $TF = \emptyset$ to the original block.

4.4   Tensorial Expression of P

In order to describe the matrix $\mathbf{P}$ we adopt a top-down approach. First, $\mathbf{P} = \mathbf{DE} \cdot (\mathbf{P}^{one})^{n_g} \cdot \mathbf{C}$ where $\mathbf{DE}$ corresponds to the delay decrementation phase with expansion of the probability vector and $\mathbf{C}$ corresponds the contraction of this vector. We do not detail these matrices since, in the implementation, they are not stored. Rather the effect of their multiplication is coded by two algorithms whose time complexities are linear w.r.t. the size of the probability vector. The matrix $\mathbf{P}^{one}$ represents the change due to the firing of one fireable general transition in a state or no change if there is no such transition. Depending on the state reached at the instant $h\Delta$ and of the choice of the transitions, a sequence with variable length of general transitions may be fired. However due to our assumption about the distributions, this length is bounded by n_g. Thus our expression is sound.

Let us detail $\mathbf{P}^{one}$. We denote $\mathbf{P}^{one}(i, TF, j, TF')$ the block corresponding to the state changes from $M(TE_i) \times \left(\prod_{g=1}^{n_g} range^n(t_g, TE_i, TF)\right)$ to $M(TE_j) \times \left(\prod_{g=1}^{n_g} range^n(t_g, TE_j, TF')\right)$.

If $TF = \emptyset$ then $\mathbf{P}^{one}(i, \emptyset, i, \emptyset) = \mathbf{I}$ and for every $(j, TF') \neq (i, \emptyset)$ $\mathbf{P}^{one}(i, \emptyset, j, TF') = \mathbf{0}$.

Assume now that $TF \neq \emptyset$ then $\mathbf{P}^{one}$ corresponds to a probabilistic choice of a transition in TF and its firing. Thus we write:

$$\mathbf{P}^{one}(i, TF, j, TF') = \sum_{t \in TF} \frac{w(t)}{\sum_{t' \in TF} w(t')} \mathbf{P}_t^{one}(i, TF, j, TF')$$

Similarly to the case of $\mathbf{Q}$, $\mathbf{P}_t^{one}(i, TF, j, TF')$ can be expressed as a tensorial product:

$$\mathbf{P}_t^{one}(i, TF, j, TF') = \mathbf{PM}(t, i, j) \bigotimes \left(\bigotimes_{g=1}^{n_g} \mathbf{PG}(t_g, t, i, TF, j, TF') \right).$$

The binary matrix $\mathbf{PM}(t, i, j)$ only depends on the reachability relation between markings: the presence of a 1 for row m and column m' witnesses the fact that $m \xrightarrow{t} m'$. These matrices can be computed on the fly when building the reachability graph of the untimed net.

For $t_g \in T_G$, the matrix $\mathbf{PG}(t_g, t, i, TF, j, TF')$ expresses what happens to the delay of t_g when t is fired. Its structure is described in table 2. The only possibility for t_g to be fireable is that it was still fireable and not fired (second row of first column). If t_g is enabled but not fireable, then it was still in this situation with the same delay or was disabled and then the delay is chosen according to its distribution (see the second column). If, after the firing of t, t_g is disabled then its delay is reset (see the third column of the table).

Table 2. Structure of matrices $\mathbf{PG}(t_g, t, i, TF, j, TF')$

	$t_g \in TF'$	$t_g \in TE_j \setminus TF'$	$t_g \in T_G \setminus TE_j$
$t_g = t \vee t_g \in T_G \setminus TE_i$	$[0]$	$\Phi^n(t_g)$	$[1]$
$t_g \in TF \setminus \{t\}$	$[1]$	$[0, 0, \ldots, 0]$	$[1]$
$t_g \in TE_i \setminus TF$	$\begin{bmatrix} 0 \\ 0 \\ \vdots \\ 0 \end{bmatrix}$	I	$\begin{bmatrix} 1 \\ 1 \\ \vdots \\ 1 \end{bmatrix}$

5 Implementation Details

Our implementation is coded in the Python language [22] completed with the Numerical package [23] for better linear algebra computations and the sparse package [24] for efficient handling of sparse matrices. All our computations were done on a Pentium-PC 2.6Ghz, 512MB.

Let us explain some important features of this implementation. The SPN is described by means of a Python function called during the initial phase of the computation. During this step all data structures which are independent of the "precision" parameter n (or Δ) are built. This includes the reachability graph, the sets TE_i, TF for a given TE_i and the state spaces $M(TE_i)$. We also store the matrices $\mathbf{RX}_{i,j}$ and the $\mathbf{PM}(t, i, j)$ matrices in a sparse format. The $\mathbf{R}'_{i,j}(t_g)$ matrices are encoded in a symbolic way, that is to say we store what kind of matrix we are dealing with, among the ones given in Table 1. Each of these matrices will be "expanded" for a given n when needed during the probability vector computation.

The main part of the algorithm is to iteratively compute the approximate (either steady-state or transient) probability vector for a given precision n. Accordingly with the introduction of the tensorial description of the matrices, we never compute global matrices but in contrast vector-matrix products $\mathbf{V} \bigotimes_{k=1}^{K} \mathbf{X}_k$ where $\mathbf{V}$ and $\mathbf{X}_k$ have compatible dimensions. Theses products are implemented following the so-called Shuffle algorithm adapted to non square matrices (see the synthesis [25]) and taking into account the symbolic representation of the $\mathbf{R}'_{i,j}(t_g)$ matrices. Such products are involved during the CTMC computation and during computation at times $h\Delta$. Note that, in the present work, we do not exploit anymore the efficient implementation of the product $\mathbf{VX}$ (where $\mathbf{X}$ is a sparse matrix) provided by the Pysparse package, since we always deal with a tensorial expression of the matrix $\mathbf{X}$.

A tricky point is the expansion phase of the state space during the computation at times $h\Delta$. We chose to store our data structure as multilevel lists so that the expansion is mainly a matter of careful traversals of the list structures for which Python is well-suited.

6   Related Works

Since we deal with systems composed of general and exponential distributions, it is impossible, except for special cases, to derive analytical expressions of the transient or even steady-state distributions of the states. Thus most results are developed on so-called state based models and they involve numerical solution algorithms.

When the system exhibits complex synchronization, the Queueing Network framework becomes frequently too restrictive and in fact, many works have studied non exponential activities with the help of the non Markovian Stochastic Petri Nets (NMSPN) formalism, some of them being adapted to general distributions and other ones to deterministic distributions only. In this context, there are two main categories of works.

The first family of solutions defines conditions under which i) the underlying stochastic process is a MRGP and ii) the parameters of this MRGP can be derived from the NMSPN definition. In [8], the author introduces "Cascaded" Deterministic SPN (C-DSPN). A C-DSPN is a DSPN for which *when two or more deterministic transitions (activities) are concurrently enabled they are enabled in the same states*. With the additional constraint that the $(k + 1)$th firing time is a multiple of the kth one, it is possible to compute efficiently the probability distribution as we do. In [9], the authors derive the elements of the MRGP underlying a SPN with general finite support distributions. However, the

NMSPN must satisfy the condition that *several generally distributed transitions concurrently enabled must become enabled at the same time* (being able to become disabled at various times). The transient analysis is achieved first in the Laplace transform domain and then by a numerical Laplace inverse transformation. A simpler method is used for the steady-state solution.

The second family of solutions is based on phase-type distributions, either continuous (CPHD) or discrete (DPHD). In [10], the authors compare the qualities of fitting general distributions with DPHD or CPHD. It is shown that the time step (the scale factor) of DPHD plays a essential role in the quality of the fitting. [11] introduces the Phased Delay SPNs (PDSPN) which mix CPHD and DCPHD (general distributions must have been fitted to such distributions by the modeller). As pointed out by the authors, without any restriction, the transient or steady-state solutions of PDSPN can only be computed by stochastic simulation. However *when synchronization is imposed between firings of a CPHD transition and resamplings of DPHD transitions* the underlying stochastic process is a MRGP and its parameters can be derived from the reachability graph of the PDSPN. The approach of [12] is based on full discretization of the process. The distributions of the transitions are either DPHD or exponential (general distributions must be fitted with DPHD). For an appropriate time step, all exponential distributions are then discretized as DPHD and the solution is computed through the resulting process which is a DTMC. We note that discretization may introduce simultaneous event occurrences corresponding to achievement of continuous Markovian activities, an eventuality with zero probability in the continuous setting.

In contrast to our approach, the other approaches derive the stochastic process underlying the SPN which is then solved, possibly with approximate methods. However, restrictions on the concurrency between generally distributed activities are always imposed in order to design efficient methods for transient or steady-state solutions.

7   Conclusions and Perspectives

Main results. We have presented a new approximate method for stochastic Petri nets including non Markovian concurrent activities associated with transitions. Contrary to the other methods, we have given an approximate semantics to the net and applied an exact analysis rather than the opposite. The key factor for the quality of this approximation is that the occurrences of Markovian transitions are not approximated as it would be in a naive discretisation process. Furthermore, the design of its analysis is based on robust numerical methods (i.e., uniformization) and the steady-state and transient cases are handled similarly. Finally due to the Petri nets formalism we have been able to exploit tensorial methods which have led to significant space complexity savings.

Future work: applications. We informally illustrate here the usefulness of our method for some classes of "periodic" systems. Let us suppose that we want to analyse a database associated to a library. At any time, interactive research transactions may be activated by local or remote clients. In addition, every day at midnight, a batch transaction is performed corresponding to the update of the database by downloading remote information from a central database. In case of an overloaded database, the previous update may be still active. Thus the new update is not launched. Even if the modeller considers that all the transactions durations are defined by memoryless distributions, this non Markovian model does not admit a stationary distribution. However applying the current tools for non Markovian models will give the modeller a useless steady-state distribution with no clear interpretation. Instead we can model such a system in an exact way by considering that our approximate process of a SPN is in this case its real semantics. Then with our method, the modeller may analyse the asymptotic load of its system at different moments of the day in order to manage the additional load due to the batch transaction.

Another application area of our method is the real-time systems domain. Such systems are often composed by periodic tasks and sporadic tasks both with deadlines. With our hybrid model, we can efficiently compute the steady-state probability of deadline missing tasks.

Future work: SPNs with multiple time-scales. It is well known that stochastic systems with events having very different time scales often lead to difficulties during numerical transient or steady-state analysis. This is even worse when these events are non Markovian. These difficulties mainly arise because we need to study the stochastic process during the largest time scale but with a precision which is driven by the smallest time scale. Thus the resulting state space is generally huge. In [26], we have extended our generic method in order to efficiently deal with very different time scales. We plan to adapt this work in the framework of SPNs.

References

1. Cox, D.R.: A use of complex probabilities in the theory of stochastic processes. Proc. Cambridge Philosophical Society (1955) 313–319
2. German, R., Logothesis, D., Trivedi, K.: Transient analysis of Markov regenerative stochastic Petri nets: A comparison of approaches. In: Proc. of the 6th International Workshop on Petri Nets and Performance Models, Durham, NC, USA, IEEE Computer Society Press (1995) 103–112
3. Cox, D.R.: The analysis of non-Markov stochastic processes by the inclusion of supplementary variables. Proc. Cambridge Philosophical Society (Math. and Phys. Sciences) **51** (1955) 433–441
4. German, R., Lindemann, C.: Analysis of stochastic Petri nets by the method of supplementary variables. Performance Evaluation **20**(1–3) (1994) 317–335 special issue: Peformance'93.
5. Ajmone Marsan, M., Chiola, G.: On Petri nets with deterministic and exponentially distributed firing times. In Rozenberg, G., ed.: Advances in Petri Nets 1987. Number 266 in LNCS. Springer–Verlag (1987) 132–145

6. Lindemann, C., Schedler, G.: Numerical analysis of deterministic and stochastic Petri nets with concurrent deterministic transitions. Performance Evaluation 27–28 (1996) 576–582 special issue: Proc. of PERFORMANCE'96.
7. Lindemann, C., Reuys, A., Thümmler, A.: DSPNexpress 2.000 performance and dependability modeling environment. In: Proc. of the 29th Int. Symp. on Fault Tolerant Computing, Madison, Wisconsin (1999)
8. German, R.: Cascaded deterministic and stochastic petri nets. In B. Plateau, W.J.S., Silva, M., eds.: Proc. of the third Int. Workshop on Numerical Solution of Markov Chains, Zaragoza, Spain, Prensas Universitarias de Zaragoza (1999) 111–130
9. Puliafito, A., Scarpa, M., Trivedi, K.: K-simultaneously enable generally distributed timed transitions. Performance Evaluation 32(1) (1998) 1–34
10. Bobbio, A., Telek, A.H.M.: The scale factor: A new degree of freedom in phase type approximation. In: International Conference on Dependable Systems and Networks (DSN 2002) - IPDS 2002, Washington, DC, USA, IEEE C.S. Press (2002) 627–636
11. Jones, R.L., Ciardo, G.: On phased delay stochastic petri nets: Definition and an application. In: Proc. of the 9th Int. Workshop on Petri nets and performance models (PNPM01), Aachen, Germany, IEEE Comp. Soc. Press. (2001) 165–174
12. Horváth, A., Puliafito, A., Scarpa, M., Telek, M.: A discrete time approach to the analysis of non-markovian stochastic Petri nets. In: Proc. of the 11th Int. Conf. on Computer Performance Evaluation. Modelling Techniques and Tools (TOOLS'00). Number 1786 in LNCS, Schaumburg, IL, USA, Springer–Verlag (2000) 171–187
13. Haddad, S., Mokdad, L., Moreaux, P.: Performance evaluation of non Markovian stochastic discrete event systems - a new approach. In: Proc. of the 7th IFAC Workshop on Discrete Event Systems (WODES'04), Reims, France, IFAC (2004)
14. Gross, D., Miller, D.: The randomization technique as a modeling tool an solution procedure for transient markov processes. Operations Research 32(2) (1984) 343–361
15. Lindemann, C.: DSPNexpress: A software package for the efficient solution of deterministic and stochastic Petri nets. In: Proc. of the Sixth International Conference on Modelling Techniques and Tools for Computer Performance Evaluation, Edinburgh, Scotland, UK, Edinburgh University Press (1992) 9–20
16. Donatelli, S., Haddad, S., Moreaux, P.: Structured characterization of the Markov chains of phase-type SPN. In: Proc. of the 10th International Conference on Computer Performance Evaluation. Modelling Techniques and Tools (TOOLS'98). Number 1469 in LNCS, Palma de Mallorca, Spain, Springer–Verlag (1998) 243–254
17. Sidje, R., Stewart, W.: A survey of methods for computing large sparse matrix exponentials arising in Markov chains. Computational Statistics and Data Analysis 29 (1999) 345–368
18. Stewart, W.J.: Introduction to the numerical solution of Markov chains. Princeton University Press, USA (1994)
19. German, R.: Iterative analysis of Markov regenerative models. Performance Evaluation 44 (2001) 51–72
20. Ciardo, G., Zijal, R.: Well defined stochastic Petri nets. In: Proc. of the 4th Int. Workshop on Modeling, Ananlysis and Simulation of Computer and Telecommunication Systems (MASCOTS'96), San Jose, CA, USA, IEEE Comp. Soc. Press (1996) 278–284
21. Scarpa, M., Bobbio, A.: Kronecker representation of stochastic Petri nets with discrete PH distributions. In: International Computer Performance and Dependability Symposium - IPDS98, Duke University, Durham, NC, IEEE Computer Society Press (1998)

22. Python team: Python home page: `http://www.python.org` (2004)
23. Dubois, P.: Numeric Python home page: `http://www.pfdubois.com/numpy/` (2004) and the Numpy community.
24. Geus, R.: PySparse home page: `http://www.geus.ch` (2004)
25. Buchholz, P., Ciardo, G., Kemper, P., Donatelli, S.: Complexity of memory-efficient kronecker operations with applications to the solution of markov models. IN-FORMS Journal on Computing **13**(3) (2000) 203–222
26. Haddad, S., Moreaux, P.: Approximate analysis of non-markovian stochastic systems with multiple time scale delays. In: Proc. of the 12th Int. Workshop on Modeling, Analysis and Simulation of Computer and Telecommunication Systems (MASCOTS 2004), Volendam, The Netherlands (2004) 23–30

Nested Nets for Adaptive Systems

Kees M. van Hee[1], Irina A. Lomazova[2,*], Olivia Oanea[1,**],
Alexander Serebrenik[1], Natalia Sidorova[1], and Marc Voorhoeve[1]

[1] Department of Mathematics and Computer Science
Eindhoven University of Technology
P.O. Box 513, 5600 MB Eindhoven, The Netherlands
{k.m.v.hee, o.i.oanea, a.serebrenik, n.sidorova, m.voorhoeve}@tue.nl
[2] Program Systems Institute of Russian Academy of Science,
Pereslavl-Zalessky, 152020, Russia
irina@lomazova.polnet.botik.ru

Abstract. We consider nested nets, i.e. Petri nets in which tokens can be Petri nets themselves. We study the value semantics of nested nets rather than the reference semantics, and apply nested nets to model adaptive workflow, i.e. flexible workflow that can be modified during the execution. A typical domain with a great need for this kind of workflow is health care, from which domain we choose the running example. To achieve the desired flexibility we allow transitions that create new nets out of the existing ones. Therefore, nets with completely new structure can be created at the run time. We show that by careful selection of basic operations on the nets we can obtain a powerful modeling formalism that enforces correctness of models. Moreover, the formalism can be implemented based on existing workflow engines.

Keywords: Petri nets; modeling; workflow, adaptivity.

1 Introduction

In this paper we consider nested nets, i.e. Petri nets in which tokens can be Petri nets themselves. This means that processes are considered as objects that can be manipulated by other processes. We apply nested nets to construct more flexible workflow management systems. In classical workflow management systems the process structure is determined at design time. During execution no structural changes are possible. This implies that designers need to take into account all possible executions, exceptional situations and combinations of them. In case of so-called *ad hoc workflow* [4, 5, 14, 34] the algorithm for processing cases is not known at design time, so it is impossible to use a classical workflow management system and so-called *case handling systems* are used instead. These systems have no formal process semantics which makes testing and verification difficult.

* The work has been partly supported by the Presidium of the Russian Academy of Sciences, program "Intellectual computer systems" (project 2.3) and Russian Fund for Basic Research (project 06-01-00106). Some of this work was done during the visit to the TU/e in July 2006 funded by NWO.
** Supported by the NWO Open Competitie project MoveBP, Project number 612.000.315.

S. Donatelli and P.S. Thiagarajan (Eds.): ICATPN 2006, LNCS 4024, pp. 241–260, 2006.

In this paper we propose a solution with more flexibility for adaptation than classical workflow systems and more structure than ad hoc workflow systems. We call them *adaptive workflow systems*. By adaptivity we understand an ability to modify processes in a structured way, for instance by replacing a subprocess or extending it. We assume a given library of protocols to be used as basic building blocks for constructing more complex protocols. A natural way to model these complex protocols in an adaptive way is by means of nested Petri nets. Nested Petri nets are Petri nets in which tokens can be Petri nets themselves, called *token nets*. The ability to modify a token net rather than a part of the net itself accounts for the following advantages:

- an ability to update the library of protocols at runtime;
- an ability to modify the ongoing processes at runtime;
- an ability to model decisions taken by different parties (separation of concerns).

Traditionally workflow is modeled by workflow nets [1, 2, 3, 6]. We extend this notion by introducing a mechanism for exception handling. Exception handling is recognized as a critical challenge for workflow management systems [22]. In our extended workflow nets we introduce final transitions whose firings reflect exceptional situations. When used as token nets, these firings are normally synchronized with the firings of the higher-level net, which terminates the execution of the lower-level net.

A typical domain with a great need for this kind of adaptive workflow systems is healthcare. Today the medical protocols have the form of guidelines that involve doctors, nurses and paramedical personnel. They are often combinations of lower-level protocols where the way of combining may be determined at runtime depending, e.g., on the patient's state. Therefore, our adaptive workflow nets are well-suited for modeling this kind of systems.

To make the results of this paper applicable in practice, we were looking for an approach that could be implemented by standard workflow engines with only slight modifications. In fact, to execute a transition for which consumed tokens are nets, we have to invoke the workflow engine again for the token nets, however this can be fresh instantiations of the workflow engine.

Related research. Net in nets are extensively studied in the Petri net literature: [10, 23, 24, 25, 28, 29, 35, 36]. In all these works, the goal was to extend the expressive power or the expressive comfort of Petri nets. So, in object Petri nets, the authors aim at making Petri nets suitable for modeling according to the object-oriented style [36]. Unlike our work, the stress is on mobility and not on constructing new nets from the existing ones. In [10] the authors consider an object-oriented approach in defining synchronization between objects of different classes. For instance, an object might call methods of other objects using the operators: parallel composition, sequential composition and alternative composition. This corresponds to our approach of creating a new net token using these operators.

Nested nets [26] define nets as tokens on top of colored Petri net semantics [21]. Nested nets have the value semantics, i.e., independent copies (multisets) of token nets are considered. The framework we propose is partially inspired by this work, extending it by introducing operations on token nets and including data in the synchronization mechanism.

The idea of controlled modification of token nets is also considered for high-level net and rule (HLNR) systems in [20, 13]. Unlike our approach that easily supports arbitrary (but fixed) nesting level and synchronization between different levels of a nested net, the previous results considered nesting of depth one only. Moreover, [20] carries structural modification of P/T token nets by means of rule tokens, whereas our approach uses predefined and well-known operations, such as sequential and parallel composition.

The idea of combining workflow and "nets in nets" is going back to [8], where the authors consider object Petri nets with workflow nets as token nets. Therefore, the differences between our work and [36] are also applicable to this work.

Effective response in presence of exceptions is recognized as a critical challenge for workflow management systems [22, 30]. We believe that exceptions form an essential part of the process, and therefore, they should be included in the model. Alternatively, as suggested in [7, 9, 22] a workflow system can be extended by an exception handling mechanism. However, in such a case no formal reasoning involving exceptions is possible.

Petri nets have been used for modeling of healthcare workflow, also known as careflow [19, 33, 32, 30]. The guideline execution system GUIDE [33] translates formalized guidelines to a hierarchical timed colored Petri net. The resulting net can be run to simulate implementation of the guideline in clinical setting. However, this formalism misses adaptivity and separation of concerns.

The rest of the paper is organized as follows. Section 2 introduces a motivating example from the healthcare world. In Section 3 we present the notion of an extended workflow net and in Section 4 we discuss operations on extended workflow nets. Section 5 describes nested nets and adaptive workflow nets. Finally, in Section 6 we discuss the results presented and possible directions for future work.

2   Example: A Medical Protocol

A typical domain with a great need for adaptive workflow systems is *health care*, in particular patient treatment. In this domain the classical way to support the medical process is to use an EPR-system (EPR stands for Electronic Patient Record). An EPR-system is a data-oriented system in which for each patient a data structure is created that is updated at each step of the medical process. Normally the data structure is stored in a relational database system. On top of this database, rule engines are used to support the medical decision making to determine a diagnosis or a treatment. In recent years the *protocol-based medicine* became a standard approach. In this approach for many kinds of medical problems there are protocols defined to derive a diagnosis or a treatment. These *protocols* are in fact process descriptions. Medical experts have to follow the steps of these protocols and to document the decisions they make. This is not only important for the quality of care but also for avoiding claims when things go wrong: medical experts can prove that they treated their patient according to the best practice that is reflected in the protocol. For each patient a specific workflow is created and an EPR-system augmented with workflow system functionality is an important improvement of the process of patient treatment.

Today the medical protocols have the form of *guidelines*. A guideline is not limited to doctors but also cover the workspace of nurses and paramedical personnel. There have been several attempts to formalize guidelines as *flowcharts* and *decision diagrams* and incorporate them into medical decision support systems (MDSS). For example, GLIF [31] and GUIDE [32, 33] support medical processes by enacting guidelines, and integrating them in EPR-systems.

As a motivating example we consider the process of diagnosis and treatment of a small-cell lung cancer (SCLC). The example is inspired by the practice guidelines for SCLC [15] created by the National Comprehensive Cancer Network (SUA) and modified by The University of Texas M.D. Anderson Cancer Center for their patient population.

The guideline is built on the basis of a library of standard protocols. A protocol may describe the decision process for establishing a diagnosis, treatment schemes and tests used in the process of diagnosis or after the treatment. Each protocol has an initial point, a final point and it handles exceptional situations by terminating the process. For instance, the protocol used for mandatory tests in the initial diagnosis stage describes the process of executing two tests (computer tomography of brain and the bone scan) and evaluating their results. An exceptional situation is considered when the result of at least one of the tests turns out positive. We model such a protocol as a workflow net extended with special transitions, which are transitions without output places. These transitions carry an exception label indicating the exceptional situation. Figure 1 shows the extended workflow net modeling the mandatory test protocol.

The interfaces of protocols in the library are given as $Prot\langle ex_1, \ldots, ex_n \rangle$, where *Prot* is the name of the protocol and $ex_1, \ldots, ex_n$ is the list of exception labels. In case the protocol has no exceptions, it is denoted solely by its name.

Figure 2 shows another protocol net (*STests*), modeling the test protocol used in the surveillance stage. The process performs some tests such as CXR test, creatine test and liver function test and registers their results in the variables x, y, z, u, w. The evaluation of the tests employs a function that combines the results of the tests. According to constraints specified in predicates, the progress of the cancer and the response to the treatment are evaluated: the cancer can relapse (modeled by the transition labeled *relapse*) or the treatment can have a partial response or good response (modeled by transitions labeled accordingly).

Protocols can make use of other protocols, create new protocols from the existing ones or modify them. For this purpose, we consider nets as colored tokens. The

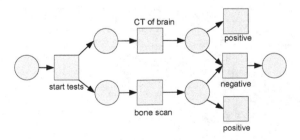

Fig. 1. Mandatory test protocol (*MandT*⟨*positive*⟩)

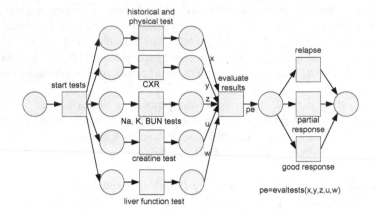

Fig. 2. Surveillance test protocol (*STests*)

protocols can be built of more primitive ones by using the operations such as choice (+), sequential (·) and parallel (∥) compositions. The operation *init* initializes an extended workflow net with its initial marking. Consider the surveillance protocol in Figure 3. The protocol makes use of other therapy and test protocols, namely *Radiotherapy, STests, Cisplatin, Etoposide, ProphT* and *RadCon⟨radiation scarring⟩*. The protocol iterates the surveillance treatment scheme until the results of the surveillance tests show signs of relapse. A regular surveillance treatment is started at runtime by creating a process to-ken (denoted by the constant *str* = *init*(*Radiotherapy · STests*) on the outgoing arc of the transition *start surveillance*). The process of this token consists of the *Radiotherapy* protocol followed by the *STests* protocol. The transitions labeled by *relapse, partial response, good response* in the surveillance test protocol are synchronized with transitions having the same labels in the upper level net (*Surveillance*).

In case of a partial response, a new token is created, namely *init*(((*Cisplatin· Etoposide*)∥*Radiotherapy*)·*STests*). This process token corresponds to radiotherapy

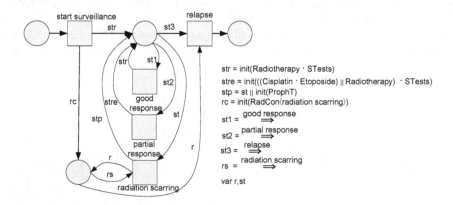

Fig. 3. Surveillance protocol (*Surveillance*)

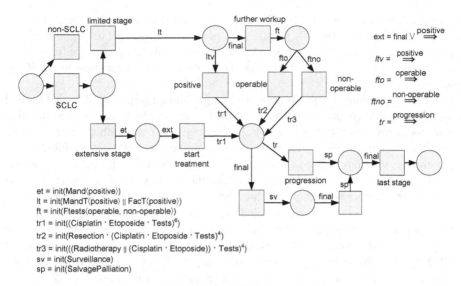

$$et = init(Mand\langle positive\rangle)$$
$$lt = init(MandT\langle positive\rangle \parallel FacT\langle positive\rangle)$$
$$ft = init(Ftests\langle operable, non\text{-}operable\rangle)$$
$$tr1 = init((Cisplatin \cdot Etoposide \cdot Tests)^6)$$
$$tr2 = init((Resection \cdot (Cisplatin \cdot Etoposide \cdot Tests)^4)$$
$$tr3 = init(((Radiotherapy \parallel (Cisplatin \cdot Etoposide)) \cdot Tests)^4)$$
$$sv = init(Surveillance)$$
$$sp = init(SalvagePalliation)$$

Fig. 4. Main SCLC protocol

performed in parallel with chemotherapy — cisplatin treatment followed by an etoposide treatment — followed by the *STests* protocol.

At the same time with the application of a surveillance treatment, a radiation control protocol (*RadCon⟨radiation scarring⟩*) is used to monitor the radiation effect on the patient. In case the patient shows signs of scars due to the radiation (the exception *radiation scarring* is signaled), a prophylactic treatment (*ProphT*) is conducted in parallel with the actual surveillance protocol (*st∥init(ProphT)*).

The protocol from the library describing the main process is depicted in Figure 4. The process starts with the decision on whether the patient has small-cell lung cancer (SCLC) or non-small cell lung cancer (Non-SCLC). The transition *Non-SCLC* models an exception that allows finishing the guideline.

Once the diagnosis has been established, the stage of the illness needs to be assessed. The stage is determined by the extent of spread of the cancer basing on the test results. Here, the protocol makes use of and combines some standard test procedures depending on the preliminary diagnosis of the phase of the cancer. In case the patient shows signs of the extensive stage, a procedure with mandatory tests *MandT⟨positive⟩* (Figure 1) is created. Once the protocol *MandT⟨positive⟩* has terminated or a transition with exception label *positive* indicating that the tests are finished is fired (the arc inscription $ext = final \vee \overset{positive}{\Rightarrow}$ is true), a specific treatment protocol can be started.

In case the preliminary diagnosis shows signs of the limited stage (*limited stage* transition) more tests are needed than in the extensive case. The protocol executing these tests combines two existing test protocols taken from the library, i.e. *MandT⟨positive⟩* and *FacT⟨positive⟩*, by performing them in parallel. Once one of the tests has a positive outcome (the transition labeled *positive* is fired), it must be synchronized with the respective transition in the main protocol since this is a symptom for the extensive stage cancer. If none of the results turns out positive, the test protocol terminates properly (this is indicated by the arc inscription *final*).

The patient with limited-stage SCLC is further tested to determine whether or not she/he can be operated. For this purpose, the protocol *FTests⟨operable, non-operable⟩* is instantiated, and one of the exceptional outcomes of this test procedure (the patient is non-operable or operable) is synchronized with the respective transition in the SCLC protocol.

For each of the three diagnosis (extensive stage, limited stage operable, and limited stage non-operable), a special treatment scheme is created, which is actually an iteration of chemotherapy, radiotherapy and tests. For example, for the limited stage operable diagnosis, a treatment scheme token considering the resection *Resection*, followed by four iterations of chemotherapy (*Cisplatin* treatment followed by *Etoposide* treatment) and tests (*Tests⟨progression⟩*) is created. Once there is a sign of progression of the cancer after performing the tests at each of the cycles signaled by the occurrence of an exception handled by the transition labeled by *progression*), the treatment is interrupted and the patient goes in the final stage, where only the Palliative/Salvage treatment can be applied. In case the initial treatment has been successfully completed (the transition labeled by *last stage* is fired) a surveillance protocol (*Surveillance*) is created and its completion determines the patient to enter the palliation/salvage stage.

3   Extended Workflow Nets

Preliminaries. We briefly introduce the notation and the basic concepts that will be used in the remainder of the paper.

$\mathbb{N}$ denotes the set of natural numbers. The set of all subsets of a set P is 2^P. A relation between sets P and Q is a subset of $P \times Q$. If $R \subseteq P \times Q$ is a relation, then we use the infix notation $p \, R \, q$ for $(p,q) \in R$ and the inverse R^{-1} of R is defined as $\{(q,p) \mid p \, R \, q\}$. Functions are a special kind of relations. Given a function F we write $dom(F)$ to denote the domain of F and $ran(F)$ to denote the range of F. If the domain of a function is finite we also write $\{d_1 \mapsto v_1, \ldots, d_n \mapsto v_n\}$. The operator $\circ$ denotes function composition, $(f \circ g)(x)$ is $f(g(x))$. A *bag (multiset)* m over P is a mapping $m : P \to \mathbb{N}$. The set of all bags over P is also denoted by $\mathbb{N}^P$. We use $+$ and $-$ for the sum and the difference of two bags and $=, <, >, \leq, \geq$ for comparisons of bags, which are defined in the standard way. We overload the set notation, writing $\emptyset$ for the empty bag and $\in$ for the element inclusion. We abbreviate bags by writing, for example, $2[p] + [q]$ instead of $\{p \mapsto 2, q \mapsto 1\}$.

Let F be a function, such that $ran(F)$ is a set of sets. Then *the generalized product* of F, written as $\Pi(F)$, is the set of all functions f with $dom(f) = dom(F)$ and $f(x) \in F(x)$ holds for all $x \in dom(F)$. Note that $\Pi(F) = \{\emptyset\}$ if $dom(F) = \emptyset$. We further extend this notion and given $S \subseteq dom(F)$ write $\Pi_{x \in S}(F)$ to denote the set of all functions f with $dom(f) = S$ such that for any $x \in S$, $f(x) \in F(x)$.

A *transition system* is a tuple $E = \langle S, \Sigma, T \rangle$ where S is a set of *states*, Σ is a set of *actions* and $T \subseteq S \times \Sigma \times S$ is a set of transitions.

Colored nets. In Section 5 we are going to introduce nested nets that are considered as a special kind of colored nets. To this end we define colored Petri nets parameterized with a value universe $\mathcal{U}$. We assume that $\mathcal{U}$ contains the value *black*, which we use as an object carrying no information. Our definition does not essentially differ from the

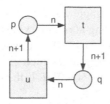

Fig. 5. Colored Petri net

existing ones [21] but it has been adapted for our purpose, notably by the addition of transition labels.

In a colored net each place is mapped to a type, which is a subset of $\mathcal{U}$. We also assume a set L of labels for transitions such that $\tau \notin L$. Every label is associated with a unique natural number, called the arity of the label. Then we define the set of transition labels $\Sigma = \{\tau\} \cup \{a(x_1, \ldots, x_n) \mid a \in L, n = \mathrm{arity}(a), x_1, \ldots, x_n \in \mathcal{U}\}$. The label τ is the special "silent" label.

Definition 1. *A colored net over the universe $\mathcal{U}$ is a 6-tuple $(P, T, F, \upsilon, \rho, \ell)$, where*

- *P and T are disjoint finite sets of places, respectively transitions;*
- *$F \subseteq (P \times T) \cup (T \times P)$ is a set of arcs,*
- *υ is a place typing function with $\mathrm{dom}(\upsilon) = P$ such that $\upsilon(p) \subseteq \mathcal{U}$ for all $p \in P$;*
- *ρ is a transition function with $\mathrm{dom}(\rho) = T$ such that $\rho(t) \subseteq \Pi(\varphi_t) \times \Pi(\psi_t)$ for all $t \in T$, where $\varphi_t = \{p \mapsto \upsilon(p) \mid (p, t) \in F\}$ and $\psi_t = \{p \mapsto \upsilon(p) \mid (t, p) \in F\}$;*
- *ℓ is a labeling function with $\mathrm{dom}(\ell) = \{(t, \gamma, \delta) \mid t \in T \wedge (\gamma, \delta) \in \rho(t)\}$ and $\mathrm{ran}(\ell) \subseteq \Sigma$.*

The labeling function is used to report information on transition firings. In this way we can distinguish between internal firings that are labeled with τ and externally observable firings labeled with labels different from τ. Note that the labeling function ℓ allows for the association of several different labels to a transition and the same label to different transitions. We illustrate the definition above by means of an example.

Example 2. Assume that $\mathcal{U}$ contains the set of natural numbers $\mathbb{N}$. Consider the net in Figure 5. It contains two places, p and q, and two transitions, t and u, with the following set of arcs $\{(p, t), (t, q), (q, u), (u, p)\}$. Both places are typed by $\mathbb{N}$, i.e., $\upsilon(p) = \upsilon(q) = \mathbb{N}$. The transition relation is given by $\rho(t) = \{(\{p \mapsto n\}, \{q \mapsto n + 1\}) \mid n \in \mathbb{N}\}$ and $\rho(u) = \{(\{q \mapsto n\}, \{p \mapsto n + 1\}) \mid n \in \mathbb{N}\}$. The labeling function ℓ is defined by $\ell(t, \gamma, \delta) = \tau$ and $\ell(u, \gamma, \delta) = a(\gamma(q))$. $\qquad\square$

Next we introduce the notion of *marking* of a colored net over a given set of types.

Definition 3. *Given a colored net $N = (P, T, F, \upsilon, \rho, \ell)$ over the universe $\mathcal{U}$, a marking of N is a function $M : P \times \mathcal{U} \to \mathbb{N}$, such that for any $p \in P$ and any $u \in \mathcal{U}$, $M(p, u) > 0$ implies $u \in \upsilon(p)$. The set of all markings of a colored net N is denoted by $\mu(N)$.*

A marked colored net over $\mathcal{U}$ is a pair (N, M), where N is a colored net over $\mathcal{U}$ and $M \in \mu(N)$ is a colored marking of N.

To illustrate this notion we return to our previous example. Some possible markings for the net in Figure 5 are $M_n^p = \{(p, n) \mapsto 1\}$ and $M_n^q = \{(q, n) \mapsto 1\}$ for any $n \in \mathbb{N}$. So

M_n^p, respectively M_n^q, consists of a single token with value n in place p, respectively q. To simplify the notation when talking about markings we use the bag notation $[(p,n)]$ instead of the functional one $\{(p,n) \mapsto 1\}$. If n coincides with *black* then we write $[p]$.

A colored net defines a transition system which gives the observable behavior of the net. To formalize this notion we denote by $\mathfrak{N}$ ($\mathfrak{M}$) the set of (marked) colored nets over the given universe.

Definition 4. *The ternary relation* $\_ \xrightarrow{} \_ \subseteq \mathfrak{M} \times \Sigma \times \mathfrak{M}$ *is defined as the smallest relation such that* $(N,M+\gamma) \xrightarrow{\ell(t,\gamma,\delta)} (N,M+\delta)$ *for all* $(N,M) \in \mathfrak{M}$, $t \in T$ *and* $(\gamma,\delta) \in \rho(t)$. *We also write* $(N,M) \xrightarrow{\sigma}$ *for some* $\sigma \in \Sigma$ *if and only if there exists a marking* $M' \in \mu(N)$ *such that* $(N,M) \xrightarrow{\sigma} (N,M')$. *Finally,* $(N,M) \xrightarrow{*} (N,M')$ *means that there exists a sequence* $(\sigma_1,\ldots,\sigma_n) \in \Sigma^*$ *such that* $(N,M) = (N,M_1) \xrightarrow{\sigma_1} (N,M_2) \xrightarrow{\sigma_2} \ldots \xrightarrow{\sigma_n}$ $(N,M_{n+1}) = (N,M')$. *In this case we also say that* (N,M') *is* reachable *in* (N,M). *The set of all markings reachable in* (N,M) *is denoted* $\mathcal{R}(N,M)$.

Consider net N_1 from Example 2. Then $(N_1,M_3^p) \xrightarrow{\tau} (N_1,M_4^q)$.

Extended workflow nets. Workflow nets (WF nets) are used to model workflow systems (see e.g. [1, 2, 3, 6]). Workflow nets have one initial and one final place and every place or transition is on a directed path from the initial to the final place. We extend the notion of a WF net to model *exceptions*, i.e. transitions that should terminate the execution of the current net. As mentioned in the introduction, exception handling is recognized as a critical challenge for workflow management systems.

To ensure that firings of exception transitions are observable and cannot be confused with firings of non-exception transitions, we require the set of transition labels Σ to be split into two parts: Σ^e for the firings of exception transitions and Σ^n for the firings of non-exception transitions. Formally, we require $\Sigma = \Sigma^e \cup \Sigma^n$, $\Sigma^e \cap \Sigma^n = \emptyset$ and $\tau \in \Sigma^n$.

Definition 5. *A colored net* $N = (P,T,F,\upsilon,\rho,\ell)$ *over the universe* $\mathcal{U}$ *is an* extended workflow net *(EWF net) with the initial place* $i \in P$, *and the final place* $f \in P$ *and the set of exception transitions* $T' \subseteq T$ *if*

1. $\{t \mid (t,i) \in F\} = \{t \mid (f,t) \in F\} = \emptyset$;
2. $\upsilon(i) = \upsilon(f) = \{black\}$;
3. $t \in T'$ *if and only if* $\ell(t,\gamma,\delta) \in \Sigma^e$ *for all* $(\gamma,\delta) \in \rho(t)$;
4. $\{p \mid (t,p) \in F\} = \emptyset$ *for all* $t \in T'$;
5. *For any node* $n \in (P \cup T)$ *there exists a path from* i *to* n;
6. *For any node* $n \in (P \cup T)$ *there exists a path from* n *to a node in* $T' \cup \{f\}$.

Traditional WF nets are EWF nets with the empty set of exception transitions. Since exception transitions produce no output, for any $t \in T'$ whenever $\ell(t,\gamma,\delta)$ is defined, $\delta = \emptyset$.

Example 6. Consider the nets in Figure 6 as colored nets over the universe $\{black\}$. Both nets are EWF nets: the left-hand side net with the initial place i and with the final place f, the right-hand side net with the initial place i' and with the final place f'. The exception transitions are $\{v\}$ and $\{v'\}$, respectively. All firings of these transitions are labeled by a and a', respectively, where $a,a' \in \Sigma^e$. $\qquad\square$

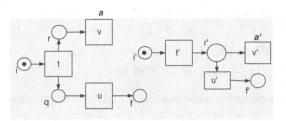

Fig. 6. EWF nets

EWF nets provide a number of advantages from the modeling point of view. First and foremost they make a clear distinction between normal termination and termination caused by an exception. Unlike traditional WF nets, where special care should be taken to remove all tokens present in the system when an exceptional situation is encountered, no similar overhead is incurred by EWF nets. We define the following concepts regarding initiality and finality of EWF nets.

Definition 7. *Let N be an EWF net with initial place i and final place f and let M ∈ μ(N). The marked net (N,M) is called* initial, *resp.* final *if and only if M = [i], resp. M = [f]. The initialization* init(N) *of N is the marked net (N,[i]).*

An important correctness property for EWF nets is *soundness*. Classical WF nets are called sound if one can reach the final marking from any marking reachable from the initial marking [1]. The intuition behind this notion is that no matter what happens, there is always a way to complete the execution and reach the final state. This soundness property is sometimes also called *proper termination* and corresponds to 1-soundness in [18].

Definition 8. *An EWF net $N = (P,T,F,\upsilon,\rho,\ell)$ with the initial place i and the final place f over the universe $\mathcal{U}$ is called* sound *if and only if for all (N,M) such that $[i] \xrightarrow{\sigma} M$ for some $\sigma \in (\Sigma^n)^*$ the following two conditions hold:*

1. *either $(N,M) \xrightarrow{\gamma} (N,[f])$ for some $\gamma \in (\Sigma^n)^*$*
   or there exists $(N,M') \in \mathcal{R}(N,M)$ such that $(N,M') \xrightarrow{a}$ for some $a \in \Sigma^e$;
2. *$(N,M) \xrightarrow{*} (N,m+[f])$ implies $m = \emptyset$.*

The intuition behind the notion of soundness formalized in Definition 8 is that for any possible behavior there is always a possibility to complete the execution and reach the final state, or to report an exception. The second requirement of soundness stresses that whenever the final state is reached, the execution has been completed. In case of classical WF nets, the second condition of soundness is redundant (cf. Lemma 11 in [18]). The following example shows that this is not the case for EWF nets.

Example 9. Recall the EWF nets from Figure 6. The left-hand side net is unsound. Indeed, consider $(N_l,[q]+[r])$, which is reachable from $(N_l,[i])$ since it can be obtained by firing t. Moreover, $(N_l,[q]+[r]) \xrightarrow{a}$ and $a \in \Sigma^e$. Hence, the first part of the soundness condition is satisfied. However, $(N_l,[q]+[r]) \xrightarrow{*} (N_l,[f]+[r])$, and since $[r] \neq \emptyset$ the second condition of soundness is violated and the net is unsound.

The right-hand side net is sound. Indeed, $\mathcal{R}(N_r, [i']) = \{[i'], [r'], [f'], \emptyset\}$, i.e., the only marking that contains f' is $[f']$ itself. Hence, the second condition is satisfied. The first condition is satisfied since $(N_r, [i']) \xrightarrow{*} (N_r, [r']), (N_r, [r']) \xrightarrow{*} (N_r, [r']) \xrightarrow{a'}$ on the one hand, and $(N_r, [f']) \xrightarrow{*} (N_r, [f'])$ on the other hand.   □

Note that adding an exception transition to a sound EWF net preserves soundness.

Lemma 10. *Let* $N = (P, T, F, \upsilon, \rho, \ell)$ *be a sound EWF net over the universe* $\mathcal{U}$. *Let* $Q \subseteq P \setminus \{f\}$, $t \notin T$ *and* $h : \Pi_{p \in Q}(\upsilon(p)) \to \Sigma^e$. *Let* N' *be* $(P, T \cup \{t\}, F \cup \{(p, t) \mid p \in Q\}, \upsilon, \rho', \ell')$, *where*

- $\rho'(u) = \rho(u)$ *for* $u \in T$ *and* $\rho'(t) = \Pi(\varphi_t) \times \{\emptyset\}$;
- $\ell'(u, \gamma, \delta) = \ell(u, \gamma, \delta)$ *for* $u \in T$ *and* $(\gamma, \delta) \in \rho(u)$, *and* $\ell'(t, \gamma, \emptyset) = h(\gamma)$ *for any* $\gamma \in \Pi(\varphi_t)$.

Then, the net N' *is a sound EWF net over* $\mathcal{U}$.

Note that since t is an exception transition we have $\delta = \emptyset$ for any $(\gamma, \delta) \in \rho'(t)$.

Proof. The proof is done by checking Definition 8.   □

Lemma 10 allows us to use an incremental approach to modeling by modeling the normal course of events first, and adding exceptions afterward.

We define bisimilarity of EWF nets, by extending the standard notion with preservation of initiality and finality of states, cf. [17].

Definition 11. *Let* N_1, N_2 *be EWF nets. A relation* $R \subseteq \mu(N_1) \times \mu(N_2)$ *is called a* simulation *if and only if for all* $M_1, M_1' \in \mu(N_1)$, $M_2 \in \mu(N_2)$ *and* $\sigma \in \Sigma$ *such that* $(N_1, M_1) \xrightarrow{\sigma} (N_1, M_1')$ *there exists* $M_2' \in \mu(N_2)$ *such that* $(N_2, M_2) \xrightarrow{\sigma} (N_2, M_2')$ *and* $M_1' R M_2'$.
The marked EWF nets (N_1, M_1) *and* (N_2, M_2) *with initial and final places* i_1 *and* i_2, f_1 *and* f_2 *respectively, are called* EWF-bisimilar, *denoted* $(N_1, M_1) =_e (N_2, M_2)$ *if and only if there exists a relation* $R \subseteq \mu(N_1) \times \mu(N_2)$ *such that* $M_1 R M_2$, *both* R *and* R^{-1} *are simulations, and whenever* $x R y$, *the following holds:*

- $x = [i_1]$ *if and only if* $y = [i_2]$;
- $x = [f_1]$ *if and only if* $y = [f_2]$;
- $x > [f_1]$ *if and only if* $y > [f_2]$;

The EWF nets N_1 *and* N_2 *are EWF-bisimilar, denoted* $N_1 =_e N_2$, *if and only if their initializations are EWF-bisimilar.*

Note that EWF-bisimilarity is based on the traditional notion of strong bisimilarity. Further notions of bisimilarity, such as branching bisimilarity, can be adapted for EWF nets. EWF-bisimilarity preserves soundness of EWF nets. The requirement $> [f]$ to be preserved by the simulation relation is required to ensure that the second part of Definition 8 is satisfied. It is not needed for traditional workflow nets. We define the *behavior* of a EWF net as its equivalence class modulo EWF-bisimilarity. The next lemma states that soundness is a behavioral property.

Lemma 12. *Let N_1, N_2 be EWF-bisimilar nets. Then N_1 is sound if and only if N_2 is sound.*

Proof (Sketch). We prove the "only if" side; the "if" side then follows by symmetry. Assume that N_1 is sound. We should prove that N_2 is sound. Let R be a bisimulation between $init(N_1)$ and $init(N_2)$. Then, $[i_1] \; R \; [i_2]$. Let M_2 be such that $(N_2, [i_2]) \xrightarrow{*} (N_2, M_2)$. Since R^{-1} is a simulation, there exists M_1 such that $M_1 \; R \; M_2$ and $(N_1, [i_1]) \xrightarrow{*} (N_1, M_1)$. Since N_1 is sound, either $(N_1, M_1) \xrightarrow{*} (N_1, [f_1])$ or $(N_1, M_1) \xrightarrow{*} (N_1, M_1') \xrightarrow{\sigma}$ for some $M_1' \in \mu(N_1)$ and $\sigma \in \Sigma^e$. Since R is a simulation, either $(N_2, M_2) \xrightarrow{*} (N_2, [f_2])$ or $(N_2, M_2) \xrightarrow{*} (N_2, M_2') \xrightarrow{\sigma}$ for some $M_2' \in \mu(N_2)$.

Finally, we need to prove that $(N_2, [i_2]) \xrightarrow{*} (N_2, [f_2] + m)$ implies $m = \emptyset$. Since $(N_2, [i_2]) \xrightarrow{*} (N_2, [f_2] + m)$, there exists M_1 such that $M_1 \; R \; [f_2] + m$ and $(N_1, [i_1]) \xrightarrow{*} (N_1, M_1)$. If $m > \emptyset$, then by the properties of R we have $M_1 > [f_1]$. This contradicts the soundness of M_1, so in fact $m = \emptyset$.   □

4 Operations on EWF Nets

In this section we discuss predicates and operations that can be applied to EWF nets and marked EWF nets. In Section 5 we will introduce nested nets and we will see how these operations can serve as basic building blocks that can be used in defining the transition function of a higher-level net. We try and keep the set of operations as simple as possible so that they can be performed by existing workflow engines.

From here on in this section, all nets are EWF nets. The set of all (marked) EWF nets is denoted $\mathfrak{N}^w$ ($\mathfrak{M}^w$). We consider a number of predicates and operations on nets and marked nets. Using Definition 7 we convert a net into a marked net by adding to it a corresponding initial marking, and for any marked net we can check whether its marking is the initial one or the final one.

Next, two nets can be combined to produce a new net by means of sequential ($\cdot$) and parallel ($\|$) composition and choice ($+$). Moreover, parallel composition can be applied to marked nets and sequential composition to a marked net and a net. In the past, similar operations have been defined for workflow nets [6]. We adapt these notions for EWF nets.

One of the most natural operations on EWF nets is *sequential composition*. For instance, a doctor can prescribe a resection followed by a medication. Hence, the treatment prescribed by the doctor is the sequential composition of two nets: resection and medication. Other common operations on nets are parallel composition[1] and choice. The three operations are depicted in Figure 7. N_1 and N_2 are unmarked EWF nets with the initial places i_1 and i_2, and the final places f_1 and f_2, respectively. The sign of equality denotes place fusion: for sequential composition the final place of the first net and the initial place of the second net have to be fused, for choice the initial places have to be fused as well as the final ones. Firing the transitions ti, tf is labeled by τ. The operations can be formalized by writing out the net structure, which is trivial but tedious.

---

[1] It should be noted that the parallel composition we define is not exactly the well-known process-algebraic operation. For instance, after the initialization, $x + (y\|z)$ can make an internal step to a state where x cannot be executed anymore.

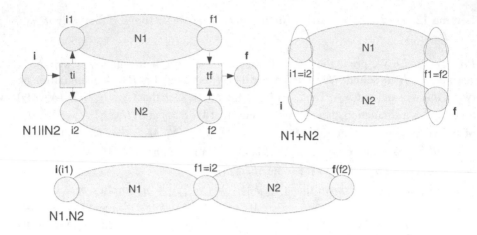

Fig. 7. Parallel composition $N1\|N_2$, Choice $N_1 + N_2$, Sequential composition $N1 \cdot N_2$

Lemma 13. *For any $N_1, N_2 \in \mathfrak{N}^w$, $N_1 \cdot N_2, N_1\|N_2, N_1 + N_2 \in \mathfrak{N}^w$. If N_1 and N_2 are sound, then $N_1 \cdot N_2, N_1\|N_2, N_1 + N_2$ are sound as well. Moreover, $\cdot$ and $+$ are associative and $\|$ and $+$ are commutative.*

One of the typical operations one might like to consider is *iteration*. For instance, the doctor can prescribe a certain treatment to be performed three times. Associativity of $\cdot$ allows us to model this by writing $N \cdot N \cdot N$, where N is the unmarked EWF net modeling the treatment. Formally we define N^k as N if $k = 1$, and as $N \cdot N^{k-1}$ if $k > 1$.

Next, consider marked nets. We start with the *sequential composition*. The intention of the sequential composition is to extend a running process by a new functionality. Therefore, we define the sequential composition as an operation on a marked net (a running process) and a net (additional functionality). Formally, we define $(N_1, M) \cdot N_2 = (N_1 \cdot N_2, M)$. Similarly, the *parallel composition* of two marked EWF nets (N_1, M_1) and (N_2, M_2) is defined as $(N_1, M_1)\|(N_2, M_2) = (N_1\|N_2, M_1 + M_2)$.

A *choice operation* on marked nets does not seem useful for our purpose.

Operations on the marked nets satisfy the following lemma.

Lemma 14. *For any $(N_1, M_1), (N_2, M_2) \in \mathfrak{M}^w$ and $N \in \mathfrak{N}^w$, $(N_1, M_1) \cdot N \in \mathfrak{M}^w$, and $(N_1, M_1)\|(N_2, M_2) \in \mathfrak{M}^w$. Moreover, $\cdot$ is associative and $\|$ is commutative.*

Parallel composition and choice are congruences w.r.t. EWF-bisimilarity and sequential composition is a congruence if the first operand is sound. Its proof can be achieved by standard meta-theory of process algebra [16].

Theorem 15. *Let N_1, N_2, N_1', N_2' be EWF nets such that $N_1 =_e N_1'$ and $N_2 =_e N_2'$. Then $N_1 \theta N_2 =_e N_1' \theta N_2'$, for $\theta \in \{+, \|, \cdot\}$.*

Let $(N_3, M_3), (N_3', M_3'), (N_4, M_4), (N_4', M_4')$ be marked EWF nets with $(N_3, M_3) =_e (N_3', M_3')$ and $(N_4, M_4) =_e (N_4', M_4')$. Then $(N_3, M_3)\|(N_4, M_4) =_e (N_3', M_3')\|(N_4', M_4')$. and $(N_3, M_3) \cdot N_2 =_e (N_3', M_3') \cdot N_2'$.

5 Nested Nets

In this section we introduce *nested nets*. In Section 2 we showed how nested nets can be used to model adaptive systems. We start by defining general nested nets and then move to nested EWF nets.

Nested nets form an extension of colored nets over a special universe. The only extension is that color of tokens can be changed without firing transitions. We recursively define value universes, nets and marked nets of any depth n, for $n = 0 \ldots \omega$, assuming the existence of an initial universe $\mathcal{U}_0$ containing basic values such as integers and composed values like pairs, lists and sets of basic or composed values. Subsequent universes and sets of nets are defined recursively as follows.

Definition 16. *The sets $\mathfrak{N}_0, \mathfrak{M}_0$ of nets and marked nets of depth zero are defined as the sets of colored nets and marked colored nets over the universe $\mathcal{U}_0$, respectively. For each $n > 0$ the value universe $\mathcal{U}_n$ and the sets $\mathfrak{N}_n, \mathfrak{M}_n$ of nets and marked nets of depth n are recursively defined by $\mathcal{U}_n = \mathcal{U}_{n-1} \cup \mathfrak{M}_{n-1}$ and $\mathfrak{N}_n$ and $\mathfrak{M}_n$ as the set of colored nets and marked colored nets over $\mathcal{U}_n$. We set $\mathfrak{N}_\omega = \bigcup_{n \geq 0} \mathfrak{N}_n$, $\mathfrak{M}_\omega = \bigcup_{n \geq 0} \mathfrak{M}_n$ and $\mathcal{U}_\omega = \mathcal{U}_0 \cup \mathfrak{M}_\omega$.*

For the sake of uniformity we call elements of $\mathfrak{N}_0$ ($\mathfrak{M}_0$) (marked) nested nets of depth 0. Observe that the recursive definition of a notion of a marked nested net of depth n allows tokens in it to be colored by nested nets of depth $n - 1$.

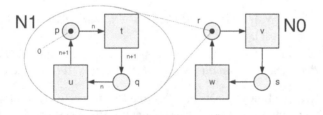

Fig. 8. Nested net

Example 17. To illustrate Definition 16 we first assume that the only elements of $\mathcal{U}_0$ are natural numbers. Then, net N_1 we introduced in Example 2 belongs to $\mathfrak{N}_0$. We can use N_1 as a token in a higher-level net as illustrated by Figure 8. We call this net N_0. It has two places, r and s, and two transitions v and w. The places r and s are of the "net type", i.e., the place typing function υ_0 should map them to subsets of $\mathfrak{M}_\omega$. For instance, we can take $\upsilon : \{r, s\} \rightarrow \{(N_1, M) \mid M \in \mu(N_1)\}$. $\square$

The firing relation $\longrightarrow$ for nested nets containing token nets may depend on the firing relation in the tokens themselves. We can thus achieve, for example, vertical synchronization: a transition in a lower-level net and another one in a higher-level net must fire simultaneously. The transition relation $\rho(t)$ may also react to initiality and finality of input tokens.

Example 18. Example 17, continued. We define the transition function and allow v and w to fire only if a transition labeled $a(n)$ for some $n \in \mathbb{N}$ fires in N_1. Formally, we write this as follows

$$\rho_1(v) = \{(\{r \mapsto (N_1,M)\}, \{s \mapsto (N_1,M')\}) \mid \exists n \in \mathbb{N} :: (N_1,M) \xrightarrow{a(n)} (N_1,M')\},$$

$$\rho_1(w) = \{(\{s \mapsto (N_1,M)\}, \{r \mapsto (N_1,M')\}) \mid \exists n \in \mathbb{N} :: (N_1,M) \xrightarrow{a(n)} (N_1,M')\}.$$

By defining ρ_1 as above, we require that firings of both v and w must synchronize with firings of u.

To complete the definition of the net we introduce the labeling function ℓ_1. For any γ and δ, $\ell_1(v,\gamma,\delta) = \tau$. For $\ell_1(w, s \mapsto (N_1,M), r \mapsto (N_1,M')) = b(n)$, where $n \in \mathbb{N}$ is such that $(N_1,M) \xrightarrow{a(n)} (N_1,M')$. □

We have seen that a token net can fire synchronously with the net containing it. However, we also want to allow a token net to proceed on its own, synchronizing only when some specific events occur or special states are reached. For this reason, we allow τ-labeled firing of token nets, thus causing a color change, without firing a transition in the higher-level net. Of course, in $\mathfrak{M}_0$ this cannot occur. We formalize a new firing relation $\Longrightarrow$ in the following definition.

Definition 19. *Let the ternary relation* $\_ \Longrightarrow \_ \subseteq \mathfrak{M}_\omega \times \Sigma \times \mathfrak{M}_\omega$ *be the smallest relation satisfying*

- $(N, M + \gamma) \xRightarrow{\ell(t,\gamma,\delta)} (N, M + \delta)$ *for all* $(N,M) \in \mathfrak{M}_\omega$ *such that* $N = (P,T,F,\upsilon,\rho,\ell)$, $t \in T$ *and* $(\gamma,\delta) \in \rho(t)$;
- $x \xRightarrow{\tau} x'$ *implies* $(N, M + [(p,x)]) \xRightarrow{\tau} (N, M + [(p,x')])$, *for all* $(N,M) \in \mathfrak{M}_\omega$ *such that* $N = (P,T,F,\upsilon,\rho,\ell)$, *and all* $p \in P$ *such that* $\upsilon(p) \subseteq \mathfrak{M}_\omega$, *and* $x \in \upsilon(p)$,

The classes of (marked) nested nets are $\mathfrak{M}_\omega$ and $\mathfrak{N}_\omega$ respectively with $\Longrightarrow$ as firing relation. This firing relation defines the nested net transition system. Note that $\Longrightarrow$ and $\longrightarrow$ coincide in $\mathfrak{M}_0$. For $\mathfrak{M}_\omega$ we let the $\Longrightarrow$ firing relation replace the $\longrightarrow$ relation and define EWF-bisimilarity and soundness with respect to this relation $\Longrightarrow$.

We illustrate the firing relation introduced by means of Example 2.

Example 20. Given the markings of N_1 we observe that for any natural number n, four kinds of markings are possible for N_0: the token in N_0 is either in r or in s and the token in N_1 has value n and is either in p or in q. We denote the four combinations by $M_n^{r(p)}$, $M_n^{r(q)}$, $M_n^{s(p)}$, and $M_n^{s(q)}$. Then, the following sequence of firings is possible starting from $(N_0, M_0^{r(p)})$. First, an autonomous firing can take place in N_1, moving the token in it from p to q and increasing its numerical value, i.e., $(N_0, M_0^{r(p)}) \xRightarrow{\tau} (N_0, M_1^{r(q)})$. Next, u can fire in N_1. This transition increments the value of the parameter to 2. The firing of u is labeled by $a(1)$ and by definition of ρ_1 v fires as well. It should be noticed that by means of the labeling and the transition function synchronization is achieved. Observe also that since any firing of v is labeled by ℓ_1 as τ, this synchronized step is also silent, i.e. $(N_0, M_1^{r(q)}) \xRightarrow{\tau} (N_0, M_2^{s(p)})$. Next, N_1 can again fire autonomously and again the value of the parameter is incremented: $(N_0, M_2^{s(p)}) \overset{\tau}{\Longrightarrow} (N_0, M_3^{s(q)})$. Finally, a synchronized firing of u and w occurs. The firing of u increments the parameter further. The firing is labeled by $a(3)$. This time the firing of N_0 is observable since ℓ_1 labels all firings of w by $b(n)$, where n is the value of the parameter in the labeling of the firing of u, i.e., $(N_0, M_3^{s(q)}) \overset{b(3)}{\Longrightarrow} (N_0, M_4^{r(p)})$.   □

The class of (marked) nested EWF nets are (marked) EWF nets over the universe $\mathcal{U}_\omega$ and the firing relation $\Longrightarrow$, which we represent by $\mathfrak{M}_\omega^w$ and $\mathfrak{N}_\omega^w$ respectively.

The healthcare workflow net discussed in detail in Section 2 is an example of a nested EWF net.

Since nested nets are derived from colored nets, and nested EWF nets are derived from EWF nets, operations and predicates introduced in Section 4 are readily applicable to nested nets. The example in Section 2 uses operators to combine processes modeled as nets. For instance, radiotherapy should be ministered in parallel with a cisplatine treatment followed by an etoposide treatment. After each session of treatment tests are carried out and the entire process is repeated four times. Using the operations we have introduced, this medical scenario can be written as $init(((Radiotherapy\|(Cisplatin \cdot Etoposide)) \cdot Tests)^4)$. This is a sound EWF net, provided the operand nets are sound.

Theorem 21. *Let* $\mathbb{L} \subseteq \mathfrak{N}_\omega^w$ *be a library of sound nested EWF nets. Then any term denoting a (marked) nested EWF net, obtained from the nets in* $\mathbb{L}$ *by application of* $\cdot$, $\|$, $+$ *and init is sound.*

In order to apply the theory we need a language to define the transition function. The language should be expressive enough to model protocols such as the one described in Section 2. However, we restrict the set of operations to concentrate on the behavior of the token nets rather than the way they are modeled. We call this class of nested EWF nets *adaptive* workflow nets.

We assume the existence of a library $\mathbb{L}$ of sound adaptive workflow nets. We assume that a countable set of variables $\mathbb{V}$ is given. The set of variables appearing in a term t is denoted $Vars(t)$.

Definition 22. *Let* $A, B \subseteq \mathbb{V}$, *and let* t *be a term constructed from constants from* $\mathbb{L}$, *and variables from* $A \cup B$ *by means of application of* $\cdot$, $\|$, $+$ *and init. Then, the following are called* basic net predicates with respect to A and B:

- $initial(t)$, $final(t)$, $t \overset{l}{\longrightarrow}$, *where* $l \in \Sigma \setminus \{\tau\}$, *and for any* $v \in Vars(t)$ *it holds that* $v \in A$;
- $t \overset{l}{\longrightarrow} u$, *where* $u \in B$, $l \in \Sigma \setminus \{\tau\}$, *and for any* $v \in Vars(t)$ *it holds that* $v \in A$;
- $u = t$, *where* $u \in B$, *and for any* $v \in Vars(t)$ *it holds that* $v \in A$;

Definition 23. *A nested EWF net* $(P, T, F, \upsilon, \rho, \ell)$ *is called* adaptive *if for each* $t \in T$ *we have* $\rho(t)$ *is defined as follows. Let* $\alpha, \beta : P \to \mathbb{V}$ *be one-to-one mappings with* $ran(\alpha) \cap ran(\beta) = \emptyset$, *and* ϕ *is a first order logic formula constructed by means of predicates associated with* $\mathcal{U}_0$ *and the basic net predicates with respect to* $ran(\alpha)$ *and* $ran(\beta)$ *such that*

- $\exists$ *and* $\forall$ *are applied only to variables of type* $\mathcal{U}_0$;
- *for any free variable* x *of* $\phi(\alpha, \beta)$, *it holds that* $x \in ran(\alpha) \cup ran(\beta)$.

Let $\theta : \mathbb{V} \to \mathcal{U}_\omega$. Let α^* be the maximal subset of $\theta \circ \alpha$ such that for any x, $\alpha^*(x) \in v(x)$ and let β^* be the maximal subset of $\theta \circ \beta$ such that for any x, $\beta^*(x) \in v(x)$. Then, $\rho(t)$ should be $\{(\gamma, \delta) \mid \gamma \in \alpha^* \wedge \delta \in \beta^* \wedge \phi(\gamma, \delta)\}$.

Note that $\overset{l}{\Longrightarrow}$ coincides with $\overset{l}{\longrightarrow}$ whenever $l \neq \tau$. So the predicates $t \overset{l}{\longrightarrow}$ and $t \overset{l}{\longrightarrow} u$ in Definition 23 coincide with $t \overset{l}{\Longrightarrow}$ and $t \overset{l}{\Longrightarrow} u$ respectively.

The following theorem states a basic fact about adaptive nets. Its proof uses the fact that the unary net predicates only address behavior, i.e., they are true for a net x if and only if they are true for a net y that is EWF-bisimilar to x, and binary predicates always contain a variable in $\mathfrak{M}_\omega$. In other words, test predicates such as $x = C$ for $x \in ran(\gamma)$ and $C \in \mathbb{L}$ are excluded.

Theorem 24. *Let $x \in \mathfrak{M}_\omega^w$ be EWF-bisimilar to an adaptive workflow net y. Let $N = (P, T, F, v, \rho, \ell)$ in $\mathfrak{N}_\omega^w$ and $p \in P$ such that $v(p) = \mathfrak{M}_\omega^w$. Then $(N, M + [(p,x)]) \in \mathfrak{M}_\omega^w$ is EWF-bisimilar to $(N, M + [(p,y)])$.*

To illustrate the definition above we formalize a number of transitions from the motivating example discussed in Section 2. We start with the transition labeled *operable* in Figure 4. For the sake of simplicity, assume that this transition is called t. It has one input place, say p, and one output place, say q. Transition t can fire only if a firing labeled *operable* occurs in the token net of the input place. In this case, resection should be performed followed by four iterations of chemotherapy (cisplatin and then etoposide) and testing. Let $\rho(t) = \{(\gamma, \delta) \mid \phi(\gamma, \delta)\}$. Since p is the only input place and q is the only output place, γ should have a form $\{p \mapsto x\}$, and δ should have a form $\{q \mapsto y\}$. To restrict the input we require $x \overset{operable}{\longrightarrow}$. To produce the output we write $y = init(Resection \cdot (Cisplatin \cdot Etoposide \cdot Tests)^4)$. Hence,

$$\rho(t) = \{(\{p \mapsto x\}, \{q \mapsto y\}) \mid x \overset{operable}{\longrightarrow} \wedge$$
$$y = init(Resection \cdot (Cisplatin \cdot Etoposide \cdot Tests)^4)\}.$$

The second example is the transition labeled *radiation scarring* in Figure 3. Whenever radiation scarring is observed by the doctor, prophylactic treatment should be administered in parallel with the ongoing process. We model this part of the guideline as a transition that can fire whenever *radiation scarring* is observed in a lower-level net. This transition produces the parallel composition of one of its inputs and *ProphT*, on the one hand, and returns the radiation control net to its original place. Since *ProphT* is a library net it should be first initialized and then used in the parallel composition. Assuming that the transition is called u, its input and output places are *surv* and *radc*, we write

$$\rho(u) = \{(\{surv \mapsto x_1, radc \mapsto x_2\}, \{surv \mapsto y_1, radc \mapsto y_2\}) \mid$$
$$x_2 \overset{radiation\ scarring}{\longrightarrow} y_2 \wedge y_1 = x_1 \| init(ProphT)\}.$$

6   Conclusion

Inspired by the existing work on the "nets in nets" paradigm [20, 27, 36], we introduced the class of nested nets to model processes with an adjustable structure. We also introduced EWF nets that extend classical workflow nets with an exception handling mechanism. Adaptive workflow nets are nested EWF nets whose behavior depends on the behavior of the token nets rather than on the way the token nets are modeled. Adaptive workflow nets are more flexible than classical workflow nets and have more structure than ad hoc systems.

To manipulate token nets in an adaptive workflow net we have identified a number of operations on EWF nets, including sequential composition, parallel composition and choice.

We have shown that the formalism presented allows to model realistic adaptive systems such as medical guidelines.

Future work. Our framework is obviously Turing complete. Hence, properties like soundness are in principle undecidable. Therefore, we would like to investigate subsets of adaptive nets where behavioral properties are decidable. It should be noted that coverability is decidable for the class of nested nets considered in [26], which gives us the background for finding suitable subclasses of EWF nets.

The fact that nets can be built or modified at runtime makes it mandatory to develop patterns of nets for which soundness can be derived (soundness by construction). Another line of research is the implementation of our framework within an existing workflow engine.

References

1. W. M. P. van der Aalst. Verification of workflow nets. In P. Azéma and G. Balbo, editors, *Application and Theory of Petri Nets 1997, ICATPN'1997*, volume 1248 of *Lecture Notes in Computer Science*. Springer-Verlag, 1997.
2. W. M. P. van der Aalst. The Application of Petri Nets to Workflow Management. *The Journal of Circuits, Systems and Computers*, 8(1):21–66, 1998.
3. W. M. P. van der Aalst. Workflow verification: Finding control-flow errors using Petri-net-based techniques. In W. M. P. van der Aalst, J. Desel, and A. Oberweis, editors, *Business Process Management: Models, Techniques, and Empirical Studies*, volume 1806 of *Lecture Notes in Computer Science*, pages 161–183. Springer-Verlag, 1999.
4. W. M. P. van der Aalst and T. Basten. Inheritance of workflows: an approach to tackling problems related to change. *Theor. Comput. Sci.*, 270(1-2):125–203, 2002.
5. W. M. P. van der Aalst, T. Basten, H. M. W. E. Verbeek, P. A. C. Verkoulen, and M. Voorhoeve. Adaptive workflow-on the interplay between flexibility and support. In *ICEIS*, pages 353–360, 1999.
6. W. M. P. van der Aalst and K. M. van Hee. *Workflow Management: Models, Methods, and Systems*. MIT Press, 2002.
7. W. M. P. van der Aalst, Arthur H. M. ter Hofstede, B. Kiepuszewski, and A. P. Barros. Workflow patterns. *Distributed and Parallel Databases*, 14(1):5–51, 2003.
8. W. M. P. van der Aalst, D. Moldt, R. Valk, and F. Wienberg. Enacting Interorganizational Workflows Using Nets in Nets. In J. Becker, M. Mühlen, and M. Rosemann, editors, *Proceedings of the 1999 Workflow Management Conference Workflow-based Applications, Münster, Nov. 9th 1999*, Working Paper Series of the Department of Information Systems, pages 117–136, University of Münster, Department of Information Systems, Steinfurter Str. 109, 48149 Münster, 1999. Working Paper No. 70.

9. M. Adams, A. H. M. ter Hofstede, D. Edmond, and W. M. P. van der Aalst. Facilitating flex-ibility and dynamic exception handling in workflows through worklets. In O. Belo, J. Eder, J. Falcão e Cunha, and O. Pastor, editors, *CAiSE Short Paper Proceedings*, volume 161 of *CEUR Workshop Proceedings*. CEUR-WS.org, 2005.

10. O. Biberstein, D. Buchs, and N. Guelfi. Object-oriented nets with algebraic specifications: The CO-OPN/2 formalism. In *Concurrent Object-Oriented Programming and Petri Nets*, volume 2001 of *Lecture Notes in Computer Science*, pages 73–130, 2001.

11. G. Ciardo and P. Darondeau, editors. *Applications and Theory of Petri Nets 2005, 26th Inter-national Conference, ICATPN 2005, Miami, USA, June 20-25, 2005, Proceedings*, volume 3536 of *Lecture Notes in Computer Science*. Springer, 2005.

12. J. Desel, W. Reisig, and G. Rozenberg, editors. *Lectures on Concurrency and Petri Nets, Advances in Petri Nets*, volume 3098 of *Lecture Notes in Computer Science*. Springer, 2004.

13. H. Ehrig and J. Padberg. Graph grammars and Petri net transformations. In Desel et al. [12], pages 496–536.

14. A. Fent, H. Reiter, and B. Freitag. Design for change: Evolving workflow specifications in ULTRAflow. In A. B. Pidduck, J. Mylopoulos, C. C. Woo, and M. T. Özsu, editors, *Advanced Information Systems Engineering, 14th International Conference, CAiSE 2002, Toronto, Canada, May 27-31, 2002, Proceedings*, volume 2348 of *Lecture Notes in Com-puter Science*, pages 516–534. Springer, 2002.

15. F. V. Fossela, R. Komaki, and G. L. Walsh. Small-cell lung cancer. Practice Guideline, Available at `http://utm-ext01a.mdacc.tmc.edu/mda/cm/CWTGuide.nsf/LuHTML/SideBar1?OpenDocument` by following `Thoracic` and `Small Cell Lung Cancer` links, 2000.

16. J. Groote and F. Vaandrager. Structured operational semantics and bisimulation as a congru-ence. *Information and Computation*, 100(2):202–260, 1992.

17. K. M. van Hee, N. Sidorova, and M. Voorhoeve. Soundness and separability of workflow nets in the stepwise refinement approach. In W. M. P. van der Aalst and E. Best, editors, *ICATPN*, volume 2679 of *Lecture Notes in Computer Science*, pages 337–356. Springer, 2003.

18. K. M. van Hee, N. Sidorova, and M. Voorhoeve. Generalised soundness of workflow nets is decidable. In J. Cortadella and W. Reisig, editors, *Application and Theory of Petri Nets 2004, ICATPN'2004*, volume 3099 of *Lecture Notes in Computer Science*, pages 197–216. Springer-Verlag, 2004.

19. K. Hoffman. Run time modification of algebraic high level nets and algebraic higher order nets using folding and unfolding construction. In G. Hommel, editor, *Proceedings of the 3rd Internation Workshop Communication Based Systems*, pages 55–72. Kluwer Academic Publishers, 2000.

20. K. Hoffmann, H. Ehrig, and T. Mossakowski. High-level nets with nets and rules as tokens. In Ciardo and Darondeau [11], pages 268–288.

21. K. Jensen. *Coloured Petri Nets - Basic Concepts, Analysis Methods and Practical*. Springer-Verlag, 1992.

22. M. Klein and C. Dellarocas. A knowledge-based approach to handling exceptions in work-flow systems. *Comput. Supported Coop. Work*, 9(3-4):399–412, 2000.

23. M. Köhler and H. Rölke. Reference and value semantics are equivalent for ordinary object petri nets. In Ciardo and Darondeau [11], pages 309–328.

24. C. Lakos. From coloured Petri nets to object Petri nets. In *ICATPN*, volume 935 of *Lecture Notes in Computer Science*, pages 278–297, 1995.

25. I. A. Lomazova. Nested Petri nets: Multi-level and recursive systems. *Fundam. Inform.*, 47(3-4):283–293, 2001.

26. I. A. Lomazova. Modeling dynamic objects in distributed systems with nested Petri nets. *Fundam. Inform.*, 51(1-2):121–133, 2002.

27. I. A. Lomazova. *Nested Petri nets: modeling and analysis of distributed systems with object structure*. Moscow: Nauchny Mir, 2004. in Russian.

28. I. A. Lomazova and P. Schnoebelen. Some decidability results for nested Petri nets. In D. Bjørner, M. Broy, and A. V. Zamulin, editors, *Ershov Memorial Conference*, volume 1755 of *Lecture Notes in Computer Science*, pages 208–220. Springer, 1999.

29. D. Moldt and F. Wienberg. Multi-agent-systems based on coloured Petri nets. In P. Azéma and G. Balbo, editors, *ICATPN*, volume 1248 of *Lecture Notes in Computer Science*, pages 82–101. Springer, 1997.

30. S. Panzarasa, S. Maddè, S. Quaglini, C. Pistarini, and M. Stefanelli. Evidence-based careflow management systems: the case of post-stroke rehabilitation. *Journal of Biomedical Informatics*, 35(2):123–139, 2002.

31. M. Peleg, A. Boxwala, S. Tu, D. Wang, O. Ogunyemi, and Q. Zengh. Guideline interchange format 3.5 technical specification. InterMed Project, 2004.

32. S. Quaglini, S. Panzarasa, A. Cavallini, G. Micieli, C. Pernice, and M. Stefanelli. Smooth integration of decision support into an existing electronic patient record. In S. Miksch, J. Hunter, and E. T. Keravnou, editors, *AIME*, volume 3581 of *Lecture Notes in Computer Science*, pages 89–93. Springer, 2005.

33. S. Quaglini, M. Stefanelli, A. Cavallini, G. Micieli, C. Fassino, and C. Mossa. Guideline-based careflow systems. *Artificial Intelligence in Medicine*, 20(1):5–22, 2000.

34. S. Rinderle, M. Reichert, and P. Dadam. Correctness criteria for dynamic changes in workflow systems - a survey. *Data Knowl. Eng.*, 50(1):9–34, 2004.

35. R. Valk. Nets in computer organization. In Brauer, W., Reisig, W., and Rozenberg, G., editors, *Lecture Notes in Computer Science: Petri Nets: Applications and Relationships to Other Models of Concurrency, Advances in Petri Nets 1986, Part II, Proceedings of an Advanced Course, Bad Honnef, September 1986*, volume 255, pages 218–233. Springer-Verlag, 1987. NewsletterInfo: 27.

36. R. Valk. Object Petri nets: Using the nets-within-nets paradigm. In Desel et al. [12], pages 819–848.

Analyzing Software Performance and Energy Consumption of Embedded Systems by Probabilistic Modeling: An Approach Based on Coloured Petri Nets

Meuse N.O. Junior[1], Silvino Neto[1], Paulo Maciel[1], Ricardo Lima[2], Angelo Ribeiro[1], Raimundo Barreto[1], Eduardo Tavares[1], and Frederico Braga[3]

[1] Centro de Informática (CIn) Universidade Federal de Pernambuco (UFPE)-Brazil
{mnoj, svvn, prmm, arnpr, rsb, eagt}@cin.ufpe.br
[2] Departamento de Sistemas Computacionais Universidade de Pernambuco (UPE)- Brazil
ricardo@upe.poli.br
[3] ITPE Instituto Tecnológico do Estado de Pernambuco -Brazil
fcabraga@itpe.br

Abstract. This paper presents an approach for analyzing embedded systems' software energy consumption and performance based on probabilistic modeling. Such an approach applies Coloured Petri Net modeling language (CPN) for supporting simulation and analysis. The proposed approach offers three basic contributions for embedded system analysis field:(i)code modeling, a probabilistic model for scenarios exploration being presented, (ii) formalism, a formal and widespread modeling language (CPN) being applied, with previously validated engines and algorithms for simulation;(iii) flexibility, the proposed approach enabling modeling of different micro-controllers at different abstraction levels.

1 Introduction

Technological advances bring new techniques and paradigms for field of embedded system. In order to improve designs in such novel paradigms, methods for design space exploration play an important role, by seeking the optimization of parameters such as performance and energy consumption. A crucial point for such exploration is defining the typical operation scenarios[1] and worst-case scenario. Classic approaches use model scenarios as test vectors where sets of events and data loads are gathered to perform simulation analysis [20]. In the context of processor performance exploration, such vectors are approached by executing standard benchmark code. In this way, it is possible to estimate the typical execution time and energy consumption for a specific processor applied to a specific domain. However, design space exploration implies evaluating a specific code in its target application. How can the execution time and energy consumption possibilities for a specific processor running a specific code within a system with a non-specific set of scenarios be evaluated? Defining test vectors according to possible scenarios is normally very difficult. This paper approaches such a question by extending our previous works [10, 11], where Coloured Petri Nets (CPN) [7] are applied to describing an application (code + processor architecture), focusing on probabilistic modeling to describe scenarios. The proposed modeling allows control flow

---

[1] In this text, the term scenario is restricted to *set of events*.

S. Donatelli and P.S. Thiagarajan (Eds.): ICATPN 2006, LNCS 4024, pp. 261–281, 2006.

exploration to statistically cover a large range of load and event possibilities. From this approach, a framework for analyzing embedded system software is presented. The application of the CPN modeling language is motivated by: (i) formalism, a consolidated set of rules being applied to describe models, (ii) graphic representation, offering more cognitive model reasoning and representation, (iii) hierarchical modeling capability, allowing different abstraction levels, (iv) behavioral description supported by programing language, guaranteeing an accurate executable model, (iv) simulation resources, tools and validated simulation algorithms being available. Thus, a formal model can be constructed and validated with minimal effort, if compared with *ad hoc* construction based on general program language. Once the model being constructed, a specific and efficiency centered tool comes directly from application of software engineering rules.

This research is centered on the micro-controller domain, the power model implemented describing the consumption of instruction by measuring entire micro-controller consumption. In this way, the code- and data-memory consumption is naturally aggregated to the instruction power-model. In this work, results concerning pipeline-less architectures are presented, the AT89S8252[2] micro-controller being chosen as a target. This paper is organized as follows: Section 2 presents related works, Section 3 presents instruction power characterization, Section 4 introduces the architecture description model, Section 5 presents a framework and analysis environment, Section 6 shows the model validation, Section 7 presents experimental results and Section 8 presents the paper's conclusions.

2   Related Works

In the context of digital systems, the concept of statistical simulation and probabilistic analysis has been applied to hardware power estimation [17]. This concept attacks *the pattern-dependence problem*, namely the dependence between results and entry test vector. From the concept of probabilistic analysis, circuit events are modeled as random processes with certain statistical characteristics. The probabilistic analysis is applied to estimate the circuit activity level, which is directly related to energy consumption [20]. Burch *et al* [5] have proposed an alternative approach mixing the premises of probabilistic analysis with a simulation-based technique. The method consist of applying randomly-generated input patterns to the circuit, simulating and monitoring the power value result. The result is evaluated until it reaches a desired *accuracy*, at a specified *confidence level*, according to the Monte Carlo simulation technique. In the context of software performance analysis, Bernat *et al* [2] have proposed a framework for performing probabilistic worst-case execution time(WCET) analysis, termed pWCET. The aim of pWCET is to estimate the probability distribution of the WCET of a given code fragment. Based on a syntax tree model, execution probabilities are associated with basic blocks (sequences of instructions that have no branch instruction). Such probabilities are captured either by measurement (with code instrumentation) or by cycle-accurate simulation, both based on a large set of test vectors (scenarios). The WCET distribution is obtained by analyzing the resulting trace[3] against the syntax tree model. pWCET is

---

[2] An 8051-like architecture with code and data memory integrated on a chip.

[3] A list of tuples (instruction,timestamp).

an interesting and functional approach but *the pattern-dependence problem*[20] remains due to the test-vector mechanism. Marculescu *et al* [15] have presented a probabilistic modeling technique for system-level performance and energy consumption analysis based on a formal model termed *Stochastic Automata Network* (SAN). Their work is based on a *process graph* that represents a net of automata. The proposed *process graph* is characterized by *execution rates* that applied under exponentially distributed activity durations, serve as a Markov chain. SAN thus, provides a formal way for supporting probabilistic evaluation. As a result Marculescu's approach makes it possible to obtain the steady-state probability distribution of system element utilization for a given scenario. In the context of CPN application to embedded systems and power estimation, Burns *et al*[6] applied CPN as a formalism for modeling a generic superscalar processor in order to perform real-time analysis, Murugavel [16] proposed a CPN extension, termed as *Hierarchical Colored Hardware Petri Net*, for dealing with gates and interconnections delays for hardware systems power estimation. More recently, Jørgensen *et al* [8] have demonstrated the strength of CPN centered frameworks as estimating the memory usage of smartphones software.

In our work, differently from [5], the randomly generated pattern is generated during code model simulation according to a probabilistic model of system behavior, as will be clarified later. The generation mechanism is performed by assumptions about the probabilities of an event occurring in a system. Such assumptions can be captured from (i) *ad-hoc* designer knowledge, (ii) a more abstract system model and/or (iii) a system behavior statistical database. Modeling formally such probabilities comprise a research addressed as future works. Currently, this research is focused on *ad-hoc* designer inferences. Since an event can be related to a set of conditional branches in the code, branch's probabilities are inserted in the assembly source-code by annotations. The analysis is currently based only on assembly codes, but the mechanism for extracting such annotations from C code is being developed. The code is modeled as a CPN so that a formal and extensible representation is built. As a result of CPN-model simulation, execution time and energy consumption are automatically estimated. Note that the scenario is being addressed by mapping from a set of branch probabilities onto a set of system events.

3   Instructions Power Characterization

In order to characterize the instruction consumption in the target micro-controller , a measurement setup was implemented, taking works [14] and [18] as references. An *Application for Automatic Power-Characterization* (AAPC) was developed, covering: target microcontroller code download, data capture and database generation. For 8051 architecture, a set of 589 characterization instruction-codes[4] were written. Each instruction-code represents an instruction in a specific situation concerning operands. For example, instruction **ADD A,#Imm** generates various instruction-codes where the **A-value** and **Imm-value** are explored. The instruction-code performs continuous loops, generating hardware outputs to synchronize measurements of execution time and energy consumption. Figure 1 illustrates the automatic measurement cycle and operation of AAPC internal modules.

---

[4] Codes that perform target instructions in a repetitive way.

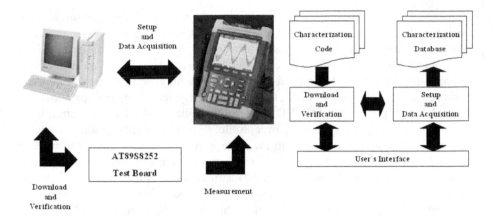

Fig. 1. Automatic power-characterization mechanism

From a SPI (Serial Peripheral Interface) link, between a PC computer and the test-board, the AAPC is able to download and verify characterization codes inside the target micro-controller. A digital oscilloscope is linked with the test-board so that measurements of consumption and execution time are captured as oscilloscope waveforms, the specific instruction execution interval being identified, due to characterization-code *markers*. The AAPC is composed of three basic modules: user interface, download-verification module and acquisition module (See Figure 1). The user interface supports

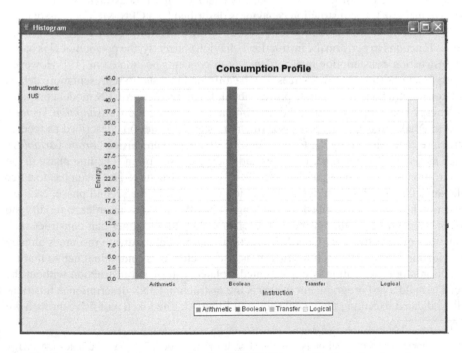

Fig. 2. Average energy consumption for AT89S8252 according to instruction classes

commands for defining target instruction-codes and output database directories. Additionally, AAPC implements analysis resources by showing histograms of measurements and average values related to the class of instructions. For each measurement, the acquisition module saves a file into the characterizations database. When an error occurs the acquisition module and download-verification module re-synchronize the process. Database files are formatted in XML, thus allowing easy access and manipulation.

The AAPC applied to the AT89S8252 micro-controller resulted in two main conclusions about the device's instruction-power profile: (i) there is small variation[5] in energy consumption among instructions with the same execution time within an instruction class and (ii) the intensity of energy consumption is clustered according to instruction classes. Resulting from this, the AT89S8252 instruction-power model is constructed by a small set of values according to execution time and class of instruction (See Figure 2). It is important to emphasize that such an evaluation is specific to the AT89S8252 micro-controller, different devices and families possibly displaying very distinct instruction power characteristics. This argument is supported by comparing works such as [12], [4] and [3].

4   Architecture Description Model

The current model extends our previous architecture CPN-Model [11] so that it performs a simulation based on probabilistic alternatives of code flow. The model presented in [11] is constructed in a way that any code can be built from a set of instruction-CPN model. This set of instruction-CPN models describes the processor ISA (*Instruction Set Architecture*). The instruction behavior is described by a CPN-ML code, associated with transitions in the instruction-CPN model (code transition). The CPN-ML code invokes functions to perform the instruction behavior. Currently, the power model is based on instruction consumption in accordance with concepts postulated in [19]. However, these functions (hardware resources) can be unfolded in order to represent more details for constructing a more accurate power model. Each instruction-CPN model computes its energy consumption during the net simulation, feeding the *Consumption Vector*[6]. The internal context of the processor (memories and registers) is described as record-structure encapsulated by the token. Additionally, the token carries *"probe variables"* such as accumulated number of cycles and energy consumption. The first phase in the description process consist of defining the token data structure. The token has to carry elements that are modified due to instruction execution. In the second phase, each instruction behavior is described by invoking CPN-ML functions in order to modify the token. Depending on the abstraction level intended, the designer can construct such functions or directly apply CPN-ML functions that model hardware resources already provided in the model. Each instruction is constructed as an individual net so that instruction model validation is implemented isolatedly, for each instruction without the need to build a test program. For instance, the instruction CJNE description is functionally validated executing the model shown in Figure 3. The token will flow through the

---

[5] Less than 1% variation, that can be associated with measurements error.

[6] A vector where each position is associated with an instruction. The position holds the energy consumed by the instruction during the simulation.

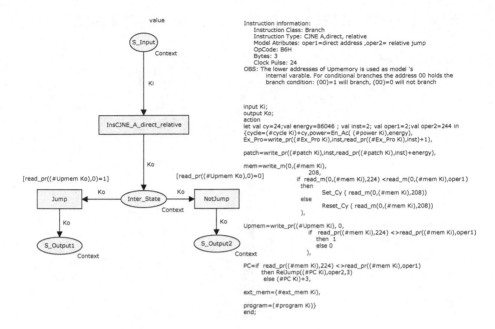

value

S_Input

Context

Ki

Instruction information:
    Instruction Class: Branch
    Instruction Type: CJNE A,direct, relative
    Model Attributes: oper1=direct address ,oper2= relative jump
    OpCode: B6H
    Bytes: 3
    Clock Pulse: 24
OBS: The lower addresses of Upmemory is used as model 's
    internal varable. For conditional branches the address 00 holds the
    branch condition: (00)=1 will branch, (00)=0 will not branch

InsCJNE_A_direct_relative

input Ki;
output Ko;
action
let val cy=24;val energy=86046 ; val inst=2; val oper1=2;val oper2=244 in
{cycle=(#cycle Ki)+cy,power=En_Ac((#power Ki),energy),
Ex_Pro=write_pr((#Ex_Pro Ki),inst,read_pr((#Ex_Pro Ki),inst)+1),

patch=write_pr((#patch Ki),inst,read_pr((#patch Ki),inst)+energy),

mem=write_m(0,(#mem Ki),
        208,
        if read_m(0,(#mem Ki),224) <read_m(0,(#mem Ki),oper1)
        then
            Set_Cy (read_m(0,(#mem Ki),208))
        else
            Reset_Cy (read_m(0,(#mem Ki),208))
),

Upmem=write_pr((#Upmem Ki), 0,
        if read_pr((#mem Ki),224) <>read_pr((#mem Ki),oper1)
        then 1
        else 0
),

PC=if read_pr((#mem Ki),224) <>read_pr((#mem Ki),oper1)
    then RelJump((#PC Ki),oper2,3)
    else (#PC Ki)+3,

ext_mem=(#ext_mem Ki),

program=(#program Ki)}
end;

Ko

[read_pr((#Upmem Ko),0)=1]   [read_pr((#Upmem Ko),0)=0]

Jump   Ko   Inter_State   Ko   NotJump

Context

Ko   Ko

S_Output1   S_Output2

Context   Context

Fig. 3. A branch instruction description in the CPN-deterministic model

instruction model until it reaches place *S_Output1* or *S_Output2*, depending on the code segment evaluation. Note that the decision process is implemented by a net structure (transition Jump and NotJump) according to an internal flag generated by code segment in transition. Each instruction model computes its clocks cycle. Therefore, the CPN simulation implements a cycle-accurate processor simulation. In order to automatically generate a CPN-Model application, a Binary-CPN compiler is applied. Given a code of interest and the processor description, a Binary-CPN compiler constructs a CPN model of the application (processor + code). The CPN structure *per si* models the possible program flows, where each possible program state is modeled as a place. Each transition is a substitution-transition encapsulating the instruction-CPN model. Figure 4 shows the CPN model in the CPNTools environment [1]. On the left-hand side, hardware resources are modeled by CPN-ML functions, implementing operation associated with hardware, and on the right-hand side the application model is represented as a Coloured Petri Net. In order to perform the behavior of CALL-RET instructions and interruptions, a special structure is implemented, the *token distributor*. When a CALL- or RET-type instruction is executed, the token is sent to the *token distributor* structure (Figure 4). From there, the token is driven to output *afterPC* place by firing transition *CalculatingPC* (Figure 4). From *afterPC* place, the token is driven to a net place (code address) according to the PC value by firing *PCAFTrans_n*, where *n* is the place addressed by the PC value. The simulation stops when a RET instruction is detected without a previous CALL instruction being executed. In this case, after being evaluated by *calculatingPC* transition, the token is held in *afterPC* place of the *token distributor*. In order to provide exit points in the code under evaluation, isolated-RET instructions are inserted into the code. Such an expedient is applied as the only code-instrumentation mechanism.

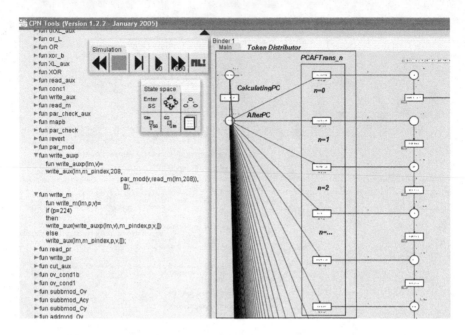

Fig. 4. Application model in CPNTools edition and simulation interface

4.1 Probabilistic Model

In the proposed probabilistic model, the behavioral description is removed from the instruction-CPN model. Thus, the instruction-CPN model does not simulate instruction operation any longer, only register its time and energy consumption. Only unconditional branch instructions, such as CALL, RET and JMP, remain with their original behavior. For conditional branch instructions, a probabilistic behavioral model is implemented (See Figure 5). All others keeps only their computation of cycle and energy consumption. This implies: (i) improvement on performance simulation and (ii) more independence of model concerning processor instruction set. The *token distributor* was modified in order to process the stop-criteria evaluation. Specific functions for statistic manipulation and stop-criteria evaluation support were implemented. Figure 6 shows the CPN model in the CPNTools environment. The binary-CPN compiler was modified in order to (i) insert parameters related to the probabilistic model, (ii) insert the stop-criteria evaluation code into *calculatingPC* (See Figure 7) transition in the *token distributor* structure and (iii) model basic blocks as a transition (Basic Block Modeling). The parameters related to probabilistic modeling and stop-criteria evaluation are captured from annotations in the source code. There are two type of annotation: header-annotation and instruction-annotation. Header-annotation informs simulation parameters associated with stop-criteria. Instruction-annotation informs simulation parameters associated with probabilistic instruction behavior.

4.2   Simulation Parameters

In order to simulate the code flow according to event probabilities, two parameters have been inserted into the model: *prob* and *wloop*(See Figure 5). Given a conditional branch instruction, its probability to branch is defined by *prob*'s value. The branch condition is evaluated by comparing the *prob* value against a random real number generated in a [0.0,1.0] interval, through a uniform distribution function. If the generated number is equal or less than *prob*'s value, the instruction branches. In this way, the probability of instruction branching is equal to *prob*'s value. *wloop* restricts the number of consecutive branches. It has two important roles in the model:(i) allowing the modeling of limited iterative-deterministic branches (loops) and (ii) restricting consecutive branches in probabilistic branch instructions. Note that deterministic modeling is implemented by assigning value 0.0 or 1.0 to *prob*. A deterministic limited loop structure is modeled as a branch instruction with *prob=1.0* and *wloop* equal to number of iterations intended. Additionally, the model supports a sweep simulation mode. In the sweep mode, the simulation explores a set of chosen instructions by evaluating them considering probability intervals. For each branch instruction, a fixed probability to be applied in the normal simulation mode and a probability interval to be applied in the sweep mode are defined. Sweep mode is useful in systems or code fragments where particular external events can be directly mapped to branch instructions. Additionally, the combination of every data captured from all sweep simulation provides a general vision of the entire set of scenarios. The probability interval is constructed according to the format:

$$[P_i, P_f, st]$$

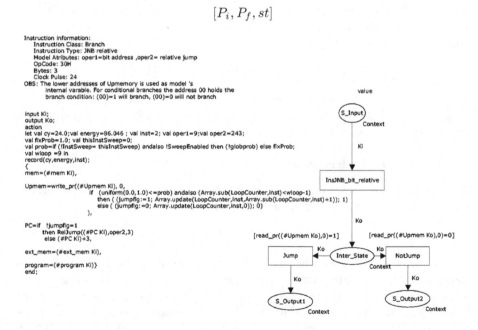

Fig. 5. Probabilistic model of branch instructions

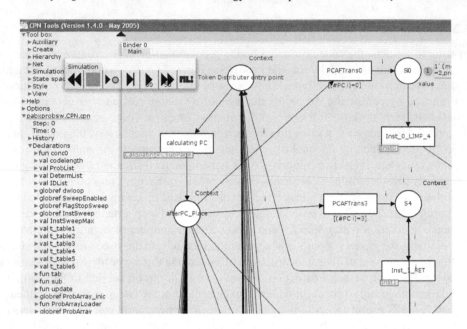

Fig. 6. Probabilistic model in CPNTool environment.

where P_i is the initial probability value, P_f is the final probability value and st is the probability step to be performed during the sweep. Thus, the instruction-annotation is constructed according to the format:

$$< @[P_i, P_f, st]|sweepmodefree|fixprob|wloop@ > .$$

Where $[P_i, P_f, st]$ is the sweep interval, $sweepmodefree$ is a boolean value that informs whether the instruction operates only with fixed probability, and $fixprob$ is the fixed probability associated with the instruction.

4.3 Stop Criteria Evaluation

The Stop-Criteria parameters are captured from the header-annotation. The header-annotation is constructed according to the format:

$$< \$ConfInterv|E_{max}|TarMet|N_s\$ > .$$

Where $ConfInterv$ is the confidence interval identifier, E_{max} is the maximum error intended, $TarMet$ is the target metric and N_s is the minimum number of simulation cycles. $ConfInterv$ informs which distribution table is being applied in accordance with the confidence interval intended. Currently, the model supports six confidence interval setups, from 90% ($ConfInterv = 1$) to 99% ($ConfInterv = 6$). The $TarMet$ sets the target metric for stop criteria evaluation, power ($TarMet = 1$), energy ($TarMet = 2$) or execution-time ($TarMet = 3$). In order to guarantee enough data for the estimation of probability distribution, N_s has priority over other stop criteria. For instance, the

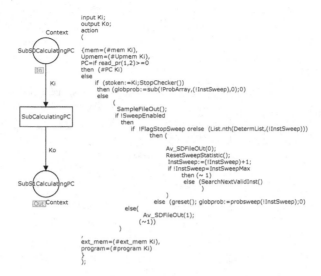

Fig. 7. Mechanism for Stop Criteria Evaluation

stop criterion is checked only if a minimum of N_s simulations have been performed. Note that the designer can tune the quantity of samples from the header-annotation. The stop criteria are evaluated by specific code inserted in the *calculatingPC* transition (See Figure 7). When an isolated RET instruction is executed, the *StopChecker()* function is invoked and, according to return value, a new simulation cycle is started or the simulation is ended.

4.4   Basic Block Modeling

The Binary-CPN compiler improves the final CPN-model based on the fact that probabilistic models do not implement the code behavior. Basic blocks are modeled as only one transition. Additionally, unconditional loops implemented as delays, such as self-loop and loop with internal-NOP instructions are modeled as only one transition. This operation reduces the number of simulation steps, hence speeding up the simulation. For architectures with pipelines, basic block modeling has to take into account pipeline-data and -structural conflicts so as to compute their time penalty and energy cost. The method presented in this work is focused on a pipeline-less model, in which simple-scalar architecture can be evaluated since it can be modeled regardless of pipeline conflicts. At the present stage of the research, a probabilistic model for simplescalar architecture is under construction.

5   Framework and Analysis Environment

Figure 8 shows the methodology for model construction, simulation and analysis. The framework consists of (i) insertion of probabilistic (instruction-annotation), (ii) generation

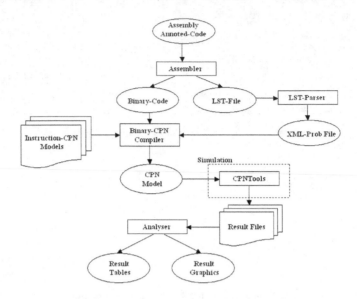

Fig. 8. Framework for probabilistic modeling and evaluation

of a binary code and a LST-code[7] file, (iii) generation of an XML-Prob file (iv) generation of a CPN-Model, (v) simulation of the CPN-Model and (vi) analysis of results.

There are two basic aspects to take into account while inserting instruction-annotation. Firstly, defining whether the instruction operates as a sensor of an external event (probabilistic behavior) or as a countable and predictable branch (deterministic behavior). Finding code structures that represent abstract control structures such as **FOR** and **WHILE(*Constant*)** is a straightforward means of identifying branch instructions with deterministic behavior. In order to identify branch instructions with probabilistic behavior, the designer should analyze those that evaluate variables associated with the system sensors and actuators. Secondly, the header-annotation and probability interval should be correctly specified in order to allow fast and consistent simulation. A high confidence level associated with high accuracy (low maximum error) implies a longer simulation time. Therefore, the designer has to explore values for header-annotation taking into account: code size[8], time simulation and intended accuracy. On the other hand, the probability interval should be carefully specified so as not to insert nonsense scenarios.

The annotated-code is assembled and an executable code (binary code) with a LST code file are obtained. The LST-code file is an ASCII file, where instruction memory-addresses are inserted into instruction lines. Additionally, LST-code files preserve the code's comments, thus preserving the code's annotations. In this way, the LST-code file provides a table mapping code-memory address to instruction-annotation.

In order to extract parameters from instructions-annotations, the LST-code file is processed by a LST-Parser. The LST-Parser generates XML file as output, termed

---

[7] Text file generated during source code assembly.

[8] It is naturally associated with time simulation.

XML-Prob. XML-Prob holds information present in the code-annotation as XML file elements. The next stage is concerned with the generation of the CPN-Model. The Binary-CPN compiler constructs a CPN-Model from binary code, XML-prob and instruction-CPN model files. The CPN-Model is generated in a CPNTools input file format, a well-defined description in XML format. Additionally, the Binary-CPN compiler allows disabling basic block modeling in order to generate detailed profiles as in the deterministic model. Currently, the CPNTools environment has been used as a platform for modeling, editing and simulation. CPNTools offers an efficient environment based on the widespread CPN semantics [13]. Finally, it is important to highlight that the CPN-Model is a computation model, so more specific and efficient engines can be implemented by focusing on specific aspects of the simulation process.

In the last stage, the partial results are processed to provide tables and graphs, characterizing scenarios. Probability distributions representing execution time, energy consumption and power are drawn for each scenario explored during the simulation. A frequency profile is constructed according to frequency classes in captured samples. Applying the conceptual identity between frequency and probability, a probability distribution of classes is constructed, namely a probability distribution of metric value intervals. The length of interval associated with classes is adjustable, the designer being able to adjust it, so as to explore different resolutions for distribution of probability. In order to promote operation flexibility, all the above stages are gathered in an environment for Petri net model integration termed EZPetri [9].

The EZPetri environment is an integration framework combining Petri net tools and new applications in a single environment. The EZPetri was extended in order to support the proposed framework, and was termed EZPetri-PCAF (*Power Cost Analysis Framework*). EZPetri-PCAF operates with three menus: editing, compiling and analyzing. In the editing menu, the user can edit code and insert annotations, the CPN-Model is generated from the compiling menu, and the analysis option is set in the analysis menu. Probability distribution is visualized in a histogram, where the class interval is adjustable from the analysis menu. Additionally, the profile of average and standard deviation is presented.

6   Model Validation

In order to validate the probabilistic model, a BubbleSort code was evaluated. The BubbleSort algorithm was chosen due to its analytical predictability, where best and worst execution times can be obtained from a specific data load. Figure 9 illustrates the validation framework. The validation was performed in two phases: (i) the BubbleSort code was modeled and evaluated according to framework described in section 5, comparing results with measures on hardware, and (ii) a set of 1000 test vectors was generated and automatics measures of energy consumption and execution times were performed applying the system described in section 3. The set of test vectors was generated from a "*seed*" code, each vector consist of the *seed* code with a randomly generated data-vector. Thus, the test vector set represents samples of BubbleSort operation scenarios. The validation was performed evaluating phases (i) results and comparing them with results of phases (ii).

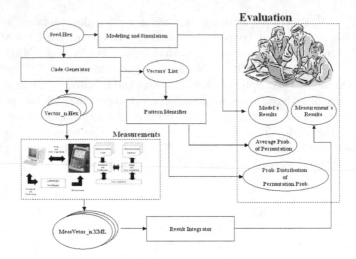

Fig. 9. Framework for model validation

```
1 ;*********BubbleSort for Probilistic Evaluation*************
2 ;************* Header-Annotation ***********************
3              ;<$1|0.05|3|1000$> ;
4 begin:
5       mov r4,#00H
6 loop:
7       mov a,r4
8       add a,#30H
9       mov r4,a
10      lcall swaap
11      mov a,r4
12      subb a,#30H
13      mov r4,a
14      inc r4
15      mov a,r4                    ; Instruction-Annotation
16      cjne a,#9H,loop            ;<@[0.0,1.0,0.1]|true|1.0|9@>
17      ret                        ; Exit point
18 swaap:
19      mov 00H,r4
20      mov r1,00H
21 inner_loop:
22      clr c
23      inc r1
24      mov a,@r0
25      subb a,@r1                 ; Instruction-Annotation
26      jc exit                    ;<@[0.0,1.0,0.1]|false|1.0|256000@>
27      inc r5
28      mov b,@r0
29      mov a,@r1
30      mov @r0,a
31      mov @r1,b
32 exit:
33      mov a,r1                   ; Instruction-Annotation
34      cjne a,#39H,inner_loop     ;<@[1.0,1.0,0.1]|true|1.0|5@>
35      ret
```

Fig. 10. BubbleSort code for model validation

Figure 10 shows a BubbleSort code (seed code) implementation for ordering a vector with 10 elements. Note that all the program's flow variance is defined for three branch-instructions at lines 16 (*inst16*), 26 (*inst26*) and 34 (*inst34*). Instructions *inst16* and *inst34* are vector-length dependent, in which a deterministic behavior is performed. On the other hand, instruction *inst26* has a probabilistic behavior, depending on the ordering level of vectors. Analyzing the BubbleSort algorithm, it is verified that:(i) instruction *inst16* branches $vector\_length - 1$ times , (ii) for each *inst16* branch,

Table 1. Model Validation: Deterministic Behavior of BubblSort's code

Metric	Estimated. Worst-Case	Estimated. Best-Case	Measured. Worst-Case	Measured. Best-Case	Error(%) Worst-Case	Error(%) Best-Case
Average Power(mW)	49,99	49,75	47,86	47,06	4,45	5,72
Total Energy(μJ)	47,59	31,69	44,13	29,98	7,84	5,50
Execution-Time(μs)	952	637	952	637	0	0

identified by index from $n = 1$ to $n = vector\_length - 1$, *inst34* branches b_n times, and (iii) b_n is defined in an arithmetic-progressive function as follows:

$$b_n = (vector\_length - 1) + (n - 1) * (-1).$$

Thus, the total number of *inst34* branches is deterministic and given by:

$$T_{branches}^{inst34} = \sum_n b_n = (b_1 + b_{(vector-length-1)}) * (vector - length - 1)/2.$$

For a vector with 10 elements, $T_{branches}^{inst34} = 45$. Note that *inst34* branches 45 times as *inst16* branches 9 times. Hence, the average *inst34*'s number of branches is 5. As the model intends to estimate the total number of branches, the total branches can be calculated based on average values. From theses analytical considerations, *inst16* and *inst34* are modeled as deterministic (*fixprob*=1.0, see subsection 4.2) and limited to 9 and

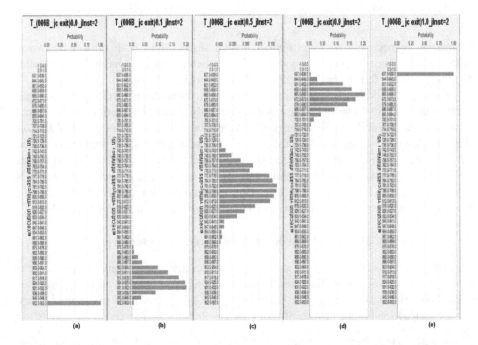

Fig. 11. Probability distribution for execution time depending on scenarios (a) P(*inst26*)=0.0 (worst-case)(b)P(*inst26*)=0.1 (c)P(*inst26*)=0.5 (d)P(*inst26*)=0.9(e)P(*inst26*)=1.0 (best-case)

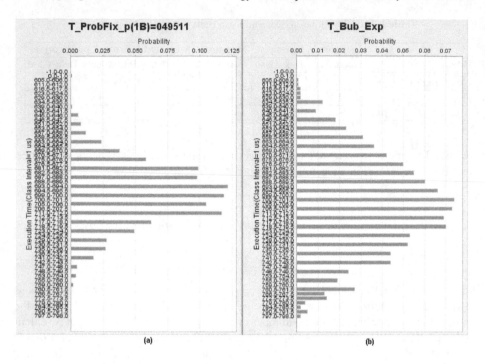

Fig. 12. Consistency validation toward execution in hardware:(a) Probability distribution esti-mated for P(*inst26*)=0.49511 and (b) Probability distribution from hardware experiment

5 branches respectively. As deterministic branch instructions, both are free of sweep-mode effects (*sweepmode free = true*). The probability interval has to be declared to avoid LST-Parser failure, but it is not effective during simulation.

In order to validate the deterministic modeling, instruction *inst26* is set to a de-terministic behavior. The worst case is calculated by setting instruction *inst26* branch probability to 0.0, and for calculating the best case, branch probability is set to 1.0. Con-sidering these two scenarios, the best and worst case for execution time are obtained. Table 1 shows some results obtained from the model evaluation and measured values, where the respective errors are highlighted. The exact accuracy reached in execution time, validates the behavioral modeling. Due to some constraints of the measurement process[9], errors associated with power and energy estimation are overestimated.

6.1 Consistency of Sweep-Mode Results as a Scenario Inferencer

As explained in Section 4.2, this work simulates scenarios by mapping from a set of probabilities to a set of system events. The simulation is steady-state, allowing two oper-ation modes: normal mode (instruction branch probability fixed) and sweep mode(vary-ing the instruction branch probability). If event probabilities are fixed, there is only one

---

[9] There are some problems concerning the oscilloscope memory limitation that impact on sam-pling rate, and consequently on measurement accuracy, but this issue is outside the scope of this paper.

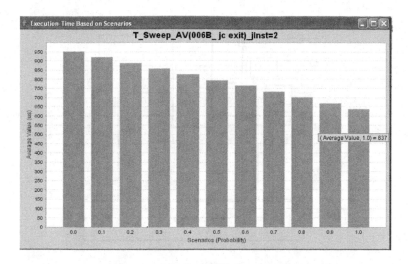

Fig. 13. Execution-Time average value according to scenario

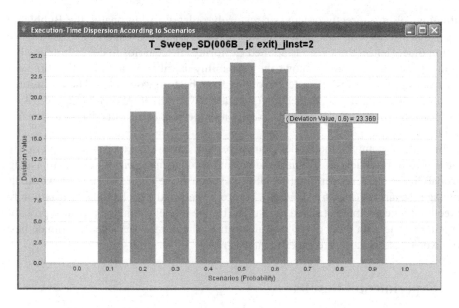

Fig. 14. Execution-Time dispersion according to scenario

scenario to be evaluated, the simulation should operate in normal mode, evaluating average values and the probability distribution of metrics. However, for dynamics systems, event probability can change according to the environmental context or data loading, these being more than one scenario to be evaluated. For events directly related to a branch-instruction, the model allows the exploration of different scenarios through the sweep simulation mode. Analyzing the BubbleSort algorithm, can be deduced that the probability of instruction *inst26* depends on the data pattern, there being a scenario associated with each probability. All scenarios are defined clearly between two extreme

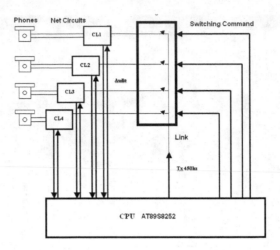

Fig. 15. Case study: Office Communicator

situations: vector directly ordered or reversely ordered. Additionally, the BubbleSort algorithm also informs that such extreme scenarios are the best and worst case for execution time. Hence, the sweep is applied exploring all scenarios to cover from the best to the worst case. Figure 11 shows the probability distribution for execution time during five stage (scenarios) of sweep, evolving from best- to worst-case. The evolution of average values and sample dispersion (standard deviation) are shown in Figure 13 and 14 respectively. The model behaves exactly as expected, being effective as a scenarios inferencer.

In order to compare the model estimative with hardware experiment of phase (ii), the BubbleSort model was evaluated in normal mode using P(*inst26*) equals to average probability of elements permutation into test vectors (See Figure 9), in other world, the average scenario (P(*inst26*)=0.49511). Figure 12 shows the probability distribution for execution time estimated (a) and measured (b), the model estimated the most probable execution time with error of 0.99%. In addition, the model was able to estimate best- and worst-case as hardware experiment was not.

7   Experimental Results

In order to evaluate the proposed framework, an office communicator was developed. Figure 15 illustrates this system. The system is normally in idle mode[10], there being code execution only when a link is closed (request for call) or opened (end of call). During the conversation, and ring timing, the CPU remaining in idle mode. Hence, the goal of analysis is to explore different scenarios that drive the system, closing and opening links. The system's characteristics are depicted as follows: (i)there are only four phones, (ii) only one conversation can be performed at each time, (iii)the ring tone consists of 1 second sending a 450Hz tone and 4 seconds in silence,(iv) if the

---

[10] Low consumption mode of AT89S8252, being assumed here as zero consumption.

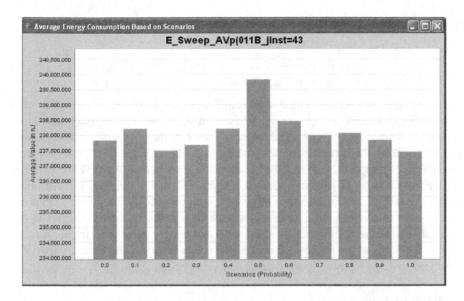

Fig. 16. Average energy consumption per scenario

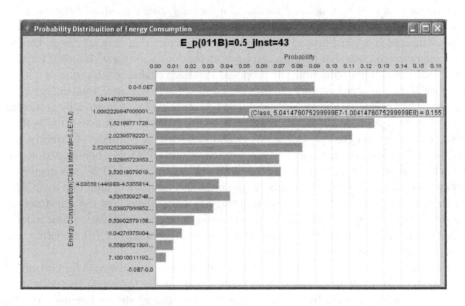

Fig. 17. Energy consumption probability distribution for P(event2)=0.5

receiver-phone does not respond after 10 ring tones, a busy tone is generated to sender-phone for 20 seconds or until the sender-phone hangs up,(v) if a phone hangs up during conversation, a busy tone is generated to the other phone during 20 seconds or until it hangs up. The assembly code comprising 94 instruction-annotations was evaluated, where 30 scenarios have been considered. Each scenario was defined based on three events, for each event 10 probabilities being explored through sweep-mode simulation.

The events are listed below.

Event 1: Link circuit busy.
Event 2: Sender-phone calls a wrong receiver-phone number (calling itself).
Event 3: Sender-phone calls a wrong receiver-phone number (calling invalid number).

The evaluations were performed with a confidence level of 90%, error within 5%, at least 1000 simulations, and stop-criteria based on the power metric. Each simulation value was obtained after 1001 simulation runs, therefore the total number of simulation runs, considering the various parameters steps, was 30030. Figure 16 shows the average energy consumption of **event 2** for 11 scenarios generated through event probability varying from 0.0 to 1.0, with step set to 0.1. Figure 17 shows the energy consumption probability distribution for scenarios of **event 2** with probability 0.5. These graphs elucidate the system consumption pattern. Note that such profiles help the designer to perceive the relation between operational scenario and battery autonomy. In this way, the proposed model and the framework implement a step toward a resource for analyzing embedded systems' software performance and energy consumption.

8 Conclusion

The framework presented offers a mechanism for analyzing embedded systems' software performance and energy consumption. Additionally, a well-known formal modeling language is applied instead of a general purpose programming language. Due to this, many micro-controller families can be modeled considering a widespread formalism, and simulation mechanism. The proposed approach enables exploration of a large number of scenarios considering possible execution of events. Power, total energy consumption and execution time distributions, can be constructed and graphically presented free of pattern dependence. Given a set of scenarios related to an event, the system response can be analyzed providing power and performance estimates. Currently, the EZPetri-PCAF environment operates in conjunction with CPNTools, being restricted to CPNTools simulation speed. Running into Athlon3000 processor, CPNTools displayed an average speed of 34767 transitions per second. It is important to highlight that CPN-Tools is a tool for general CPN model construction and analysis, and not a customized engine for simulating the proposed model. In order to improve simulation time a dedicated simulation engine is under construction. The formal nature of CPN modeling will guarantee the quality of software engineering.

In conclusion, the proposed approach offers three basic contributions for embedded system analysis field:(i)code modeling, a probabilistic model for scenarios exploration being presented, (ii) formalism, a formal and widespread modeling language (CPN) being applied, with previously validated engines and algorithms for simulation;(iii) flexibility, the proposed approach enabling modeling of different micro-controllers at different abstraction levels.

Acknowledgments

The authors would like to thank the anonymous reviewers whose comments helped to improve this paper.

References

1. Cpntools, version 1.4.0. *http://wiki.daimi.au.dk/cpntools/cpntools.wiki.*
2. G. Bernat, A. Colin, and S. M. Petters. pwcet, a tool for probabilistic wcet analysis of real-time systems. In *WCET*, pages 21–38, 2003.
3. A. Bona, M. Sami, D. Sciuto, V. Zaccaria, C. Silvano, and R. Zafalon. Energy estimation and optimization of embedded vliw processors based on instruction clustering. In *Proceedings of the 39th conference on Design automation*, pages 886–891. ACM Press, 2002.
4. A. Bona, M. Sami, D. Sciuto, V. Zaccaria, C. Silvano, and R. Zafalon. An instruction-level methodology for power estimation and optimization of embedded vliw cores. In *Proceedings of the conference on Design, automation and test in Europe*, page 1128. IEEE Computer Society, 2002.
5. R. Burch, F. Najm, and T. T. P. Yang. A monte carlo approach for power estimation. ieee transactions on vlsi systems, vol. 1, march 1993, pp. 63-71.
6. F. Burns, A. Koelmans, and A. Yakovlev. Wcet analysis of superscalar processor using simulation with coloured petri nets. *International Journal of Time-Critical Computing Systems,*.
7. K. Jensen. Coloured petri nets. basic concepts, analysis methods and practical use. *Volume 1, Basic Concepts. Monographs in Theoretical Computer Science*, (2), 1997.
8. J. Jørgensen, S. Christensen, A.-P. Tuovinen, and J. Xu. Tool support for estimating the memory usage of mobile phone software. *International Journal on Software Tools for Technology Transfer (STTT)*, 2006.
9. A. A. Jr, G. A. Jr, R. Lima, P. Maciel, M. O. Jr, and R. Barreto. Ezpetri: A petri net interchange framework for eclipse based on pnml. In *In First Int. Symp. on Leveraging Applications of Formal Method (ISoLA'04)*, 2004.
10. M. N. O. Junior, P. Maciel, R. Barreto, and F. Carvalho. Towards a software power cost analysis framework using colored petri net. In *PATMOS 2004*, volume 3254, pages 362–371. LNCS Kluwer Academic Pubishers, September 2004.
11. M. N. O. Júnior, P. R. M. Maciel, R. M. F. Lima, A. Ribeiro, C. Oliveira, A. Arcoverde, R. S. Barreto, E. Tavares, and L. Amorin. A retargetable environment for power-aware code evaluation: An approach based on coloured petri net. In *PATMOS 2005*, volume 3728, pages 49–58. LNCS Kluwer Academic Pubishers, September 2005.
12. B. Klass, D.Thomas, H. Schmit, and D. Nagle. Modeling inter-instruction energy effects in a digital signal processor. In *Power-Driven Microarchitecture Workshop In conjunstion with ISCA98*, 1998.
13. L. Kristensen, S. Christensen, and K. Jensen. The practitioner's guide to coloured petri nets. *International Journal on Software Tools for Technology Transfer: Special section on coloured Petri nets*, 2(2):98–132, 1998.
14. T. Laopoulos, P. Neofotistos, C. Kosmatopoulos, and S. Nikolaidis. Current variations measurements for the estimation of software-related power consumption. *IEEE Instrumentation and Measurement Technology Conference*, May 2002.
15. R. Marculescu and A. Nandi. Probabilistic application modeling for system-level perfromance analysis. In *DATE '01: Proceedings of the conference on Design, automation and test in Europe*, pages 572–579, Piscataway, NJ, USA, 2001. IEEE Press.
16. A. K. Murugavel and N. Ranganathan. Petri net modeling of gate and interconnect delays for power estimation. In *Proc. of the 39th conf. on Design automation*, pages 455–460. ACM Press, 2002.
17. F. N. Najm. A survey of power estimation techniques in vlsi circuits. *IEEE Trans. Very Large Scale Integr. Syst.*, 2(4):446–455, 1994.
18. S. Nokolaidis, N. Kavvadias, T. Laopoulos, L. Bisdoumis, and S. Blionas. Instruction level energy modeling for pipelined processors. In *PATMOS 2003*, pages 279–288. LNCS Kluwer Academic Pubishers, September 2003.

19. V. Tiwari, S. Malik, and A. Wolfe. Power analysis of embedded software: A first step towards software power minimization. *IEEE Transactions on Very Large Scale Integration Systems*, 2(4):437–445, December 1994.
20. G. Yeap. *Practical Low Power Digital VLSI Design*. Kluwer Academic Publishers, 1998.

Infinite Process Semantics of Inhibitor Nets

H.C.M. Kleijn[1] and M. Koutny[2]

[1] LIACS, Leiden University
P.O. Box 9512, NL-2300 RA Leiden, The Netherlands
kleijn@liacs.nl
[2] School of Computing Science, University of Newcastle
Newcastle upon Tyne, NE1 7RU, United Kingdom
Maciej.Koutny@ncl.ac.uk

Abstract. This paper is concerned with defining causality semantics for infinite executions of Petri nets with inhibitor arcs. We first show how one can deal with infinite step sequences and the corresponding occurrence nets (processes) and causal structures. We then discuss how to improve the succinctness of both finite and infinite processes generated from step sequences. In the latter case, the proposed constructions avoid infinite branching in the case of bounded nets.

Keywords: theory of concurrency, Petri nets, weighted inhibitor arcs, causality semantics, infinite processes, occurrence nets, step sequences.

1 Introduction

Petri nets are a formal model of concurrent computation that has been the subject of extensive development in the past few decades (see [6, 13]). In the standard formalisation, a Petri net consists of places (local states) and transitions (actions). The latter can be executed if a specified set of local states is currently active (or *marked*). Such a model is usually referred to as *Place/Transition nets* (PT-nets). In the case of Petri nets with inhibitor arcs (PTI-nets), executing a transition can also depend on some specific local states being *unmarked*. PTI-nets are well suited to model situations involving testing for a specific condition, rather than producing and consuming resources, and proved to be useful in areas such as communication protocols [2] and performance analysis [4]. Particularly attractive from a modelling point of view are PTI-nets supporting weighted inhibitor arcs which can be used for testing whether a place does not contain more than a certain threshold number of tokens [10].

This paper is a continuation of the work of [7] on elementary net systems with inhibitor arcs, which has been further developed in [10]. Its key aspect is to use so-called *stratified order structures*, generalising partial orders in order to provide a causality semantics consistent with the operational semantics defined in terms of *finite* step sequences. In order to obtain such a semantics, one unfolds a given net into an occurrence net with additional *activator* arcs reflecting the role of inhibitor arcs. The resulting activator occurrence net (or *process*) is acyclic in a

S. Donatelli and P.S. Thiagarajan (Eds.): ICATPN 2006, LNCS 4024, pp. 282–301, 2006.

sense which also includes the activator arcs, allowing one to extract a (labelled) stratified order structure which describes precisely the causality relationships between the events in the given run. An axiomatic characterisation of processes that can be obtained in this way can also be provided.

The work on causality semantics of PTI-nets in [7, 10] assumed that all executions are finite, and so are the corresponding processes and structures. In this paper, we relax this assumption by considering infinite step sequences and infinite activator occurrence nets. Such an extension is needed for the definition of branching processes of PTI-nets in the style of [5], and their subsequent application in the development of efficient model checking techniques for PTI-nets (in the style of, e.g., [12] and [9]).

We will proceed by adopting those notions which were effective in the finite case. In particular, we will take advantage of the so-called *semantical framework* developed in [10] providing a uniform platform on which issues relating to executions, processes and causality can be expressed, and their mutual consistency evaluated. A crucial technical aspect which needs to be addressed is related to the well-foundedness of stratified order structures and processes. As it turns out a suitable treatment for the former has already been proposed in [8].

The paper consists of two parts. In the first one (Sections 3–6), we show how one can instantiate the semantical framework of [10] to deal with infinite step sequences and infinite activator occurrence nets generated by PTI-nets (and, using a different technique, also by a subclass of PTI-nets with complemented inhibitor places, or PTCI-nets). In doing so, we assume basic familiarity with the technical development in [10] and avoid repeating explanations and discussions which can be found there. Proofs for this part can be found in [11].

Activator occurrence nets corresponding to finite and infinite step sequences may and usually do exhibit a certain degree of redundancy in the number of arcs and/or nodes they contain. This is clearly not desirable if, for instance, one is interested in using them as a basis for model checking algorithms. Moreover, the limit constructions used in the definitions of processes can lead to infinite branching. Therefore, in the second part of this paper (Section 7), we address the problem of excessive branching in the processes of PTI-nets, and show how one can improve the succinctness of both finite and infinite processes. The proposed constructions avoid infinite branching in the case of bounded underlying nets.

2   Preliminaries

We use the standard mathematical notation. In particular, $\uplus$ denotes disjoint set union, $\mathbb{N}$ the set of natural numbers (including 0) and $\mathbb{N}^+$ the set of positive natural numbers. The powerset of a set X is denoted by $\mathbb{P}(X)$.

Functions. The standard notation for the composition of functions is used also in the special case of two functions, $f : X \to \mathbb{P}(Y)$ and $g : Y \to \mathbb{P}(Z)$, for which $(g \circ f) : X \to \mathbb{P}(Z)$ is defined by $g \circ f(x) \stackrel{\mathrm{df}}{=} \bigcup_{y \in f(x)} g(y)$, for all $x \in X$. The restriction of a function $f : X \to Y$ to a set $Z \subseteq X$ is denoted by $f|_Z$.

Binary relations. We will sometimes use an infix notation and write xPy rather than $(x,y) \in P$. Moreover, $dom_P \stackrel{\text{df}}{=} \{x \mid (x,y) \in P\}$ and $codom_P \stackrel{\text{df}}{=} \{y \mid (x,y) \in P\}$. The composition of two relations, $P \subseteq X \times Y$ and $Q \subseteq Y \times Z$, is given by $P \circ Q \stackrel{\text{df}}{=} \{(x,z) \mid \exists y \in Y : (x,y) \in P \wedge (y,z) \in Q\}$. The restriction of a relation $P \subseteq X \times Y$ to a set $Z \subseteq X \times Y$ is denoted by $P|_Z$. By id_X we denote the identity relation on a set X. Relation $P \subseteq X \times X$ is reflexive if $id_X \subseteq P$; irreflexive if $id_X \cap P = \varnothing$; and transitive if $P \circ P \subseteq P$. The transitive closure of P is denoted by P^+, and the transitive and reflexive closure by $P^\star$.

Multisets. A multiset over a set X is a function $\mathsf{m} : X \to \mathbb{N}$, and a subset of X may be viewed through its characteristic function as a multiset over X. m is finite if there are finitely many $x \in X$ such that $\mathsf{m}(x) \geq 1$; the cardinality of m is then defined as $|\mathsf{m}| \stackrel{\text{df}}{=} \sum_{x \in X} \mathsf{m}(x)$. The sum of two multisets over X, m and m', is the multiset given by $(\mathsf{m} + \mathsf{m}')(x) \stackrel{\text{df}}{=} \mathsf{m}(x) + \mathsf{m}'(x)$ for all $x \in X$.

Sequences. We use the notation $\sigma = \langle x_i \rangle_{\mathcal{I}}$ to represent an infinite $x_1 x_2 \ldots$ or finite $x_1 x_2 \ldots x_n$ sequence σ, including the empty one ε. In the former case $\mathcal{I} = \mathbb{N}^+$ and in the latter $\mathcal{I} = \{1, 2, \ldots, n\}$ or $\mathcal{I} = \varnothing$, respectively. For example, $\langle xyz \rangle_{\mathbb{N}^+} = xyzxyzxyz \ldots$. We will also denote $\mathcal{I}_0 \stackrel{\text{df}}{=} \mathcal{I} \cup \{0\}$. If all the x_i's are sets then $\bigcup \sigma \stackrel{\text{df}}{=} \bigcup_{i \in \mathcal{I}} x_i$. If each x_i is a finite multiset, then σ is a *step sequence*.

Petri nets. A *net* is a triple $N \stackrel{\text{df}}{=} (P, T, W)$ such that P and T are disjoint sets, and $W : (T \times P) \cup (P \times T) \to \mathbb{N}$. The elements of P and T are respectively the *places* and *transitions*, and W is the *weight function*. In diagrams, places are drawn as circles, and transitions as rectangles. If $W(x,y) \geq 1$ for some $(x,y) \in (T \times P) \cup (P \times T)$, then (x,y) is an *arc* leading from x to y. As usual, an arc is annotated with its weight if the latter is greater than 1.

The *pre-* and *post-multiset* of a transition (or place) x are multisets of places (resp. transitions), $\text{PRE}_N(x)$ and $\text{POST}_N(x)$, respectively given by $\text{PRE}_N(x)(y) \stackrel{\text{df}}{=} W(y, x)$ and $\text{POST}_N(x)(y) \stackrel{\text{df}}{=} W(x, y)$. We assume that $\text{PRE}_N(x)$ and $\text{POST}_N(x)$ are non-empty for every transition x.

A *marking* is a multiset M of places[1]. In diagrams, it is represented by drawing in each place p exactly $M(p)$ tokens (small black dots).

A *step* is a finite non-empty multiset U of transitions. It is *enabled* at a marking M if $M(p) \geq \sum_{t \in T} U(t) \cdot \text{PRE}_N(t)(p)$ for all $p \in P$. In such a case, U can be *executed* leading to the marking M' given by

$$M'(p) \stackrel{\text{df}}{=} M(p) - \sum_{t \in T} U(t) \cdot \text{PRE}_N(t)(p) + \sum_{t \in T} U(t) \cdot \text{POST}_N(t)(p)$$

for all $p \in P$. We also write $M[U\rangle M'$.

A (possibly infinite) sequence $\sigma = \langle U_i \rangle_{\mathcal{I}}$ of non-empty steps is a step sequence from a marking M_0 if there are markings $\langle M_i \rangle_{\mathcal{I}}$ satisfying $M_{i-1}[U_i\rangle M_i$ for every $i \in \mathcal{I}$. For a finite $\mathcal{I}$, if $\mathcal{I} = \varnothing$ then $\sigma = \varepsilon$ is a step sequence from M_0 to M_0; otherwise σ is a step sequence from M_0 to M_n, where n is the largest index in $\mathcal{I}$.

---

[1] For technical reasons, we do not require that M be finite.

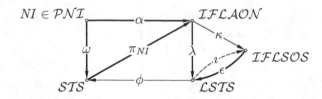

Fig. 1. The semantical setup (bold arcs indicate mappings to powersets and the dashed arc indicates a partial function). The meaning of various semantical domains is as follows: $\mathcal{STS}$ are executions of inhibitor nets in $\mathcal{PNI}$; $\mathcal{IFLAON}$ are nets used to define processes of inhibitor nets; $\mathcal{LSTS}$ are executions of nets in $\mathcal{IFLAON}$; and $\mathcal{IFLSOS}$ are structures capturing the causality in nets in $\mathcal{IFLAON}$.

If σ is a step sequence from M we write $M[\sigma\rangle$, and if σ is a step sequence from M to M' we write $M[\sigma\rangle M'$, calling M' *reachable* from M. If we want to make it clear which net we are dealing with, we may add a subscript N and write $[\cdot\rangle_N$ rather than $[\cdot\rangle$.

3   Semantical Framework

In this section, we instantiate the generic semantical framework of [10], which aims at a systematic presentation of processes and the causality semantics for different classes of Petri nets. The concrete setup is shown in Figure 1, and its various components are described in the rest of this section.

Inhibitor nets $\mathcal{PNI}$ and their executions $\mathcal{STS}$. The family $\mathcal{PNI}$ of nets we consider consists of *PT-nets with inhibitor arcs* (or PTI-nets). A PTI-net is a tuple $NI \overset{\mathrm{df}}{=} (P, T, W, I, M_0)$ such that $\mathrm{UND}(NI) \overset{\mathrm{df}}{=} (P, T, W)$, its underlying net, is *finite* (i.e., P and T are finite sets), $I : P \times T \to \mathbb{N} \cup \{\infty\}$, and M_0 is the initial marking. If $I(p, t) = k \in \mathbb{N}$, then p is an *inhibitor place* of t, and t can only be executed if p does not contain more than k tokens. In diagrams, we draw an arrow from p to t with a small circle as arrowhead and annotated with its weight k whenever $k > 0$. The notations introduced above for transitions, places and markings are defined for NI through its underlying net.

We also consider the class $\mathcal{PNCI} \subset \mathcal{PNI}$ of PT-nets with *complemented inhibitor places* (or PTCI-nets). This means that every inhibitor place p has a designated complement place, denoted by p^{cpl}, such that $\mathrm{PRE}_{NI}(p) = \mathrm{POST}_{NI}(p^{cpl})$ and $\mathrm{POST}_{NI}(p) = \mathrm{PRE}_{NI}(p^{cpl})$. In this case $\mathrm{BND}_{NI}(p) = \mathrm{BND}_{NI}(p^{cpl}) \overset{\mathrm{df}}{=} M_0(p) + M_0(p^{cpl})$ is a common bound on the number of tokens in both p and p^{cpl}.

In NI, a step U is *enabled* at a marking M if it is enabled at M in $\mathrm{UND}(NI)$ and, in addition, there is no transition t and place p such that $U(t) \geq 1$ and $M(p) > I(p, t)^2$. Step sequences and related notions are defined as for ordinary nets, using the modified notion of enabledness.

---

[2] This definition of enabledness is based on an *a priori* condition: the inhibitor places of transitions occurring in a step should obey the inhibitor constraints *before* the step is executed. In the *a posteriori* approach [3], the condition is strengthened so that the inhibitor place property must also be true *after* executing U.

We use *step sequences* STS to model the operational (or behavioural) semantics of a PTI-net NI. The set of step sequences of NI is the set $\omega(NI)$ comprising all step sequences starting from the initial marking M_0.

Causal structures $\mathcal{IFLSOS}$. These are *initially finite labelled stratified order structures* (or ifso-structures — in accordance with [10]) defining an abstract causality semantics of PTI-nets. An ifso-structure is a tuple $ifsos \stackrel{\mathrm{df}}{=} (X, \prec, \sqsubset, \ell)$, where: X is a countable set (the domain); ℓ is a labelling for X; and $\prec$, $\sqsubset$ are two binary relations over X such that for all $x, y, z \in X$ the following hold: (C0) there are only finitely many y such that $y \sqsubset x$; (C1) $\sqsubset$ is irreflexive; (C2) $\prec$ is included in $\sqsubset$; (C3) $x \sqsubset y \sqsubset z$ and $x \neq z$ implies $x \sqsubset z$; and (C4) $x \sqsubset y \prec z$ or $x \prec y \sqsubset z$ implies $x \prec z$. In diagrams, $\prec$ is represented by normal arcs, and $\sqsubset$ by dashed arcs. We sometimes omit arcs that can be deduced using (C1)-(C4) (see Figure 2(f)).

The $\Diamond$–closure ([7]) is an operation which constructs an ifso-structure from a relational structure $rs = (X, \prec, \sqsubset, \ell)$, where: X is a countable set; ℓ is a labelling for X; and $\prec$, $\sqsubset$ are two binary relations over X. The $\Diamond$–*closure* of such a structure is $rs^{\Diamond} = (X, \prec', \sqsubset', \ell)$, where

$$\prec' \stackrel{\mathrm{df}}{=} (\prec \cup \sqsubset)^* \circ \prec \circ (\prec \cup \sqsubset)^* \quad \text{and} \quad \sqsubset' \stackrel{\mathrm{df}}{=} (\prec \cup \sqsubset)^* \backslash id_X .$$

We say that rs is $\Diamond$–*acyclic* if $\prec'$ is irreflexive, and $\Diamond$–*initially finite* if for every x in X there are only finitely many y such that $y \sqsubset' x$.

Proposition 1. $rs^{\Diamond}$ *is an ifso-structure iff* rs *is* $\Diamond$–*acyclic and* $\Diamond$–*initially finite.*

Activator occurrence nets $\mathcal{IFLAON}$ **and their causal structures.** The acyclic nets $\mathcal{IFLAON}$ underpinning abstract processes of inhibitor nets are *initially finite labelled activator occurrence nets* (or ifao-nets — in accordance with [10]). An ifao-net is a tuple $AON \stackrel{\mathrm{df}}{=} (B, E, R, Act, \ell)$ such that:

- UND$(AON) \stackrel{\mathrm{df}}{=} (B, E, R)$ is a countable underlying net (i.e., B and E are countable sets) such that $R \subseteq (B \times E) \cup (E \times B)^3$, and $Act \subseteq B \times E$ is the set of *activator arcs* drawn with small black dots as arrowheads.
- For every $b \in B$, $|\text{PRE}_{AON}(b)| \leq 1$ and $|\text{POST}_{AON}(b)| \leq 1$.
- The structure $rs_{AON} \stackrel{\mathrm{df}}{=} (E, \prec_{loc}, \sqsubset_{loc}, \ell|_E)$ is $\Diamond$–acyclic and $\Diamond$–initially finite, where $\prec_{loc}$ and $\sqsubset_{loc}$ are relations respectively given by $(R \circ R)|_{E \times E} \cup (R \circ Act)$ and $Act^{-1} \circ R$.
- ℓ is a labelling for $B \cup E$.

The various notations introduced above for transitions in E (called *events*), places in B (called *conditions*) and markings are defined for AON through its underlying net. Moreover, the set $\text{MIN}_{AON} \stackrel{\mathrm{df}}{=} \{b \in B \mid |\text{PRE}_{AON}(b)| = 0\}$ is called the *implicit initial marking* of AON.

The relations $\prec_{loc}$ and $\sqsubset_{loc}$ represent *local* information about the causal relationships between the events contained in AON, and Figure 2(a,b,c) shows how

---

[3] The weight function R is treated as a binary relation which always returns 0 or 1.

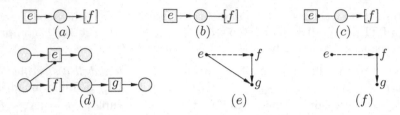

Fig. 2. Two cases defining $e \prec_{loc} f$ (a, b); one case defining $e \sqsubset_{loc} f$ (c); an ifao-net (d) with identity labelling; the ifso-structure it generates (e); and an abbreviated drawing of this ifso-structure where an arc has been omitted as it could be deduced using C4 (f). Notice that $e \prec_{AON} g$, whereas $e \prec^*_{loc} g$ does not hold.

they are derived. These causal relationships can be extracted by the mapping κ using the notion of $\diamond$–closure. This mapping associates with each ifao-net an ifso-structure and is defined by: $\kappa(AON) = (E, \prec_{AON}, \sqsubset_{AON}, \ell|_E) \stackrel{\mathrm{df}}{=} rs_{AON}{}^{\diamond}$ the ifso-structure *generated* by AON (see Figure 2(d,e)).

Proposition 2. $\kappa(AON)$ *is an ifso-structure.*

If $(b, e) \in Act$, then e can only be executed if b contains a token whose presence is tested without the implication of it being consumed by e. Formally, in AON, a step U is *enabled* at a marking M if it is enabled at M in $\mathrm{UND}(AON)$ and, in addition, there is no event e and condition b such that $U(e) \geq 1$, $(b, e) \in Act$ and $M(b) = 0$.[4] Step sequences and related notions are defined as for ordinary nets, using the modified notion of enabledness.

It is easy to check that the underlying net $\mathrm{UND}(AON)$ of an ifao-net AON is a standard occurrence net [1, 14], and so one can import several of its well-established properties; in particular, it is the case that if $\langle E_i \rangle_{\mathcal{I}}$ is a step sequence of $\mathrm{UND}(AON)$ from the marking MIN_{AON}, then the E_i's are mutually disjoint finite sets. We can easily relate the step sequences generated by the two nets.

Proposition 3. *Let* $\sigma = \langle E_i \rangle_{\mathcal{I}}$ *be a step sequence.*

1. σ *is a step sequence of* AON *from* MIN_{AON} *iff* σ *is a step sequence of* $\mathrm{UND}(AON)$ *from* MIN_{AON} *and for all* $i \in \mathcal{I}$ *and* $e \in E_i$, $f \prec_{loc} e$ *implies* $f \in \bigcup_{j<i} E_j$, *and* $f \sqsubset_{loc} e$ *implies* $f \notin \bigcup_{j>i} E_j$.
2. *If* σ *is a step sequence of* AON *from* MIN_{AON} *then the* E_i's *are mutually disjoint finite sets.*

This provides a useful characterisation of step sequences executed by ifao-nets; however, a fuller account of their operational semantics will also include the labelling ℓ, as described next.

Labelled executions $\mathcal{LSTS}$ of acyclic nets and causal structures. We use *labelled step sequences $\mathcal{LSTS}$* to model the operational semantics of ifao-nets.

---

[4] This definition of enabledness is again based on the *a priori* condition. For alternative definitions see [3, 15].

A *labelled step sequence* is a pair $\varpi \stackrel{\text{df}}{=} (\sigma, \ell)$ such that $\sigma = \langle X_i \rangle_{\mathcal{I}}$ is a step sequence consisting of mutually disjoint finite sets, and ℓ is a labelling for the set $\bigcup \sigma$ called the *domain* of ϖ. Moreover, for every x in this domain, $ind(\varpi, x)$ is the index of the unique set X_i such that $x \in X_i$. With ϖ we also associate the step sequence $\phi(\varpi) \stackrel{\text{df}}{=} \langle \ell \langle X_i \rangle \rangle_{\mathcal{I}}$, where $\ell \langle \{x_1, \ldots, x_k\} \rangle \stackrel{\text{df}}{=} \{\ell(x_1)\} + \cdots + \{\ell(x_k)\}$ thus defining the function ϕ of Figure 1.

For an ifao-net $AON = (B, E, R, Act, \ell)$, the set of labelled step sequences $\lambda(AON)$ comprises all $\varpi = (\sigma, \ell|_E)$ such that σ is a step sequence of AON from MIN_{AON} and $E = \bigcup \sigma$. Note that λ is well-defined due to Proposition 3.

Similarly, the set $\epsilon(ifsos)$ of labelled step sequences of an ifso-structure $ifsos = (X, \prec, \sqsubset, \ell)$ comprises all $\varpi = (\sigma, \ell)$ such that $X = \bigcup \sigma$ and for all $x, y \in X$: $x \prec y$ implies $ind(\varpi, x) < ind(\varpi, y)$, and $x \sqsubset y$ implies $ind(\varpi, x) \le ind(\varpi, y)$.

Related to ϵ is a partial mapping which allows one to associate a causal structure with a set of labelled step sequences. The *ifso-structure intersection* of a non-empty set $LSTS$ of labelled step sequences with the same domain X and labelling ℓ is defined as $\imath(LSTS) \stackrel{\text{df}}{=} (X, \prec, \sqsubset, \ell)$, where $\prec$ and $\sqsubset$ are binary relations on X such that for all $x, y \in X$: $x \prec y$ if $ind(\varpi, x) < ind(\varpi, y)$ for all $\varpi \in LSTS$, and $x \sqsubset y$ if $ind(\varpi, x) \le ind(\varpi, y)$ for all $\varpi \in LSTS$.

Aims and properties. The arcs in Figure 1 indicate mappings defining and relating three different views on the semantics of a PTI-net *NI*. Our overall goal is to show that these different semantics agree in the sense that processes ($\mathcal{IFLAON}$) and causal structures ($\mathcal{IFLSOS}$) describe relations between events consistent with the chosen operational semantics ($\mathcal{STS}$). In particular, the (as yet undefined) mapping α associates with *NI* a non-empty set of ifao-nets (*processes*) satisfying certain *axioms*, and an ifao-net is given an operational semantics through the mapping λ which associates with it a non-empty set of *labelled step sequences*. Labelled step sequences can be interpreted as ordinary step sequences (of the original PTI-net *NI*) by *forgetting* some irrelevant information through the total function ϕ. Finally, the (also not yet defined) partial function π_{NI} gives, for each step sequence of *NI*, a non-empty set of ifao-nets which can be viewed as operationally defined processes of *NI*.

Three aims can be formulated which, when fulfilled, mean that the axiomatic and behavioural process definition as well as the operational semantics of nets in $\mathcal{PNI}$ are in full agreement: the axiomatic processes of *NI* (defined through α) coincide with the operational processes of *NI* (defined through $\pi_{NI} \circ \omega$); the operational semantics of *NI* (defined through ω) coincides with the operational semantics of the processes of *NI* (defined through $\phi \circ \lambda \circ \alpha$); and the causality in a process of *NI* (defined through κ) coincides with the causality structure implied by its operational semantics (defined through $\imath \circ \lambda$). Formally, we have:

Aim 1 $\alpha = \pi_N \circ \omega$.

Aim 2 $\omega = \phi \circ \lambda \circ \alpha$.

Aim 3 $\kappa = \imath \circ \lambda$.

The two corollaries below provide further justification of the consistency of the process and abstract causality semantics of the PTI-net NI with its operational semantics given by the function ω (which captures the dynamics of the nets in $\mathcal{PNI}$ given through step sequences).

Corollary 1. $\omega = \phi \circ \lambda \circ \pi_N \circ \omega.$

Corollary 2. $\omega = \phi \circ \epsilon \circ \kappa \circ \alpha.$

As shown in [10], the above consistency characteristics hold whenever the five properties formulated below are satisfied.

Property 1. The following functions are total: (i) ω, (ii) α, (iii) λ, (iv) ϕ, and (v) $\pi_{NI}|_{\omega(NI)}$. Moreover, the following functions never return the empty set: (vi) ω, (vii) α, (viii) λ, and (ix) $\pi_{NI}|_{\omega(NI)}$.

Property 2. For all $\sigma \in \mathcal{STS}$ and $AON \in \mathcal{IFLAON}$,

$$\sigma \in \omega(NI) \wedge AON \in \pi_{NI}(\sigma) \text{ iff } AON \in \alpha(NI) \wedge \sigma \in \phi(\lambda(AON)) .$$

Property 3. The following functions are total: (i) κ, (ii) ϵ, and (iii) $\imath|_{\lambda(\mathcal{IFLAON})}$. Moreover, the following function never returns the empty set: (iv) ϵ.

Property 4. $\imath \circ \epsilon = id_{\mathcal{IFLSOS}}$.

Property 5. $\lambda = \epsilon \circ \kappa$.

In [11] we show that the above five behavioural properties are indeed satisfied for the concrete set-up in Figure 1 with two different instantiations for the pairs of mappings α and π_{NI}. Extending [10] to the potentially infinite case, one instantiation is specifically for PTCI-nets and the other for general PTI-nets. These missing functions will be defined in Sections 5 and 6, respectively.

4   Marking Reachability in ifao-nets

For the standard occurrence nets, marking reachability from their implicit initial markings can be treated using the notions of a slice and configuration. In the case of ifao-nets, the situation is much more complex even in the finite case, as reported in [10]. Below we show how the results of [10] can be extended to ifao-nets.

Let $AON \stackrel{\text{df}}{=} (B, E, R, Act, \ell)$ be an ifao-net and $ON = \text{UND}(AON)$ be its underlying occurrence net. We will say that a set S of conditions has a *finite past* if there are only finitely many events e such that $(e, b) \in R^+$ for some $b \in S$.

We first recall a few notions and results taken from the theory of occurrence nets [14]. A *configuration* of ON is a finite set $D \subseteq E$ which comprises events together with their causal predecessors, i.e., $e \in D$, $f \in E$ and $(f, e) \in R^+$ implies $f \in D$. We denote this by $D \in \text{CNF}(ON)$. A *slice* of ON is a maximal w.r.t. set inclusion set $S \subseteq B$ with finite past, such that the conditions it comprises are causally unrelated, i.e., $(S \times S) \cap R^+ = \varnothing$. We denote this by $S \in \text{SL}(ON)$. It can be seen that $\text{SL}(ON)$ coincides with the set of all markings reachable in ON from MIN_{AON}; and that $\text{CNF}(ON)$ coincides with the sets of events which can be executed by finite step sequences starting from MIN_{AON}. For ifao-nets, however, we have to distinguish in addition two different kinds of configurations and slices, described next.

A finite set $D \subseteq E$ is a *strong configuration* of AON, denoted by $D \in \text{SCNF}(AON)$, if $e \in D$ and $(f, e) \in \prec^+_{loc}$ implies $f \in D$; similarly, D is a *weak configuration* of AON, if $e \in D$ and $(f, e) \in (\prec_{loc} \cup \sqsubset_{loc})^+$ implies $f \in D$. Clearly, $\text{CNF}(ON) \supseteq \text{SCNF}(AON) \supseteq \text{WCNF}(AON)$, and if $Act = \varnothing$ then both inclusions become equalities and all three classes coincide.

Proposition 4. *If σ is a finite step sequence of AON from MIN_{AON}, then $\bigcup \sigma$ is a strong configuration of AON.*

The two new kinds of slices for ifao-nets are defined using two relations generalising the idea of causally related conditions: $\text{SLIN}(AON) \stackrel{\text{df}}{=} (R \circ \prec^*_{loc} \circ R)|_{B \times B}$ and $\text{WLIN}(AON) \stackrel{\text{df}}{=} (R \circ (\prec_{loc} \cup \sqsubset_{loc})^* \circ R)|_{B \times B}$. Clearly, $R^+|_{B \times B} \subseteq \text{SLIN}(AON) \subseteq \text{WLIN}(AON)$, and if $Act = \varnothing$ then both inclusions become equalities. A *strong (weak) slice* of AON is a maximal w.r.t. set inclusion set $S \subseteq B$ with finite past, such that the conditions it comprises are incomparable w.r.t. $\text{SLIN}(AON)$ (resp. $\text{WLIN}(AON)$), i.e., $(S \times S) \cap \text{SLIN}(AON) = \varnothing$ (resp. $(S \times S) \cap \text{WLIN}(AON) = \varnothing$). We denote this by $S \in \text{SSL}(AON)$ (resp. $S \in \text{WSL}(AON)$). By using a similar argument as in [10] for the finite case, one can show that $\text{WSL}(AON) \subseteq \text{SSL}(AON) \subseteq \text{SL}(ON)$, and if $Act = \varnothing$ then both inclusions become equalities.

As an example, consider the ifao-net in Figure 2(d). In this net, the marking M reached after executing f is a strong slice. It is however not a weak slice. Moreover there is no marking reachable from M at which e could still be executed. In general, we have the following result.

Proposition 5. *Let σ be a finite step sequence of AON from MIN_{AON} to a marking M. Then the following statements are equivalent:*

1. *There exists a step sequence σ' of AON from M such that $\bigcup \sigma \sigma' = E$.*
2. $\bigcup \sigma \in \text{WCNF}(AON)$.
3. $M \in \text{WSL}(AON)$.

A crucial property which can be used in the reachability analysis of ifao-nets is that strong slices of AON coincide with the set $\mathcal{M}$ of all markings reachable from MIN_{AON} in AON, and the weak ones with the set $\mathcal{M}'$ of all markings from $\mathcal{M}$ which have the additional property that all the 'unused' events can be executed. Thus $M \in \mathcal{M}'$ iff $M \in \mathcal{M}$ and there is a step sequence σ' of AON from M such that $\sigma \sigma'$ comprises all the events of AON.

Proposition 6. *We have the following:*

$$\mathcal{M} = \text{SSL}(AON) = \{\text{MAR}(D) \mid D \in \text{SCNF}(AON)\}$$

$$\mathcal{M}' = \text{WSL}(AON) = \{\text{MAR}(D) \mid D \in \text{WCNF}(AON)\},$$

where $\text{MAR}(D) \stackrel{\text{df}}{=} \text{MIN}_{AON} \cup \{b \mid \exists e \in D : (e, b) \in R\} \setminus \{b \mid \exists e \in D : (b, e) \in R\}$.

5 Process Semantics of PTCI-Nets

Let $NCI = (P, T, W, I, M_0)$ be a PTCI-net, fixed for the rest of this section. In this section, we will define the mappings α and π_{NI} of Figure 1 for this kind of net, which will be denoted by α^{cpl} and π_{NCI}^{cpl}, respectively.

We first provide the operational process definition which takes a step sequence and constructs a corresponding ifao-net.

Definition 1. *Let $\sigma = \langle U_i \rangle_{\mathcal{I}}$ be a step sequence of NCI. A complement activator process (ca-process) generated by σ is a labelled net with activator arcs*

$$AON = (B, E, R, Act, \ell) \stackrel{\text{df}}{=} \left(\bigcup_{k \in \mathcal{I}_0} B_k, \bigcup_{k \in \mathcal{I}_0} E_k, \bigcup_{k \in \mathcal{I}_0} R_k, \bigcup_{k \in \mathcal{I}_0} Act_k, \bigcup_{k \in \mathcal{I}_0} \ell_k \right)$$

obtained as the limit of a sequence $\langle N_k \rangle_{\mathcal{I}_0}$ of nets, where for $k \in \mathcal{I}_0$:

$$N_k = (B_k, E_k, R_k, Act_k, \ell_k) \stackrel{\text{df}}{=} \left(\biguplus_{i=0}^{k} B^i, \biguplus_{i=0}^{k} E^i, \biguplus_{i=0}^{k} R^i, \biguplus_{i=0}^{k} Act^i, \biguplus_{i=0}^{k} \ell^i \right)$$

is constructed in the following way (in this, and a similar definition later on, it is assumed that the sets of conditions, events and arcs do not contain any elements other than those specified explicitly).

- *For $i \in \mathcal{I}_0$, $\ell^i : B^i \cup E^i \to P \cup T$ is a labelling defined below.*
- *$E^0 = \varnothing$ and for $i \in \mathcal{I}$, E^i comprises a distinct event for each transition occurrence in U_i. The event corresponding to the j-th occurrence of t in U_i is t–labelled and denoted by $t^{i,j}$.*
- *B^0 comprises a distinct condition for each place occurrence in M_0. The condition corresponding to the j-th occurrence of s in M_0 is s–labelled and denoted by s^j.*
- *For $i \in \mathcal{I}$ and for every $e \in E^i$, B^i comprises a distinct condition for each place occurrence in $\text{POST}_{NCI}(\ell_i(e))$. The condition corresponding to the j-th occurrence of p in $\text{POST}_{NCI}(\ell_i(e))$ is p–labelled and denoted by $p^{e,j}$.*
- *$R^0 = \varnothing$, and for $i \in \mathcal{I}$ and every $e \in E^i$:*
  - *We add an arc $(e, p^{e,j})$ to R^i for each $p^{e,j} \in B^i$.*
  - *We choose a disjoint (i.e., $B_f \cap B_g = \varnothing$ whenever $f \neq g$) set of conditions $B_e \subseteq B_{i-1} \backslash \text{dom}_{R_{i-1}}$ such that $\ell_i \langle B_e \rangle = \text{PRE}_{NCI}(\ell_i(e))$ and add an arc (b, e) to R^i for each $b \in B_e$.*
- *$Act^0 = \varnothing$, and for $i \in \mathcal{I}$ and every $e \in E^k$, if p is an inhibitor place of $\ell_k(e)$ then we choose a set A_e of exactly $\text{BND}_{NCI}(p) - \text{INH}_{NCI}(\ell_k(e))(p)$ conditions in $B_{k-1} \backslash \text{dom}_{R_{k-1}}$ labelled by p^{cpl}. After that we add an activator arc (b, e) to Act^i for each $b \in A_e$.*

We will denote this by $AON \in \pi_{NCI}^{cpl}(\sigma)$.

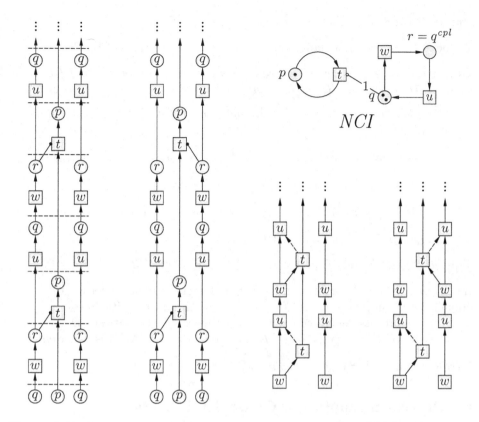

Fig. 3. A PTCI-net, two ca-processes generated by $\sigma = \langle\{w,w\}\{t\}\{u,u\}\rangle_{\mathbb{N}+}$, and ifso-structures generated by these ca-processes

Note that the above definition (as well as the further three process definitions) is a conservative extension of that defined in [10] for the finite nets. Furthermore, as observed there, complements for bounded places of a PTI-net (or even a PT-net) cannot just be added since this may lead to new processes. In case no reachable marking of *NCI* assigns more than one token to a place, there is only one candidate set B_e of conditions in the last part of the definition. It may happen that $A_e \cap A_f \neq \varnothing$ for $e \neq f$. Moreover, as the next result indicates, the required sets A_e can always be (effectively) found.

Proposition 7 ([10]). *Let $k \in \mathcal{I}$, $e \in E^k$ and p be an inhibitor place of $\ell(e)$. Then $|\{b \in B_{k-1} \backslash dom_{R_{k-1}} \mid \ell(b) = p^{cpl}\}| \geq \mathrm{BND}_{NCI}(p) - \mathrm{INH}_{NCI}(\ell(e))(p)$.*

Figure 3 shows a PTCI-net *NCI* and two different ca-processes for the infinite step sequence $\sigma = \langle\{w,w\}\{t\}\{u,u\}\rangle_{\mathbb{N}+}$. Note that $\mathrm{BND}_{NI}(q) = \mathrm{BND}_{NI}(r) = 2$ and $r = q^{cpl}$. The horizontal dashed lines indicate the stages in which the first of the two process was constructed. Moreover, we also show the ifso-structures generated by these ifao-nets. It is easy to check that ca-processes generated by step sequences are ifao-nets.

Proposition 8. *If* $\sigma \in \omega(NCI)$ *then* $\pi_{NCI}^{cpl}(\sigma) \subseteq \mathcal{IFLAON}$.

We next provide an axiomatic definition for the ca-processes of PTCI-nets.

Definition 2. *A* complement activator process *(or ca-process) of NCI is an ifao-net* $AON = (B, E, R, Act, \ell)$ *such that*

- ℓ *is a labelling function for* $B \cup E$ *such that* $\ell(B) \subseteq P$ *and* $\ell(E) \subseteq T$.
- *For all* $e \in E$,

$$\text{PRE}_{NCI}(\ell(e)) = \ell\langle\text{PRE}_{AON}(e)\rangle \quad \text{and} \quad \text{POST}_{NCI}(\ell(e)) = \ell\langle\text{POST}_{AON}(e)\rangle .$$

- *If* $e \in E$ *and* p *is an inhibitor place of* $\ell(e)$ *then*

$$|\{b \in \ell^{-1}(p^{cpl}) \mid (b,e) \in Act\}| = \text{BND}_{NCI}(p) - \text{INH}_{NCI}(\ell(e))(p) .$$

- MIN_{AON} *is finite and* $M_0 = \ell\langle\text{MIN}_{AON}\rangle$.

We will denote the set of ca-processes of NCI by $\alpha^{cpl}(NCI)$.

Intuitively, the third condition means that if event e is enabled then there are enough tokens in p^{cpl} to ensure that p does not inhibit transition $\ell(e)$.

Every ca-process generated by a step sequence of NCI satisfies Definition 2 and is therefore a ca-process of NCI. Consequently, $\alpha^{cpl}(NCI)$ is never empty.

Proposition 9. *If* $\sigma \in \omega(NCI)$ *then* $\pi_{NCI}^{cpl}(\sigma) \subseteq \alpha^{cpl}(NCI)$.

6   Process Semantics of General PTI-Nets

In this section, we will define a process semantics for general PTI-nets by extending that given in [10]. Since we cannot rely on complements of inhibitor places, another feature is needed to test that an inhibitor place does not contain too many tokens. The solution in [10] is to add 'on demand' new artificial conditions (labelled by the special symbol λ) with activator arcs to fulfill this role.

Let $NI = (P, T, W, I, M_0)$ be a PTI-net fixed for the rest of this section. If $p \in P$ and $t, w \in T$ are such that $\text{INH}_{NI}(t)(p) \neq \infty$ and $\text{PRE}_{NI}(w)(p) + \text{POST}_{NI}(w)(p) \neq 0$, then we denote $w \xrightarrow{p} t$, or simply $w \multimap t$. Similarly, for an ifao-net $AON = (B, E, R, Act, \ell)$, if $b \in B$ and $e, f \in E$ are such that $(b, e) \in Act$ and $\text{PRE}_{AON}(f)(b) + \text{POST}_{AON}(f)(b) \neq 0$, then we denote $f \xrightarrow{b} e$, or simply $f \multimapdot e$.

The main idea behind the construction presented next is to ensure that if $w \multimap t$ then any two occurrences, f of w and e of t, are adjacent to a common condition so that $f \multimapdot e$, and thus are related in the corresponding causal structure. Actually, whether or not it is necessary to enforce these relations depends on the current number of tokens in the inhibitor places involved. The approach we follow here is a uniform strategy based on 'local' structural relations and ultimately leads to an abstract causality semantics in agreement with the operational semantics of PTI-nets (see also the discussion in [10]). First we define the operational process semantics and demonstrate how to construct an ifao-net for a given step sequence of NI.

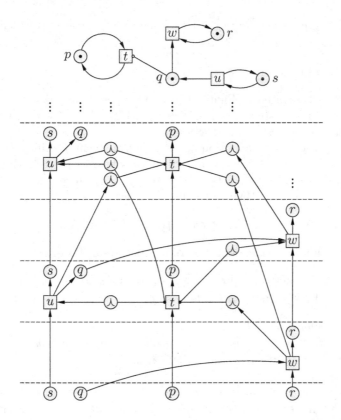

Fig. 4. A PTI-net and an a-process generated by $\sigma = \langle\{w\}\{t,u\}\rangle_{\mathbb{N}^+}$

Definition 3. *Let $\sigma = \langle U_i \rangle_{\mathcal{I}}$ be a step sequence of NI. An* activator process *(or* a-process*) generated by σ is a labelled net with activator arcs*

$$AON = (B, E, R, Act, \ell) \stackrel{\mathrm{df}}{=} \left(\bigcup_{k \in \mathcal{I}_0} B_k \cup \widetilde{B}_k, \bigcup_{k \in \mathcal{I}_0} E_k, \bigcup_{k \in \mathcal{I}_0} R_k, \bigcup_{k \in \mathcal{I}_0} Act_k, \bigcup_{k \in \mathcal{I}_0} \ell_k \right)$$

obtained as the limit of a sequence $\langle N_k \rangle_{\mathcal{I}_0}$ of nets, where for $k \in \mathcal{I}_0$:

$$N_k = (B_k \uplus \widetilde{B}_k, E_k, R_k, Act_k, \ell_k) \stackrel{\mathrm{df}}{=} \left(\biguplus_{i=0}^{k} B^i \uplus \biguplus_{i=0}^{k} \widetilde{B}^i, \biguplus_{i=0}^{k} E^i, \biguplus_{i=0}^{k} R^i, \biguplus_{i=0}^{k} Act^i, \biguplus_{i=0}^{k} \ell^i \right)$$

is constructed as in Definition 1, except that $\widetilde{B}^0 = Act^0 \stackrel{\mathrm{df}}{=} \varnothing$ and, for $k \in \mathcal{I}$:

- $\ell^k(b) \stackrel{\mathrm{df}}{=} \lambda$ *for all $b \in \widetilde{B}^k$.*
- *If $e \in E^k$ and $f \in E^j$ (for $j < k$) are such that $\ell_k(f) \multimapinv \ell_k(e)$ then we create exactly one condition $b \in \widetilde{B}^k$ and add two arcs: $(f, b) \in R^k$ and $(b, e) \in Act^k$.*

- If $f \in E^k$ and $e \in E^j$ (for $j \leq k$) are such that $\ell_k(f) \multimap \ell_k(e)$ then we create exactly one condition $b \in \widetilde{B}^k$ and add two arcs: $(b, f) \in R^k$ and $(b, e) \in Act^k$.

We will denote this by $AON \in \pi_{NI}(\sigma)$.

Definition 3 is illustrated in Figure 4 for a PTI-net (which is not a PTCI-net) and one of its infinite step sequences, $\sigma = \langle\{w\}\{t, u\}\rangle_{\mathbb{N}^+}$. As before, horizontal lines indicate the stages of construction.

In the construction of Definition 3, whenever an event f is introduced before an event e and $\ell(f) \multimap \ell(e)$, then this will *always* lead to $f \prec e$ in the generated ifso-structure. Similarly, whenever an event e is introduced not later than an event f and $\ell(f) \multimap \ell(e)$, then this will *always* lead to $e \sqsubset f$.

Proposition 10. *If $\sigma \in \omega(NI)$ then $\pi_{NI}(\sigma) \subseteq \mathcal{IFLAON}$.*

We next have an axiomatic definition of a-processes of PTI-nets.

Definition 4. *An* activator process *(or a-process) of NI is an ifao-net $AON = (B \uplus \widetilde{B}, E, R, Act, \ell)$ satisfying the following:*

1. $\ell(B) \subseteq P$ and $\ell(E) \subseteq T$.
2. *The conditions in $\widetilde{B} = dom_{Act}$ are labelled by the special symbol $\curlywedge$.*
3. $\text{MIN}_{AON} \cap B$ *is finite and* $M_0 = \ell\langle\text{MIN}_{AON} \cap B\rangle$.
4. *For all $e \in E$,*
   $\text{PRE}_{NI}(\ell(e)) = \ell\langle\text{PRE}_{AON}(e) \cap B\rangle$ and $\text{POST}_{NI}(\ell(e)) = \ell\langle\text{POST}_{AON}(e) \cap B\rangle$.
5. *For all $b \in \widetilde{B}$, there are unique $g, h \in E$ such that $(b, h) \in Act_{AON}$, $\ell(g) \multimap \ell(h)$ and $\text{PRE}_{AON}(b) + \text{POST}_{AON}(b) = \{g\}$.*
6. *For all $e, f \in E$, if $\ell(f) \multimap \ell(e)$ then there is exactly one $c \in \widetilde{B}$ such that $f \stackrel{c}{\multimapdot} e$.*
7. *For all $e \in E$ and $S \in \text{SSL}(AON)$, if $\text{PRE}_{AON}(e) \cup \{b \mid (b, e) \in Act_{AON}\} \subseteq S$ then $\ell\langle S \cap B\rangle \leq \text{INH}_{NI}(\ell(e))$ (one can easily see that $S \cap B$ is a finite set).*

We will denote the set of a-processes of NI by $\alpha(NI)$.

Definition 4(1,3,4) guarantees that $\text{UND}(AON)$, after deleting $\widetilde{B}$ and the adjacent arcs, is a process of $\text{UND}(NI)$. Definition 4(5) describes the immediate neighbourhood of a $\curlywedge$–labelled condition, which has to correspond to an inhibitor arc in NI. Conversely, Definition 4(6) requires that whenever events in AON represent transitions related through an inhibitor place, there should be a $\curlywedge$–labelled condition relating these events. Finally, Definition 4(7) refers to Proposition 6, and requires that the strong slices of AON (i.e., markings reachable from MIN_{AON}) properly reflect the inhibitor constraints present in NI: an event can only occur at a strong slice if there are not too many conditions corresponding to tokens in the inhibitor places of its counterpart in NI.

Every a-process generated by a step sequence of NI satisfies Definition 4 and thus is an a-process of NI. Consequently, $\alpha(NI)$ is never empty.

Proposition 11. *If $\sigma \in \omega(NI)$ then $\pi_{NI}(\sigma) \subseteq \alpha(NI)$.*

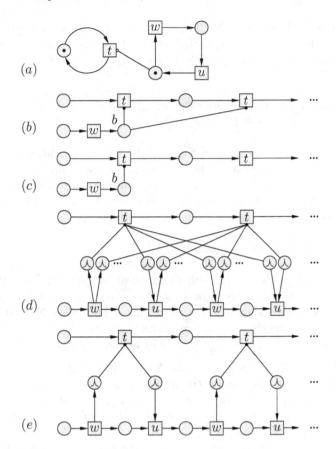

Fig. 5. A PTCI-net (a); ca-process generated by $\{w\}\langle\{t\}\rangle_{\mathbb{N}+}$ (b); modified ca-process (c); a-process generated by $\langle\{w\}\{t\}\{u\}\rangle_{\mathbb{N}+}$ (d); and modified a-process (e)

7   Reducing Branching in Process Nets

We now look at the constructions described in Definitions 1 and 3 in the context of the PTCI-net shown in Figure 5 together with a ca-process and a-process. Focus is on the feature that some of the events and conditions in such processes may exhibit infinite branching (with respect to ordinary arcs or activator arcs in either direction) something which is not true of occurrence nets generated by infinite step sequences of PT-nets. In other words, all infinite branching is related to dealing with inhibitor arcs since the underlying occurrence nets are always finitely branching and have, in fact, non-branching conditions. We will now propose two modifications which attempt to rectify this problem.

Let us consider a ca-process $AON = (B, E, R, Act, \ell)$ constructed in Definition 1 for a PTCI-net $NCI = (P, T, W, I, M_0)$, with the view of deleting as many of its activator arcs as possible without changing the induced ifso-structure. We first introduce an auxiliary notation.

For every $b \in B$, let Act_b be the set of all events such that $(b, e) \in Act$. Moreover, let Act_b^{min} (Act_b^{max}) be the set of all $e \in Act_b$ for which there is no $f \in Act_b$ such that fR^+e (resp. eR^+f). We propose the following modification to AON.

Modification 1. *Delete each activator arc $(b, e) \in Act$ such that $e \notin Act_b^{min} \cup Act_b^{max}$, denoting the result by AON^{cmod}.*

When applied to the ca-process in Figure 5(b), the modification results in the ifao-net shown in Figure 5(c), where the previously infinitely branching ca-process has become finitely branching (note that in this case $Act_b^{max} = \varnothing$). The modification works in the sense that the causality relationships generated by the original and modified constructions are the same.

Proposition 12. $\kappa(AON) = \kappa(AON^{cmod})$.

Proof. Clearly, all causal relationships generated by AON^{cmod} are included in those generated by AON. To show that the reverse also holds, suppose that $(b, e) \in Act$ is a deleted activator arc and consider two cases. Note that the two cases directly relate to the way the relations $\prec_{loc}$ and $\sqsubset_{loc}$ were introduced in order to define the ifso-structure generated by an ifao-net (see also Figure 2(b,c)).

Case 1: $(f, b) \in R$. Then, since $\kappa(AON)$ satisfies (C0), there is $g \in Act_b^{min}$ such that $(g, e) \in R^+$. Moreover, the activator arc $(b, g) \in Act$ is not deleted by Modification 1. Hence the direct causal relationship induced by $(f, b) \in R \wedge (b, e) \in Act$ in AON is represented in AON^{cmod} by $(f, b) \in R \wedge (b, g) \in Act \wedge (g, e) \in R^+$.

Case 2: $(b, f) \in R$. Then, since $b \in dom_R$ the set Act_b is finite. Hence there is $g \in Act_b^{max}$ such that $(e, g) \in R^+$. Moreover, the activator arc $(b, g) \in Act$ is not deleted by Modification 1. Hence the direct causal relationship induced by $(b, f) \in R \wedge (b, e) \in Act$ in AON is represented in AON^{cmod} by $(b, f) \in R \wedge (b, g) \in Act \wedge (e, g) \in R^+$. $\square$

The above modification typically results in a reduction in branching; moreover, we can show that infinite branching disappears provided that the underlying net with the initial marking taken from NCI is *bounded* (i.e., there is an upper bound on the size of markings reachable in $\text{UND}(NCI)$ from M_0).

Proposition 13. *If $\text{UND}(NCI)$ with the initial marking M_0 is bounded, then AON^{cmod} is finitely branching.*

Proof. If AON^{cmod} is not finitely branching, then there is $b \in B$ such that $Act_b^{min} \cup Act_b^{max}$ is infinite. Thus at least one of Act_b^{min} and Act_b^{max} is an infinite set of events. Moreover, no two events in this set are related by R^+. Hence, from the standard properties of occurrence nets of PT-nets, it follows that there is no upper bound on the size of markings reachable in $\text{UND}(NCI)$ from M_0, a contradiction. $\square$

Let us now consider the construction of an a-process for a PTI-net $NI = (P, T, W, I, M_0)$ as described in Definition 3. We now also intend to discuss the expectation that, when constructing an a-process for NI, only a 'small' set of the nodes generated so far (call it a *frontier*) would be necessary to construct new events and conditions. Clearly, for PTCI-nets and ca-processes constructed in Definition 1, a suitable frontier consists only of the conditions without outgoing arcs. However, in Definition 3 the frontier can be much bigger, as it also must include some of the events already generated. Indeed, the construction requires that *all* (and only) the events labelled by transitions adjacent to inhibitor places be included in the frontier. This is, in many cases, excessive and a possible remedy is directly motivated by the transitivity properties (C4).

Assuming the notation as in Definition 3 and $k \in \mathcal{I}$, let $\mathcal{F}_k$ be the set of all $e \in E_k$ such that there is no $f \in E_k$ such that $\ell_k(f) = \ell_k(e)$ and $eR_k^+ f$.

Modification 2. *Replace the last two items of Definition 3 by the following:*

- *If $e \in E^k$ and $f \in \mathcal{F}_{k-1}$ are such that $\ell_k(f) \multimap \ell_k(e)$ and it is not the case that $(f, e) \in R_{k-1} \circ Act_{k-1} \circ R_k^+$, then we create exactly one condition $b \in \widetilde{B}^k$ and add two arcs: $(f, b) \in R^k$ and $(b, e) \in Act^k$.*
- *If $f \in E^k$ and $e \in E^k \cup \mathcal{F}_{k-1}$ are such that $\ell_k(f) \multimap \ell_k(e)$ and it is not the case that $(e, f) \in Act_{k-1}^{-1} \circ R_{k-1} \circ R_k^+$, then we create exactly one condition $b \in \widetilde{B}^k$ and add two arcs: $(b, f) \in R^k$ and $(b, e) \in Act^k$.*

Denote the result by AON^{mod}.

The above is illustrated in Figure 5(e), which in contrast to the previously infinitely branching a-process shown in Figure 5(d), is finitely branching.

Note that with every modified a-process AON^{mod} constructed as just described corresponds a normally constructed a-process (denoted AON). Their only difference is that AON^{mod} may have less λ-labelled conditions and hence less activator arcs and ordinary arcs connected to these places.

As before, the causality relationships after the modification are unchanged.

Proposition 14. $\kappa(AON) = \kappa(AON^{mod})$.

Proof. Clearly, all causal relationships generated by AON^{mod} are included in those generated by AON. To show that the reverse also holds, we consider two cases (below, the primed components are related to AON^{mod}). Again, the two cases directly relate to the way the relations $\prec_{loc}$ and $\sqsubset_{loc}$ were introduced in order to define the ifso-structure generated by an ifao-net (see also Figure 2(b,c)). Below we use the fact that for events g, h, if $(g, h) \in R^\star$ then $(g, h) \in (R')^\star$.

Case 1: $(f, e) \in R \circ Act$. Suppose that e has been added in the k-th step of both constructions. To start with, $f \in E_{k-1}$ and so, since N_k is finite, there is $f' \in \mathcal{F}_{k-1}$ such that $(f, f') \in R^\star$ and f, f' have the same label. If $(f', e) \in R'_{k-1} \circ Act'_{k-1} \circ (R'_k)^+$ then we have $(f, e) \in (R')^+ \circ Act' \circ (R')^+$. Otherwise, we have $(f', e) \in R'_k \circ Act'_k$, and so $(f, e) \in (R')^+ \circ Act'$.

Case 2: $(e, f) \in Act^{-1} \circ R$. Suppose that f has been added in the k-th step of both constructions. If $e \in E^k$ then one can easily see that $(e, f) \in (Act')^{-1} \circ R'$. Otherwise, $e \in E_{k-1}$ and, similarly as before, there is $e' \in \mathcal{F}_{k-1}$ such that $(e, e') \in R^\star$ and e, e' have the same label. If $(e', f) \in Act_{k-1}^{-1} \circ R_{k-1} \circ R_k^+$ then we have $(e, f) \in (R')^\star \circ (Act')^{-1} \circ (R')^+$. Otherwise, we have $(e', f) \in Act_k^{-1} \circ R_k'$, and so $(e, f) \in (R')^\star \circ (Act')^{-1} \circ R'$. $\qquad\square$

As before, infinite branching disappears for bounded underlying nets.

Proposition 15. *If* UND(NI) *with the initial marking M_0 is bounded, then AON^{mod} is finitely branching.*

Proof. If AON^{mod} is not finitely branching then there are events $e, e_1, e_2, \ldots$ such that $(e, e_i) \in R \circ Act$, for all $i \geq 1$; or $(e, e_i) \in Act^{-1} \circ R$, for all $i \geq 1$. Since UND(NI) with the initial marking M_0 is bounded, there are e_k and e_l such that $(e_k, e_l) \in R^+$. We then obtain a contradiction with Modification 2. $\qquad\square$

We can also estimate the number of the frontier events in the bounded case, which follows from the fact that in any occurrence net generated from UND(NI) with the initial marking M_0, there can be no more causally unrelated instances of a given transition than the maximal size of the reachable markings.

Proposition 16. *If* UND(NI) *with the initial marking M_0 is bounded, then $|\mathcal{F}_k| \leq |T| \cdot \beta$ for every k, where β is the maximal size of markings reachable from M_0 in* UND(NI).

Therefore for NI the size of a frontier, which also includes all conditions without outgoing directed arcs, is bounded by $(1 + |T|) \cdot \beta$. A tighter upper bound can be obtained in the safe case, due to the fact that in an occurrence net generated from a safe PT-net, any two occurrences of the same transition are causally related. Since there are at most $|P|$ conditions without outgoing directed arcs, this also implies that there are never more than $|P|$ events in a frontier.

Proposition 17. *If* UND(NI) *with the initial marking M_0 is safe (i.e., each marking reachable from M_0 in* UND(NI) *is a set) then $|\mathcal{F}_k| \leq min(|T|, |P|)$.*

Hence the size of a frontier is bounded by $2|P|$.

8   Final Remarks

In this paper we have introduced and discussed infinite processes of PTI-nets. It is our intention to use the results obtained here in the definition of branching processes of PTI-nets, leading to an extension of the theory developed in [5]. It is worth mentioning that the two ways of deriving processes of PT-nets with inhibitor arcs are not too distant, and in essence overlap for a wide range of nets. For example, using a result proven in [10], it is immediate that if a PTCI-net NCI is an ordinary inhibitor net (i.e., $I(p, t) \in \{0, \infty\}$ for all p and t),

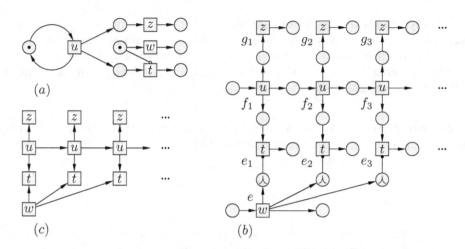

Fig. 6. A PTI-net (a); its modified a-process generated by $\{w\}\langle\{u\}\{z\}\{t\}\rangle_{\mathbb{N}+}$ (b); and the corresponding causality relationship (c)

then $\kappa(\alpha^{cpl}(NCI)) = \kappa(\alpha(NCI))$. This means that the two semantical mappings result in identical descriptions of the causalities in the net NCI.

We will now discuss some issues related to the two modifications proposed in the last section. To start with, Modification 1 might look somewhat unsatisfactory since it involves deleting redundant activator arcs rather than not generating them in the first place. This seems to be in general unavoidable since it may be impossible to predict whether a newly generated event will eventually belong to Act_b^{max}. However, if the underlying net is safe, it is always possible to check this by looking at the transitions occurring in the original step sequence.

Let us now consider Modification 2 for unbounded PTI-nets. In such a case the frontier can grow indefinitely, as it can for PT-nets [1] (there is no upper bound on the size of reachable markings). Moreover, we lose the finite branching property which holds for infinite processes of PT-nets. Consider the example in Figure 6. The a-process there has been obtained using the modified version of our construction (which in this particular case does not have any effect), yet it still contains an infinitely branching event e. One might, of course, hope that the situation could be improved in some way (notice that gluing all λ-labelled conditions together only shifts the source of infinite branching), but one can see that there is no finitely branching ifao-net which would generate the same ifso-structure as that in Figure 6. Indeed, to generate the causalities $e \prec e_i$, we cannot use (as an intermediary) the f_i's since this would introduce causalities between e and (some of) the g_i's. Moreover, no e_l can be used to generate causalities between e and another e_j since the e_i's are unrelated. We therefore conclude that infinite branching is in general unavoidable when generating unfoldings of unbounded PTI-nets. This somewhat pessimistic remark should however not impact on the application to model checking of the constructions proposed here, as the latter typically applies only to bounded Petri nets (see, e.g., [9,12]).

Finally, the above argument essentially implies that the ifso-structure in Figure 6 can only be generated by an ifao-net exhibiting infinite branching. Since the PTI-net in Figure 6 has exactly the same *a priori* and *a posteriori* step sequences, what we have just observed would not be changed by adopting a different operational semantics for PTI-nets.

Acknowledgments. We are grateful to the referees for their constructive criticism. This research was supported by the EPSRC project CASINO and the EC IST project 511599 RODIN.

References

1. Best, E., Devillers, R.: Sequential and Concurrent Behaviour in Petri Net Theory. Theoretical Computer Science 55 (1988) 87–136
2. Billington, J.: Protocol Specification Using P-Graphs, a Technique Based on Coloured Petri Nets. In: Part II of [13] (1998) 331–385
3. Busi, N., Pinna, G.M.: Process Semantics for Place/Transition Nets with Inhibitor and Read Arcs. Fundamenta Informaticae 40 (1999) 165–197
4. Donatelli, S., Franceschinis, G.: Modelling and Analysis of Distributed Software Using GSPNs. In: Part II of [13], (1998) 438–476
5. Engelfriet, J.: Branching Processes of Petri Nets. Acta Inf. 28 (1991) 575–591
6. Ehrig, H., Juhás, G., Padberg, J., Rozenberg, G. (eds.): Unifying Petri Nets. Springer-Verlag, Lecture Notes in Computer Science 2128 (2001)
7. Janicki, R., Koutny, M.: Semantics of Inhibitor Nets. Information and Computation 123 (1995) 1–16
8. Janicki, R., Koutny, M.: Order Structures and Generalisations of Szpilrajn's Theorem. Acta Inf. 34 (1997) 367–388
9. Khomenko, V., Schröter, C.: Parallel LTL-X Model Checking of High-Level Petri Nets Based on Unfoldings. In: R.Alur and D.A.Peled (eds.): CAV'04. Springer-Verlag, Lecture Notes in Computer Science 3114 (2004) 109-121
10. Kleijn, H.C.M., Koutny, M.: Process Semantics of General Inhibitor Nets. Information and Computation 190 (2004) 18–69
11. Kleijn, H.C.M., Koutny, M.: Infinite Process Semantics of Inhibitor Nets. Technical Report 932, University of Newcastle upon Tyne (2005)
12. McMillan, K.L.: A Technique of State Space Search Based on Unfoldings. Formal Methods in System Design 6 (1995) 45-65
13. Reisig, W., Rozenberg, G. (eds.): Lectures on Petri Nets. Springer-Verlag, Lecture Notes in Computer Science 1491,1492 (1998)
14. Rozenberg, G., Engelfriet, J.: Elementary Net Systems. In: W.Reisig and G.Rozenberg (eds.) Advances in Petri Nets. Lectures on Petri Nets I: Basic Models. Springer-Verlag, Lecture Notes in Computer Science 1491. (1998) 12–121
15. Vogler, W.: Partial Order Semantics and Read Arcs. Theoretical Computer Science 286 (2002) 33–63

Towards Synthesis of Petri Nets from Scenarios

Robert Lorenz[1] and Gabriel Juhás[2]

[1] Lehrstuhl für Angewandte Informatik
Katholische Universität Eichstätt-Ingolstadt, Eichstätt, Germany
robert.lorenz@ku-eichstaett.de
[2] Faculty of Electrical Engineering and Information Technology
Slovak University of Technology, Bratislava, Slovakia
gabriel.juhas@stuba.sk

Abstract. Given a set of scenarios, we answer the question whether this set equals the set of all executions of a Petri net.

Formally, scenarios are expressed by (isomorphism classes of) labelled partial orders (LPOs), also known as pomsets or partial words. An LPO is an execution of a Petri net if it is a sequentialization of an LPO generated by a process of the net. We propose a definition of regions for a set of LPOs, i.e for a partial language. Given a partial language of scenarios, we prove a necessary and sufficient condition (based on regions) for the partial language of scenarios to be the partial language of executions of a place/transition Petri net. Finally, we prove our notion of regions to be consistent with the notion of regions of trace languages.

1 Introduction

Scenario based specifications of information systems become recently very popular both in theory and practice, see e.g. [16]. There are approaches which concentrate on scenario mining [1, 2]. There are also several methods for the synthesis of a system model from a set of scenarios (see e.g. [23, 15, 3, 28, 7, 8] for some of them), the scenario based verification of the system model ([21, 9]), and for test generation and validation purposes ([22]). They differ mostly in the formalisms used for scenarios and the system model. In this paper, we consider place/transition Petri nets (p/t-nets) as system models.

The simplest way to describe scenarios for p/t-nets is to use sequences of fired transitions. Thus, the set of scenarios specifying the system can be described by a formal language over the set of transitions or by a transition system. For this kind of specifications the synthesis problem is solved using the theory of regions [3, 6, 7, 8, 5, 4, 25]. However, sequences lack any information about independence between occurring events and Petri nets are popular mainly because they allow to easily describe their non-sequential behavior and concurrency of occurring events.

There are several different ways how to formalize the non-sequential behavior of Petri nets, most of them based on a partial order between events labelled by transitions. In such a partial order, an ordering of events a and b expresses that the occurrence of a precedes the occurrence of b. The absence of an ordering expresses their concurrent (independent) occurrence.

S. Donatelli and P.S. Thiagarajan (Eds.): ICATPN 2006, LNCS 4024, pp. 302–321, 2006.

The probably most common way to express the partial order based behavior of Petri nets is by labelled causal (occurrence) nets and processes [12, 13]. Dropping the conditions and keeping the events in a process leads to a labelled partial order (LPO), called *run*. Runs capture the causal ordering of events. Naturally, events which are independent can occur sequentially in any order. Thus, adding order to a run still leads to a possible execution. For example, any occurrence sequence of transitions, which can be seen as a labelled total order, sequentializes a run. Generalizing this relationships, any LPO which sequentializes a run, is a possible *execution* of the net.

A partial order based semantics can also be defined through sequences of (concurrent) steps of transitions. A step of transitions is a multi-set of transitions. It is enabled in a marking, if there are enough tokens to fire all transitions concurrently, where a transition can be contained in a step more than once (auto-concurrency). The non-sequential behavior of p/t-nets can be described by step transition systems. A region based characterization of those step transition systems describing the non-sequential behavior of p/t-nets was given in [24]. The notion of enabled steps of transitions leads to the notion of *enabled LPOs*: Given a partial order of events labelled by transitions, it is said to be enabled to occur if the following condition is satisfied for every cut (i.e. for every maximal set of independent events) [14]: If all the events before the cut have occurred, then in the reached marking the step of transitions determined by the cut is enabled to occur. It was proven in [20, 29] that an LPO is an execution if and only if it is enabled.

There are also other descriptions of the non-sequential behavior of Petri nets. For example, one can take step sequences and relate them via an equivalence, getting so called *traces*. A region based characterization of trace languages which describe the non-sequential behavior of p/t-nets was given in [17].

Because the identity of vertices (events) of LPOs is not important, it is usual to distinguish LPOs only up to isomorphism. An isomorphism class of LPOs can be also understood as a *partial word* (over a set of labels), similarly as a labelled total order (a sequence) is understood as a (total) word [14]. A set of LPOs (over a set of labels) is called a *partial language* (over a set of labels). Another view is to understand isomorphism classes of LPOs as *pomsets* (partially ordered multi-sets (of labels)) [27].

In [14] Grabowski characterizes partial languages of executions of 1-safe Petri nets. In case of p/t-nets, there is no characterization of partial languages of executions similar to those based on regions for total languages or trace languages. In this paper, we fill this gap.

In [18] we gave a characterization of p/t-net executions based on so called *token flow functions*. Roughly speaking, a token flow function labels any arc connecting an event a with an event b in an LPO by the number of tokens produced by the occurrence of a which are consumed by the occurrence of b in a fixed place p. For each event the sum of token numbers annotated to its outgoing arcs defines the *outgoing token flow* (w.r.t. p). The sum of token numbers annotated to ingoing arcs defines its *ingoing token flow* (w.r.t. p). Given a p/t-net, we formulated a necessary and sufficient condition for the executability of an LPO, called the *token flow property*. An LPO fulfills the token flow property w.r.t. to a marked p/t-net if and only if for every place p there exists a token flow function of the LPO satisfying the following: for every event, its outgoing token flow (w.r.t. p) does not exceed the number of tokens produced by the corresponding transition, provided that for every event its ingoing token flow (w.r.t. p) equals the number of tokens consumed by the corresponding transition of the p/t-net.

The notion of regions presented in this paper is based on the extension of the notion of token flow functions from single LPOs to partial languages of LPOs. Regions are exactly those extended token flow functions, for which there is an upper bound for the outgoing token flows of events and equally labelled events have equal ingoing token flows. Given a set of scenarios in form of a partial language of LPOs, its regions are used to define its so called *associated p/t-net*: The regions form the set of places and the event labels form the set of transitions. Given a region r and a transition t, the ingoing token flow of events labelled by t defines the weight of the arc connecting r with t, and the maximum of all outgoing token flows of events labelled by t defines the weight of the arc connecting t with r.

As the main result of the paper, we show that a partial language $\mathcal{L}$ of LPOs equals the partial language of executions of a p/t-net if and only if all executions of the associated p/t-net belong to $\mathcal{L}$. We also prove that our notion of regions is consistent with the notion of regions for trace languages as given in [17].

The main result of this paper is a step towards the synthesis of p/t-nets from sets of scenarios. It is also a step towards the synthesis of p/t-nets from incomplete sets of scenarios $\mathcal{L}$, since the p/t-net associated to $\mathcal{L}$ can be considered as the net with the "smallest" behavior including $\mathcal{L}$. The necessary steps to derive synthesis algorithms, namely a finite representation of an infinite set of scenarios and a necessary and sufficient condition for existence of a finite representation of the (possibly) infinite set of regions are discussed in the conclusion (Section 4).

The rest of the paper is organized as follows: In Section 2 we state basic mathematical notations and give the definitions of LPOs and of syntax and semantics of p/t-nets. In Section 3 we briefly restate the definitions and results from [18], define regions of partial languages, prove a characterization of partial languages of executions of p/t-nets based on this notion of regions and finally show that our notion of regions is consistent with the notion of regions of trace languages as presented in [17].

2   Preliminaries

2.1   Mathematical Notations

By $\mathbb{N}$ we denote the *nonnegative integers*. Given a function f from A to B and a subset C of A we write $f|_C$ to denote the *restriction* of f to the set C. By id_A we denote the *identity function* on a set A, and by 1_X we denote the *characteristic function* of a subset $X \subseteq A$ (given by $1_X(a) = 1$ for $a \in X$ and $1_X(a) = 0$ otherwise). Given a finite set A, the symbol $|A|$ denotes the *cardinality* of A. The set of all subsets of a set A we denote by 2^A. The set of all *multi-sets* over a set A is the set $\mathbb{N}^A$ of all functions $f : A \to \mathbb{N}$. We do not distinguish between a subset $X \subseteq A$ and the multi-set 1_X. Addition $+$ on multi-sets is defined as usual by $(m + m')(a) = m(a) + m'(a)$. We also write $\sum_{a \in A} m(a)a$ to denote a multi-set m over A. Given a binary relation $R \subseteq A \times A$ over a set A, the symbol R^+ denotes the *transitive closure* of R, and R^* the *reflexive transitive closure* of R.

2.2   Labelled Partial Orders

In this subsection we recall the definition of labelled partial orders (LPOs). It is based on the notion of directed graphs. A *directed graph* is a pair $(V, \rightarrow)$, where V is a finite *set of nodes* and $\rightarrow \subseteq V \times V$ is a binary relation over V called the *set of arcs*. As usual, given a binary relation $\rightarrow$, we write $a \rightarrow b$ to denote $(a, b) \in \rightarrow$.

Definition 1   ((Labelled) partial order)
A partial order *is a directed graph* po $= (V, <)$, *where* $<$ *is an irreflexive and transitive binary relation on* V. *A labelled partial order (LPO) is a triple* lpo $= (V, <, l)$, *where* $(V, <)$ *is a partial order, and* $l : V \rightarrow T$ *is a* labelling function *with set of labels* T.

Two nodes $v, v' \in V$, $v \neq v'$, of a partial order $(V, <)$ are called *independent* if $v \not< v'$ and $v' \not< v$. By co $\subseteq V \times V$ we denote the set of all pairs of independent nodes of V. A *co-set* is a subset $C \subseteq V$ fulfilling: $\forall x, y \in C : x \operatorname{co} y$. A *cut* is a maximal co-set. If co is transitive, then the partial order $(V, <)$ is called *stepwise linear*.

   For a co-set C of a partial order $(V, <)$ and a node $v \in V \setminus C$ we write $v < C$, if $v < s$ for an element $s \in C$, and $v \operatorname{co} C$, if $v \operatorname{co} s$ for all elements $s \in C$. For two co-sets C', C we write $C' < C$, if $s' < C$ for all elements $s' \in C'$.

   For each co-set C of $(V, <)$ the partial order $(V', < |_{V' \times V'})$ with $V' = \{v \in V \mid v < C\} \cup C$ is called *prefix* of $(V, <)$.

Definition 2   (Sequentialization, step-linearization)
Given two partial orders po$_1$ $= (V, <_1)$ *and* po$_2$ $= (V, <_2)$, *we say that* po$_2$ *is a* sequentialization *of* po$_1$ *if* $<_1 \subseteq <_2$, *and a* proper sequentialization *if additionally* $<_1 \neq <_2$.

   If po$_2$ *is stepwise linear and a sequentialization of* po$_1$, *the* po$_2$ *is called a* step-linearization *of* po$_1$.

We use the above notations defined for partial orders also for labelled partial orders. If X is the set of labels of lpo $= (V, <, l)$, i.e. $l : V \rightarrow X$, then for a set $V' \subseteq V$, we define the multi-set $|V'|_l \subseteq \mathbb{N}^X$ by $|V'|_l(x) = |\{v \in V \mid v \in V' \wedge l(v) = x\}|$.

   In the case, an LPO lpo $= (V, <, l)$ is stepwise linear, the relation co $\cup \{(v, v) \mid v \in V\}$ is an equivalence relation on V. By $[v]_{\operatorname{co}}$ we denote the equivalence classes of this equivalence relation. Let $\{[v]_{\operatorname{co}} \mid v \in V\} = \{\nu_1, \dots, \nu_k\}$ such that $i < j \Longrightarrow \nu_i < \nu_j$. Then $|\nu_1|_l \dots |\nu_k|_l$ is the *step sequence* of lpo.

   We will often consider LPOs only up to isomorphism. As usual, two LPOs $(V, <, l)$ and $(V', <', l')$ are called *isomorphic*, if there is a bijective mapping $\psi : V \rightarrow V'$ such that $l(v) = l'(\psi(v))$ for $v \in V$, and $v < w \Longleftrightarrow \psi(v) <' \psi(w)$ for $v, w \in V$. By [lpo] we will denote the set of all LPOs isomorphic to lpo. The LPO lpo is said to *represent* the isomorphism class [lpo].

Definition 3   (Partial language). *Let* T *be a set. A subset* $L \subseteq \{[\text{lpo}] \mid \text{lpo is an LPO with set of labels } T\}$ *is called* partial language *over* T.

2.3   Place/Transition-Nets

In this subsection we give the definitions of p/t-nets and their semantics based on processes and labelled partial orders (also known as partial words [14] or pomsets [27]).

The syntax of p/t-nets is based on the notion of nets. A *net* is a triple (P, T, F), where P is a (possibly infinite) set of *places*, T is a finite set of *transitions* satisfying $P \cap T = \emptyset$, and $F \subseteq (P \times T) \cup (T \times P)$ is a *flow relation*.

Let $x \in P \cup T$. The *preset* $^\bullet x$ *of* x is the set $\{y \in P \cup T \mid (y, x) \in F\}$. The *postset* $x^\bullet$ *of* x is the set $\{y \in P \cup T \mid (x, y) \in F\}$. Given a set $X \subseteq P \cup T$, this notation is extended by $^\bullet X = \bigcup_{x \in X} {}^\bullet x$ and $X^\bullet = \bigcup_{x \in X} x^\bullet$. For technical reasons, we consider only nets in which every transition has a nonempty preset and postset.

Definition 4 (Place/transition net)
A place/transition-net *(shortly* p/t-net*)* N *is a quadruple* (P, T, F, W), *where* (P, T, F) *is a net, and* $W : F \to \mathbb{N}^+$ *is a weight function.*

We extend the weight function W to pairs of net elements $(x, y) \in (P \times T) \cup (T \times P)$ with $(x, y) \notin F$ by $W((x, y)) = 0$.

A *marking* of a net $N = (P, T, F, W)$ is a function $m : P \to \mathbb{N}$, i.e. a multi-set over P. A *marked p/t-net* is a pair (N, m_0), where N is a p/t-net, and m_0 is a marking of N, called *initial marking*.

We omit the semantics of a p/t-net $N = (P, T, F, W)$ based on occurrence sequences and step sequences. The semantics of p/t-nets based on processes is defined using occurrence nets. An *occurrence net* is a net $O = (B, E, G)$ such that $|{}^\bullet b|, |b^\bullet| \leqslant 1$ for every $b \in B$ (i.e. places are *unbranched*), and O is *acyclic* (i.e. G^+ is a partial order). Places of an occurrence net are called *conditions* and transitions of an occurrence net are called *events*.

The set of conditions of an occurrence net $O = (B, E, G)$ which are minimal (maximal) according to G^+ is denoted by $\mathrm{Min}(O)$ ($\mathrm{Max}(O)$). Clearly, $\mathrm{Min}(O)$ and $\mathrm{Max}(O)$ are cuts w.r.t. G^+ (recall that events have nonempty pre- and postsets by assumption).

Definition 5 (Process)
Let (N, m_0) *be a marked p/t-net,* $N = (P, T, F, W)$. *A process of* (N, m_0) *is a pair* $K = (O, \rho)$, *where* $O = (B, E, G)$ *is an occurrence net and* $\rho : B \cup E \to P \cup T$ *is a labelling function, satisfying:*

(i) $\rho(B) \subseteq P$ *and* $\rho(E) \subseteq T$,
(ii) $\forall e \in E, \forall p \in P : |\{b \in {}^\bullet e \mid \rho(b) = p\}| = W((p, \rho(e)))$ *and* $\forall e \in E, \forall p \in P :$
$|\{b \in e^\bullet \mid \rho(b) = p\}| = W((\rho(e), p))$, *and*
(iii) $\forall p \in P : |\{b \in \mathrm{Min}(O) \mid \rho(b) = p\}| = m_0(p)$.

Definition 6 (Run, Execution)
Let $K = (O, \rho)$, $O = (B, E, G)$, *be a process of a marked p/t-net* (N, m_0). *Then* $\mathrm{lpo}_K = (E, G^+|_{E \times E}, \rho|_E)$ *is called* run *of* (N, m_0) *representing* K.

A run lpo *of* (N, m_0) *is said to be* minimal, *if* lpo *is not a proper sequentialization of a run* lpo'.

A sequentialization lpo *of a run* lpo' *of* (N, m_0) *is called an* execution *of* (N, m_0).

If lpo *is an execution of* (N, m_0), *then the isomorphism class* $[\mathrm{lpo}]$ *is also called* execution *of* (N, m_0), . *Denote* $\mathfrak{Lpo}(N, m_0) = \{[\mathrm{lpo}] \mid \mathrm{lpo}$ *execution of* $(N, m_0)\}$ *the* partial language of executions *of a marked p/t-net* (N, m_0).

Another definition of the non-sequential semantics of p/t-nets is given by the notion of *enabled LPOs*:

Definition 7 (Enabledness). *Let* (N, m_0) *be a marked p/t-net,* $N = (T, P, F, W)$. *An LPO* lpo $= (V, <, l)$ *with* $l : V \to T$ *is called* enabled *(to occur) w.r.t.* (N, m_0) *if for every cut* C *of* lpo *and every* $p \in P$:

$$m_0(p) + \sum_{v \in V \wedge v < C} (W((l(v), p)) - W((p, l(v)))) \geq \sum_{v \in C} W((p, l(v))).$$

Its occurrence *leads to the marking* m' *given by* $m'(p) = m(p) + \sum_{v \in V} (W((l(v), p)) - W((p, l(v))))$ *for* $p \in P$.

An isomorphism class [lpo] is called *enabled* w.r.t. a marked p/t-net (N, m_0), if lpo is *enabled* w.r.t. (N, m_0). An important result relating the notions of *enabled LPOs* and *executions* was proven in [20, 29].

Theorem 1. *Let* (N, m_0) *be a marked p/t-net. An LPO* lpo *is enabled w.r.t.* (N, m_0) *if and only if* lpo *is an execution of* (N, m_0).

3   Synthesis of p/t-Nets

In this section, we first recall an alternative characterization of LPOs to be an execution called *token flow property*, we presented in [18]. We then discuss the formal problem setting of p/t-net synthesis from a given partial language. After that, we propose a notion of regions of partial languages and derive a characterization of those partial languages which allow a p/t-net synthesis. Finally, we establish the relationship between our notion of regions and the definition of regions of a trace language as defined in [17].

3.1   Token Flow Property

There is another characterization of LPOs to be executions of a marked p/t-net called *token flow property*. For motivation purposes we shortly recall the basic definitions and results.

Fix a marked p/t-net (N, m_0) and a place p of $N = (P, T, F, W)$. Given an LPO lpo $= (V, \prec, l)$ with $l(V) = T$ we assign non-negative integers to its edges through a so called *token flow function*. The aim is to find a token flow function x assigning values $x((v, v'))$ to edges (v, v') in such a way that there is a process with exactly $x((v, v'))$ post-conditions of v labelled by p which are also pre-conditions of v'. Thus, such a token flow function of lpo abstracts from the individuality of conditions of a process and encodes the flow relation of this process by natural numbers. Clearly, finding such a token flow function for every place means that lpo is a sequentialization of the run representing this process. In order to simplify the formal definition of the token flow property, let us define an extension of lpo $= (V, \prec, l)$ by adding an initial node which is smaller than all nodes from V and is labelled by a new label. It represents a transition producing the initial marking and helps to avoid several case differentiations in the following definitions.

Definition 8 (0-extension, of an LPO). *Let* lpo $= (V, \prec, l)$ *be a LPO. Then an LPO* lpo$^0 = (V^0, \prec^0, l^0)$, *where* $V^0 = (V \cup \{v_0\})$ *with* $v_0 \notin V$, $\prec^0 = \prec \cup(\{v_0\} \times V)$, *and* $l^0(v_0) \notin l(V)$ *and* $l^0|_V = l$, *is called 0-extension of* lpo.

We denote for an LPO lpo $= (V, <, l)$, a function $x :< \to \mathbb{N}$ and $v \in V$:

- $In(v, x) = \sum_{v' < v} x((v', v))$.
- $Out(v, x) = \sum_{v < v'} x((v, v'))$.

Definition 9 (Token flow function, of an LPO). *Let* lpo $= (V, \prec, l)$ *be an LPO and* lpo$^0 = (V^0, \prec^0, l^0)$ *be a 0-extension of* lpo. *A function* $x :\prec^0 \to \mathbb{N}$ *is called* token flow function *of* lpo, *if it satisfies the following property:*

(Tff) x *is* consistent *with the labelling l in the following sense:*

$$\forall v, v' \in V^0 : l(v) = l(v') \implies In(v, x) = In(v', x).$$

We denote $In(v, x)$ the intoken flow *of v w.r.t. x and $Out(v, x)$ the* outtoken flow *of v w.r.t. x for $v \in V$.*

Example 1. Figure 1 shows a p/t-net, a process of this p/t-net and a 0-extension lpo^0 of the run representing this process. The LPO lpo^0 has annotated a token flow function. Its intoken flow (outtoken flow) equals 0 (2) for the a- and b-labelled node, resp. 2 (0) for the c-labelled nodes.

This definition differs from that in [18]. While in [18] token flow functions were defined as general as possible, we here additionally require property (**Tff**). This is more intuitive and does not restrict the setting or change the argumentations, since (**Tff**) is implicitly contained in the token flow property defined below. Each process of a marked p/t-net defines a token flow function of the run representing this process in the following way:

Let $K = (O, \rho)$ be a process of a marked p/t-net (N, m_0) with $O = (B, V, G)$ and let lpo $= (V, <, l)$ be the run representing K. Let lpo$^0 = (V^0, <^0, l^0)$ be a 0-extension of lpo. Denote $v_0^\bullet = \text{Min}(O)$. We define for every place $p \in P$ the *canonical token flow function* $x_p :<^0 \to \mathbb{N}$ of lpo w.r.t. p as follows:

$$x_p((v, v')) = |\{b \in B \mid \rho(b) = p \land b \in v^\bullet \cap {}^\bullet v'\}|.$$

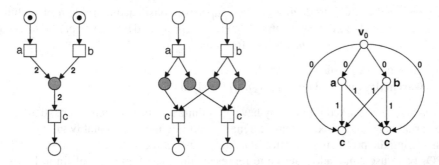

Fig. 1. A p/t-net (left picture), a process of this p/t-net (middle picture) and a 0-extension of the run representing this process together with annotated canonical token flow function w.r.t. the grey place (right picture)

Example 2. In Figure 1 there is shown the canonical token flow function of the shown run w.r.t. the grey place.

Observe, that by definition 5 (ii) the canonical token flow function of a run lpo w.r.t. a place p fulfills (**Tff**) (what justifies the above definition). By definition, each canonical token flow function respects the weight function and the initial marking of (N, m_0) in the sense that the following two properties are fulfilled:

- (**IN**) The intoken flow of an event v equals the number of tokens consumed from place p by the occurrence of transition $l(v)$.
- (**OUT**) The outtoken flow of an event v is less or equal to the number of tokens which are produced by the occurrence of transition $l(v)$ in place p. In particular, the outtoken flow of the source event v_0 is less or equal to the number of tokens in place p of the initial marking m_0.

In general, we say that an arbitrary LPO, whose labels are transitions of (N, m_0), fulfills the token flow property w.r.t. (N, m_0), if for every place there exists a token flow function which fulfills the properties (**IN**) and (**OUT**).

Definition 10 (Token flow property). *Let* $\text{lpo} = (V, \prec, l)$ *be an LPO with* $l(V) = T$, $\text{lpo}^0 = (V^0, \prec^0, l^0)$ *be a 0-extension of* lpo *and let* (N, m_0) *be a marked p/t-net,* $N = (P, T, F, W)$. *Denote* $W((l(v_0), p)) = m_0(p)$ *for each place* $p \in P$. *We say that* lpo *fulfills the token flow property w.r.t.* (N, m_0) *if the following statement holds: For all* $p \in P$ *there is a token flow function* $x_p : \prec^0 \rightarrow \mathbb{N}$ *with*

(**IN**) $In(v', x_p) = W((p, l(v')))$*for every* $v' \in V$,
(**OUT**) $Out(v', x_p) \leqslant W((l(v'), p))$ *for every* $v' \in V^0$.

In [18] we showed:

Theorem 2. *An LPO is an execution of a marked p/t-net if and only if it fulfills the token flow property w.r.t. this marked p/t-net.*

3.2   Problem Setting

As mentioned in the introduction, in this paper we consider the problem of finding a marked p/t-net whose non-sequential behavior is represented by a given partial language $\mathcal{L}$. The *p/t-net synthesis problem* reads formally:

> **Given:**   A partial language $\mathcal{L}$.
> **Searched:** A marked p/t-net (N, m_0) satisfying $\mathcal{L} = \mathfrak{Lpo}(N, m_0)$.

Obviously, a marked p/t-net may have an infinite number of (finite) runs, that means the partial language of executions of a marked p/t-net may be countably infinite. When considering the problem in practice, one has to restrict the setting to finite sets of finite LPOs or to use some adequate finite representation of infinite sets of finite LPOs. A detailed discussion of this topic is out of scope of this paper, but we will present some thoughts in the conclusion.

A partial language $\mathcal{L}$ of executions of a marked p/t-net satisfies the following immediate properties (induced by the definition of executions):

($\mathcal{L}_1$) The set of labels of $\mathcal{L}$ is finite.
($\mathcal{L}_2$) $\forall$[lpo] $\in \mathcal{L}$: If lpo$'$ is a prefix of lpo, then [lpo$'$] $\in \mathcal{L}$.
($\mathcal{L}_3$) $\forall$[lpo] $\in \mathcal{L}$: If lpo$'$ is a sequentialization of lpo, then [lpo$'$] $\in \mathcal{L}$.

Therefore, in the following, we will only consider partial languages which satisfy these conditions and denote them as *LPO-specifications*.

Definition 11 (LPO-specification). *A partial language $\mathcal{L}$ is called* LPO-specification, *if it satisfies the properties ($\mathcal{L}_1$) - ($\mathcal{L}_3$).*

For the following notions we will need a set L of concrete LPOs representing a given LPO-specification $\mathcal{L}$, i.e. satisfying [lpo] $\in \mathcal{L} \iff \exists$lpo$' \in L :$ [lpo] $=$ [lpo$'$]. We denote $\mathcal{L}(L) = \{$[lpo] $\mid$ lpo $\in L\}$. For technical reasons we will require such a representation L of $\mathcal{L}$ to fulfill the following properties.

(L1) The set of labels of L is finite.
(L2) $\forall$lpo $\in L$: If lpo$'$ is prefix of lpo, then lpo$' \in L$.
(L3) $\forall$lpo $\in L$: If lpo$'$ is sequentialization of lpo, then lpo$' \in L$.
(L4) $\forall(V, <, l), (V', <', l') \in L: v \in V \cap V' \Rightarrow l(v) = l'(v)$.
(L5) For each two LPOs lpo $= (V, <, l)$, lpo$' = (V', <', l') \in L$ with $V \cap V' \neq \emptyset$ there is lpo$'' \in L$ such that lpo and lpo$'$ both are sequentializations of prefixes of lpo$''$.

That means, L should be prefix and sequentialization closed and should represent alternative executions by LPOs with disjoint sets of nodes (condition (L5) separates the node sets of executions which are in conflict). It is straightforward to observe that an LPO-specification always allows such a representation.

Remark 1. In particular, (L5) allows to decompose L into subsets X of LPOs with pairwise disjoint node sets, each representing one (infinite) run. Namely, each such set X is defined as a maximal prefix and sequentialization closed subset of L satisfying:

(Seq) Each two LPOs lpo, lpo$' \in X$ are sequentializations of prefixes of another LPO lpo$'' \in X$.

For a set of BPOs L we denote $W_L = \bigcup_{(V,<,l) \in L} V$, $E_L = \bigcup_{(V,<,l) \in L} <$ and $l_L = \bigcup_{(V,<,l) \in L} l$ and write for an LPO lpo $= (V, <, l) \in L$, a function $x : E_L \to \mathbb{N}$ and $v \in V$:

- $In_{\text{lpo}}(v, x) = \sum_{v' < v} x((v', v))$.
- $Out_{\text{lpo}}(v, x) = \sum_{v < v'} x((v, v'))$.

In examples we will always give such L by a set of minimal LPOs, such that each LPO in L is a sequentialization of some prefix of one of these minimal LPOs.

Example 3. Figure 2 shows the two different LPO-specifications $L_1 = \{$lpo$_1$, lpo$_2\}$ and $L_2 = \{$lpo$_{3,n} \mid n \in \mathbb{N}\}$.

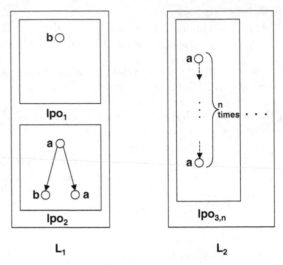

Fig. 2. Two LPO-specifications

3.3 Regions

For the rest of the paper we consider each LPO-specification given by a set of LPOs L satisfying (L1)-(L5). In subsection 3.1 we established that a token flow function of a partially ordered multi-set of transitions which fulfills the properties (**IN**) and (**OUT**) for some place defines a distribution of the tokens produced by a transition in this place onto the following transition in the given order, such that all transitions in the LPO are enabled w.r.t. this place. This gives rise to the idea to consider token flow functions of LPOs as regions, that define places of a p/t-net. For this, we define, as for LPOs, the notions of 0-*extension* and *token flow function* of L.

Definition 12 (0-extension). *Let* $v_0 \notin W_L$ *and* $\mathrm{lpo}^0 = (V^0, <^0, l^0)$ *be a 0-extension of each* $\mathrm{lpo} \in L$ *such that* $V^0 = V \cup \{v_0\}$, *and* $l_1^0(v_0) = l_2^0(v_0)$ *for each two* $(V_1, <_1, l_1), (V_2, <_2, l_2) \in L$. *Then the set* $L^0 = \{\mathrm{lpo}^0 \mid \mathrm{lpo} \in L\}$ *is called* 0-extension of L. *We denote* $W_L^0 = W_{L^0}$, $E_L^0 = E_{L^0}$ *and* $l_L^0 = l_{L^0}$.

Example 4. Figure 3 shows 0-extensions of the LPO-specifications L_1 and L_2 shown in Figure 2.

Definition 13 (Token flow function). *Let* L^0 *be a 0-extension of* L. *A function* $x : E_L^0 \to \mathbb{N}$ *is called* token flow function of L, *if it satisfies the following property:*

($\mathfrak{I}\mathrm{ff}$) x *is consistent over all LPOs:*

$$\forall \mathrm{lpo} = (V, <, l), \mathrm{lpo}' = (V', <', l') \in L^0,$$
$$\forall v \in V^0, \forall v' \in (V')^0 :$$
$$l(v) = l'(v') \implies In_{\mathrm{lpo}^0}(v, x) = In_{(\mathrm{lpo}')^0}(v', x).$$

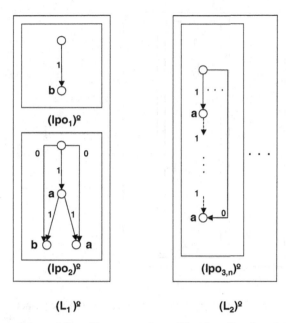

Fig. 3. 0-extensions of L_1 and L_2 with annotated possible token flow functions x_1 of L_1 and x_2 of L_2

Example 5. Figure 3 shows possible token flow functions x_1 of L_1 and x_2 of L_2 given by arc annotations.

Observe that according to $(\mathfrak{T}\mathfrak{ff})$ for lpo $= (V, <, l) \in L$ and $v \in V$ the value of $In_{\text{lpo}^0}(v, x)$ is independent of the choice of lpo. It is interpreted as the number of tokens consumed by transition $l(v)$. On the other hand, the value $Out_{\text{lpo}^0}(v, x)$ is dependent on lpo and can even differ for equally labelled nodes. It is interpreted as the number of tokens produced by transition $l(v)$ which are consumed by other transitions. Not consumed tokens remain in the final marking of the considered lpo. Of course, a token flow function can represent a place of a p/t-net only if the number of produced tokens represented by $Out_{\text{lpo}^0}(v, x)$ is bounded over all LPOs lpo $\in L$:

Definition 14 (Region). *Let L^0 be a 0-extension of L. A token flow function $r : E_L^0 \to \mathbb{N}$ of L is called a region of L if there is $M \in \mathbb{N}$ such that*

$$(\mathbf{R}) \quad |\{Out_{\text{lpo}^0}(v, r) \mid \text{lpo} = (V, <, l) \in L, v \in V^0\}| \leqslant M.$$

We denote

- $\text{Pre}_r(v) = In_{\text{lpo}^0}(v, r)$ *for* lpo $= (V, <, l) \in L$ *with* $v \in V$.
- $\text{Post}_r(v) = \max\{Out_{\text{lpo}^0}(v', r) \mid \text{lpo} = (V, <, l) \in L, v' \in V^0, l_L^0(v') = l_L^0(v)\}$
  for $v \in W_L^0$.
- $m_0(r) = \text{Post}_r(v_0)$.

A region which does not equal the 0-function is called non-trivial.

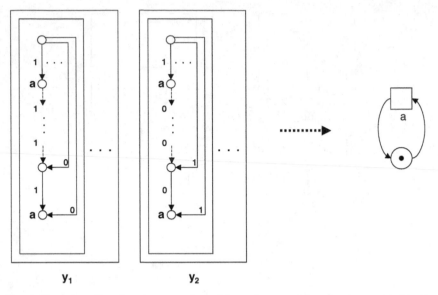

y_1 y_2

Fig. 4. Two token flow function y_1 and y_2 of L_2 and a part of the p/t-net associated to L_2

For a label a, a node v with $l_L(v) = a$ and a region r of L we also denote $\mathrm{Pre}_r(a) = \mathrm{Pre}_r(v)$ *and* $\mathrm{Post}_r(a) = \mathrm{Post}_r(v)$.

Example 6. The token flow functions x_1 and x_2 shown in Figure 3 are both regions. It holds: $\mathrm{Pre}_{x_1}(a) = \mathrm{Pre}_{x_1}(b) = 1$, $\mathrm{Post}_{x_1}(a) = 2$, $\mathrm{Post}_{x_1}(b) = 0$ and $m_0(x_1) = 1$.

Consider the token flow functions y_1 and y_2 of L_2 shown in Figure 4. While y_1 defines a region with $\mathrm{Pre}_{y_1}(a) = \mathrm{Post}_{y_1}(a) = m_0(y_1) = 1$, y_2 does not because $Out_{\mathrm{lpo}_{3,n}^0}(v_0, y_2) = n$ for $n \in \mathbb{N}$.

By the definition of regions, we can associate a p/t-net to L by considering regions as places. By construction, each LPO of L is an execution of the associated p/t-net, but in general the associated p/t-net has more executions as specified.

Definition 15 (associated p/t-net). *We denote $\mathcal{R}_L$ the set of non-trivial regions of L. Denote $P = \{p_r \mid r \in \mathcal{R}_L\}$, T the set of labels of L, $W((p_r, l_L(v))) = \mathrm{Pre}_r(v)$ and $W((l_L(v), p_r)) = \mathrm{Post}_r(v)$ for $l_L(v) \in T$ and $p_r \in P$, $F = \{(x, y) \mid W((x, y)) > 0\}$, and $m_0(p_r) = m_0(r)$ for $p_r \in P$. Then the p/t-net (N_L, m_L), $N_L = (P, T, F, W)$, we call the p/t-net associated to L.*

Example 7. The left part in Figure 5 shows some (of the infinite many) regions of L_2. The right part shows (a part of) the p/t-net associated to L_2 with the places $p_i = p_{z_i}$, $i = 1, 2, 3$.

Of course, the associated p/t-net (N_L, m_L) is a candidate for a p/t-net satisfying $\mathcal{L}(L) = \mathfrak{Lpo}(N_L, m_L)$. Observe that $P = \emptyset$ (i.e. $\mathcal{R}_L = \emptyset$) is possible. In this case the associated p/t-net would have an unrestricted behavior and trivially $\mathcal{L}$ is part of this behavior.

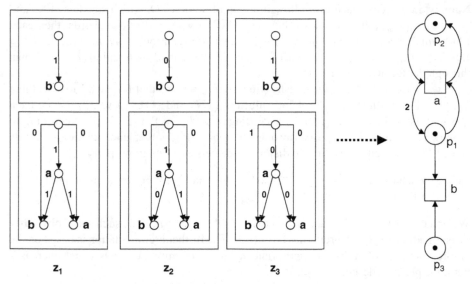

Fig. 5. Some regions of L_2 with corresponding places of the associated p/t-net

In particular, if $^\bullet t = \emptyset$ for a transition t of (N_L, m_L), (N_L, m_L) does not satisfy $\mathcal{L}(L) \supseteq \mathfrak{Lpo}(N_L, m_L)$. The following Lemma tells us that (N_L, m_L) always satisfies $\mathcal{L}(L) \subseteq \mathfrak{Lpo}(N_L, m_L)$.

Lemma 1. *Let* $\mathrm{lpo} \in L$ *and* (N_L, m_L), $N_L = (P, T, F, W)$, *be the p/t-net associated to* L. *Then* $[\mathrm{lpo}]$ *is an execution of* (N_L, m_L).

Proof. We can assume without loss of generality that $P \neq \emptyset$. According to Theorem 2, it is enough to show that $\mathrm{lpo} = (V, \prec, l)$ fulfills the token flow property w.r.t (N_L, m_L). In fact, for each place $p_r \in P$ the token flow function $r|_{\prec^0}$ fulfills the properties (**IN**) and (**OUT**) according to ($\mathfrak{Tff}$) and (**R**).   □

The main theorem reads:

Theorem 3. *There is a marked p/t-net* (N, m_0) *with* $\mathcal{L}(L) = \mathfrak{Lpo}(N, m_0)$ *if and only if* $\mathcal{L}(L) = \mathfrak{Lpo}(N_L, m_L)$ *for the marked p/t-net* (N_L, m_L) *associated to* L.

We prove the theorem in the next subsection.

3.4   Proof of the Main Result

The "if"-part is obvious. To prove the "only if"-part, let (N, m_0), $N = (P, T, F, W)$, be a p/t-net satisfying $\mathcal{L}(L) = \mathfrak{Lpo}(N, m_0)$. It is enough to show that for each $p \in P$ there is a region $r = r(p)$ of L such that for each transition t

(i) $\mathrm{Pre}_r(t) = W((p, t))$,
(ii) $\mathrm{Post}_r(t) \leqslant W((t, p))$,
(iii) $m_0(r) \leqslant m_0(p)$.

Namely, in this case each token flow function of some LPO, which satisfies (**IN**) and (**OUT**) w.r.t. p_r, also satisfies (**IN**) and (**OUT**) w.r.t. p. In other words, then each LPO which is an execution of (N_L, m_L) is also an execution of (N, m_0), i.e. $\mathcal{L}(L) \subseteq \mathfrak{Lpo}(N_L, m_L) \subseteq \mathfrak{Lpo}(N, m_0) = \mathcal{L}(L)$. The construction of such regions $r(p)$ is shown through the following Lemmata.

Fix a place $p \in P$. For the construction of $r(p)$ we use that for each lpo $\in L$ there is a token flow function x_p^{lpo} of lpo fulfilling (**IN**) and (**OUT**) w.r.t. p (since each lpo $\in L$ is enabled w.r.t. (N, m_0) according to $\mathcal{L}(L) = \mathfrak{Lpo}(N, m_0)$). According to remark 1, L can be decomposed (through an appropriate equivalence relation) into sets X of LPOs , which have pairwise disjoint node sets and satisfy property

(Seq) Each two LPOs lpo, lpo$' \in X$ are sequentializations of prefixes of another
  LPO lpo$'' \in X$.

We first construct $r(p)$ for each such maximal subset X of L. Each such set X obviously satisfies additionally the properties (L1)-(L5). The further argumentation is based on the "representation" of X by an appropriate sequence of minimal LPOs which increases w.r.t. the prefix ordering:

Lemma 2. *Let X be a set of LPOs satisfying (L1) - (L5) and (Seq). Then there is a sequence of LPOs $(\mathrm{lpo}_n)_{n \in \mathbb{N}} \subseteq X$, such that*

(i) lpo_n is prefix of lpo_{n+1},
(ii) lpo_n is minimal in X, and
(iii) each LPO in X is a sequentialization of a prefix of some lpo_n.

Proof. Observe that each two LPOs lpo, lpo$' \in X$ are sequentializations of prefixes of another LPO lpo$'' \in X$. Order the set of all nodes of LPOs in X inductively as follows:

  – As node v_1 choose an arbitrary node which is minimal for one arbitrary LPO in X.
  – Suppose the nodes $v_1, \ldots, v_n$ are chosen. Remove the nodes $v_1, \ldots, v_n$ from all LPOs in X and denote X_n the set of all those modified LPOs. As node v_{n+1} choose an arbitrary node which is minimal for one arbitrary LPO in X_n.

Note that for each $n \in \mathbb{N}$ there are LPOs in X with node set $\{v_1, \ldots, v_n\}$ (according to the chosen ordering and (L5)), and that there is in fact a unique minimal one (since two different LPOs with the same node set are sequentializations of prefixes of some other LPO). Let lpo_n be this minimal LPO with node set $\{v_1, \ldots, v_n\}$. Obviously, lpo_n is then a prefix of lpo_{n+1}, since lpo_{n+1} restricted to the node set $\{v_1, \ldots, v_n\}$ is also in X. Clearly, every LPO in X is a sequentialization of a prefix of some lpo_n. □

If the LPOs of all such $X \subseteq L$ have "consistent" token flow functions, we can construct a region of L:

Lemma 3. *Let $\mathrm{Pre}, \mathrm{Post} : T \to \mathbb{N}$ and $(\mathrm{lpo}_n)_{n \in \mathbb{N}}$ be a sequence of LPOs such that*

(i) The set of labels of each lpo_n is a subset of T.
(ii) lpo_n is prefix of lpo_{n+1},
(iii) for each n there is a token flow function x of $\mathrm{lpo}_n = (V_n, <_n, l_n)$ with

(IN) $In(v', x) = \text{Pre}(a)$ *for every* $v' \in V_n^0$ *with* $l_n(v') = a$,
(OUT) $Out(v', x) \leqslant \text{Post}(a)$ *for every* $v' \in V_n^0$ *with* $l_n(v') = a$.

Then for each $n \in \mathbb{N}$ *there is a token flow function* x_n *of* lpo_n *satisfying* **(IN)** *and* **(OUT)**, *such that* $x = \lim_{n \to \infty} x_n$ *is a region of the set* X *of all sequentializations of prefixes of some* lpo_n.

Proof. Choose the token flow functions x_n inductively as follows:

- Let x_1 be an arbitrary token flow function of lpo_1 satisfying **(IN)** and **(OUT)**.
- Choose a token flow function x_{n+1} of lpo_{n+1} satisfying **(IN)** and **(OUT)** such that m is maximal with $x_{n+1}|_{<_m} = x_n|_{<_m}$.

We first show that the sequence of functions $(x_n)_{n \in \mathbb{N}}$ converges pointwise, i.e. on every edge. For this, we fix $k \in \mathbb{N}$ and show that there is $n(k) \in \mathbb{N}$ such that $\forall n, m \geqslant n(k) : x_n|_{<_k} = x_m|_{<_k}$. That means that the sequence $(x_n)_{n \in \mathbb{N}}$ becomes stationary on every lpo_k. This can be seen as follows: The set of all token flow functions of lpo_k fulfilling **(IN)** and **(OUT)** is finite, since there are only finitely many arcs and the value of such a token flow function on an arc is bounded through **(IN)** and **(OUT)** by $\max(\max\{\text{Pre}(a) \mid a \in T\}, \max\{\text{Post}(a) \mid a \in T\})$. The functions $y_n = x_n|_{<_k}$ are token flow function of lpo_k fulfilling **(IN)** and **(OUT)** for each $n \geqslant k$. Assume that $y_i \neq y_{i+1}$ and $y_i = y_j$ for $k \leqslant i < j$. Then $z_{i+1} = x_j|_{<_{i+1}}$ is a token flow function of lpo_{i+1} fulfilling **(IN)** and **(OUT)** with $z_{i+1}|_{<_k} = y_j = y_i = x_i|_{<_k}$. This contradicts the choice of x_{i+1} since $x_{i+1}|_{<_k} = y_{i+1} \neq y_i = x_i|_{<_k}$. Thus, y_n can change only finitely often.

Thus, x is well-defined. Assuming $x_n = 0$ on every edge not in $<_n$, we get that x is a token flow function of X. To see this, let $\text{lpo} = (V, <, l), \text{lpo}' = (V', <', l') \in X$ and $v \in V^0$ and $v' \in (V')^0$ with $l(v) = l'(v')$, lpo be sequentialization of a prefix of lpo_k and lpo' be sequentialization a prefix of lpo_m. Then, since all x_n fulfill **(IN)**:

$$
\begin{aligned}
In_{\text{lpo}^0}(v, x) &= \lim_{n \to \infty} In_{\text{lpo}^0}(v, x_n) \\
&= \lim_{n \to \infty} In_{\text{lpo}_k^0}(v, x_n) \overset{\textbf{(IN)}}{=} \text{Pre}(l(v)) \\
&\overset{\textbf{(IN)}}{=} \lim_{n \to \infty} In_{\text{lpo}_m^0}(v', x_n) = In_{(\text{lpo}')^0}(v', x).
\end{aligned}
$$

Observe here that $x_n|_{<_k}$ is a token flow function of lpo_k for $n \geqslant k$. Then x fulfills **($\mathfrak{T}$ff)**, i.e. is a token flow function. Finally we get that x is even a region: Let $\text{lpo} = (V, <, l) \in X$ be a prefix of a sequentialization of lpo_k. It follows for $v \in V$, since all x_n fulfill **(OUT)**:

$$
\begin{aligned}
Out_{\text{lpo}^0}(v, x) &= \lim_{n \to \infty} Out_{\text{lpo}^0}(v, x_n) \\
&\leqslant \lim_{n \to \infty} Out_{\text{lpo}_k^0}(v, x_n) \overset{\textbf{(OUT)}}{\leqslant} \text{Post}(l(v)).
\end{aligned}
$$

That means x fulfills **(R)**, i.e. is a region of X. $\qquad \square$

For each maximal subset $X \subseteq L$ there is a sequence $(\text{lpo}_n)_{n \in \mathbb{N}} \subseteq X$ satisfying the properties stated in Lemma 2 and fulfilling the preconditions of Lemma 3 with $\text{Pre}(l(v)) = W((p, l(v)))$ and $\text{Post}(a) = W((l(v), p))$ by construction. By Lemma 3 there is an appropriate region for each such X, Moreover, BPOs in different such sets X have disjoint node sets. Therefore we can construct a searched region $r(p)$ as the union of all regions of such sets X.

3.5   Relationship to Other Definitions of Regions

In this section we establish the relationship between regions of LPO-specifications and other definitions of regions. For this one has to distinguish between regions of languages and regions of graphs. Whereas in the case of languages it is required that the synthesized p/t-net produces the given language, in the case of graphs it is required that state-graph of the synthesized p/t-net and the given graph are isomorphic, what is a substantial stronger requirement. The regions presented in this paper are language based, since we distinguish different executions by LPOs with disjoint node sets (no matter whether they have a common prefix or not). Another definition of regions which is based on languages and allows the specification of concurrent events is that for trace languages. Therefore, we exemplarily compare our definition of regions to that one for trace languages as presented in [17].

We recall the definition of trace semantics of a p/t-net as given in [17]. A *(generalized) concurrency alphabet* is a pair (A, I), where A is a finite alphabet, and $I \subseteq (\mathbb{N}^A)^+ \times \mathbb{N}^A$ is an *independence relation*. For a concurrency alphabet (A, I) we denote $\rho \sim_I \rho'$ if there are $\rho_1, \rho_2 \in (\mathbb{N}^A)^+$ and $u, v, u', v' \in \mathbb{N}^A$, such that $\rho = \rho_1 uv\rho_2$ and $\rho' = \rho_1 u'v'\rho_2$, $u + v = u' + v'$ and $(\rho_1, u + v) \in I$, $\equiv_I = (\sim_I)^*$, and $[\rho]_I = \{\rho' \mid \rho \equiv_I \rho'\}$. $L \subset (\mathbb{N}^A)^+$ is called *consistent* (w.r.t. (A, I)), if $\forall \rho \in L : [\rho]_I \subseteq L$. A *trace language* (w.r.t. (A, I)) is a triple (L, A, I), where L is consistent (w.r.t. (A, I)).

For a trace language $Tr = (L, A, I)$, a *region* of Tr is a function $r : L \cup A \rightarrow \mathbb{N} \cup (\mathbb{N} \times \mathbb{N})$ satisfying the following conditions for $\rho \in L$, $a \in A$ and $u \in \mathbb{N}^A$:

(TR1) $r(\rho) \in \mathbb{N}$ and $r(a) = (\text{Pre}_r(a), \text{Post}_r(a)) \in \mathbb{N} \times \mathbb{N}$
(TR2) If $\rho u \in L$ then $r(\rho) \geqslant \sum_{a \in A} u(a)\text{Pre}_r(a)$, and
(TR3) If $\rho u \equiv \rho'$ then $r(\rho') = r(\rho) + \sum_{a \in A} u(a)(\text{Post}_r(a) - \text{Pre}_r(a))$.

The set of all regions is denoted by R_{Tr}. We can now associate a p/t-net (N, m_0) to Tr with $N = (R_{Tr}, A, F, W)$ via $W(r, a) = \text{Pre}_r(a)$, $W(a, r) = \text{Post}_r(a)$ and $m_0(r) = r(\underline{0})$, where $\underline{0}$ denotes the empty step sequence.

A trace language $Tr = (L, A, I)$ is called *PN trace language*, if the following axioms are satisfied: (PN1) $L \neq \emptyset$, (PN2) $\rho u \in L \Rightarrow \rho \in L$, (PN3) $\rho u \in L \Rightarrow (\rho, u) \in I$ and (PN4) $\rho \in L \Rightarrow (\forall r \in R_{Tr} : r(\rho) \geqslant \sum_{a \in A} u(a)\text{Pre}_r(a) \Rightarrow \rho u \in L)$. If a trace language $Tr = (L, A, I)$ satisfies (PN1) - (PN4), then the associated p/t-net can be shown to have L as its set of step sequences. On the other hand, the trace language of a given p/t-net always fulfills (PN1) - (PN4) (see [17]).

To each LPO-specification it can associated a trace language in a natural way:

Definition 16 (Trace language, of an LPO-specification). *Let L' represent an LPO-specification. Define the trace language (L, A, I) associated to L' as follows: L is the* set of all step sequences of step-linearizations of LPOs in L', $A = l_{L'}(W_{L'})$ and $(\rho, u) \in I \iff \rho u \in L$.

Obviously such L is consistent.

Lemma 4. *Let L' represent an LPO-specification and let r be a region of L'. Let further (L, A, I) be the trace language associated to L'. Define $r_L : L \cup A \to \mathbb{N} \cup (\mathbb{N} \times \mathbb{N})$ by $r_L(\underline{0}) = \text{Post}_r(v_0)$, $r_L(a) = (\text{Pre}_r(a), \text{Post}_r(a))$ and $r_L(u_1 \ldots u_n) = r_L(0) + \sum_{i=1}^{n} \sum_{a \in A} u_i(a)(\text{Post}_r(a) - \text{Pre}_r(a))$. Then r_L is a region of (L, A, I).*

Proof. We first show (TR2), i.e. that for $\rho \in L$ and $u \in \mathbb{N}^A$:

$$(*) \quad \rho u \in L \Rightarrow r_L(\rho) \geqslant \sum_{a \in A} u(a)\text{Pre}_r(a).$$

Let $\text{lpo} \in L'$ such that ρu is the step sequence of a step-linearization of lpo. Denote $\rho = u_1 \ldots u_n$. Since lpo is enabled w.r.t. the p/t-net $(N_{L'}, m_{L'})$, $N_{L'} = (P, T, F, W)$, associated to L', it holds for the place p_r corresponding to the region r and each cut C of lpo:

$$(**) \quad m_{L'}(p_r) + \sum_{v \in V \wedge v < C} (W((l(v), p_r)) - W((p_r, l(v)))) \geq \sum_{v \in C} W((p_r, l(v))).$$

Choose $S \subseteq V$ with $|C| = u$ and $|\{v \mid v < C\}| = u_1 + \ldots + u_n$ and remember that $W((p_r, l(v))) = \text{Pre}_r(v)$ and $W((l(v), p_r)) = \text{Post}_r(v)$. Then $(**)$ translates to $(*)$.

This implies moreover $r_L(\rho) \in \mathbb{N}$ for $\rho \in L$. Since obviously $r_L(\underline{0}) \in \mathbb{N}$ and $r(a) \in \mathbb{N} \times \mathbb{N}$ for $a \in A$, we deduce (TR1).

Finally, we show (TR3). For $\rho u = \rho'$ it holds by definition that $r_L(\rho') = r_L(\rho) + \sum_{a \in A} u(a)(\text{Post}_r(a) - \text{Pre}_r(a))$. To show this equation for the general case, it is enough to show $\rho \equiv_I \rho' \Rightarrow r_L(\rho) = r_L(\rho')$. It is even enough to consider the case $\rho \sim_I \rho'$, i.e. $\rho = \rho_1 vw\rho_2$ and $\rho' = \rho_1 v'w'\rho_2$ with $v+w = v'+w'$ and $(\rho_1, v+w) \in I$. The statement follows then from $\sum_{a \in A}(v+w)(a)(\text{Post}_r(a) - \text{Pre}_r(a)) = \sum_{a \in A}(v' + w')(a)(\text{Post}_r(a) - \text{Pre}_r(a))$. □

Lemma 5. *Let L' represent an LPO-specification and (L, A, I) be the trace language associated to L'. Let further $r : L \cup A \to \mathbb{N} \cup (\mathbb{N} \times \mathbb{N})$ be a region of (L, A, I). Then there is a region r' of L' satisfying $\text{Post}_{r'}(v_0) = r(\underline{0})$, $\text{Pre}_{r'}(a) = \text{Pre}_r(a)$ and $\text{Post}_{r'}(a)) \leqslant \text{Post}_r(a)$.*

Proof. It is enough to show that for each $\text{lpo} = (V, \prec, l) \in L'$ there is a token flow function x fulfilling $In(v', x) = \text{Pre}_r(a)$ and $Out(v', x) \leqslant \text{Post}_r(a)$ for every $v' \in V^0$. Then analogously as in the proof of Theorem 3 the statement follows.

Assume there is $\text{lpo} = (V, \prec, l) \in L'$ such that there is no token flow function x satisfying (**IN**) and (**OUT**). Then, by an argumentation in our paper [18] (Lemma 7), there is a cut C of lpo for which $r(\underline{0}) + \sum_{v < C} \text{Post}_r(v) - \sum_{v < C} \text{Pre}_r(v) - \sum_{v \in C} \text{Pre}(v) < 0$. This contradicts (TR2) for the step sequence of an adequate step linearization of lpo. □

4   Conclusion

As already mentioned in the introduction, we consider the presented results as a step towards solving the p/t-net synthesis problem from non-sequential specifications. To this end, the problem and the results must be transformed in two ways. First, it is necessary to represent the possibly infinite set of all regions of an LPO-specification by a finite subset. By this, a finite p/t-net could be assigned to an LPO-specification L with the same behavior as the (possibly infinite) p/t-net associated to L. Second, one needs a finite representation of LPO-specifications. When these problems are solved, finally an effective test of $L = \mathfrak{Lpo}(N', m_0')$ for the marked p/t-net (N', m_0') associated to L has to be developed. We are currently working on these topics and have already found partial solutions. These are not presented in detail due to lack of space. Instead, we shortly give some hints in this subsection.

Concerning the finite representation of LPO-specifications let us first consider the case, where simply the LPO-specification L is finite. Then, of course, the associated p/t-net cannot have an infinite behavior, i.e. it's marking graph cannot contain cycles (see the LPO-specifications from Figure 2 as an example). That means, in a finite representation of an infinite LPO-specification, the LPOs of a representation are only considered as prefixes of runs of the searched p/t-net, and one has additionally to specify desired cycles. This could be done for example by specifying cut-off events of LPOs. If no desired cycles are specified, it is nevertheless possible to detect such regions (places of the associated p/t-net) which prohibit cycles (which are, loosely spoken, those with initial marking different to the final marking of some sub-LPO).

Concerning finite representations of p/t-nets corresponding to LPO-specifications L, it is well known (see e.g. [24]) that there are p/t-nets with an infinite set of places which cannot be represented by p/t-nets with a finite set of places (see also the example below). That means we need a further characterization of LPO-specifications equaling the partial language of executions of p/t-nets with finite set of places. However, if L is finite then also the set of states (represented by L) is finite. In this case one can construct a p/t-net with finite set of places having L as the partial language of its executions. Observe moreover that for the synthesis of elementary nets from LPO-specifications there are only finitely many possible regions.

Example 8.   Consider the marked p/t-net $N = (P, T, F, W, m_0)$ defined by $T = \{a, b\}$, $P = \{p_n \mid n \in \mathbb{N}\}$, $W((a, p_n)) = n + 1$ and $W((p_n, b)) = n$, and $m_0(p_n) = n - 1$ for $n \in \mathbb{N}$. After one occurrence of a, the concurrent step $2b$ is enabled. Then *each* further occurrence of a enables transition b *exactly once*, since for increasing n the places p_n more and more restrict the behavior. This is the reason, why P cannot be replaced by a finite subset, say $P_k = \{p_1, \ldots, p_k\}$ for some $k \in \mathbb{N}$. Namely, in this case after each k occurrences of a the transition b would be enabled twice (not only once). In fact, the places p_n "converge" to the place p with $W((a, p)) = 1$, $W((p, b)) = 1$ and $m_0(p) = 1$. This place can be added to P without further restricting the behavior. On the other side, since b is not enabled in the initial marking, P can also not be equivalently replaced by the place set $P' = \{p\}$ only containing p, since w.r.t. P' the transition b is enabled in the initial marking. The problem is the increasing number of initial tokens in p_n for increasing n. If this number would be bounded (that means $m_0(p) \leqslant k$ for a fixed $k \in \mathbb{N}$), *each* occurrence of a (from the beginning) would enable transition b *exactly once*. The same behavior could be achieved by the finite place set $P'' = \{q\}$ with $W((a,q)) = 1$, $W((q,b)) = 1$ and $m_0(q) = 0$, i.e. in this case there would be a finite representation.

When considering finite LPO-specifications L with some representation $\mathfrak{Lpo}(L)$, there is moreover the following effective test of $L = \mathfrak{Lpo}(N', m_0')$ for the marked p/t-net (N', m_0') associated to L. In fact it must be shown that LPOs not in L are no executions of (N', m_0'). For this, first compute a finite set of regions representing the set of all regions. Then test the following LPOs to satisfy the token flow property w.r.t. all these regions: All LPOs constructed from minimal LPOs of maximal length in $\mathfrak{Lpo}(L)$ by deleting some edge, and all LPOs not in $\mathfrak{Lpo}(L)$ constructed from minimal LPOs in $\mathfrak{Lpo}(L)$ by adding a unique maximal node with some label. Each LPO not in L is also no execution of (N', m_0') if and only if none of these LPOs fulfills the token flow property w.r.t. all of the finite many regions.

References

1. W.M.P. van der Aalst, A.K.A. de Medeiros and A.J.M.M. Weijters. *Genetic Process Mining.* LNCS 3536, pages 48-69, 2005.
2. W.M.P. van der Aalst, T. Weijters and L. Maruster. *Workflow Mining: Discovering Process Models from Event Logs.* IEEE Trans. Knowl. Data Eng. 16/9, pages 1128-1142, 2004.
3. E. Badouel and P. Darondeau. *Theory of Regions.* LNCS 1491, pages 529-586, 1998.
4. J. Cortadella, M. Kishinevsky, L. Lavagno and A. Yakovlev. *Deriving Petri Nets for Finite Transition Systems.* IEEE Trans. Computers 47/8, pages 859-882, 1998.
5. P. Darondeau. *Unbounded Petri Net Synthesis.* LNCS 3098, pages 413-438, 2004.
6. J. Desel and W. Reisig. *The Synthesis Problem of Petri Nets.* Acta Inf. 33/4, pages 297-315, 1996.
7. A. Ehrenfeucht and G. Rozenberg. *Partial (Set) 2-Structures. Part I: Basic Notions and the Representation Problem.* Acta Inf. 27/4, pages 315-342, 1989.
8. A. Ehrenfeucht and G. Rozenberg. *Partial (Set) 2-Structures. Part II: State Spaces of Concurrent Systems.* Acta Inf. 27/4, pages 343-368, 1989.
9. J. Esparza and K. Heljanko *Implementing LTL Model Checking with Net Unfoldings.* LNCS 2057, pages 37-56, 2001.
10. L.R. Ford, Jr. and D.R. Fulkerson. Maximal Flow Through a Network. *Canadian Journal of Mathematics* 8, pp. 399–404, 1955.
11. A. Goldberg and S. Rao. Beyond the Flow Decomposition Barrier. *Journal of the ACM* 45/5, pp. 783–797, 1998.
12. U. Goltz and W. Reisig. The Non-Sequential Behaviour of Petri Nets. *Information and Control*, 57(2-3), pp. 125-147, 1983.
13. U. Goltz and W. Reisig. Processes of Place/Transition Nets. LNCS 154, pp. 264-277, 1983.
14. J. Grabowski. On Partial Languages. *Fundamenta Informaticae* IV.2, pp. 428–498, 1981.
15. D. Harel, H. Kugler and A. Pnueli. *Synthesis Revisited: Generating Statechart Models from Scenario-Based Requirements.* LNCS 3393, pages 309-324, 2005.
16. D. Harel and R. Marelly. *Come, Let's Play: Scenario-Based Programming Using LSCs and the Play-Engine* Springer, 2003.
17. P.W. Hoogers and H.C.M. Kleijn and P.S. Thiagarajan. A trace semantics for Petri nets. *Information and Computation* 117/1, pp. 98–114, 1995.

18. G. Juhás and R. Lorenz and J. Desel. Can I execute my scenario in your net? LNCS 3536, pp. 289–308, 2005.

19. A.V. Karzanov. Determining the Maximal Flow in a Network by the Method of Preflows. *Soviet Math. Doc.* 15, pp. 434–437, 1974.

20. A. Kiehn. On the Interrelationship between Synchronized and Non-Synchronized Behavior of Petri Nets. *Journal Inf. Process. Cybern. EIK* 24, pp. 3 – 18, 1988.

21. J. Klose and H. Wittke. *An Automata Based Interpretation of Live Sequence Charts.* LNCS 2031, pages 512-527, 2001.

22. M. Lettrari and J. Klose. *Scenario-Based Monitoring and Testing of Real-Time UML Models.* LNCS 2185, pages 317-328, 2001.

23. N. Mansurov. *Automatic synthesis of SDL from MSC and its applications in forward and reverse engineering.* Comput. Lang. 27/1, pages 115-136, 2001.

24. M. Mukund. *Petri Nets and Step Transition Systems.* Int. J. Found. Comput. Sci. 3/4, pages 443-478, 1992.

25. M. Nielsen, G. Rozenberg and P.S. Thiagarajan. *Elementary Transition Systems.* Theor. Comput. Sci. 96/1, pages 3-33, 1992.

26. M. Nielsen, G. Rozenberg and P.S. Thiagarajan. *Transition systems, event structures and unfoldings.* Information and Computation 118/2, pages 191-207, 1995.

27. V. Pratt. Modelling Concurrency with Partial Orders. *Int. Journal of Parallel Programming* 15, pp. 33–71, 1986.

28. A. Roychoudhury, P.S. Thiagarajan, T. Tran and V.A. Zvereva. *Automatic Generation of Protocol Converters from Scenario-Based Specifications.* Proceedings of the 25th IEEE Real-Time Systems Symposium (RTSS 2004), pages 447-458, 2004.

29. W. Vogler. Modular Construction and Partial Order Semantics of Petri Nets. *LNCS* 625, 1992.

Designing Reversibility-Enforcing Supervisors of Polynomial Complexity for Bounded Petri Nets Through the Theory of Regions

Spyros A. Reveliotis[1] and Jin Young Choi[2]

[1] School of Industrial & Systems Engineering
Georgia Institute of Technology
spyros@isye.gatech.edu
[2] Digital Communications Infra Division
Samsung Networks Inc.
jin_young.choi@samsung.com

Abstract. This paper proposes an analytical method for the synthesis of reversibility-enforcing supervisors for bounded Petri nets. The proposed me-thod builds upon recent developments from (i) the theory of regions, that enables the design of Petri nets with pre-specified behavioral requirements, and (ii) the theory concerning the imposition of generalized mutual exclusion constraints on the net behavior through monitor places. The derived methodology takes the form of a Mixed Integer Programming formulation, which is readily solvable through canned optimization software. The last part of the paper discusses extensions of the presented method so that it accommodates uncontrollable behavior and any potential complications arising from the large-scale nature of the underlying plant nets and their behavioral spaces. Finally, the relevance and the efficacy of the proposed approach is demonstrated through its application in the synthesis of liveness-enforcing supervisors for process-resource nets.

1   Introduction

Reversibility is a well-characterized and important property in many contemporary technological applications and it implies the ability of the studied system to retrieve its initial state from any state that this system can reach during its operation. Clearly, under this basic definition, reversibility applies to the entire range of systems that can be modelled through dynamical system theory; however, in this work we deal with the concept of reversibility in the more restricted class of systems that can be modelled by bounded Petri nets (PN) [5]. For systems modelled in this representational framework, we seek to develop a methodology that will support the design of controllers (i) enforcing the reversibility of the underlying system and (ii) possessing an "on-line" computational cost that is polynomial with respect to the size of this system. The pursued approach is a combination of (i) Petri net supervisory control based on the theory of monitor places [3, 4] and (ii) the design of Petri nets with

S. Donatelli and P.S. Thiagarajan (Eds.): ICATPN 2006, LNCS 4024, pp. 322–341, 2006.

a desired / pre-specfied topology for their reachability space through the theory of regions [1]. In this sense, our work presents considerable similarity with the works of [2, 9], which also sought to develop monitor-based supervisors for some PN sub-classes modelling sequential resource allocation, while exploiting insights and results coming from the theory of regions. However, the main differentiator of our work from those past efforts is the emphasis that we place on the (polynomial) complexity of the derived solutions. More specifically, in the previous works, the authors sought to derive a set of monitor places that would encode the *maximally permissive* supervisor for the considered application context, where the latter was initially computed through state space-based approaches (typically, Ramadge & Wonham's supervisory control theory [7]). Two significant implications of that approach were that (i) the derived supervisor might employ a number of monitor places that was a super-polynomial function of the size of the underlying Petri net, and (ii) there might be cases that the approach would fail to return a supervisor since it might not be possible to enforce the target behavior through a set of monitor places. Contrary to those past works, in our approach we define *a priori* the maximum number of monitor places that we want to use in the derived solution, and we seek to develop a supervisor that will guarantee "best performance" under this size restriction. The applied performance criterion can be quite general; for the purposes of the subsequent discussion, we shall assume that it can be expressed by a sum of values / weights taken over the set of states that are admitted by the derived supervisor. By restricting the number of the considered monitor places to be a polynomial function of the underlying system size, our approach can guarantee the polynomial "on-line" complexity of the derived solution. Furthermore, as it will be established in the following, the overall design problem reduces to the solution of a mathematical programming (MP) formulation consisting of the aforementioned performance objective and a set of constraints expressing the requirement for reversibility of the controlled system. This formulation essentially constitutes an implicit search for a best supervisor over the entire set of supervisors that can enforce the system reversibility while observing the imposed size constraint, and therefore, it is richer in feasible solutions than the earlier developed approaches. In addition, the explicit parameterization of the proposed approach through the maximum number of the employed monitor places allows the systematic study of the trade-off between the computational complexity of the applied supervisor and the attained performance. Finally, in principle, this approach can still enable the computation of the maximally permissive supervisor – assuming that this supervisor is implementable by a set of monitor places – by setting the number of the provided monitor places to a sufficiently large value.

From a more historical standpoint, this research falls within a broader initiative of ours, seeking to derive polynomial-complexity, monitor-based, reversibility-enforcing supervisors for a class of PN's modelling sequential resource allocation. This class of PN's and the currently available results on its reversibility-enforcing supervision through monitor-based approaches, are presented in [8, 11][1]. Yet, one of the open research questions raised in [8] is the analytical characterization of the entire set of supervisors that can enforce the reversibility of any given instance of the considered PN sub-class, while employing a pre-specified number of monitor places. This question is resolved in this paper through the constraint set of the aforementioned MP formulation.

In the light of the above introduction of the presented work, the rest of the paper is organized as follows: Section 2 first reviews the basic PN concepts and results that are necessary for the development of this work, and subsequently it summarizes the key elements of the theory of regions, according to the perspective adopted in [2], and of the monitor-based Petri net control theory, developed in [4]. Section 3 develops the supervisor design approach proposed in this work, while Section 4 demonstrates the modelling and analytical power of this approach through a detailed example. Section 5 discusses some enhancements and extensions of the basic methodology presented in Section 3, and, finally, Section 6 concludes the paper and highlights directions for future work.

2   Preliminaries

2.1   Petri Net Fundamentals

Petri net Definition [5]. A *(marked) Petri net (PN)* is defined by a quadruple $\mathcal{N} = (P, T, W, M_0)$, where

- P is the set of *places*,
- T is the set of *transitions*,
- $W : (P \times T) \cup (T \times P) \rightarrow Z_0^+$ is the *flow relation*,[2] and
- $M_0 : P \rightarrow Z_0^+$ is the net *initial marking*, assigning to each place $p \in P$, $M_0(p)$ *tokens*.

Also, for the purposes of the subsequent analysis, the *size* of PN $\mathcal{N} = (P, T, W, M_0)$ is defined as $|\mathcal{N}| \equiv |P| + |T| + \sum_{p \in P} M_0(p)$.

The first three items in the above definition essentially constitute a *weighted bipartite digraph* representing the system *structure* that governs its underlying dynamics. The last item determines the system *initial state*. A conventional graphical representation of the net structure and its marking depicts nodes corresponding to places by empty circles, nodes corresponding to transitions by bars, and the tokens located at the various places by small filled circles. The flow relation W is depicted by directed edges that link every nodal pair for which the corresponding W-value is non-zero. These edges point from the first node of the corresponding pair to the second, and they are also labelled – or,

---

[1] In fact, one of the main results for this PN sub-class is that the net liveness and reversibility are equivalent concepts; hence, in [8, 11], the aforementioned supervisory control problem is characterized as *liveness* rather than reversibility-enforcing supervision.

[2] In this work, Z_0^+ denotes the set of nonnegative integers, and $\Re$ denotes the set of reals.

"weighed" – by the corresponding W-value. By convention, absence of a label for any edge implies that the corresponding W-value is equal to unity.

Some structure-related PN concepts. For computational purposes, the net flow relation W is encoded by two $|P| \times |T|$ matrices, Θ^+ and Θ^-, with $\Theta^+(p, t) = W(t, p)$ and $\Theta^-(p, t) = W(p, t)$. The difference $\Theta^+ - \Theta^-$ is known as the net *flow matrix* and it is denoted by Θ. A PN is said to be *pure* if and only if (*iff*) $\forall p \in P, \forall t \in T, \Theta^-(p, t)\Theta^+(p, t) = 0$. Notice that for pure PN's, the net flow relation, W, is completely characterized by the net flow matrix, Θ.

Given a transition $t \in T$, the set of places p for which $(p, t) > 0$ (resp., $(t, p) > 0$) is known as the set of *input* (resp., *output*) places of t. Similarly, given a place $p \in P$, the set of transitions t for which $(t, p) > 0$ (resp., $(p, t) > 0$) is known as the set of *input* (resp., *output*) transitions of p. It is customary in the PN literature to denote the set of input (resp., output) transitions of a place p by $\bullet p$ (resp., $p \bullet$). Similarly, the set of input (resp., output) places of a transition t is denoted by $\bullet t$ (resp., $t \bullet$). This notation is also generalized to any set of places or transitions, X, e.g. $\bullet X = \bigcup_{x \in X} \bullet x$.

The ordered set $X = < x_1 \ldots x_n > \in (P \cup T)^*$ is a *path* iff $x_{i+1} \in x_i^\bullet, i = 1, \ldots, n-1$. Furthermore, a path X is characterized as a *circuit* iff $x_1 \equiv x_n$.

The particular class of PN's with a flow relation W mapping onto $\{0, 1\}$ are characterized as *ordinary*. An ordinary PN with $|t^\bullet| = |\bullet t| = 1, \forall t \in T$, is characterized as a *state machine*, while an ordinary PN with $|p^\bullet| = |\bullet p| = 1$, $\forall p \in P$, is characterized as a *marked graph*.

Some dynamics-related PN concepts. In the PN modelling framework, the system state is represented by the net *marking* M, i.e., a function from P to Z_0^+ that assigns a *token* content to the various net places. The net marking M is initialized to marking M_0, introduced in the PN definition provided at the beginning of this section, and it subsequently evolves through a set of rules summarized in the concept of *transition firing*. A concise characterization of this concept has as follows: Given a marking M, a transition t is *enabled iff* for every place $p \in \bullet t$, $M(p) \geq W(p, t)$, or equivalently, $M \geq \Theta^-(\cdot, t)$, and this fact is denoted by $M[t\rangle$. $t \in T$ is said to be *disabled* by a place $p \in \bullet t$ at M iff $M(p) < W(p, t)$, or, equivalently, $M(p) < \Theta^-(p, t)$. Given a marking M, a transition t can be *fired* only if it is enabled in M, and firing such an enabled transition t results in a new marking M', which is obtained from M by removing $W(p, t)$ tokens from each place $p \in \bullet t$, and placing $W(t, p')$ tokens in each place $p' \in t^\bullet$. The marking evolution incurred by the firing of a transition t can be concisely expressed by the *state equation*:

$$M' = M + \Theta \cdot 1_t \tag{1}$$

where 1_t denotes the unit vector of dimensionality $|T|$ and with the unit element located at the component corresponding to transition t.

Given a PN $\mathcal{N}$, a sequence of transitions, $\sigma = t_1 t_2 \ldots t_n$, is *fireable* from some marking M iff $M[t_1\rangle M_1[t_2\rangle M_2 \ldots M_{n-1}[t_n\rangle M_n$; we shall also denote this fact by $M \xrightarrow{\sigma} M_n$. The *length* of σ is defined by the number of transitions in it, and it will be denoted by $|\sigma|$. Also, the *Parikh vector* of σ is a $|T|$-dimensional vector, $\bar{\sigma}$, with each component $\bar{\sigma}(t)$, $t \in T$, stating the number of appearances of transition t in σ.

The set of markings reachable from the initial marking M_0 through any *fireable* sequence of transitions is denoted by $R(\mathcal{N}, M_0)$ and it is referred to as the net *reachability space*. Equation 1 implies that a necessary condition for $M \in R(\mathcal{N}, M_0)$ is that the following system of equations is feasible in z:

$$M = M_0 + \Theta z \tag{2}$$

$$z \in (Z_0^+)^{|T|} \tag{3}$$

The *reachability graph*, $\mathcal{G}(\mathcal{N}, M_0)$, of $\mathcal{N}$, is a labelled directed graph with its node set being equal to $R(\mathcal{N}, M_0)$, and its edge set being defined by the nodal pairs $(M, M') \in R(\mathcal{N}, M_0) \times R(\mathcal{N}, M_0)$ for which there exists $t \in T$ such that $M[t\rangle M'$; the edges of $\mathcal{G}(\mathcal{N}, M_0)$ are labelled by the corresponding transitions. A PN $\mathcal{N} = (P, T, W, M_0)$ is said to be *bounded iff* all markings $M \in R(\mathcal{N}, M_0)$ are bounded. $\mathcal{N}$ is said to be *structurally bounded iff* it is bounded for any initial marking M_0. $\mathcal{N}$ is said to be *reversible iff* $M_0 \in R(\mathcal{N}, M)$, for all $M \in R(\mathcal{N}, M_0)$, and any marking $M \in R(\mathcal{N}, M_0)$ such that $M_0 \in R(\mathcal{N}, M)$ is a *co-reachable* marking of $\mathcal{N}$. A transition $t \in T$ is said to be *live iff* for all $M \in R(\mathcal{N}, M_0)$, there exists $M' \in R(\mathcal{N}, M)$ such that $M'[t\rangle$; non-live transitions are said to be *dead* at those markings $M \in R(\mathcal{N}, M_0)$ for which there is no $M' \in R(\mathcal{N}, M)$ such that $M'[t\rangle$. PN $\mathcal{N}$ is *quasi-live iff* for all $t \in T$, there exists $M \in R(\mathcal{N}, M_0)$ such that $M[t\rangle$; it is *weakly live iff* for all $M \in R(\mathcal{N}, M_0)$, there exists $t \in T$ such that $M[t\rangle$; and it is *live iff* for all $t \in T$, t is live.

PN semiflows. PN semiflows provide an analytical characterization of various concepts of *invariance* underlying the net dynamics. Generally, there are two types, p and t-semiflows, with a *p-semiflow* formally defined as a $|P|$-dimensional vector y satisfying $y^T \Theta = 0$ and $y \geq 0$, and a *t-semiflow* formally defined as a $|T|$-dimensional vector x satisfying $\Theta x = 0$ and $x \geq 0$. In the light of Equation 2, the invariance property expressed by a p-semiflow y is that $y^T M = y^T M_0$, for all $M \in R(\mathcal{N}, M_0)$. Similarly, Equation 2 implies that for any t-semiflow x, $M = M_0 + \Theta x = M_0$.

2.2 Petri Net Design Through the Theory of Regions

In this section we overview an interpretation of the *theory of regions* provided in [2]. According to this interpretation, the problem addressed by the theory of regions can be stated as follows: Given a directed graph, $G = (N, E)$, with its edges labelled by elements from some set T, and containing a node $n_0 \in N$ such that there exists a path from n_0 to any other node $n \in N$, find a *pure* PN $\mathcal{N} = (P, T, W, M_0)$, such that its reachability graph $\mathcal{G}(\mathcal{N}, M_0)$ is expressed by G, when setting $M_0 \equiv n_0$. Since the net $\mathcal{N}$ is required to be pure, it can be fully defined by specifying the row $\Theta(p, \cdot)$ of the net flow matrix Θ and the initial marking $M_0(p)$, for each place $p \in P$. These parameters can be subsequently obtained through a system of equations derived from the structure of the target graph G and the logic underlying Equations 1–3.

In particular, Equation 1 implies that, for any undirected cycle, γ, in graph G:

$$\forall p \in P, \quad \Theta(p, \cdot) \cdot \bar{\gamma} = 0 \tag{4}$$

Equation 4 is known as the *"cycle"* equation of the theory of regions, and the parameter $\bar{\gamma}$ appearing in it is a vector of dimensionality $|T|$, and with component $\bar{\gamma}(t)$ denoting the difference between the number of times that t is encountered in γ labelling an edge pointing in the direction of the traversal of γ, and the number of times that t is encountered in γ labelling an edge pointing in the opposite direction.

Similarly, the reachability of a node $n \in N$ from node n_0 through some path $\xi(n)$, implies that

$$\forall p \in P, \quad M_0(p) + \Theta(p, \cdot) \cdot \bar{\xi}(n) \geq 0 \tag{5}$$

Equation 5 is known as the *"reachability condition"* associated with node n, and the parameter $\bar{\xi}(n)$ appearing in it is a vector of dimensionality $|T|$ and with component $\bar{\xi}(n; t)$ indicating the number of appearances of transition t in path $\xi(n)$. For nodes n reachable from n_0 through more than one paths, only one of the corresponding reachability conditions should be included in the considered system of equations, since the reachability conditions corresponding to the remaining paths can be derived from the included condition and the cycle equations discussed above.

On the other hand, for every node $n \in N$ and transition $t \in T$ such that there is no edge emanating from n that is labelled by t, there must exist a place $p \in P$ that disables the firing of transition t at the marking M corresponding to node n. This requirement is imposed by the following equation:

$$\exists p \in P, \quad M_0(p) + \Theta(p, \cdot) \cdot \bar{\xi}(n) + \Theta(p, t) \leq -1 \tag{6}$$

Equation 6 is known as the *"event separation condition"* associated with node-transition pair (n, t), and the parameter $\bar{\xi}(n)$ appearing in it is the same with that appearing in Equation 5. Also, the node-transition pairs, (n, t), such that there is no edge emanating from n that is labelled by t, are characterized as the *"event separation instances"*.

Finally, a last requirement is that the various nodes $n \in N$ of graph G correspond to different markings of the PN $\mathcal{N}$; i.e., for any given nodal pair (n, n'),

$$\exists p \in P, \quad \Theta(p, \cdot) \cdot \bar{\xi}(n) \neq \Theta(p, \cdot) \cdot \bar{\xi}(n') \tag{7}$$

Equation 7 in known as the *"state separation condition"*, and the parameter $\bar{\xi}(n)$ appearing in it is defined as in Equations 5 and 6.

In the light of the above characterizations, the theory of regions is epitomized by the following theorem:

Theorem 1. *[2] Consider a directed graph, $G = (N, E)$, with its edges labelled by elements from some set T, and containing a node $n_0 \in N$ such that there* *exists a path from n_0 to any other node $n \in N$. Then, there exists a pure PN $\mathcal{N} = (P, T, W, M_0)$ with graph G as its reachability graph and with node n_0 corresponding to its initial marking M_0, iff (i) for each place $p \in P$, the flow vector $\Theta(p, \cdot)$ satisfies (a) the cycle equation corresponding to each undirected cycle γ of G and (b) the reachability condition corresponding to each node n of G, where the latter is stated with respect to some arbitrary path from n_0 to n; (ii) the net flow matrix Θ satisfies the state separation condition for every nodal pair (n, n') with $n \neq n'$; and (iii) for every event separation instance in G, there exists a place $p \in P$ with its flow vector $\Theta(p, \cdot)$ satisfying the corresponding event separation condition.*

In Section 3 we employ this result towards the development of a methodology that will support the design of reversibility-enforcing supervisors for bounded PN's.

2.3   Petri-Net Supervisory Control Based on Generalized Mutual Exclusion Constraints and "monitor" Places

In many PN control applications, one seeks to impose a set of constraints on the marking, M, of a plant net, $\mathcal{N} = (P, T, W, M_0)$, that are expressed as a set of linear inequalities of the type

$$A \cdot M \leq b \tag{8}$$

where the elements of matrix A and the right-hand-side (rhs) vector b are non-negative integers. Marking constraints of the type expressed by Equation 8 are known as *Generalized Mutual Exclusion (GME)* constraints. Consider the GME constraint of Equation 8 that is defined by the row $A(i, \cdot)$ of matrix A and the component $b(i)$ of the rhs vector b. Then, according to the theory of [4], this constraint can be imposed on the plant net $\mathcal{N}$ by super-imposing on it a single *"monitor"* place $p_c(i)$; this place must be connected to the rest of the network according to the flow vector:

$$\Theta(p_c(i), \cdot) = -A(i, \cdot) \cdot \Theta \tag{9}$$

and its initial marking must be set to:

$$M_0(p_c(i)) = b(i) \tag{10}$$

Under the aforementioned configuration, $p_c(i)$ enforces the constraint

$$A(i, \cdot) \cdot M \leq b(i) \tag{11}$$

on the markings, M, of the original net, by essentially establishing the invariant

$$A(i, \cdot) \cdot M + M(p_c(i)) = b(i) \tag{12}$$

Equation 12 indicates that the token content, $M(p_c(i))$, of place $p_c(i)$ expresses the "slack" of Constraint 11 under marking M, and justifies the characterization of the control place $p_c(i)$ as a *"monitor"* place.

We conclude this brief discussion on GME constraints and their enforcing monitor places, by establishing the following result, that will be useful in the developments of Section 3:

Lemma 1. *Consider a monitor place $p_c(i)$ that enforces the GME constraint of Equation 11 on a plant net $\mathcal{N}$. Then, every t-semiflow, x, of $\mathcal{N}$ is also a t-semiflow for place $p_c(i)$.*

Proof: We need to show that $\Theta(p_c(i), \cdot) \cdot x = 0$. But this is an immediate implication of Equation 9 and the fact that x is a t-semiflow of the original net $\mathcal{N}$. □

3   A Formal Statement of the Considered Problem and the Proposed Supervisor Design Methodology

Having established in the previous section all the concepts and results that are necessary for the formal development of this work, we can now proceed to the detailed statement of the undertaken problem and the systematic exposition of the methodology proposed for its solution. We start with the formal problem statement.

A formal statement of the problem considered in this work. The problem considered in this work can be formally defined as follows: Given a non-reversible, bounded PN $\mathcal{N}$, identify a set of GME constraints

$$A \cdot M \leq b \tag{13}$$

such that

i. when imposed on the plant net $\mathcal{N}$, will incur the reversibility of the controlled system.
ii. Furthermore, the cardinality of the imposed constraint set must not exceed a pre-specified parameter K.
iii. In addition,

$$\forall i, j, \ A(i,j) \in \{0, 1, \ldots, \bar{A}(i,j)\} \text{ and } \forall i, \ b(i) \in \{0, 1, \ldots, \bar{b}(i)\}, \tag{14}$$

where $\bar{A}(i,j)$ and $\bar{b}(i)$ are *finitely* valued, externally provided parameters.
iv. Finally, assuming that every reachable marking $M_i \in R(\mathcal{N}, M_0)$ of $\mathcal{N}$ is associated with some value w_i, the developed supervisor must maximize the total value of the admissible markings, over the set of supervisors satisfying the aforementioned requirements.

In the sequel, a PN supervisor that is defined by Equation 13 for some pricing of matrix A and vector b, will be referred to as the supervisor $\mathcal{S}(A, b)$.

Overview of the proposed solution. Next, we provide a *Mixed Integer Programming (MIP)* formulation for the aforestated problem. The objective function of this formulation will express the optimality requirement stated in item (iv) above. Requirement (ii) will be captured by the structure of the decision variables of the presented formulation, while requirements (i) and (iii) will be explicitly encoded in its constraints. More specifically, given a pricing of the matrix A and the rhs vector b, the constraint set must check whether this pricing abides to requirement (iii) and it must also assess the ability of this pricing to satisfy requirement (i), i.e., establish the reversibility of the controlled system. This last requirement further implies that all the markings $M \in R(\mathcal{N}, M_0)$ that remain reachable under the considered GME constraints, are also co-reachable under these constraints. Hence, the constraint set of the proposed formulation must be able to assess the reachability and co-reachability of the markings $M \in R(\mathcal{N}, M_0)$ under the net supervision by any tentative GME constraint set, $A \cdot M \leq b$, and it must also be able to validate that all reachable markings are also co-reachable. The rest of this section proceeds to the detailed derivation of a formulation that possesses the aforementioned qualities.

Characterizing the net transition firing under supervision by a GME constraint-based supervisor $\mathcal{S}(A, b)$. In order to be able to assess the reachability and co-reachability of the various markings $M \in R(\mathcal{N}, M_0)$ under supervision by a supervisor $\mathcal{S}(A, b)$, it is necessary to characterize how the various transitions, $t \in T$, of the plant net $\mathcal{N}$, retain their fireability in the controlled system. Next, we introduce a set of variables and constraints that will achieve this purpose. The main issue to be addressed is whether a transition t that was fireable in some marking $M_i \in R(\mathcal{N}, M_0)$, leading to another marking $M_j \in R(\mathcal{N}, M_0)$, will remain fireable under supervision by $\mathcal{S}(A, b)$. For this to be true, t must be enabled at M_i by all the monitor places, $p_c(k)$, $k = 1, \ldots, K$, that implement the supervisor $\mathcal{S}(A, b)$. Testing whether transition t is enabled at marking M_i by a monitor place $p_c(k)$ can be done through the employment of a *binary* variable z_{ij}^k, that will be priced to one, if this condition is true, and to zero, otherwise. A set of constraints that will enforce the pricing of z_{ij}^k according to the aforementioned scheme is the following:

$$M_0(p_c(k)) + \sum_{(u,v) \in \xi(i)} \Theta(p_c(k), t(u, v)) + \Theta(p_c(k), t(i, j)) + (z_{ij}^k - 1)L_{ij}^k \geq 0 \quad (15)$$

$$M_0(p_c(k)) + \sum_{(u,v) \in \xi(i)} \Theta(p_c(k), t(u, v)) + \Theta(p_c(k), t(i, j)) - z_{ij}^k U_{ij}^k \leq -1 \quad (16)$$

The parameter $\xi(i)$ appearing in Equations 15 and 16 denotes any path in $R(\mathcal{N}, M_0)$ leading from M_0 to M_i. (u, v) denotes an edge of $\xi(i)$ leading from node M_u to node M_v, and $t(u, v)$ denotes its labelling transition. L_{ij}^k denotes a lower bound for the quantity $M_0(p_c(k)) + \sum_{(u,v) \in \xi(i)} \Theta(p_c(k), t(u, v)) + \Theta(p_c(k), t(i, j))$, and U_{ij}^k denotes an upper bound for the quantity $M_0(p_c(k)) + \sum_{(u,v) \in \xi(i)} \Theta(p_c(k), t(u, v)) + \Theta(p_c(k), t(i, j)) + 1$. Then, it is clear, that, when $M_0(p_c(k)) + \sum_{(u,v) \in \xi(i)} \Theta(p_c(k), t(u, v)) + \Theta(p_c(k), t(i, j)) \geq 0$ – i.e., when transition $t(i, j)$ is enabled by monitor place $p_c(k)$ in marking M_i – the above set of constraints is satisfied by setting $z_{ij}^k = 1$. On the other hand, when $M_0(p_c(k)) + \sum_{(u,v)\in\xi(i)} \Theta(p_c(k), t(u,v)) + \Theta(p_c(k), t(i,j)) < 0$ the above constraint set is satisfied by setting $z_{ij}^k = 0$.

It remains to connect the variables $M_0(p_c(k))$ and $\Theta(p_c(k), \cdot)$ to the primary problem variables, A, b, and explain how to compute the bounds L_{ij}^k and U_{ij}^k employed in the above equations. Connecting $M_0(p_c(k))$ and $\Theta(p_c(k), \cdot)$ to the variables A, b can be done straightforwardly through Equations 9 and 10; the corresponding substitutions respectively transform Equations 15 and 16 to:

$$b(k) - \sum_{(u,v)\in\xi(i)} A(k, \cdot) \cdot \Theta(\cdot, t(u,v)) - A(k, \cdot) \cdot \Theta(\cdot, t(i,j)) + (z_{ij}^k - 1)L_{ij}^k \geq 0 \quad (17)$$

$$b(k) - \sum_{(u,v)\in\xi(i)} A(k, \cdot) \cdot \Theta(\cdot, t(u,v)) - A(k, \cdot) \cdot \Theta(\cdot, t(i,j)) - z_{ij}^k U_{ij}^k \leq -1 \quad (18)$$

Finally, it should be clear from the structure of Constraints 17 and 18 that the bound L_{ij}^k (resp., U_{ij}^k), defined above, can be obtained by minimizing (resp., maximizing) the quantity $b(k) - \sum_{(u,v)\in\xi(i)} A(k, \cdot) \cdot \Theta(\cdot, t(u,v)) - A(k, \cdot) \cdot \Theta(\cdot, t(i,j))$ over the space defined by the admissible ranges of the involved variables $A(k, \cdot)$ and $b(k)$ (c.f., item (iii) in the formal problem statement provided at the beginning of this section).

Once variables z_{ij}^k have been properly priced for all k, the feasibility of $M_i[t(i,j)\rangle M_j$ can be assessed by introducing another *real* variable, z_{ij}, that is priced according to the following constraints:

$$z_{ij} \leq z_{ij}^k, \quad \forall k \in \{1, \ldots, K\} \quad (19)$$

$$z_{ij} \geq \sum_{k=1}^{K} z_{ij}^k - K + 1 \quad (20)$$

$$0 \leq z_{ij} \leq 1 \quad (21)$$

To understand the pricing logic behind Constraints 19–21, first notice that Constraint 21 restricts the variable z_{ij} within the interval $[0, 1]$. Then, Constraint 19 sets it to zero, as long as any of the variables z_{ij}^k is priced to zero – and therefore, the corresponding monitor place $p_c(k)$ disables $t(i,j)$. On the other hand, when all variables z_{ij}^k are priced to one, Constraint 20 forces variable z_{ij} to its extreme value of one.

Characterizing the reachability of the markings $M_i \in R(\mathcal{N}, M_0)$ under supervision by a GME constraint-based supervisor $S(A, b)$. The availability of the variables z_{ij}, defined above, subsequently enables the characterization of the reachability of the various markings $M_i \in R(\mathcal{N}, M_0)$ under supervision by the GME constraint-based supervisor $S(A, b)$. This can be done by introducing the *real* variables y_i^l, $0 \leq i \leq |R(\mathcal{N}, M_0)|$, $0 \leq l \leq \bar{l}$, and pricing them so that

$y_i^l = 1$ indicates that marking M_i is reachable from the initial marking M_0 under supervision by $\mathcal{S}(A, b)$ and the minimum length of any transition sequence leading from M_0 to M_i is l; if M_i is not reachable from M_0 under supervision by $\mathcal{S}(A, b)$, y_i^l should be set to zero, for all l. Clearly, in order to satisfy this definition of y_i^l, $\bar{l}$ must be set to the length of the maximum path in $\mathcal{G}(\mathcal{N}, M_0)$ that starts from M_0 and contains no cycles. Then, a set of constraints that achieves the pricing of y_i^l described above, is as follows:

$$y_i^0 = \begin{cases} 1, i = 0 \\ 0, i \neq 0 \end{cases} \tag{22}$$

$$0 \leq y_i^l, \quad \forall i \in \{1, \ldots, |R(\mathcal{N}, M_0)|\}, \; l \in \{1, \ldots, \bar{l}\} \tag{23}$$

$$\sum_{l=0}^{\bar{l}} y_i^l \leq 1 \tag{24}$$

$$\delta_{ji}^l \leq y_j^{l-1}, \quad \forall j : (M_j, M_i) \in \mathcal{G}(\mathcal{N}, M_0) \tag{25}$$

$$\delta_{ji}^l \leq z_{ji}, \quad \forall j : (M_j, M_i) \in \mathcal{G}(\mathcal{N}, M_0) \tag{26}$$

$$y_i^l \leq \sum_j \delta_{ji}^l \tag{27}$$

$$y_i^l \geq y_j^{l-1} + z_{ji} - 1 - \sum_{q=0}^{l-1} y_i^q, \quad \forall j : (M_j, M_i) \in \mathcal{G}(\mathcal{N}, M_0) \tag{28}$$

Constraint 22 expresses the fact that marking M_0 is reachable from itself in zero steps, under supervision by $\mathcal{S}(A, b)$, and this is the only marking in $R(\mathcal{N}, M_0)$ possessing this property. Constraint 23 states the nonnegative real nature of variables y_i^l, $i > 0$, $l > 0$, while Constraint 24 expresses the fact that, according to the pricing scheme discussed above, only one of the variables y_i^l, $0 \leq l \leq \bar{l}$, can be priced to one. Constraints 25, 26 and 27 express the fact that, under supervision by $\mathcal{S}(A, b)$, there is a minimal path from marking M_0 to marking M_i of length l, only if there is a minimal path of length $l - 1$ from M_0 to some marking M_j such that (i) $(M_j, M_i) \in \mathcal{G}(\mathcal{N}, M_0)$ and (ii) this transition remains feasible under $\mathcal{S}(A, b)$. In particular, variables δ_{ji}^l is a set of auxiliary *real* variables that are used to force y_i^l to zero every time that the aforestated condition is violated for all the markings $M_j \in R(\mathcal{N}, M_0)$ such that $(M_j, M_i) \in \mathcal{G}(\mathcal{N}, M_0)$. On the other hand, Constraint 28 tends to price variable y_i^l to one every time that there exists a marking M_j such that (i) $(M_j, M_i) \in \mathcal{G}(\mathcal{N}, M_0)$, (ii) this transition remains feasible under $\mathcal{S}(A, b)$, and (iii) M_j is reachable from M_0 under supervision by $\mathcal{S}(A, b)$ through a minimal path of length $l-1$; however, this pricing is enforced only when the quantity $\sum_{q=0}^{l-1} y_i^q$ appearing in the right-hand-side of this constraint is equal to zero – i.e., only when the marking M_i cannot be reached from the initial marking M_0 through a path of smaller length.

Characterizing the co-reachability of the markings $M_i \in R(\mathcal{N}, M_0)$ under supervision by a GME constraint-based supervisor $\mathcal{S}(A, b)$. It is well-known that the co-reachability of a marking $M_i \in R(\mathcal{N}, M_0)$ is equivalent to the reachability of the same marking in the graph $\mathcal{G}^R(\mathcal{N}, M_0)$, obtained from $\mathcal{G}(\mathcal{N}, M_0)$ by reversing all its arcs. In the light of this observation, the set of constraints characterizing the co-reachability of the markings $M_i \in R(\mathcal{N}, M_0)$, under supervision by a GME constraint-based supervisor $\mathcal{S}(A, b)$, can be obtained through a straightforward modification of the constraint set 22–28, characterizing the reachability of these markings. More specifically, let ψ_i^l be a *real* variable that will be priced to one, if $M_i \in R(\mathcal{N}, M_0)$ is co-reachable under supervision by $\mathcal{S}(A, b)$, and a minimal transition sequence leading from M_i to M_0 has a length equal to l; otherwise, ψ_i^l should be priced to zero. By following a logic similar to that employed in the previous paragraph for the pricing of variables y_i^l, we obtain the following set of constraints for the pricing of variables ψ_i^l:

$$\psi_i^0 = \begin{cases} 1, i = 0 \\ 0, i \neq 0 \end{cases} \tag{29}$$

$$0 \leq \psi_i^l, \quad \forall i \in \{1, \ldots, |R(\mathcal{N}, M_0)|\}, \ l \in \{1, \ldots, \tilde{l}\} \tag{30}$$

$$\sum_{l=0}^{\tilde{l}} \psi_i^l \leq 1 \tag{31}$$

$$\eta_{ij}^l \leq \psi_j^{l-1}, \quad \forall j : (M_i, M_j) \in \mathcal{G}(\mathcal{N}, M_0) \tag{32}$$

$$\eta_{ij}^l \leq z_{ij}, \quad \forall j : (M_i, M_j) \in \mathcal{G}(\mathcal{N}, M_0) \tag{33}$$

$$\psi_i^l \leq \sum_j \eta_{ij}^l \tag{34}$$

$$\psi_i^l \geq \psi_j^{l-1} + z_{ij} - 1 - \sum_{q=0}^{l-1} \psi_i^q, \quad \forall j : (M_i, M_j) \in \mathcal{G}(\mathcal{N}, M_0) \tag{35}$$

The parameter $\tilde{l}$, appearing in Equations 30 and 31, denotes the length of the maximum path in $\mathcal{G}^R(\mathcal{N}, M_0)$ that leads from node M_0 to node M_i and contains no cycles, and the auxiliary variables η_{ij}^l, that appear in Constraints 32 and 33, play a role identical to that played by variables δ_{ji}^l in Constraints 25 and 26.

Characterizing the closure of the sub-space that is reachable and co-reachable under supervision by a GME constraint-based supervisor $\mathcal{S}(A, b)$. Let x_i be a *real* variable that will be priced to one when the marking $M_i \in R(\mathcal{N}, M_0)$ is reachable and co-reachable under supervision by $\mathcal{S}(A, b)$, and it will be priced to zero, otherwise. Then, in the light of the above characterizations of reachability and co-reachability, the desired pricing of x_i can be enforced by the following constraints:

$$x_i \leq \sum_{l=0}^{\bar{l}} y_i^l \tag{36}$$

$$x_i \leq \sum_{l=0}^{\bar{l}} \psi_i^l \tag{37}$$

$$x_i \geq \sum_{l=0}^{\bar{l}} y_i^l + \sum_{l=0}^{\bar{l}} \psi_i^l - 1 \tag{38}$$

$$0 \leq x_i \leq 1 \tag{39}$$

Constraint 39 restricts x_i in the interval $[0, 1]$. Then, Constraints 36 and 37 force it to zero, when marking M_i is not reachable or co-reachable. On the other hand, if M_i is both reachable and co-reachable, Constraint 38 forces x_i to its extreme value of one.

Finally, the availability of variables x_i allows us to express the requirement for closure of the sub-space of $R(\mathcal{N}, M_0)$ that is reachable and co-reachable under supervision by $\mathcal{S}(A, b)$, through the following constraint:

$$(1 - x_i) + x_j \geq z_{ij}, \quad \forall i, j : (M_i, M_j) \in \mathcal{G}(\mathcal{N}, M_0) \tag{40}$$

When $x_i = 1$ and $x_j = 0$ – i.e., when x_i belongs to the target space of markings that are reachable and co-reachable under supervision by $\mathcal{S}(A, b)$, but x_j does not belong to this set – Constraint 40 forces variable z_{ij} to zero – i.e., it requires that the corresponding transition $M_i[t(i, j)\rangle M_j$ is disabled by $\mathcal{S}(A, b)$. In any other case, the left-hand-side of Constraint 40 is greater than or equal to one, and therefore, the constraint becomes inactive.

The objective function of the proposed formulation. The objective function of the considered formulation is straightforwardly expressed as follows:

$$\max \sum_i w_i x_i \tag{41}$$

Proving the correctness of the proposed formulation. Next, we state and prove the correctness of the derived formulation.

Theorem 2. *The formulation of Equations 14,17–41 returns an optimal solution to the problem stated at the beginning of this section, provided that such a solution exists; otherwise, this formulation will be infeasible.*

Proof: First, let us suppose that the aforementioned formulation returns a feasible solution. Then, it is clear from the earlier discussion of the various constraints of the considered formulation, that the set of markings $M_i \in R(\mathcal{N}, M_0)$

with $x_i = 1$ and the edges (M_i, M_j) of $\mathcal{G}(\mathcal{N}, M_0)$ with $z_{ij} = 1$, in the returned solution, define a strongly connected subgraph of $\mathcal{G}(\mathcal{N}, M_0)$ containing
the initial marking M_0; let us denote this subgraph by $\mathcal{G}^C(\mathcal{N}, M_0)$. Next we
show that $\mathcal{G}^C(\mathcal{N}, M_0)$ is the reachability graph of the net $\mathcal{N}^C$, that is obtained
from the plant net $\mathcal{N}$ by super-imposing on it the monitor places that imple-
ment the GME constraint set $A \cdot M \leq b$, where A, b have the values returned
by the considered formulation. To establish this result, it is sufficient to show that
the net $\mathcal{N}^C$ satisfies the conditions of Theorem 1 with respect to $\mathcal{G}^C(\mathcal{N}, M_0)$.
This can be shown as follows: First notice that the state separation condition
over $\mathcal{G}^C(\mathcal{N}, M_0)$ is immediately satisfied by $\mathcal{N}^C$, since its marking subsumes
the marking of the original net $\mathcal{N}$. The satisfaction of the reachability condition
for every node of $\mathcal{G}^C(\mathcal{N}, M_0)$ is a consequence of Constraints 17–28. For event
separation instances of $\mathcal{G}^C(\mathcal{N}, M_0)$ that were already present in $\mathcal{G}(\mathcal{N}, M_0)$, there
must be a place p in the original net $\mathcal{N}$ that satisfies the corresponding event
separation condition. For the remaining event separation instances, the defini-
tion of $\mathcal{G}^C(\mathcal{N}, M_0)$ implies that the relevant variables z_{ij} were priced to zero
in the returned solution, and therefore, for each of them, there exists a place
$p_c(k)$, $k \in \{1, \ldots, K\}$, that satisfies the corresponding event separation condition. For places p in the original net $\mathcal{N}$, the cycle equations for the various
cycles of $\mathcal{G}^C(\mathcal{N}, M_0)$ are immediately satisfied by the fact that $\mathcal{G}^C(\mathcal{N}, M_0)$ is a
subgraph of $\mathcal{G}(\mathcal{N}, M_0)$. For the monitor places $p_c(k)$, $k \in \{1, \ldots, K\}$, the satisfaction of the cycle equations for the cycles of $\mathcal{G}^C(\mathcal{N}, M_0)$ is guaranteed by
Lemma 1. Finally, the optimality of the supervisor $\mathcal{S}(A, b)$, that is returned by
the considered formulation, is guaranteed by the specification of the objective
function (c.f. Equation 41).

On the other hand, if the considered formulation is infeasible, then it is impossible to identify a strongly connected subgraph of $\mathcal{G}(\mathcal{N}, M_0)$ that contains
the initial marking M_0 and can be separated from $\mathcal{G}(\mathcal{N}, M_0)$ by using K GME
constraints with the corresponding matrix A and rhs vector b priced in the pre-
specified ranges. Hence, it can be concluded that the supervisor design problem
defined at the beginning of this section, is infeasible.   □

4   Example

In this section, we demonstrate the implementation and the efficacy of the design
methodology developed in Section 3, by applying it to the design of a liveness-
enforcing supervisor for the PN depicted in Figure 1.

Interpreting the PN of Figure 1 as a process-resource net. The PN in Figure 1
models a Resource Allocation System (RAS), consisting of three resource types,
R_1, R_2, and R_3, with respective capacities $C_1 = C_3 = 1$, and $C_2 = 2$, and
supporting two process types, JT_1 and JT_2. The process plans of these two
process types are respectively modelled by the paths $< t_{10}p_{11}t_{11}p_{12}t_{12}p_{13}t_{13} >$
and $< t_{20}p_{21}t_{21}p_{22}t_{22}p_{23}t_{23} >$; thus, it can be seen that (i) each process consists
of three consecutive stages, (ii) the execution of each processing stage by some

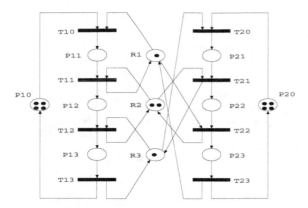

Fig. 1. The process-resource net considered in the example of Section 4

process instance requires the exclusive allocation of a single unit from a certain resource type, and (iii) a process instance can release the resource currently allocated to it and advance to the next processing stage only when it has secured the allocation of the next required resource. Finally, the places p_{10} and p_{20} are characterized as the "idle places" of the corresponding processes, and their initial marking, $M_0(p_{i0})$, $i = 1, 2$, establishes an upper bound to the number of instances of process type JT_i that can be simultaneously loaded into the system.

Liveness-enforcing supervision of process-resource nets based on Generalized Mutual Exclusion constraints and monitor places. [3] The reachability space, $R(\mathcal{N}, M_0)$, for the PN depicted in Figure 1 is provided in Figure 2, while the detailed characterization of the markings corresponding to the various nodes of the graph of Figure 2 can be found in Table 1[4]. It can be seen in Figure 2 that the considered net is not reversible. In particular, there is a class of states depicted by the darker-shaded nodes in Figure 2 such that every time that the net transitions to one of these states, there is no path to the initial state M_0; for further reference, this class of markings will be characterized as *unsafe*. From a more conceptual standpoint, the net non-reversibility can be interpreted by the development of a RAS deadlock, i.e., the entanglement of a subset of the running processes in a circular waiting pattern, where each process in this subset waits upon some other process of this set to release its currently allocated resource. Furthermore, the net non-reversibility implies that the underlying RAS might not be able to complete the currently loaded processes, under normal operation[5].

---

[3] We remind the reader that, in the considered class of process-resource nets, reversibility and liveness are equivalent concepts, and that the term "liveness-enforcing supervision (LES)" has prevailed over the term "reversibility-enforcing supervision".

[4] Table 1 provides only the markings of the places corresponding to the various processing stages, since the markings of the remaining places can be easily obtained from the net invariants corresponding to (i) the reusability of the system resources and (ii) the circuits established by the introduction of the process idle places.

[5] i.e., without external intervention to resolve the developed deadlock.

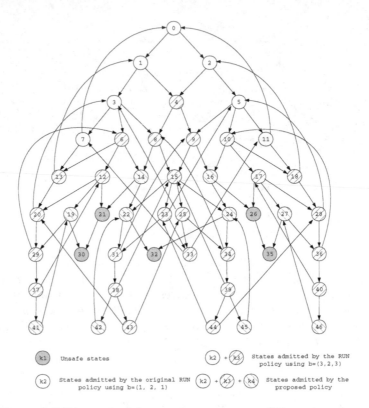

Fig. 2. The reachability graph of the process-resource net of Figure 1, and a comparison of the sub-spaces admitted by the supervisors of Equations 42, 43 and 45

The last fifteen years have seen the development of an extensive body of research seeking to develop supervisors that will enforce the reversibility of the considered class of process-resource nets. Generally speaking, these supervisors seek to constrain the behavior of the underlying process-resource net within a strongly connected component of its safe sub-space, that also contains the initial marking M_0.[6] Furthermore, one particular class of such supervisors, known as *algebraic*, seeks to establish the reversibility of the underlying process-resource net by restricting the number of process instances that can execute simultaneously various subsets of processing stages; hence, each such restriction can be expressed as a GME constraint involving the subset of places in the underlying process-resource net that correspond to the repsective processing stages. For the process-resource net of Figure 1, two such algebraic reversibility-enforcing

---

[6] Ideally, one would like to obtain the *maximally permissive* supervisor that will admit the entire safe sub-space of the underlying process-resource net, but it has been shown that, for the considered PN sub-class, the recognition of the net safe markings is an NP-complete problem [8].

Table 1. The markings of the reachability space depicted in Figure 2

State	p_{11}	p_{12}	p_{13}	p_{21}	p_{22}	p_{23}	State	p_{11}	p_{12}	p_{13}	p_{21}	p_{22}	p_{23}
0	0	0	0	0	0	0	24	0	1	0	1	1	0
1	1	0	0	0	0	0	25	0	1	0	0	0	1
2	0	0	0	1	0	0	26	1	0	0	0	2	0
3	0	1	0	0	0	0	27	0	0	0	1	2	0
4	1	0	0	1	0	0	28	0	0	0	0	1	1
5	0	0	0	0	1	0	29	1	1	1	0	0	0
6	1	1	0	0	0	0	30	1	2	0	1	0	0
7	0	0	1	0	0	0	31	1	0	1	0	1	0
8	0	1	0	1	0	0	32	1	1	0	1	1	0
9	1	0	0	0	1	0	33	0	0	1	0	0	1
10	0	0	0	1	1	0	34	0	1	0	1	0	1
11	0	0	0	0	0	1	35	1	0	0	1	2	0
12	0	2	0	0	0	0	36	0	0	0	1	1	1
13	1	0	1	0	0	0	37	0	2	1	0	0	0
14	1	1	0	1	0	0	38	0	1	1	0	1	0
15	0	1	0	0	1	0	39	0	1	0	0	1	1
16	1	0	0	1	1	0	40	0	0	0	0	2	1
17	0	0	0	0	2	0	41	1	2	1	0	0	0
18	0	0	0	1	0	1	42	1	1	1	0	1	0
19	1	2	0	0	0	0	43	0	1	1	0	0	1
20	0	1	1	0	0	0	44	0	0	1	0	1	1
21	0	2	0	1	0	0	45	0	1	0	1	1	1
22	1	1	0	0	1	0	46	0	0	0	1	2	1
23	0	0	1	0	1	0							

supervisors have been developed in [6]. They are respectively expressed by the following GME constraint sets:

$$
\begin{bmatrix} 1 & & 1 & 1 & 1 \\ 1 & 1 & & 1 & 1 \\ 1 & 1 & 1 & 1 \end{bmatrix} \cdot \hat{M} \le \begin{bmatrix} 1 \\ 2 \\ 1 \end{bmatrix} .
\tag{42}
$$

$$
\begin{bmatrix} 1 & & 1 & 1 & 1 \\ 1 & 1 & & 1 & 1 \\ 1 & 1 & 1 & 1 \end{bmatrix} \cdot \hat{M} \le \begin{bmatrix} 3 \\ 2 \\ 3 \end{bmatrix} .
\tag{43}
$$

where $\hat{M} = (M(p_{11}), M(p_{12}), M(p_{13}), M(p_{21}), M(p_{22}), M(p_{23}))^T$. Notice that the Constraint set 43 is a relaxation of the Constraint set 42 since $A_1 = A_2$ and $b_1 \le b_2$. Therefore, the supervisor established by the Constraint set 43 is expected to be more permissive than the supervisor established by the Constraint set 42, and this is indeed reflected in Figure 2 that also depicts the sub-spaces admitted by each of these two supervisors.

Obtaining a more permissive supervisor for the net of Figure 1. In this work, we employed the formulation of Equations 14,17–41 in order to compute an algebraic reversibility-enforcing supervisor for the net of Figure 1 that possesses the same computational complexity with the supervisors of Equations 42 and 43, but it is *maximally permissive*. In other words, we sought to obtain a pair

$$
A = \begin{bmatrix} a_{11} & a_{12} & a_{13} & a_{14} & a_{15} & a_{16} \\ a_{21} & a_{22} & a_{23} & a_{24} & a_{25} & a_{26} \\ a_{31} & a_{32} & a_{33} & a_{34} & a_{35} & a_{36} \end{bmatrix} , b = \begin{bmatrix} b_1 \\ b_2 \\ b_3 \end{bmatrix} ,
\tag{44}
$$

such that (i) the supervisor $\mathcal{S}(A,b) \equiv A \cdot \hat{M} \le b$ will accept a strongly connected component of the safe sub-space depicted in Figure 2 containing the initial marking M_0, and furthermore, (ii) the number of markings accepted by this supervisor is the maximal possible that can be accepted by any algebraic supervisor possessing the aforementioned structure.

Foregoing the straightforward implementational details, for the sake of brevity, we proceed to the presentation of the results of our computation. The supervisor returned by the proposed formulation is:

$$\mathcal{S}(A,b) = \mathcal{S}\left(\begin{bmatrix} 1 & 0 & 0 & 0 & 3 & 0 \\ 0 & 1 & 0 & 2 & 0 & 0 \\ 2 & 2 & 0 & 2 & 3 & 0 \end{bmatrix}, \begin{bmatrix} 6 \\ 3 \\ 8 \end{bmatrix}\right). \tag{45}$$

The sub-space admitted by the supervisor of Equation 45 is also depicted in Figure 2. As it can be seen in this figure, the obtained supervisor manages to recognize the entire safe space of the considered process-resource net, and therefore, it is optimal. Hence, this example corroborates the efficacy and analytical power of the proposed methodology.

5   Enhancements and Extensions of the Proposed Approach

In this section we consider some enhancements of the basic formulation developed in Section 3, and some modifications of the underlying supervisor design methodology, that will allow the accommodation of additional considerations, like the uncontrollability of certain transitions of the plant net $\mathcal{N}$, the potentially prohibitive computational cost resulting from the very large size of the underlying reachability space $\mathcal{G}(\mathcal{N}, M_0)$, and the imposition of additional costs and/or restrictions on the elements of the matrix A. We deal with each of these issues in a separate paragraph.

Accommodating the uncontrollability of the plant transitions. In certain cases it is possible that some of the plant transitions $t \in T$ cannot have their firing controlled by an external supervisor, but they will fire spontaneously any time that they are enabled by the plant. Their presence partitions the transition set T to the subset T^U of *uncontrollable* transitions, and its complement T^C of the *controllable* ones. Clearly, a reversibility-enforcing supervisor $\mathcal{S}(A,b)$ should not try to disable the fireability of any transition $t \in T^U$, whenever such a transition is enabled by the places of the original plant net $\mathcal{N}$. This requirement can be easily introduced in the MIP formulation of Equations 14,17–41, by adding to it the following constraint:

$$\forall i,j : (M_i, M_j) \in \mathcal{G}(\mathcal{N}, M_0) \ \wedge \ t(i,j) \in T^U, \quad z_{ij} \ge x_i \tag{46}$$

Constraint 46 essentially requests that the uncontrollable transition between markings M_i and M_j is enabled by all the monitor places implementing the supervisor $\mathcal{S}(A, b)$, but this request is enforced only for the transitions emanating from markings $M_i \in R(\mathcal{N}, M_0)$ that remain accessible during the operation of the controlled net.

Dealing with complexity considerations. It is clear from the structure of the formulation of Equations 14,17–41, that it involves a number of variables and constraints that is polynomially related to the size of the original reachability graph $\mathcal{G}(\mathcal{N}, M_0)$ of the underlying plant net $\mathcal{N}$. It is well-known, though, that, in general, the size of $\mathcal{G}(\mathcal{N}, M_0)$ is a super-polynomial function of the size of the net $\mathcal{N}$, and therefore, there might be cases where the generation and solution of the proposed formulation will be a computationally intractable task. In these cases, the approach proposed in Section 3 can still be pursued on a judiciously selected subspace of $\mathcal{G}(\mathcal{N}, M_0)$. The main requirements imposed on this subspace are that (i) it includes a strongly connected component containing the initial state M_0, and (ii) there are no uncontrollable transitions leading from this subspace to the rest of the graph $\mathcal{G}(\mathcal{N}, M_0)$; otherwise, its selection is left to the jurisdiction of the designer. Let $\mathcal{C}$ denote the *cut* from the subspace of $\mathcal{G}(\mathcal{N}, M_0)$ to be considered during the application of the proposed methodology, to the rest of $\mathcal{G}(\mathcal{N}, M_0)$. Then, the only modification required in the formulation of Equations 14,17–41 so that it effectively applies on the subgraph of $\mathcal{G}(\mathcal{N}, M_0)$ mentioned above, is that it must also contain a set of variables z_{ij}^k, $k \in \{1, \ldots, K\}$, z_{ij}, for every edge $(M_i, M_j) \in \mathcal{C}$, priced according to the constraint set 17–21, and the additional constraint:

$$\forall i, j : (M_i, M_j) \in \mathcal{C}, \quad z_{ij} \leq 1 - x_i \tag{47}$$

Constraint 47 requests the disabling by the developed supervisor of the transitions in the cut $\mathcal{C}$, but, similar to Constraint 46, it enforces this requirement only for those transitions of $\mathcal{C}$ that emanate from markings that remain reachable in the operation of the controlled net.

Restricting the elements of matrix A. In certain cases, it might be pertinent, for computational or more general implementational purposes, to put a cost structure on the elements of matrix A. As a case in point, it will be generally desirable to keep the elements of matrix A as small as possible. Of course, this additional requirement should not compromise the primary objective of the design process, which is stated in Equation 41. Hence, these additional concerns can be addressed through a *hierarchical goal programming* [10] approach. According to this approach, the formulation of Equations 14,17–41 is initially solved to optimality without any consideration of the extra concerns, in order to obtain the value of an optimal solution to the original problem stated in Section 3. Subsequently, the entire formulation is resolved with a new objective that expresses the cost criterion imposed on the elements of matrix A, while the desired value of the derived solution with respect to the original objective is fixed to the earlier computed optimal value and it is communicated to the new problem as an additional constraint; we leave to the reader the implementational details of this idea.

6   Conclusions

This paper proposed an analytical method for the synthesis of reversibility-enforcing supervisors for bounded Petri nets. The proposed method was based upon recent developments from (i) the theory of regions, that enables the design of Petri nets with pre-specified behavioral requirements, and (ii) the theory concerning the imposition of generalized mutual exclusion constraints on the net behavior through monitor places. The derived methodology takes the form of a Mixed Integer Programming formulation, which is readily solvable through canned optimization software. A small example borrowed from the theory of liveness-enforcing supervision for process-resource nets demonstrated the efficacy of the proposed approach, while the last part of the paper discussed extensions of the presented method so that it accommodates uncontrollable behavior and any potential complications arising from the large-scale nature of the underlying plant nets and their behavioral spaces.

References

1. E. Badouel and P. Darondeau. Theory of regions. In W. Reisig and G. Rozenberg, editors, *LNCS 1491 – Advances in Petri Nets: Basic Models*, pages 529–586. Springer-Verlag, 1998.
2. A. Ghaffari, N. Rezg, and X. Xie. Design of a live and maximally permissive petri net controller using the theory of regions. *IEEE Trans. on Robotics & Automation*, 19:137–141, 2003.
3. A. Giua, F. DiCesare, and M. Silva. Generalized mutual exclusion constraints on nets with uncontrollable transitions. In *Proceedings of the 1992 IEEE Intl. Conference on Systems, Man and Cybernetics*, pages 974–979. IEEE, 1992.
4. J. O. Moody and P. J. Antsaklis. *Supervisory Control of Discrete Event Systems using Petri nets*. Kluwer Academic Pub., Boston, MA, 1998.
5. T. Murata. Petri nets: Properties, analysis and applications. *Proceedings of the IEEE*, 77:541–580, 1989.
6. J. Park and S. Reveliotis. Algebraic synthesis of efficient deadlock avoidance policies for sequential resource allocation systems. *IEEE Trans. on R&A*, 16:190–195, 2000.
7. P. J. G. Ramadge and W. M. Wonham. The control of discrete event systems. *Proceedings of the IEEE*, 77:81–98, 1989.
8. S. A. Reveliotis. *Real-time Management of Resource Allocation Systems: A Discrete Event Systems Approach*. Springer, NY, NY, 2005.
9. M. Uzam. An optimal deadlock prevention policy for flexible manufacturing systems using petri net models with resources and the theory of regions. *Intl. Jrnl of Advanced Manufacturing Technology*, 19:192–208, 2002.
10. W. L. Winston. *Introduction To Mathematical Programming: Applications and Algorithms, 2nd ed.* Duxbury Press, Belmont, CA, 1995.
11. M. Zhou and M. P. Fanti (editors). *Deadlock Resolution in Computer-Integrated Systems*. Marcel Dekker, Inc., Singapore, 2004.

On the Step Explosion Problem

Stephan Roch and Karsten Schmidt

Humboldt-Universität zu Berlin
Institut für Informatik
Unter den Linden 6
D-10099 Berlin
Germany

Abstract. In many well-known extensions of place-transition nets, including read arcs, inhibitory arcs, reset arcs, priorities and signal arcs, it is sometimes possible to reach a marking through firing a step of transitions which cannot be reached through firing a sequence of single transitions. For state space analysis, it is thus recommendable to consider, in each state, all steps of transitions for firing. Since the number of activated steps may be exponential in the number of transitions, we have, in addition to the well-known state explosion problem, another explosion which we call *step explosion*. In this paper, we present an approach for alleviating step explosion. We furthermore discuss the joint application of our method with partial order reduction and in the context of CTL model checking.

Keywords: State space exploration, step semantics, read, inhibitory, reset, signal arcs, priorities, partial order reduction.

1 Introduction

In the state space analysis of Petri nets, we usually consider a single transition firing rule. The reason for not considering larger steps (firing more than one transition jointly) is: *every marking that can be reached by a step of transitions can as well be reached through a sequence of single transition occurrences.* Thereby, the sequence consists exactly of the constituents of the considered step.

For many extensions of place-transition nets, the above assumption is not true. This means, it is possible to reach markings under the step[1] firing rule which are unreachable under the single transition firing rule. The list of such extension includes

- Read arcs (see Fig. 1),
- Inhibitory arcs (see Fig. 2),
- Reset arcs (see Fig. 3),
- Capacities under the *first consume, then produce* semantics (see Fig. 4),
- Priorities (see Fig. 5) and even
- Signal arcs (see Fig. 6);

---

[1] We mean *arbitrary*, not necessarily *maximal* steps.

S. Donatelli and P.S. Thiagarajan (Eds.): ICATPN 2006, LNCS 4024, pp. 342–361, 2006.
© Springer-Verlag Berlin Heidelberg 2006

All these extensions have in common that they lead to a non-linearity in the enabling rule such that the enabling condition for the step cannot be equally divided into enabling conditions for the constituents of the step.

Among the extensions where the considered problem does not occur, are

- Colored nets and
- Capacities under the *first produce, then consume* semantics.

These extensions can be mapped into place transition nets in a way that preserves the step semantics. For other extensions of Petri nets, we did not study the possibility of additional markings under a step firing rule.

Looking at the examples, one might be tempted to argue that simultaneous occurrence of transitions is sufficiently unlikely and may thus be ignored. However, today Petri nets are used to model systems in a huge number of application areas, and on various levels of abstraction. How likely it is that transitions occur simultaneously cannot be quantified without considering the application area and the level of abstraction on which a Petri net is built. After all, the task of verification is to discover those subtle errors which remain hidden after other kinds of inspection. We thus argue in favour of having a powerful technology available for detecting markings which can be reached only through the simultaneous occurrence of transitions. Only the modeller can judge whether these markings correspond to realistic scenarios, and whether they are hazardous or not.

The effect of several extensions of place-transition nets has been studied before, including contextual nets [MR95] which include read arcs, read arcs as such [Vog97, VSY98], and inhibitory arcs [JK91a, JK91b, JK93, JK95].

While some authors, for example [MR95] and [Vog97] propose to forbid structures that lead to additional markings, others such as [JK95] take the position of this paper and propose to accept the additional markings. Considering these markings means, though, that we need to consider, in every state, all activated steps of transitions. This leads to a problem that we call *step explosion*: the number of steps to be considered in each state may be exponential in the size of the net and may be the reason for overflow in state space verification even if the number of reachable markings would fit into memory.

Step explosion is a very unfortunate phenomenon since the occurrence of markings which are unreachable by a sequence of single transition appears to be a rare event. Nevertheless, we would like to cover them since they may represent dangerous situations like hazards, race conditions, etc.

For alleviating step explosion, we propose to consider, in every state, at least all *irreducible* steps of transitions. A step s is *reducible* if it can be divided into two steps s_1 and s_2 such that the marking reached by s can as well be reached by the sequence $s_1 s_2$ (which does not necessarily imply.

that the sequence $s_2 s_1$ is executable as well). Obviously, considering irreducible steps is sufficient for covering all markings reachable by arbitrary steps.

In this paper, we first propose an approach to compute an over-approximation of the set of activated irreducible steps. It is based on a structural relation between transitions. Then we discuss the combination of our step reduction

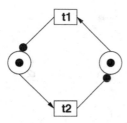

Fig. 1. Net with two read arcs. A read arc tests presence of a token but does not remove it, nor does it prevent another transition from removing or testing it simultaneously. Both transitions are enabled. Occurrence of either transition disables both in marking $(1,0)$ or $(0,1)$. Thus, the marking $(0,0)$ is unreachable under single transition firing rule but reachable by the step consisting of both transitions.

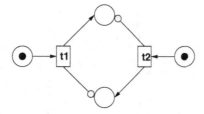

Fig. 2. Net with two inhibitory arcs. An inhibitory arc requires the absence of tokens on the pointed place but does not produce nor remove tokens. Both transitions are enabled. Occurrence of either single transition disables both and leaves one token inside the circle. Simultaneous occurrence leaves two tokens inside the circle.

technique with partial order reduction. Finally, we study the influence of step reduction on temporal logic model checking.

2   Preliminaries

In the sequel, we rely on the reader's familiarity with the usual place-transition net single transition firing rule as well as concurrent activation of steps. For simplicity, we do not consider multiple occurrence of one and the same transition. Our approach could be extended accordingly, though. We call a transition t or a step s *token-activated* in a marking m if it is activated under the single transition firing rule of place-transition nets, ignoring all extensions (including capacities). This means, there are enough tokens to feed all normal arcs from places to transitions.

We skip formal definitions of the various extensions studied in this paper. Instead, we refer to the informal explanations in the captions of Fig. 1–6. A reader who is familar with at least one of the extensions may easily understand our approach.

In the sequel, we use the term *Petri net* for any extension of place-transition nets, including place-transition nets themselves.

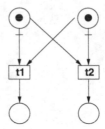

Fig. 3. Net with two reset arcs. A reset arc removes all tokens present on a place. In the figure, occurrence of one transition disables the other, due to the additional normal arc. Thus, only one place on the bottom can be marked. Occurrence of both transitions as a step can, however, be considered as possible and leaves tokens on both bottom places.

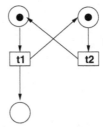

Fig. 4. ([vG05]) Net with capacities under *first consume, then produce* semantics. Assume all places have capacity 1. Then both transitions are disabled due to saturated capacity of the two topmost places. The step consisting of both transitions can nevertheless be considered to be enabled (though some readers may object) since every transition generates the required space to store the token produced by the respective other transition.

3   Step Reduction

In this section, we present the notions of *reducible* and *irreducible* steps. Firing only irreducible steps is sufficient for covering all markings which are reachable under the step firing rule. We present syntactic criteria for various Petri net extensions which over-approximate irreducible steps.

Definition 1 (Reducible and irreducible steps). *Let s be a step (i.e., a set of transitions) activated in a marking m of a Petri net. Let $m \xrightarrow{s} m'$, i.e., m' is the result of firing s in m. Then s is reducible in m if there are nonempty, disjoint steps s_1 and s_2 such that $s_1 \cup s_2 = s$ and there is a marking m'' such that $m \xrightarrow{s_1} m''$ and $m'' \xrightarrow{s_2} m'$. s is irreducible in m if it is not reducible in m. A step is reducible (irreducible) if it is reducible (irreducible) in all markings where it is activated.*

Note that we do not require that the sequence $s_2 s_1$ is firable. This means that existence of *at least one* sequence of consituents of s is sufficient for declaring s

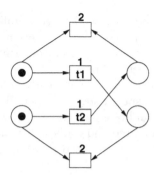

Fig. 5. Net with priorities. Under priorities, only activated transitions with highest priority may fire. Occurrence of a transition with priority 1 activates a transition with priority 2. Thus, under single transition firing rule it is impossible to have a token on both places at the right. This is, however, possible by firing both priority 1 transitions as a step.

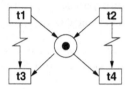

Fig. 6. Signal net. Signal nets have a step firing rule, anyway. A fired transition forces transitions pointed to by a signal arc to fire simultaneously, as long as there are sufficient tokens to enable all of them concurrently. However, considering only single *spontaneous* transitions (transitions without incoming signal arcs), it is impossible to reach another than the depicted marking. Starting with two spontaneous transitions ($t1$ and $t2$), only one of $t3$ and $t4$ is forced to fire simultaneously (since there is only one token), and two tokens are left on the place.

reducible. Observe further that, with this definition, all steps of place-transition nets (without extension) which are not single transitions, are reducible. This is actually a justification for the single transition firing rule. As Figures 1–6 show, there exist non-singular irreducible steps in many extensions of place-transition nets.

Theorem 1. *If a state space is generated by firing, in every marking m, at least all irreducible steps, then the state space contains all markings reachable under the step firing rule.*

Proof. (*Sketch*) Let m be reachable from the initial marking by a sequence of steps. If that sequence contains a reducible step s, that step can be replaced by $s_1 s_2$ as provided by Def. 1. s_1 and s_2 may be reducible again, but repetition of the argument is well-founded since s_1 and s_2 are smaller than s. Consequently, we finally arrive at a sequence of irreducible steps.   q.e.d.

For an efficient implementation of step reduction, we need an easily computable criterion. Such a criterion may over-approximate the set of irreducible steps. The criteria we are going to propose subsequently, all follow the same pattern. We introduce, for every considered extension and every marking m, a relation $\lhd(m)$ between token-activated transitions which is determined by the net structure and the current marking. Then we prove, for every criterion, that every step where the occurring transitions are not strongly connected w.r.t. the relation $\lhd(m)$, are reducible. In fact, such a step s can be divided into nonempty steps s_1 and s_2 such that $t1 \in s_1$ and $t_2 \in s_2$ implies $t_2 \ntriangleleft(m)t_1$. The criteria are such that the sequence $s_1 s_2$ can be fired and leads to the same marking as s. For read arcs, a relation similar to the one proposed here was proposed in [Vog97].

Definition 2 (Strong connectivity). *Two elements x and y are strongly connected w.r.t. a relation R iff both $[x, y]$ and $[y, x]$ are in the reflexive and transitive closure of R.*

Strong connectivity is an equivalence relation. The classes of the corresponding partition are called *strongly connected components* w.r.t. R. A set is strongly connected iff all its elements are pairwise strongly connected. Every strongly connected component is thus a strongly connevted set, but not vice versa.

Proposition 1. *Every finite set S which is not strongly connected w.r.t. a relation R can be partitioned into sets S_1 and S_2 such that $x \in S_1$ and $y \in S_2$ implies $[y, x] \notin R$.*

Proof *(Sketch)*. If S is not strongly connected, it can be partitioned into more than one strongly connected components. Among those components are *terminal* ones, i.e. such from which no elements in other components are reachable. Let S_2 be such a terminal component and S_1 consist of all remaining elements. q.e.d.

Nets with read arcs

In a net with read arcs, a transition t_1 may test a token which is consumed by another transition t_2. In such a case, putting t_2 into s_1 and t_1 into s_2 may be dangerous since t_1 might be disabled after having fired t_2. We thus define:

Definition 3 (Relation $\lhd_{Rd}(m)$). *For token-activated transitions t_1, t_2, let $t_1 \lhd_{Rd}(m)t_2$ iff there is a marked place p such that there is a read arc between p and t_1 and a normal arc from p to t_2.*

See Fig. 7 for an illustration of the defined relation.

Fig. 7. Illustration for $\lhd_{Rd}$

Theorem 2. *If, in a marking m of a net with read arcs, a step s is not strongly connected w.r.t. $\lhd_{Rd}(m)$ then s is reducible in m.*

Proof. Let $m \xrightarrow{s} m'$. Since s is not strongly connected, it can be partitioned, according to Prop. 1, into s_1 and s_2 such that $t_1 \in s_1$ and $t_2 \in s_2$ implies $t_2 \not\vartriangleleft_{Rd}(m)t_1$. Since s is activated in m, s is in particular token-activated. Thus, both s_1 and s_2 are token-activated in m. Furthermore, all places connected with read-arcs to transitions in s_1 hold a token in m. Consequently, s_1 can fire in m leading to a marking, say, m''. Since $s = s_1 \cup s_2$ was token-activated in m, s_2 is token-activated in m''. Assume, one of the transitions of s_2 is disabled in m''. Since s_2 is token-activated, there must be a place p connected a read-arc to a transition $t_2 \in s_2$ such that p does not carry a token in m''. Since t_2 was activated in m, there must be a transition $t_1 \in s_1$ which removed this token, i.e. there is a normal arc from p to t_1. Hence, $t_2 \vartriangleleft_{Rd}(m)t_1$ which contradicts the choice of s_1 and s_2. Thus, s_2 is activated in m''. It can finally be easily verified that firing s_2 in m'' leads exactly to m'.   q.e.d.

Nets with inhibitory arcs

In a net with inhibitory arcs, a transition t_1 may test absence of a token which is produced by another transition t_2. In such a case, putting t_2 into s_1 and t_1 into s_2 may be dangerous since t_1 might be disabled after having fired t_2. We thus define:

Definition 4 (Relation $\vartriangleleft_{In}(m)$). *For token-activated transitions t_1, t_2, let $t_1 \vartriangleleft_{In}(m)t_2$ iff there is an unmarked place p such that there is an inhibitory arc between p and t_1 and a normal arc from t_2 to p.*

See Fig. 8 for an illustration of the defined relation.

Fig. 8. Illustration for $\vartriangleleft_{In}$

Theorem 3. *If, in a marking m of a net with inhibitory arcs, a step s is not strongly connected w.r.t. $\vartriangleleft_{In}(m)$ then s is reducible in m.*

Proof. Follows the same line of argument as the case with read arcs.

Nets with reset arcs

In a net with reset arcs, there are two situations where transition may fire in a step but not in a sequence. First, a reset arc may remove a token which, in a step, could be used for another transition. Second, a reset arc may remove a token which, in a step, is produced later—the sequence is thus fireable but leads to a different marking than the step. We thus define:

Definition 5 (Relation $\vartriangleleft_{Rs}(m)$). *For token-activated transitions t_1, t_2, let $t_1 \vartriangleleft_{Rs}(m)t_2$ iff there is*

− a marked place p such that there is a normal arc from p to $t1$ and a reset
  arc from p to t_2, or
− a place p such that there is a reset arc from p to t_1 and a normal arc from
  t_2 to p.

See Fig. 9 for an illustration of the defined relation.

Fig. 9. Illustration for $\lhd_{Rs}$

Theorem 4. *If, in a marking m of a net with reset arcs, a step s is not strongly
connected w.r.t. $\lhd_{Rs}(m)$ then s is reducible in m.*

Proof. Let $m \xrightarrow{s} m'$. Since s is not strongly connected, it can be partitioned,
according to Prop. 1, into s_1 and s_2 such that $t_1 \in s_1$ and $t_2 \in s_2$ implies
$t_2 \ntriangleleft_{Rs}(m)t_1$. Since s is activated in m, s is in particular token-activated. Thus,
both s_1 and s_2 are token-activated in m. Reset arcs do not establish a restriction
for activation. Thus, s_1 is activated in m. Let the occurrence of s_1 in m lead to,
say, m''. s_2 is token-activated in m'' since it is token-activated in m, and by the
choice of s_1 and s_2 w.r.t. $\lhd_{Rs}(m)$ (first condition), no reset arc of s_1 removes
tokens required for token-activating s_2 in m''. Thus, s_2 can fire in m'' and leads
to a marking, say, m^*. For places without dangling reset arcs, it is immediately
clear that $m^*(p) = m'(p)$. For places with dangling reset arcs, $m'(p) = k$, where
k is the number of tokens produced on p by transitions in s itself. This is due
to the semantics of reset arcs. By the second condition, the reset arc is in s_1,
or all transitions producing tokens on p are in s_2. In either case, $m^*(p) = k$ as
well.   q.e.d.

Nets with capacities

In a net with capacities under the *first consume, then produce* semantics, a
transition that consumes tokens from a crowded place should be fired before a
transition that produces further tokens on that place. We thus define:

Definition 6 (Relation $\lhd_{Cp}(m)$). *Let $K(p)$ be the capacity of place p. Let
$W(t,p)$ be the number of tokens produced on p by t. Then $t_1 \lhd_{Cp}(m)t_2$ iff $K(p) <
m(p) + W(t_2,p)$, and there is a normal arc from p to t_1.*

See Fig. 10 for an illustration of the defined relation.

Theorem 5. *If, in a marking m of a net with capacities under the* first consume,
then produce *semantics, a step s is not strongly connected w.r.t. $\lhd_{Cp}(m)$ then
s is reducible in m.*

Fig. 10. Illustration for $\lhd_{Cp}$. Assume that the depicted place has capacity 1.

Proof. Let $m \xrightarrow{s} m'$. Since s is not strongly connected, it can be partitioned, according to Prop. 1, into s_1 and s_2 such that $t_1 \in s_1$ and $t_2 \in s_2$ implies $t_2 \not\lhd_{Cp}(m)t_1$. Since s is activated in m, s is in particular token-activated. Thus, both s_1 and s_2 are token-activated in m. Since s is activated in s, all places have a capacity which is large enough to accept all tokens produced by s, after removal of all tokens consumed by s. If a transition in s_1 produces more than $K(p) - W(t,p)$ tokens on a place p, our choice of $\lhd_{Cp}(m)$ assures that all transition of s which remove tokens from p, are in s_1, too. Consequently, s_1 is activated in m. Let its occurrence lead to a marking m''. Obviously, s_2 is token-activated in m''. Since all capacities are respected by firing s in m, the capacities are as well respected by firing s_2 in m''. Hence, s_2 is activated in m'' and its occurrence leads to m'.   q.e.d.

Nets with priorities

In a net with priorities, a transition t may activate a transition with higher priority. This way, it may deactivate a transition which can otherwise fire in a step with t. Such transitions should fire in the second part of a sequence. We thus define:

Definition 7 (Relation $\lhd_{Pr}(m)$). *Let $Pr(t)$ be the priority of transition t. Then $t_1 \lhd_{Pr}(m)t_2$ iff there is a place p and a transition t such that*

- *there is a normal arc from p to t,*
- *p does not carry as many tokens as required for firing t, and*
- *there is a normal arc from t_2 to p.*
- *$Pr(t) > Pr(t_1)$*

In other words, t_2 potentially acts towards the activation of t and thus to the deactivation of t_1.

See Fig. 11 for an illustration of the defined relation.

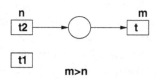

Fig. 11. Illustration for $\lhd_{Pr}$

Theorem 6. *If, in a marking m of a net with priorities, a step s is not strongly connected w.r.t. $\lhd_{Pr}(m)$ then s is reducible in m.*

Proof. Let $m \xrightarrow{s} m'$. Since s is not strongly connected, it can be partitioned, according to Prop. 1, into s_1 and s_2 such that $t_1 \in s_1$ and $t_2 \in s_2$ implies $t_2 \not\lhd_{Pr}(m) t_1$. Since s is activated in m, s is in particular token-activated. Thus, both s_1 and s_2 are token-activated in m. Furthermore, all elements of s have the same priority n, and no transition with a priority higher than n is activated. Thus, s_1 is activated in m. Let its occurrence lead to a marking m''. In m'', s_2 is token-activated. It is even activated since, through the choice of $\lhd_{Pr}(m)$, no transition with priority higher than n can be activated in m''. It is obvious that firing s_2 in m'' leads to m'. q.e.d.

Nets with signal arcs

The firing rule for signal nets is more involved than the firing rules of the extensions considered so far. Thus, we define it explicitly. In signal nets, we fire complete steps. The intuition behind complete steps is that firing a transition t emits signal pulses via its outgoing signal arcs which force the target transitions to fire instantaneously (thereby possibly emitting more signals) if they are activated concurrently with t.

Definition 8 (Complete steps in signal nets). *Consider a signal net[2]. A transition is* forced *iff it has incoming signal arcs, otherwise it is* spontaneous. *A step s is* complete *iff it is token-activated, signal-founded, and saturated. A step is* signal-founded *iff, for each $t \in s$, there is a path $t_1 t_2 \ldots t_n = t$ with $t_i \in s$ (for all i), t_1 being a spontaneous transition, and for all $i < n$, there is a signal arc from t_i to t_{i+1}. Let the* vicinity $V(s)$ *of s be the set of all transitions t such that $t \notin s$ and there is a transition $t' \in s$ with a signal arc from t' to t. s is* saturated *iff there is no t in the vicinity of s such that $s \cup \{t\}$ is token-activated.*

Due to the definition of complete steps, in particular the requirement to be signal-founded, every step must contain spontaneous transitions. Thus, steps with only one spontaneous transition are necessarily irreducible. We can therefore focus our attention on spontaneneous transitions.

As before, our goal is to separate s into s_1 and s_2 such that the disjoint union of s_1 and s_2 is s and the sequence $s_1 s_2$ can be fired and leads to the same marking as s. Consider a disjoint separation of the spontaneous transitions in s into sp_1 and sp_2. A disjoint separation of complete steps with $sp_1 \subseteq s_1$ and $sp_2 \subseteq s_2$ may be jeopardized by a transition t which is transitively forced by both sp_1 and sp_2. t could then possibly be forced to occur in both s_1 and s_2. This observation motivates the first condition below. Another problem for a sound separation is the saturation condition. Even if s is saturated, s_1 is not necessarily saturated: tokens assigned to transitions in $s \setminus s_1$ may be able to activate transitions in the vicinity of s_1. Furthermore, transitions in s_1 may produce tokens which activate transitions in the vicinity of s_2. These observations motivate the remaining conditions below. We thus define:

---

[2] Signal arcs are drawn between transitions.

Definition 9 (Relation $\lhd_{Si}(m)$). *For token-activated spontaneous transitions t_1 and t_2, let $t_1 \lhd_{Si}(m)t_2$ iff there exists a transition t_1' which is reachable via (arbitrarily many) signal arcs from t_1, and a transition t_2' reachable via (arbitrarily many) signal arcs from t_2 such that at least one of the following conditions is true:*

C1 $t_1' = t_2'$;
C2 *there is a place p with a normal arc from t_2' to p and a normal arc from p to t_1';*
C3 *there is a place p with normal arcs from p to both t_1' and t_2'.*

See Fig. 12 for an illustration of the defined relation.

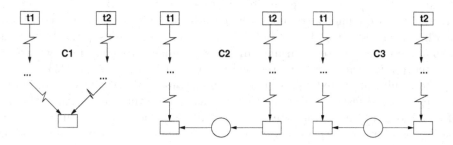

Fig. 12. Illustration for $\lhd_{Si}$

Theorem 7. *If, in a marking m of a signal net, there is a step s such that the set sp of spontaneous transitions in s is not strongly connected w.r.t. $\lhd_{Si}(m)$ then s is reducible in m.*

Proof. Let $m \xrightarrow{s} m'$. Since sp is not strongly connected, it can be partitioned, according to Prop. 1, into sp_1 and sp_2 such that $t_1 \in sp_1$ and $t_2 \in sp_2$ implies $t_2 \not\lhd_{Si}(m)t_1$. Let s_1 be the set of all transitions in s which are transitively connected with a transition in sp_1 via signal arcs. Let s_2 be the set of transitions in s which are transitively connected with a transition in sp_2 via signal arcs. Since s is signal-founded, we have $s_1 \cup s_2 = s$. By construction and condition $C1$, $s_1 \cap s_2 = \emptyset$. Both s_1 and s_2 are by construction signal-founded. Since $s_1 \subset s$, s_1 is token-activated in m. Consider an arbitrary transition t_1 in the vicinity of s_1. Assume $s_1 \cup \{t_1\}$ is token-activated. By condition C1, $t_1 \notin s_2$ and consequently $t_1 \notin s$ but $t_1 \in V(s)$. Since s is saturated, $s \cup \{t\}$ is not token-activated. Thus, in s at least one of the tokens which activate t_1 must be assigned to a transition in $s \setminus s_1 = s_2$. Hence, there must be a place p with an arc from p to t_1 and an arc from p to a transition in s_2. This contradicts condition C3, so s_1 is saturated.

Since s_1 is token-activated, signal-founded, and saturated, it is complete and can thus fire in m, leading to a marking, say, m''. Since the whole s is token-activated in m, s_2 is still token-activated in m''. Assume, there is a transition t_2 in the vicinity of s_2 such that $s_2 \cup \{t_2\}$ is token-activated in m''. By condition

C1, $t_2 \notin s_1$, so $t_2 \notin s$ but $t_2 \in V(s)$. Since s is saturated in m, it is impossible that the tokens assigned to t_2 in m'' have been already present in m: they have either been assigned to transitions in s_1 and thus been consumed, or are, in m'', still assigned to transitions in s_2. Thus, the tokens which activate t_2 in m'' must have been produced by a transition in s_1. This implies, however, the existence of a place p with and arc from p to t_2 and an arc from a transition in s_1 to p, contradicting condition C2. Hence, s_2 is saturated in m''.

As s_2 is saturated, token-activated, and founded, it is complete and can thus fire in m''. Since s is the disjoint union of s_1 and s_2, it is easy to see that the sequence $s_1 s_2$ reaches m' as well as s.   q.e.d.

In Figs. 1–6, we have $t1 \lhd t2$ and $t2 \lhd t1$ in the depicted marking. Thus, the sets $\{t1, t2\}$ are strongly connected w.r.t. the appropriate relations $\lhd$. Consequently, the steps $\{t1, t2\}$ in these figures (complete steps $\{t1, t2, t3\}$ and $\{t1, t2, t4\}$ in Fig. 6) are not qualified as reducible by our criteria (they are in fact irreducible). In Figs. 7–11, we have $t1 \lhd t2$ but not vice versa. The steps $\{t1, t2\}$ are thus reducible in the depicted marking. In these examples, the marking reached by the step $\{t1, t2\}$ can as well be reached by firing the sequence $\{t1\}\{t2\}$.

The results presented so far concern extensions of Petri nets with only one of the discussed features. For an extension with more than one feature, the list of conditions for an appropriate relation $\lhd$ needs to be extended with patterns where the different extensions interfere. Due to the mere complexity of providing such patterns for all combinations, we do not discuss those cases in this paper. For an example, the reader may refer to [SR02]. There, a relation $\lhd$ is proposed for an extension of place-transition nets with both signal and read arcs.

4   Computation of a Reduced Step List

For applying step reduction in state space exploration, we need to perform the following steps in every marking m encountered:

1. Compute the set U of token-activated (in signal nets: spontaneous) transitions;
2. Build the appropriate relation $\lhd(m)$;
3. Construct the set $\mathcal{S}$ of all subsets of U which are strongly connected w.r.t. $\lhd(m)$;
4. fire all activated elements of $\mathcal{S}$.

The first step is part of every existing state space exploration tool, so we do not need to elaborate on it. The second step is easy as the proposed conditions correspond to simple patterns in the net structure. The fourth step consists of a plain verification of the enabling conditions. Consequently, the third step in the procedure above is the most interesting one.

This step reduces to the question: given a directed graph $G = [V, E]$ with a set V of vertices and a set E of edges, compute all strongly connected *sets* (SCS) $V' \subseteq V$.

The problem hereby is that we need to compute not only strongly connected *components* (SCC) which are SCS of maximum size w.r.t. set inclusion. Nevertheless, the known linear-time algorithm for computing SCC [Tar72] can be employed to our problem. The basic idea of computing SCS is:

1. partition V into SCC;
2. for every SCC C, pick an element x;
3. compute all SCS which contain x;
4. compute all SCS which do not contain x.

The last step can be executed through recursion: partition $C \setminus \{x\}$ into SCC and so on. The third step can be implemented by declaring x "compulsory" (while other nodes are called "facultative") and restrict further computations to those sets which contain x. These ideas lead to the following procedure for the calculation of all SCS in $[V, E]$

PROCEDURE SCS($Comp, Fac$: **SET OF** Node): **SET OF SET OF** Node;

1. If $Comp \cup Fac = \emptyset$ **THEN RETURN** $\emptyset$;
2. Partition $Comp \cup Fac$ into SCC, say $V_1, \ldots, V_k$; // SCS are included in SCC
3. $Result := \emptyset$;
4. **FOR ALL** V_i such that $Comp \subseteq V_i$ **DO** // Compulsary nodes MUST occur
   (a) $F := V_i \cap Fac$; // nodes still open for inclusion/exclusion
   (b) $C := V_i \cap Comp$; // compulsary nodes
   (c) **WHILE** $F \neq \emptyset$ **DO**
      i. pick an $x \in F$;
      ii. $F := F \setminus \{x\}$;
      iii. $Result := Result \cup$ SCS(C,F); // add SCS which do NOT contain x
      iv. $C := C \cup \{x\}$; // in remaining iterations, add SCS which contain x
   (d) $Result := Result \cup \{C\}$;
5. **RETURN** $Result$;

Note that, for a non-empty $Comp$, there is at most one i such that $Comp \subseteq V_i$, so the body of statement 4 is executed at most once. It may be executed more than once if $Comp = \emptyset$.

The algorithm terminates since the recursive calls of SCS as well as the iterations of statement 4.c have arguments of decreasing size.

Theorem 8. *For every disjoint pair C^* and F^* of nodes, SCS(C^*, F^*) returns the set of all strongly connected sets included in $C^* \cup F^*$ which contain all elements of C^*.*

Proof. The empty set does not include strongly connected sets, so the first statement is correct. Otherwise, every strongly connected set of $C^* \cup F^*$ is included in at least one strongly connected component of $C^* \cup F^*$. Since we want to calculate those SCS which include C^*, the restriction in statement 2 is correct, and it remains to show that the body of statement 4 adds to $Result$ all SCS which are included in V_i and which include C^*. If $V_i = C^*$ then V_i is the only SCS that fits the specification. In this case, it is added to *Result* in statement 4.d. Otherwise, we may distinguish, among the SCS to be considered, those which do contain a certain element in $V_i \setminus C^*$ and those which do not contain that element. The second kind of SCS is computed in statement 4.c.iii, which can be shown by embedding the current argument into an induction on the recursion level. This induction is well-founded since the algorithm terminates. For the first kind of SCS, the considered element can now be treated as a compulsory one (statement 4.c.iv). By repetition of the current argument we can assert that the remaining SCS are indeed computed in the remaining iterations of the while loop.   q.e.d.

Corollary 1. *The call* $SCS(\emptyset, V)$ *returns all strongly connected sets of a graph* $[V, E]$.

The proposed algorithm may have exponential run-time, measured in the size of the given net. This cannot be avoided since it can be easily observed that there may be exponentially many irreducible steps. We are not capable of describing the complexity of the algorithm based on the number of computed steps. Experience with the implementation discussed later in this paper shows, however, that it performs well. At least, it guarantees that no SCS is enumerated twice.

5   Step Reduction and Partial Order Reduction

Explicit state space verification can only be successful if powerful reduction techniques are applied. This is even more the case if we consider nonsingular steps, as (even with step reduction) the number of steps to be fired is larger than the number of single transitions, and the number of reached markings can be larger. For some reduction techniques, joint application with step reduction is straightforward. This includes, for example, symmetry reduction [HJJJ84, Sta91, Sch00] and the sweep-line method [Mai03, Sch04]. Combination with partial order reduction is also possible. It needs, however, a few adjustments and leads to quite interesting results. Thus, we discuss this combination in more detail.

The core idea of partial order reduction [Val88, Pel93, God94] is to exploit commutativity of transitions (in our case: steps). This idea is implemented through considering, in every marking, only a subset *stubborn(m)* of transitions (in our case: steps) for firing. Through suitable conditions for the selection of the subsets, preservation of various properties can be asserted. In [Val91], the approach is presented on the abstraction level of *variable-transition systems*. This approach is well-suited for many extensions of Petri nets as well as the consideration of steps since our steps can be considered as transitions in the sense of variable-transition system. Places play the role of variables, and a marking corresponds of a valuation of the variables.

Let us, for the sake of simplicity, focus on the method that preserves deadlocks. Preserving deadlocks is one of the oldest areas partial order reduction is applied to, and it leads to fairly simple conditions. The technique in [Val91] requires the following conditions for preserving deadlocks (we formulate the conditions in terms of places and steps):

St1 If a deactivated step s is in $stubborn(m)$ then, for one condition which is sufficient for deactivation, all steps that have the capability to change that condition, have to be contained in $stubborn(m)$, too.

A sufficient condition for deactivation could, for instance, be an insufficiently marked pre-place, a marked place connected to a transition in s via an inhibitory arc, an activated transition with higher priority, or, in the case of signal nets, an activated transition in the vicinity of s. Changing the condition could consist of producing tokens on an insufficiently marked place, removing tokens from a place connected with an inhibitory arc, or firing a transition in conflict to a transition with higher priority, or to a transition in the vicinity of s, whatever is appropriate.

St2 (simplified) If an activated step s is in $stubborn(m)$, so are all steps that could potentially deactivate s through their occurrence.

This includes, for instance, steps which remove tokens from (normal) pre-places of s, put tokens on places connected with inhibitory arcs, activate transitions with higher priority, or transitions in the vicinity of s.

St3 (simplified) If an activated step s is in $stubborn(m)$, so are all steps s' such that the sequence ss' leads, in any marking, to another marking than $s's$.

St4 $stubborn(m)$ contains at least one activated step.

Applying these rules to a state space generation under (arbitrary) step semantics leads to the absence of any reduction for most systems. The reason is that, with an activated step s, many super- or subsets of s are activated, and many activated steps s_1, s_2 have an activated superset containing both. In this case, requirement St2 forces the stubborn set to contain virtually all activated steps.

This is where step reduction comes in. Since we consider only irreducible steps or a small superset thereof, the transitive connection between arbitrary steps is broken, and stubborn sets applied according to the rules adapted from [Val91] leads to substantial reduction (see the experimental results below). Since the concept of variable-transition systems is quite liberal in defining what a transition (here: a step) is, the theory can be applied almost straightforwardly to those steps considered by our step reduction. Actually, only one adjustment must be made: If we consider a step in any marking, we need to consider it in every other marking, too. This does not necessarily mean that we need to *fire* the step in every marking where it is activated. It just means that it needs to be considered in stubborn set calculation. The step reduction presented above may violate that condition since a step might be reducible in one marking, but irreducible in another.

For assuring that step reduction becomes compatible with partial order reduction, we relax the relations $\lhd$ such that they become marking-independent. As the following list of definitions shows, this relaxation is mostly straightforward, and it is immediately clear that Theorems 2–7 turn over to the relaxed case.

Definition 10 (Relation $\lhd_{Rd}$). *For transitions t_1, t_2, let $t_1 \lhd_{Rd} t_2$ iff there is a place p such that there is a read arc between p and t_1 and a normal arc from p to t_2.*

Definition 11 (Relation $\lhd_{In}$). *For transitions t_1, t_2, let $t_1 \lhd_{In} t_2$ iff there is a place p such that there is an inhibitory arc between p and t_1 and a normal arc from t_2 to p.*

Definition 12 (Relation $\lhd_{Rs}$). *For transitions t_1, t_2, let $t_1 \lhd_{Rs} t_2$ iff there is*

- *a place p such that there is a normal arc from p to t_1 and a reset arc from p to t_2, or*
- *a place p such that there is a reset arc from p to t_1 and a normal arc from t_2 to p.*

Definition 13 (Relation $\lhd_{Cp}$). *Let $t_1 \lhd_{Cp}$ iff there is a place p with finite capacity such that there is a normal arc from t_2 to p and one from p to t_1.*

Definition 14 (Relation $\lhd_{Pr}$). *Let $Pr(t)$ be the priority of transition t. Then $t_1 \lhd_{Pr} t_2$ iff there is a place p and a transition t such that there is a normal arc from p to t, and one from t_2 to p.*

Definition 15 (Relation $\lhd_{Si}$). *For spontaneous transitions t_1 and t_2, let $t_1 \lhd_{Si} t_2$ iff there exists a transition t'_1 which is reachable via (arbitrarily many) signal arcs from t_1, and a transition t'_2 reachable via (arbitrarily many) signal arcs from t_2 such that at least one of the following conditions is true:*

- *$t'_1 = t'_2$;*
- *there is a place p with a normal arc from t'_2 to p and a normal arc from p to t'_1;*
- *there is a place p with normal arcs from p to both t'_1 and t'_2.*

This way, it is possible to define partial order reduction techniques for many extensions of place-transition nets. They work well in those systems where the actual topology of the extended features leads to a large set of reducible steps.

6   Step Reduction and CTL Model Checking

So far, we have only considered the *set* of reachable markings. Step reduction preserves the set of markings reachable under the (arbitrary) step firing rule. In this section, we consider the question: which properties expressible in the temporal logic CTL are preserved by step reduction?

We start with a sketched introduction to CTL [CE82].

Definition 16 (CTL). *CTL is the logic that is built from atomic propositions (which assign a truth value to every marking), boolean connectives (with the standard interpretation), and the following operators:*

- *EX* where *EX*ϕ *is true in m iff* ϕ *is true in at least one successor marking of m;*
- *AX* where *AX*ϕ *is true in m iff* ϕ *is true in all successor markings of m;*
- *EF* where *EF*ϕ *is true in m iff* ϕ *is true in at least one marking reachable from m;*
- *AG* where *AG*ϕ *is true in m iff* ϕ *is true in all markings reachable from m;*
- *AF* where *AF*ϕ *is true in m iff* ϕ *is true at least once on every path descending from m;*
- *EG* where *EG*ϕ *is true in m iff* ϕ *is true in all markings of at least one (infinite or ending in a deadlock) path descending from m;*
- *EU* where *E(ϕUψ)* *is true in m iff there is a path descending from m where* ψ *holds in a marking m' on that path and* ϕ *holds in all markings between m (including m) and m' (not necessarily including m') on that path;*
- *AU* where *A(ϕUψ)* *is true in m iff, in all paths descending from m,* ψ *holds in a marking m' on that path and* ϕ *holds in all markings between m (including m) and m' (not necessarily including m') on that path.*

Since the semantics in CTL is defined on transition systems, we may compare the value of a CTL formula in a transition system ST under step semantics with its value in the corresponding transition system RD under step reduction. The example net in Fig. 13 teaches us that most CTL formulas are not generally preserved by step reduction. In this example, it can be easily verified that

- $EX(p3 \wedge p4)$ is true in ST but false in RD,
- $AX(\neg p3 \vee \neg p4)$ is false in ST but true in RD,
- $EG(p1 \wedge p2) \vee (p3 \wedge p4)$ is true in ST but false in RD,
- $AF(p1 \wedge p4) \vee (p2 \wedge p3)$ is false in ST but true in RD,
- $E((p1 \wedge p2)U(p3 \wedge p4))$ is true in ST but false in RD,
- $A((p1 \wedge p2)U((p1 \wedge p4) \vee (p2 \wedge p3)))$ is false in ST but true in RD.

The example is a plain place-transition net, so the problem occurs in every extension as well. After all, at least some implications hold. Since the transition

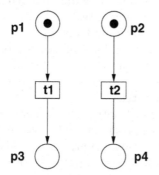

Fig. 13. In this example, several CTL formulae have change their truth value when irreducible steps (here: single transitions) are considered instead of arbitrary steps

system under step reduction is an under-approximation of the unreduced system, all purely existential (E) properties holding in RD, hold in ST as well, and all purely universal (A) properties holding in ST, hold in RD as well. Furthermore, the above list does not contain the operators AG and EF. This is due to the following theorem:

Theorem 9. *Every CTL property which is built using only atomic propositions, boolean connectives and the operators AG and EF, holds in the step reduced system if and only if it holds in the original system.*

Proof. (Sketch) Let $mm_1m_2 \ldots m_n m'$ be a sequence witnessing the validity of $EF\phi$ in m, i.e., m' satisfies ϕ. If that sequence occurs in the reduced system, it occurs in the unreduced system as well. If such a sequence occurs in the unreduced system, and involves a reducible step $m_i \xrightarrow{s} m_{i+1}$, s can be replaced by a sequence of irreducible steps. Thus, there is a sequence leading from m to m' in the reduced system as well. q.e.d.

Many traditional Petri net properties, including reachability, dead markings, dead transitions, and liveness, are expressible using only EF and AG.

7   Experimental Results

We have implemented step reduction in the tool SESA [RS02]. SESA is an adaptation of the INA [RS98] tool to an extension of Petri nets with both signal arcs and read arcs.

The first example concerns a control for a railway crossing. We verified absence of deadlocks. Full state space generation failed after more than 20 hours, due to memory overflow. At that point of time, more than 12.000.000 steps have been executed. Among them were more than 1.000.000 *different* steps. With step reduction, the state space consisting of 478.673 states and 3.867.429 transitions fit into memory. There were only 124 different irreducible steps. State space verification took 7 hours and 35 minutes.

In a second example, we took a model of a plant that produces tooth brushes [Rau96]. While full state space exploration failed after an hour with more than 10.000.000 fired steps out of more than 20.000 different steps, the exploration under step reduction succeeded after less than a minute with approx. 50.000 occurrences of 90 irreducible steps.

The third example models a control for a machinery for turning parcels. We checked reachability of a state predicate. In full state space exploration, a satisfying state has been found after having generated 48.566 states, taking 46 minutes. With step reduction, only 10.239 states were generated in 20 minutes. With joint application of step reduction and partial order reduction, only 6.351 states were generated in 20 minutes. The huge amount of time spent in full exploration is again due to the consideration of a huge number of reducible steps.

8   Conclusion

We have shown that, in many extensions of place-transition nets, there can be markings reachable under step semantics which are unreachable under a single transition firing rule. We argued that there should be verification technology to detect these markings. A naive state space generation under step semantics fails due to the step explosion problem. We proposed step reduction as a technique for alleviating step explosion. We then studied the compatibility of step reduction with partial order reduction and investigated the consequences of applying step reduction to CTL model checking. This way, we proposed a technique to discover potentially hazardous markings which can be embedded into existing state space verification technology.

Future work includes the extension of this approach to steps including auto-concurrency (simultaneous occurrence of several copies of one and the same transition), a systematic study of extensions with more than one of the studied features, and a characterisation of all extensions which lead to additional markings in the step semantics.

References

[CE82]   E.M. Clarke and E.A. Emerson. Using branching time temporal logic to synthesize synchronization skeletons. *Science of Computing 2*, pages 241–266, 1982.

[God94]   P. Godefroid. *Partial–Order Methods for the Verification of Concurrent Systems: An Approach to the State–Explosion Problem.* PhD thesis, University of Liege, Computer Science Department, 1994.

[HJJJ84] P. Huber, A. Jensen, L. Jepsen, and K. Jensen. Towards reachability trees for high–level Petri nets. In *Advances in Petri Nets 1984, Lecture Notes on Computer Science 188*, pages 215–233, 1984.

[JK91a]   R. Janicki and M. Koutny. Invariant Semantics of Nets with Inhibitor Arcs. In J. C. M. Baeten and J. F. Groote, editors, *Proceedings of the 2nd International Conference on Concurrency Theory (CONCUR'91), Amsterdam, The Netherlands, August 1991*, volume 527 of *Lecture Notes in Computer Science*, pages 317–331, Berlin, Heidelberg, 1991. Springer-Verlag.

[JK91b]   R. Janicki and M. Koutny. Invariants and Paradigms of Concurrency Theory. In E. H. L. Arts, J. van Leeuwen, and M. Rem, editors, *Parallel Architectures and Languages Europe (PARLE'91) Volume II: Parallel Languages, Eindhoven, The Netherlands, June 1991*, volume 506 of *Lecture Notes in Computer Science*, pages 59–74, Berlin, Heidelberg, 1991. Springer-Verlag.

[JK93]   R. Janicki and M. Koutny. Structure of Concurrency. *Theoretical Computer Science*, 112:5–52, 1993.

[JK95]   R. Janicki and M. Koutny. Semantics of Inhibitor Nets. *Information and Computation*, 123:1–16, 1995.

[Mai03]   T. Mailund. *Sweeping the State Space - a sweep-line state space exploration method.* PhD thesis, University of Aarhus, 2003.

[MR95]   U. Montanari and F. Rossi. Contextual Nets. *Acta Informatica*, 32(6):545–596, September 1995.

[Pel93]   D. Peled. All from one, one for all: on model–checking using representitives. *5th Int. Conf. Computer Aided Verification,Elounda, Greece, LNCS 697*, pages 409–423, 1993.

[Rau96]   M. Rausch. *Modulare Modellbildung, Syntese und Codegenerierung ereignis-diskreter Steuerungssysteme*. PhD thesis, Otto-von-Guerike Universität Magdeburg, Lehrstuhl Steuerungstechnik, 1996.

[RS98]   S. Roch and P. H. Starke. INA - Integrated Net Analyzer - Version 2.2 - Manual. Technical report, Informatik-Berichte, Humboldt-Universität zu Berlin, 1998.

[RS02]   S. Roch and P. Starke. The sesa home page. *http://www.informatik.hu-berlin.de/lehrstuehle/automaten/sesa/*, 2002.

[Sch00]   K. Schmidt. How to calculate symmetries of Petri nets. *Acta Informatica*, 36(7):545–590, 2000.

[Sch04]   K. Schmidt. Automated generation of a progress measure for the sweep-line method. In *Proc. 10th Conf. Tools and Algorithms for the Construction and Analysis of Systems (TACAS)*, volume 2988 of *Lecture Notes in Computer Science*, pages 192–204. Springer-Verlag, 2004.

[SR02]   P. H. Starke and S. Roch. Analysing Signal-Net Systems. Informatik-Berichte 162, Humboldt-Universität zu Berlin, 2002.

[Sta91]   P. Starke. Reachability analysis of Petri nets using symmetries. *J. Syst. Anal. Model. Simul.*, 8:294–303, 1991.

[Tar72]   R. E. Tarjan. Depth first search and linear graph algorithms. *SIAM J. Comput.*, 1:146–160, 1972.

[Val88]   A. Valmari. Error detection by reduced reachability graph generation. In *Proceedings of the Ninth European Workshop on Application and Theory of Petri Nets, Venice, Italy*, pages 95–112, 1988.

[Val91]   A. Valmari. Stubborn Sets for Reduced State Space Generation. In G. Rozenberg, editor, *Proc. Advances in Petri Nets 1990*, volume 483 of *Lecture Notes in Computer Science*, pages 491–515, Berlin, Germany, November 1991. Springer-Verlag.

[vG05]   Rob van Glabbeek. Personal communication. 2005.

[Vog97]   W. Vogler. Partial Order Semantics and Read Arcs. Technical Report 1997-1, Universität Augsburg, 1997.

[VSY98]   W. Vogler, A. Semenov, and A. Yakovlev. Unfolding and Finite Prefix for Nets with Read Arcs. In D. Sangiorgi and R. de Simone, editors, *Proceedings 9th International Conference on Concurrency Theory (CONCUR'98), Nice, France, September 1998*, volume 1466 of *Lecture Notes in Computer Science*, pages 501–516. Springer-Verlag, 1998.

Prospective Analysis of Cooperative Provision of Industrial Services Using Coloured Petri Nets

Katrin Winkelmann and Holger Luczak

Research Institute for Operations Management at Aachen University
Pontdriesch 14/16, D-52062 Aachen, Germany
{Katrin.Winkelmann, Holger.Luczak}@fir.rwth-aachen.de

Abstract. International customer demands in the capital goods industry force providers of industrial services to be present on a global market. Often this can only be realised efficiently by cooperating with other partners. But the process to configure a cooperation and choose an adequate cooperation alternative still lacks adequate support. Existing approaches to evaluate services or cooperation are not suitable for an assessment of different alternatives prior to their implementation. Simulation approaches for production networks on the other hand, cannot be used for industrial services. This paper deals with an approach to overcome this problem and presents a simulation model based on Petri net theory for the prospective analysis of cooperative provision of industrial services. To achieve this goal, a domain-specific conceptual model of cooperative provision of industrial services has been developed that considers the constitutive characteristics of industrial services and their cooperative provision and maps them onto the formal notation of coloured Petri nets to form an executable simulation model. As a result, machine and equipment producers will be able to assess their cooperation alternatives related to integrated service provision in advance and thus avoid cost-intensive false decisions.

1 Motivation

1.1 Trend Towards Cooperation and Need for Assessment Support

Germany's capital goods industry has an export rate of about 70 percent and its capital goods are sold around the globe. So, product-related services also have to be offered internationally [4], [45]. These product-related industrial services (e.g. repair or maintenance) are offered to (re)establish, ensure or enhance the long-term usability of industrial products and are becoming increasingly important in the global economy and also in Germany's capital goods industry: German machine and equipment producers are realising an increasing percentage of their revenues with product-related services [16]. However, international provision of high-quality services requiring expert knowledge has proved to be cost-intensive and thus threatens the profit margin. Therefore, machine and equipment producers are looking for partners to form coalitions to improve their service business through cooperative advantages such as more efficient deployment of experts, shorter order processing times or cost reduction. In other words, the process of providing services for customers is no longer

S. Donatelli and P.S. Thiagarajan (Eds.): ICATPN 2006, LNCS 4024, pp. 362–380, 2006.

owned by one single company, but by several partners. This shared process is referred to as cooperative service provision.

While many companies have gained preliminary experience with cooperative service provision, the existence of problems in finding appropriate partners and reported organisational difficulties indicate that companies lack support for the planning and configuration of service networks [46]. A variety of alternatives related to cooperation and configurations complicates a substantiated decision [10], [14], [35]. In this context, the possibility of assessing different cooperation alternatives before implementation is needed [6]. Some approaches for network and service controlling exist, but these methods are not suitable for an assessment of different alternatives prior to their implementation, i.e. a prospective analysis. Companies are not able to evaluate, for example, whether or not a certain combination of partners in the service process leads to a shorter order processing time for service orders.

To solve problems of this kind, a model for the prospective analysis of service network alternatives in the capital goods industry is needed [25]. This model has to consider the special requirements and characteristics of industrial services and their cooperative provision. In addition, it has to be represented in a formal notation that allows simulations and performance analysis.

1.2 Prospective Analysis of Service Network Alternatives Using Simulation Based on Coloured Petri Nets

The target is therefore to enable the prospective analysis of cooperative provision of industrial services. The chosen method of resolution is simulation with coloured Petri nets (CPN). Coloured Petri nets are widely used to model business processes and workflows [18], [19], [39]. The availability of many analytic techniques based on Petri nets is another advantage [13], [29], [41]. Modern systems are often so complex that system interactions can usually only be analysed using simulation techniques [1], [2]. Simulation is particularly suitable for assessing different alternatives before their actual implementation and has proven to be a successful method for prospective analysis of production network design [2], [22], [30], [36]. First applications of simulation in consumer-oriented service systems (e.g. queuing models at bank counters) have produced promising results [21], [33].

2   Relation to Other Work and Need for Action

Several existing approaches have been examined and evaluated (see Fig. 1): First, approaches based on simulation in different kinds of application have been evaluated. Petri net approaches in the workflow management and service domain have also been considered. These approaches provide a valuable contribution since they show how to model certain problems and transfer them into a simulation model. Although their objectives differ significantly from the problem at hand, these approaches provide some useful pointers for the development of a simulation model for the case described in this paper. Second, to consider the special conditions related to the cooperative provision of industrial services, approaches for assessing cooperation and industrial services were analysed. These approaches are not designed for a prospective analysis but deliver valuable information about the assessment of cooperative service processes.

First simulation applications for services were developed by [21], [27], and [33]. Although these authors examined a prospective assessment for services, their approaches do not consider cooperation aspects and characteristics of service provision in an industrial context. Advanced simulation models exist for different ranges of application, e.g. for product development [23] or for autonomous production cells [32]. Petri net approaches in the workflow domain consider consumer-oriented services in particular (e.g. [17], [39]), but they do not provide information about modelling cooperative provision of industrial services.

Approaches in the field of cooperation controlling have proved to be of little value to the problem at hand, because of their focus on a retrospective analysis and management. Concepts based on the Balanced Scorecard (e.g. [5], [26]) benefit from the need to define assessment criteria, but remain superficial about operationalisation of these criteria.

With regard to controlling and quality management of industrial services, the following approaches have proved useful: [7], [11], [15], [24], and [40]. These authors have developed assessment criteria for industrial services at different levels of detail, and their results have been taken into account in the development of the simulation model.

The analysis of relevant literature shows that some approaches exist for parts of the problem, but shortcomings prevail, especially those related to the prospective analysis of alternatives of cooperative service provision. The following deficits can be identified:

- Most of the approaches concerning prospective analysis based on simulation refer to manufacturing or other areas of application and do not consider the special characteristics of industrial services that demand significantly different approaches than manufacturing of material goods and the provision of consumer-oriented services [21]. This means that approaches that do not consider these characteristics of services cannot be used to solve the problem at hand. Initial simulation approaches for services differ importantly in terms of the underlying problem, so that they cannot be adopted to solve the problem.

- Approaches in the field of cooperation controlling also do not consider the special characteristics of services. In addition, concepts of controlling are not designed for a prospective assessment but rather for ongoing management and retrospective analysis. They are therefore not applicable for prospective analysis of service provision.

- Approaches which assess the quality of services or control services are also designed for continuous management and retrospective analysis and thus cannot be adopted for prospective analysis. In addition, these approaches focus on individual companies and neglect cooperation aspects. These aspects, however, are very important for the current case.

	Author	Year	Services	Industrial services	Captial goods industry	Cooperation	Service provision	Configration of cooperation	Configuration of service provision	Functional	Strategic	Operations research	Prospective	Simulation	Petri nets
	Sector					**Cooperation design**				**Assessment/analysis**				**Method of resolution**	
Analysis based on simulation	Banks et al.	2001	○			○		○	○			●	●	●	○
	Jansen-Vullers, Reijers	2005	●				●	○	●	●	○	●	●	●	●
	Laughery et al.	1998	●	○	○					●	○	●	●	●	
	Licht et al.	2004			○					○	○	○	●	●	●
	Mjema	1997	●	●	●				○	●	○		●	●	
	Schlick et al.	2002			○				○	●		○	●	●	●
	Seel	2002	●	○				●		●			●	●	●
	van der Aalst	2002	○					●	○	●	○	○	○	○	●
	von Steinaecker	2000			○	●		●	●	●	○	●	○	●	●
	Wurmus	2002								●				●	●
	Zimmer	2001				○		○	○	●	○	●	●	●	●
Cooperation analysis	Bauer, Kinkel	2003	●	●	●	●	●	●			●				
	Bornheim, Stüllenberg	2002				●				○	●				
	Hartel	2003	●	●	●	●	●	●	●	●	○				
	Hermann, Langhoff	2003	●	●	○	●	●	●			●				
	Hess	2001				●				○	●				
	Merkle	1999				●		○			●				
	Schuh, Güthenke	1999				●		○			●				
	Voß	2001				●		○			●				
Assessment of services	Borrmann	2003	●	●	●					●	●				
	Bruhn	2004	●	○	○						○				
	Corsten, Gössinger	2004	●	○							●				
	DIN	2002	●	○	○		●		○	●	○				
	Eichmann	2002	●	●	○		●		○	●	○				
	Hlubek et al.	2004	●	●	○		○		○	●	○				
	Kinkel	2003	●	●	○		○			●	○				
	Luczak, Drews	2005	●	●	●		○		○	●		○			
	Ulber, Elsweiler	2002	●	●	○		○		○	●	○				
	VDI	2003	●	●	○		●		○	●	○				
	Goal of presented project		●	●	●	●	●	●	●	●	○	●	●	●	●

Legend: ● covered   ○ partly covered   (blank) not covered

Fig. 1. Examination and evaluation of existing approaches

But existing approaches do provide valuable information for solving different parts of the problem at hand:

- Simulation based on Petri nets as a method of prospective analysis has proven to be of great value, especially for complex cooperative processes. First promising applications for services suggest that using Petri nets and simulation for the problem at hand would be useful.
- Coloured Petri nets in particular have proven to be a valuable modelling technique for complex cooperative processes and are applicable to cooperative service provision. Their analysis potential meets the demands of the current case.
- Assessment criteria used in service controlling and to measure service quality represent an important input for the development of a model for the assessment of

possible alternatives of cooperative service provision. Especially assessment criteria developed for industrial services are relevant for the problem at hand.

As a result of the deficits and advantages of existing approaches outlined above, the following tasks need to be addressed [44]:

- A conceptual model for the cooperative provision of industrial services has to be developed.
- It has to be implemented into a simulation model based on coloured Petri nets.
- The model has to be capable of prospective performance analysis of cooperative provision of industrial services.

3   Modelling Cooperative Provision of Industrial Services Using Coloured Petri Nets

To create a model of cooperative provision of industrial services, the special characteristics of this domain have to be taken into account. These characteristics have to be integrated into a conceptual model of cooperative provision of industrial services before this model is translated into a CPN model.

3.1   Conceptual Model of Cooperative Provision of Industrial Services

Services can be defined using constitutive attributes [9], [34]. This definition comprises three dimensions: process, structure, and outcome [9], [12].

The process dimension stresses the fact that the customer as the external factor has to be integrated into the process [9]. Another important characteristic in the process dimension is the simultaneity of production and consumption of services in respect to time and space. Together with the integration of the external factor, this implies two more specialities for industrial services that are not consumer-oriented but are rather business-to-business services in an industrial environment: Since services have to take place at the site of the product, required resources have to travel there. In addition, technicians of the client company often have to be integrated into the service process and work together with personnel of the service company. Services and especially industrial services are also very heterogeneous and standardisation of processes is not easy.

The structure dimension defines services as the ability to perform a service [9]. This shows that services are intangible and thus conventional approaches that are oriented towards tangible attributes of products are not applicable to services [34].

The simultaneity of process and production also affects the outcome dimension: It implies that there is no product whose quality can be measured in the end but the process itself is the service. Since services are not storable, this means that any delay in the process directly affects the quality of the services.

The conceptual model of cooperative provision of industrial services is based on these three dimensions and integrates the outlined characteristics of industrial services.

Process dimension. To represent the process of industrial service provision, a reference model has been developed [20]. This reference model has been adapted to fit cooperative service provision and to fulfil the modelling requirements of the Petri net notation. Furthermore, the resulting model has been structured in a hierarchy: The top level provides a process overview; all process steps of this overview are then modelled in more detail. These sub-processes are can be interpreted as separate modules that interact via interfaces. This concept provides, on the one hand, flexibility, since modules can be changed easily and locally without having to deal with the whole model. On the other hand, it allows one to model process parts in sufficient detail while keeping complexity at the top level low.

To account for the heterogeneity of the process, locally different process alternatives (e.g. necessity for problem description or revisions) have been defined. Probability functions control the flow of orders through these alternatives (for an example see chapter 3.2, Fig. 4).

Structure dimension. The structure dimension represents the resources needed to execute the processes. Mapped resources are different categories of personnel and material (e.g. service engineers, service technicians, client technicians for personnel and tools, spare parts, operating resources like gaskets or lubricants). Depending on the kind of order that has to be processed and its degree of difficulty, different categories and quantities of resources are needed and consumed.

The structure dimension also accounts for the mapping of different cooperation alternatives. The cooperation partners provide different resources and thus certain parameters change. Such parameters are the amount and structure of the provided resources and their consumption based on the degrees of difficulty of the orders. Depending on the location of the partners and spare parts suppliers, times for personnel travel and spare parts shipment vary. The times required for fulfilling a service order and the probability of the need to correct this work depend on the qualification of personnel.

Outcome dimension. The outcome of a service is the finished service order which results from the process carried out by the resources. The outcome dimension of the model therefore comprises therefore the set of all service orders and their status as they are fulfilled. This status contains different pieces of information: the degree of difficulty of the order, executed changes and revisions, and the order processing time. This information is used to calculate performance measures for the alternatives. In order to evaluate the alternatives, six performance measures in the three dimension time, cost and quality have been implemented into the model: Average time in system and average waiting time per order, average cost of personnel per order and personnel utilisation, first hit rate and variation of average time in system.

The conclusions based on the outcome of a comparison of cooperation alternatives may differ depending on the goals a company pursues in a cooperation.

3.2   Implementation of the Model Using CPN Tools

The dimensions of the conceptual model have to be translated into a CPN model. Based on the work of [38] the following analogy is presented: Fig. 2 maps the three dimensions process, structure, and outcome and connects resources, process steps, and orders to work items and activities. Using Petri net notation (*italics in brackets*), orders and resources are represented by tokens and process steps by transitions. A work item corresponds to an enabled transition and an activity to the actual firing of a transition.

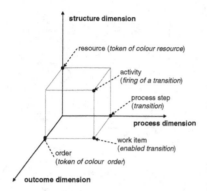

Fig. 2. Conceptual dimensions and correspondents in Petri net notation

To transfer the conceptual model into a simulation model, the software CPN Tools of the University of Aarhus was used [8]. All aspects of the overall model were mapped into a directed graph consisting of places, transitions, arcs, and markings. Place and transition names have been chosen according to the usual terminology in the area of application. The net consists of 11 pages, 47 places, and 31 transitions. Since the detailed presentation of all model elements is beyond the scope of this paper, the implementation of some of the main concepts is presented below.

Process modelling. First of all, the entire process was represented in coloured Petri net notation using places, transitions, and arcs. The main flow of the process is as follows: Requests for service orders arrive and get received. Then, the problem and the corresponding service measures have to be clarified and the request develops into an order. This order then is planned and controlled. In case changes are needed, the order moves into a loop and back to the planning process step. After successful order fulfilment the order is confirmed and invoiced.

The hierarchical concept described in section 3.1 was realised using the technique of substitution transitions. A substitution transition represents another page (subpage) in the net which starts and ends at the same places as the substitution transition, but contains a more detailed process comprising several other transitions and places. As depicted in Fig. 3, the process maps ten substitution transitions.

All time delays for process steps use random distribution functions, because the duration of the process steps varies. Since CPN Tools can handle time values only as integers, the real values of stochastic functions have been rounded, e.g. `@+round(normal(48.0,20.0))`. To be able to easily change the unit of the model time the function `myround (r) = round (r * d)` has been defined [17]. If the value of d is set to 1.0, the unit of the model time is set to minutes.

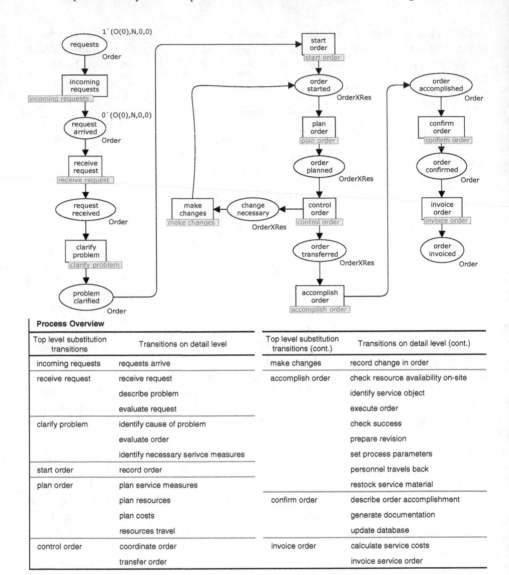

Fig. 3. Process overview

Order modelling. The primary markings in the net are orders that pass through the process. The colour set of orders (Order(i),j,k,l) is defined as follows:

```
colset Order = index Order with 0..n;
colset B = with N|Ch|Nr|Chr; with (N=normal, Ch=change,
Nr=normal revised, Chr=change revised)
colset Order = product A*B*INT*INT timed;
```

Initially, each order (respectively request, since any order enters the process as a client's request and develops into an order later on) receives a one-to-one order number (position k in Order(i),j,k,l), see first transition in Fig. 4).

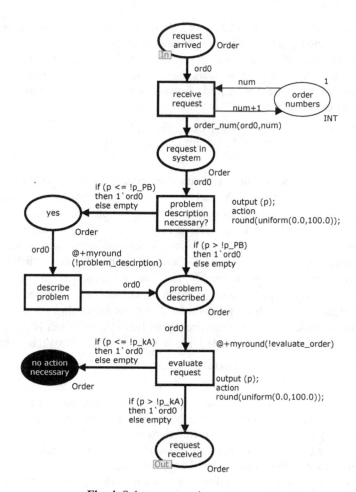

Fig. 4. Subpage `receive_request`

This is an important requirement since, in order to model several concurrent processes, order copies (with the same order number) are created. When concurrent processes become synchronous, guards like

```
getn(ord) = getn(ord_copy);
with getn(Order(i),j,k,l) = k;
```

ensure that the tokens from two input places of a transition are copies of the same order before the transition is enabled. For a more detailed description of branching and synchronisation patterns see [37].

Orders are indexed to represent different degrees of difficulty of service assignments (position i in `Order(i),j,k,l`). Depending on this degree of difficulty, different categories and quantities of resources have to be employed and fulfilment of an order takes different amounts of time.

Every order also carries information about its status regarding changes and additions in the service assignment, as well as a notation to account for conducted revisions (position j in Order(i),j,k,l), since certain process steps only apply to orders not having passed through any changes or revisions.

The last integer value (position l in Order(i),j,k,l) stores the order processing time. When the request enters the system, the corresponding time value is stored in position l. When the order is finished and invoiced, the order processing time is calculated by the function

```
fun order_time_end(Order(i),j,k,l) =
(Order(i),j,k,(intInf.toInt(time())-l));.
```

Resource modelling. The resource allocation is modelled with a separation between personnel and service material[1]. The places containing personnel and service material are made available on different subpages by using fusion places. Fusion places are duplicates of places which make the places accessible at different pages or different locations on one single page. Resource places (both personnel and service material) contain different colours of resources. Depending on the degree of difficulty of an order, different multi-sets of resources need to be allocated. A multi-set is a combination of several appearances of different colours, e.g. 1 service engineer + 2 service technicians + 3 client technicians. To be able to allocate personnel and service material multi-sets to an order, resource lists (Perslist, Matlist) have been defined and these have been added into a product colour set with the order:

```
colset OrderXRes = product Order*Perslist*Matlist
timed;
```

The process of resource allocation takes place on the subpage plan_order (see Fig. 4). After the service measures have been planned, both personnel and service material are assigned to the order[2]. The functions matmenge(Order(i),j,k,l) and persmenge(Order(i),j,k,l) calculate the resource multi-sets depending on the degree of difficulty of the order. These multi-sets are added to the two lists in the colour set OrderXRes. The following transition plan_service_measures determines the contents of the necessary service measures.

The transition plan_resources then splits the process into two parallel segments. On the right hand side, the planning process is finished by planning costs. The order is then ready to enter the control process step. On the left hand side is a copy of the order, and its allocated resources map the travelling done by personnel and the shipping times of spare parts, respectively. This transition is enabled only if the place no_change_necessary contains a copy of the same order. This is

---

[1] Since both processes are similar, resources could have been modelled in coloured Petri nets using only one place containing personnel and material as tokens of different colours. In the field of industrial services, however, personnel and material resources are commonly separated and treated quite differently. Thus, to keep comprehensibility of the model for users high, this additional complexity of having two separate places and sub-processes has been included.

[2] For both resource places, initialising transitions have been added in order to be able to use the initial values as an input parameter. This is necessary because global reference variables cannot be used directly in initial marking inscriptions.

important because changes may lead to different resource allocation. The existing allocation can only be changed (and respective tokens removed from the place `resources_planned`) when the resources are not on-site or on their way there yet. Thus, whether or not changes are necessary is mapped in the subpage `control_order`. In case of a change, the corresponding token is removed from the place `resources_planned` (which is a fusion place with another instance on the subpage `control_order`). If no change is necessary, a token is created in the place `no_change_necessary` on the subpage `control_order` (this is also a fusion place with instances on both subpages) and the transition `resources_travel` is enabled. The output place `resouces_on_site` is again a fusion place with its second instance on the subpage `accomplish_order`. The time for the transition `resources_travel` is calculated by comparing the times for personnel travel and spare parts shipment and using the larger value for both.

Parameterisation. To be able to initialise the model with company-specific starting and boundary conditions, these factors have been parameterised in the model by using global reference variables. That means that a company can specify the basic conditions of its service process by defining different values for the global reference

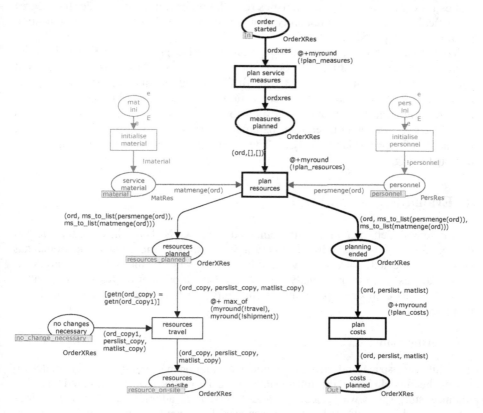

Fig. 5. Subpage `plan_order`

Parameterisation

Initialisation parameters	Cooperation alternative parameters
Probability	probability of revision necessary
probability of problem decription necessary	structure and amount of available personnel
probability of no action necessary (request is directly discarded)	personnel consumption per order
probability of no order necessary (request is not turned into an order)	cost of personnel
probability of other kind of order necessary (no service order)	travel time personnel
probability of changes necessary	time to accomplish order
probability of different degrees of difficulty of orders	
Time	
model time unit	
length of simulation runs	
time for system start-up (discarded for performance measurement)	
interarrival time of requests	
time delays for 21 different process steps	

Fig. 6. Parameterisation of the model with initialisation and cooperation alternative parameters

variables. These parameters comprise time values for the different process steps and their stochastic functions (e.g. a Poisson arrival of requests) as well as probability values for certain process characteristics (e.g. the probability that an order has degree of difficulty of 2 or the probability that revisions occur). The list of initialisation parameters is depicted in Fig. 6.

To account for the concept of cooperative service provision, the model has to change with regard to different cooperation alternatives. This means that different parts of the process may be provided by different partners and thus may differ with respect of time, cost, and quality (see section 3.1). Therefore additional parameters that are influenced by a change in responsibility of certain process steps have been added. These cooperation-relevant parameters are used to configure cooperation alternatives for performance analysis. The list of cooperation alternative parameters is depicted in Fig. 6.

4 Evaluation

The coloured Petri net model implemented using CPN Tools has been evaluated concerning its correctness by verification (i.e. testing if the conceptual model has been implemented correctly), validation (i.e. testing if the model behaves as expected), and performance analysis of two cooperation alternatives (i.e. evaluating alternatives by comparing performance measures). Details of the evaluation will now be presented.

4.1 Verification

The full state space and strongly connected components graph were calculated to verify the model. Since the timed state space is a subset of the untimed one, the untimed state space was calculated and assessed. In addition, stochastic functions which model probabilities have been replaced by variables from a small colour set that are assigned in bindings. The subpage modelling the arrival of orders and the two transitions initialising the resources have also been neglected for the state space analysis, since they are only needed for simulation. The state space analysis was carried out for one case in isolation, because competition between cases for resources is only relevant for performance analysis [38] and is considered in the validation process. The full state space and strongly connected components graph consist of 550 nodes and 1009 arcs. More properties of the net which result from the state space analysis are presented below and summarised in Fig. 7.

Boundedness Properties

Place	Upper integer bound	Lower integer bound
41 order places	1	0
3 personnel places	35	31
3 service material places	60	51

Liveness Properties

Dead marking	Token
no_action_necessary	O(0)
no_order	O(1), O(2) or O(3)
other_order	O(1), O(2) or O(3)
order_invoiced	(O(1),N), (O(2),N), (O(3),N), (O(1),Ch), (O(2),Ch), (O(3),Ch), (O(1),Nr), (O(2),Nr), (O(3),Nr), (O(1),Chr), (O(2),Chr) or (O(3),Chr)

Home Properties

Initial marking is not a home marking.

Fairness Properties

No infinite occurrence sequences.

Fig. 7. State space analysis results

Boundedness properties. In state space analysis upper and lower bounds of all places are calculated, i.e. the maximum and minimum number of tokens of a certain colour that a place can contain. A coloured Petri net is bounded, if all places have an upper and lower bound [19].

The analysis shows that all 47 places of the net are bounded. All 41 places containing orders have an upper bound of 1 and a lower bound of 0, which corresponds to the fact that one single case is considered and thus these places cannot contain more than one token. The six places containing resources are split into two groups: personnel and service material, with three fusion places each. These places have an upper bound of 35 (personnel), 60 (service material), respectively, as specified in the starting conditions for the state space analysis. The lower bounds are 31 for personnel and 51 for service material, which corresponds to the maximum number of resources consumed by an order of the highest degree of difficulty. Thus, the net is bounded.

Home properties. A home marking is a marking to which it is always possible to return [19].

In the net presented here, the initial marking is not a home marking. This is correct because once the order is received and has entered the fulfilment process, it is not possible to return to the state of the incoming order.

Liveness properties. A net is live if a set of binding elements remains active, this means that every transition will be enabled at some point [19]. In a dead state on the contrary, no transition is enabled.

The state space shows 19 dead markings for the model, because the order is supposed to reach desired end states, i.e. the process of fulfilling an order comes to an end. All 19 dead states haven been transferred to the simulation and checked with CPN Tools: In one dead marking, the place no_action_necessary contains a token of the colour order, that has not yet been evaluated: Order(0). This marking refers to the case that during the process of dealing with an incoming request, it is seen that no action has to be taken and the request does not have to be followed. In three other dead markings, the place no_order contains one order: Order(1), Order(2) or Order(3). Similarly, the place other_order contains one order in another three dead markings: Order(1), Order(2) or Order(3). These markings represent the case that after requests have been evaluated and the corresponding problems and measures are clarified, it is possible that a request (of any degree of difficulty) does not have to be followed by an order, or it does not have to be treated as a service order but as some other kind of order. In both cases, no further action within the process of the industrial service provision is necessary. The remaining 12 dead markings correspond to tokens in the place order_invoiced, when the order is successfully fulfilled and finished. Since, for an order of any degree of freedom, there are four possible modes in which it can finish (N = normal, CH = changed, Nr = normal_revised, Chr = changed_revised, see section 3.2), there are twelve different dead markings. Thus, all 19 dead markings are desired end states of the process.

Fairness properties. Fairness shows how often different binding elements occur [19].

The model is structured so that no infinite occurrence sequences exist. Although there are two loops in the net (making changes and revisions), they only occur if the order has not gone through these loops before. Thus, it is correct that the net does not contain infinite occurrence sequences.

4.2 Validation

Although validation measures are presented after the description of the model, it is stressed that model validation is not a task to be performed at the end of model development, but rather accompanies the whole process of model development [2], [3], [22], [28].

To validate the model, a panel of 13 subject matter experts was involved in the development of the model from the beginning. Three workshops and several interviews with the experts have been organised. Joint development of the model, as well as a structured walkthrough of the conceptual model [22], was used to validate the logic and assumptions of the model. The use of common expressions in names of places and transitions, as well as the graphical representation and hierarchical structure of the model in CPN Tools have proven of high value to facilitate the communication with potential users and their understanding of the model.

Simulation was then used to assess the behaviour of the model [22]. Interactive and terminating simulation runs showed that the simulation results were consistent with perceived system behaviour.

These measures, together with the evaluation of the example of performance analysis and the experimental design described in the next sections, show that the model has been validated.

4.2.1 Performance Analysis

Purpose of the model presented here is to execute a prospective analysis of cooperative provision of industrial services. Thus, now that the model has been verified and validated, an example of this prospective performance analysis is presented.

The example compares three alternatives of a fictive company for providing industrial services in the Eastern European market. The alternatives correspond to different scenarios: In alternative 1 it is assumed that the company is not cooperating and thus sending out its own engineers and technicians. This leads to high travel times, but also accounts for short order fulfilment times and a low probability of revisions, since the company's personnel is highly qualified. Alternative 2 represents the case in which the company cooperates with a partner closer to the service sites but with less qualified personnel. Finally, alternative 3 represents the cooperation with a partner right in the market with lowest travelling times but longer accomplishment times and a higher probability of revisions because of less qualified personnel. Initialisation and cooperation alternative parameters have been chosen according to input from subject matter experts. The example has been chosen to be easily understandable to be able to check output for plausibility.

Ten samples (n = 10) were collected for each alternative. For the calculation of the performance measures, a start-up phase for each simulation run was neglected to eliminate errors due to transient behaviour of the system (e.g. faster processing times because of lower resource utilisation for the first orders entering the system). The length of the start-up phase was calculated according to [1] and [42]. For the performance measures time 95%-confidence intervals were constructed. The simulation results for average time in system and average cost of personnel are presented in Fig. 8.

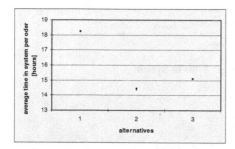 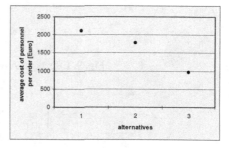

Fig. 8. Simulation results for average time in system and cost of personnel per order

The diagrams of these two performance criteria show that the conclusions drawn from the results depend on the goal a company is following and the corresponding rating of performance criteria. While the average cost of personnel per order is highest for alternative 1 and lowest for alternative 3, the average time in system is shortest for alternative 2. This shows that the entrepreneurial decision for one alternative is always linked to trade-offs between different objectives.

The example shows the effectiveness of the model in principle and the comparison of alternatives using performance analysis based on simulation.

4.2.2 Experimental Design

In order to gain more information about the model and the effects of the cooperation parameters on the performance measures, an experimental design was carried out. In this case, a 2^k factorial design was used [22], [31]. In this design all six cooperation parameters were varied systematically using two levels (- and +) for all factors, leading to $2^6 = 64$ design points. Main effects and two-factor interaction effects on two performance measures (average time in system and average personnel cost) and their 95% confidence intervals were calculated using 64 replications in the design points with five simulations runs each (320 runs in total). The results were used to reproduce expected system behaviour, as depicted in Figures 9 and 10.

Factors 5 and 6 (travel time and time to accomplish order) have the highest influence on the average time in system (see Fig. 9): Changing factor 5 from its – to its + level (i.e. a higher travel time) increases the time in system, while the same change in factor 6 (i.e. shorter time to accomplish order) decreases the time in system. Changes regarding available personnel, personnel cost, resource consumption, and the probability of revisions as well as two-factor interactions are less important compared to the effects of factors 5 and 6.

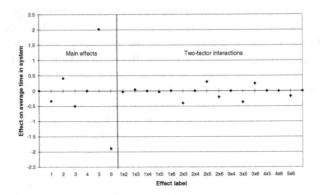

Fig. 9. Effects on Average Time in System

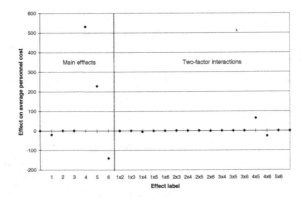

Fig. 10. Effects on Average Personnel Cost

A similar picture arises for the effects on the average personnel cost (see Fig. 10), with the difference that a change in factor 4 (cost of personnel) has the highest influence on the average personnel cost per order, as one would expect. In this case, the other main effects besides the ones of factors 4, 5, and 6 and the two-factor interactions are even less important.

The results show that the model represents a consistent behaviour since no unreasonable effects were produced and the main influencing factors are feasible.

5 Conclusions

The conceptual and simulation model of cooperative service provision presented in this paper reflects the major features of industrial services as well as their cooperative provision. It allows prospective analysis of cooperation alternatives by varying factors relevant to cooperation and comparing analysis results.

The use of coloured Petri nets as a modelling and simulation formalism provides several benefits: One the one hand, the graphical representation of the model leads to transparency, consistency, and conformity with user's expectations, so that understanding and acceptance of the model are facilitated. On the other hand, the maturity of the Petri net formalism and the availability of appropriate tools allow a direct implementation into computer models and their analysis.

It has been shown that cooperative provision of industrial services can be modelled, analysed, and simulated using coloured Petri nets and that prospective assessment of alternatives is possible. Verification and validation of the model has shown that it properly maps the cooperative provision processes of industrial services and shows consistent behaviour. The influencing factors identified in a 2k factorial design support these findings. Thus the goal of this work, to develop a CPN model capable of prospective analysis of cooperative provision of industrial services, has been achieved.

In future research the 2^k factorial design will also be used to investigate trade offs in regard to resource allocation or mechanisation, e.g. the deployment of different categories of personnel or the application of remote services.

Additional research would be helpful to further facilitate communication with model users, especially the connection to animation [43] and web based input/output facilities would be.

Since the application of simulation to the area of industrial services is still new, they still have a backlog of demand concerning the collection of precise and detailed collection of service process data, compared to production systems where simulation has already been established. But the more data will be available the more feasible a further integration of modelling and analysis of service provision and production of the corresponding investment goods will become.

References

1. Banks, J., Carson II, J. S., Nelson, B. L., Nicol, D.: Discrete-event System Simulation. Pearson Prentice Hall, Upper Saddle River, NJ (2001)
2. Banks, J.: Principles of Simulation. In: Banks, J. (ed.): Handbook of Simulation. Wiley, New York, NY (1998) 3-30

3. Bastian, M.: Modelle und Methoden in Problemlösungsprozessen. In: Luczak, H., Stich, V. (eds.): Betriebsorganisation im Unternehmen der Zukunft. Springer, Berlin (2004) 285-289
4. Bienzeisler, B., Meiren, T.: Trendstudie Dienstleistungen. IAO, Stuttgart (2005)
5. Bornheim, M., Stüllenberg, F.: Effizienz- und Effektivitätssteuerung von Kooperationen mit Hilfe der Balanced Scorecard. Controlling 14 (2002)4/5 283-289
6. Braun, J.: Veränderte Blickwinkel auf Unternehmen. In: Warnecke, H., Braun, J. (eds.): Vom Fraktal zum Produktionsnetzwerk. Springer, Berlin (1999) 43-90
7. Bruhn, M.: Qualitätsmanagement für Dienstleistungen. Springer, Berlin (2004)
8. CPN Tools: http://wiki.daimi.au.dk/cpntools/, accessed March 27, 2006
9. Corsten, H.: Dienstleistungsmanagement. Oldenbourg, Wien (2001)
10. DIHT (ed.): Services Going Abroad. Deutscher Industrie- und Handelstag, Berlin (2001)
11. DIN (ed.): Strukturmodell und Kriterien für die Auswahl und Bewertung investiver Dienstleistungen. PAS 1019. Beuth, Berlin (2002)
12. Donabedian, A.: Explorations in Quality Assessment and Monitoring. Vol. 1: The Definition of Quality and Approaches to its Assessment. Health Administration Press, Ann Arbor, Mich. (1980)
13. Girault, C., Valk, R.: Petri Nets for Systems Engineering. Springer, Tokyo (2003)
14. Gulati, R., Nohria, N., Zaheer, A.: Strategic Networks. Strategic Management Journal 21 (2000)Special Issue 203-215
15. Hlubek, W., Pötzsch, G., Kesting, J.: Certified Service. In: Luczak, H., Stich, V. (eds.): Betriebsorganisation im Unternehmen der Zukunft. Springer, Berlin (2004) 167-187
16. Hoeck, H., Kutlina, Z.: Status quo und Perspektiven im Service 2004. Verlag Klinkenberg, Aachen (2004)
17. Jansen-Vullers, M. H., Reijers, H. A.: Business Process Redesign at a Mental Healthcare Institute: A Coloured Petri Net Approach. In: Jensen, K. (ed.): Sixth Workshop and Tutorial on Practical Use of Coloured Petri Nets and the CPN Tools: Aarhus, Denmark, October 24-26, 2005. Computer Science Dept., Aarhus Univ., Aarhus (2005) 21-38
18. Jensen, K., Rozenberg, G.: High-level Petri Nets. Springer, Berlin (1991)
19. Jensen, K.: Coloured Petri Nets, Vol. 1-3. Springer, Berlin (1997)
20. Kallenberg, R.: Ein Referenzmodell für den Service in Unternehmen des Maschinenbaus. Aachen, Techn. Hochsch., Diss. (2002)
21. Laughery, R., Plott, B., Scott-Nash, S.: Simulation of Service Systems. In: Banks, J. (eds.): Handbook of Simulation. Wiley, New York, NY (1998) 629-644
22. Law, A. M., Kelton, W. D.: Simulation Modeling and Analysis. McGraw-Hill, Boston (2000)
23. Licht, T., Dohmen, L., Schmitz, P., Schmidt, L., Luczak, H.: Person-Centered Simulation of Product Development Processes Using Timed Stochastic Coloured Petri Nets. In: Geril, P. (ed.): Proceedings of the European Simulation and Modelling Conference, ESM'2004, October 25-27, 2004, Paris. EUROSIS-ETI, Ghent, Belgien (2004) 188-195
24. Luczak, H., Drews, P. (eds.): Praxishandbuch Service-Benchmarking. Service Verlag Fischer, Landsberg (2005)
25. Luczak, H., Winkelmann, K., Hoeck, H.: Internationalisierung von industriellen Dienstleistungen. In: Bruhn, M., Stauss, B. (eds.): Internationalisierung von Dienstleistungen. Gabler, Wiesbaden (2005) 389-413
26. Merkle, M.: Bewertung von Unternehmensnetzwerken. St. Gallen, Univ., Diss. (1999)
27. Mjema, E.: A Simulation Based Method for Determination of Personnel Capacity Requirement in the Maintenance Department. Aachen, Techn. Hochsch., Diss. (1997)
28. Naylor, T. H., Finger, J. M.: Verification of Computer Simulation Models. Management Science (1967)2 B-92 - B-101

29. Peterson, J. L.: Petri Net Theory and the Modeling of Systems. Prentice-Hall, Englewood Cliffs, NJ (1981)
30. Pritsker, A. A. B.: Principles of Simulation Modeling. In: Banks, J. (eds.): Handbook of Simulation. Wiley, New York, NY (1998) 31-51
31. Robinson, S.: Simulation: The Practice of Model Development and Use. Wiley, Chichester (2004)
32. Schlick, C., Reuth, R., Luczak, H.: A Comparative Simulation Study of Work Processes in Autonomous Production Cells. Human Factors and Ergonomics in Manufacturing 12 (2002)1 31-54
33. Seel, C.: Visuelle Simulation von Dienstleistungsprozessen. Saarbrücken, Univ., Diss. (2002)
34. Sontow, K.: Dienstleistungsplanung in Unternehmen des Maschinen- und Anlagenbaus. Aachen, Techn. Hochsch., Diss. (2000)
35. Sydow, J.: Network Development by Means of Network Evaluation? Human Relations 57 (2004)2
36. Zhou, M., Venkatesh, K.: Modeling, Simulation and Control of Flexible Manufacturing Systems. World Scientific, Singapore (1999)
37. van der Aalst, W. M. P., ter Hofstede, A. H. M., Kiepuszewski, B., Barros, A. P.: Workflow Patterns. Distributed and Parallel Databases 14 (2003)3 5-51
38. van der Aalst, W. M. P.: Business Process Management Demystified. Department of Technology Management, Eindhoven University of Technology, Eindhoven (2003)
39. van der Aalst, W. M. P.: Putting Petri Nets to Work in the Workflow Arena. In: van der Aalst, W. M. P., Colom, J. - M., Kordon, F., Kostis, G., Moldt, D. (eds.): Petri Net Approaches for Modelling and Validation. LINCOM Europa, München (2002) 125-143
40. VDI (ed.): Richtlinie 2893: Auswahl und Bildung von Kennzahlen für die Instandhaltung, VDI-Handbuch Betriebstechnik, Teil 4 - Betriebsüberwachung/Instandhaltung. Beuth, Berlin (2003)
41. Wang, J.: Timed Petri Nets. Kluwer Academic Publishers, Boston (1998)
42. Welch, P. D.: The Statistical Analysis of Simulation Results. In: Lavenberg, S. S. (ed.): Computer Performance Modeling Handbook. Academic Press, San Diego (1983) 267-329
43. Westergaard, M., Lassen, K. B.: Building and Deploying Visualizations of Coloured Petri Net Models Using BRITNeY Animation and CPN Tools. In: Jensen, K. (ed.): Sixth Workshop and Tutorial on Practical Use of Coloured Petri Nets and the CPN Tools: Aarhus, Denmark, October 24-26, 2005. Computer Science Dept., Aarhus Univ., Aarhus (2005) 119-135
44. Winkelmann, K.: Application of Coloured Petri Nets in Cooperative Provision of Industrial Services. In: Jensen, K. (ed.): Sixth Workshop and Tutorial on Practical Use of Coloured Petri Nets and the CPN Tools: Aarhus, Denmark, October 24-26, 2005. Computer Science Dept., Aarhus Univ., Aarhus (2005) 285-300
45. Wise, R., Baumgartner, R.: Go Downstream. Harvard Business Review 77 (1999)5 133-144
46. Zahn, E., Stanik, M.: Wie Dienstleister gemeinsam den Erfolg suchen. In: Bruhn, M., Stauss, B. (eds.): Dienstleistungsnetzwerke. Gabler, Wiesbaden (2003) 593-612

Can I Execute My Scenario in Your Net? VipTool Tells You!

Robin Bergenthum[1], Jörg Desel[1], Gabriel Juhás[2], and Robert Lorenz[1]

[1] Lehrstuhl für Angewandte Informatik
Katholische Universität Eichstätt-Ingolstadt, Eichstätt, Germany
name.surname@ku-eichstaett.de
[2] Faculty of Electrical Engineering and Information Technology
Slovak University of Technology, Bratislava, Slovakia
gabriel.juhas@stuba.sk

Abstract. This paper describes the verification module (the VipVerify Module) of the VipTool [4]. VipVerify allows to verify whether a given scenario is an execution of a system model, given by a Petri net. Scenarios can be graphically specified by means of Labeled Partial Orders (LPOs). A specified LPO is an execution of a Petri net if it is a (partial) sequentialization of an LPO generated by a process of the net. We have shown in [2] that the executability of an LPO can be tested by a polynomial algorithm. The VipVerify Module implements this algorithm. If the test is positive, the corresponding process is computed and visualized. If the test is negative, a maximal executable prefix of the LPO is computed and visualized, together with a corresponding process and the set of those following events in the LPO which are not enabled to occur after the occurrence of the prefix. Further, the VipVerify Module allows to test in polynomial time whether a scenario equals an execution with minimal causality. A small case study illustrates the verification of scenarios w.r.t. business process models.

1 Introduction

Specifications of distributed systems are often formulated in terms of scenarios. In other words, it is often part of the specification that some scenarios should or should not be executable by the system. Given the system, a natural question is whether a scenario can be executed. Answering this question can help to uncover system faults or requirements, to evaluate design alternatives and to validate the system design.

There are basically two possibilities to express single executions of distributed systems, namely as sequences of actions (that means as totally ordered sets of action names) or as partially ordered sets of action names. Since sequences lack any information about independence and causality between actions, we consider executions (and scenarios) as partially ordered sets of action names in case of distributed systems.

There exist several software packages, developed at universities or software companies, which support the design and verification of distributed systems based on scenarios. Some of them allow to compute the unfolding of a distributed system (given as a Petri net, a communicating automaton or a process algebra) in order to run LTL and

S. Donatelli and P.S. Thiagarajan (Eds.): ICATPN 2006, LNCS 4024, pp. 381–390, 2006.

CTL model checking algorithms on this unfolding (the tool PEP and the Model Checking Kit, [14, 15, 16]). Other tools use message sequence charts (MSCs) or their extension to live sequence charts (LCSs) to describe scenario-based requirements. These are used to guide the system design (the tool Mesa or the Playengine, [17, 18, 19, 20]), for test generation and validation (the tool TestConductor integrated into Rhapsody, [21, 22]), or for the synthesis of SDL or statecharts models (the tool MSC2SDL, [23, 24]). In [25, 26] a verification environment is described in which LSCs are used to express requirements that are verified against a statemate model implementation, where the verification is based on translating LSCs into automata.

Up to now, there exists no tool support to verify a given scenario to be an execution of a distributed system. One reason might be that there were no efficient verification algorithms so far for this problem. In case a scenario is given as a labeled partial order (LPO) over the set of possible actions (events) and the distributed system is given as a (place/transition) Petri net, we presented in [2] a polynomial algorithm.

The notion of *executions* of Petri nets is based on their non-sequential semantics given by occurrence nets and processes [12, 13]. Abstracting from the conditions in a process gives an LPO, called *run*. Runs capture the causal ordering of events. Events which are independent can occur sequentially in any order. Thus, adding order to a run still leads to a possible execution. For example, occurrence sequences of transitions (understood as labeled total orders) sequentialize runs. Generalizing this relationship, an LPO which (partially) sequentializes a run is an execution of the net. The process represented by such a run is called *corresponding to the specified LPO* in the following.

If a specified LPO is an execution of a given Petri net, the mentioned algorithm computes a process corresponding to the LPO. In the negative case a maximal executable prefix of the LPO is computed as well as the set of those following events in the LPO, which are not enabled to occur after the occurrence of the prefix. We further deduced a polynomial algorithm to test if a specified LPO precisely matches a process w.r.t. causality and concurrency of the events in the specification, if this process represents a minimal ordering of events among all processes.

Actually, we implemented the above described algorithms as parts of the new Vip-Verify Module of the VipTool [3, 4, 8]. The algorithms are based on computing the maximal flow in a flow network [9]. While the maximal flow algorithm presented in [9] is only pseudo-polynomial in general, there came up strict polynomial algorithms running in cubic time (see e.g. [10]) and also faster (see [11] for an overview) during the last decades. Since the basic algorithm from [9] turns out to be strict polynomial (running in cubic time) in our special case, we started with an implementation of the algorithms based on this basic algorithm. Moreover, we added a graphical interface (the VipLpoEditor module) which allows the user to graphically specify scenarios of a given Petri net in terms of LPOs over the set of transition names of the Petri net.

The paper is further organized as follows: In Section 2 we present a description of the new modules of the VipTool. A simple case study illustrates the new functionalities in Section 3. Then, in section 4, we briefly describe how the new functionalities additionally fit into the existing validation and verification concept for business process models the VipTool supports. In Section 5 we present some performance results for the implemented algorithms. Finally, the conclusion outlines the future development.

2 Description of the New VipTool Modules

To support the new functionalities, the VipEditor provides three graphical submodules: In the existing VipNetEditor Petri net models of distributed systems can be designed. In the new VipLpoEditor the user can specify scenarios in terms of LPOs. Finally, processes (computed by the VipVerify Module) are visualized in the existing VipProEditor.

The VipNetEditor is only slightly revised compared to the last version of VipTool. Very briefly, it has the following main functionalities: Drawing and painting features can be used analogously as by any standard Windows application. Size, colors, fonts can by easily changed by the user for all draw elements such as places, transitions, arcs, labels etc. Furthermore, all standard editing features such as select, move, copy, paste etc. are implemented. Beyond that, for example automatic alignment and click-and-drag-points of net arcs are supported. Usual token game simulation is also a part of the VipNetEditor. Figure 1 shows a screen-shot of the VipNetEditor with an example of a simple Petri net model of a business process.

Given a Petri net in the VipNetEditor, the user may take advantage of the VipLpoEditor. Clicking the appropriate button splits the screen and the VipLpoEditor is available. A grid makes drawing the LPO easy. An arc is automatically added between two

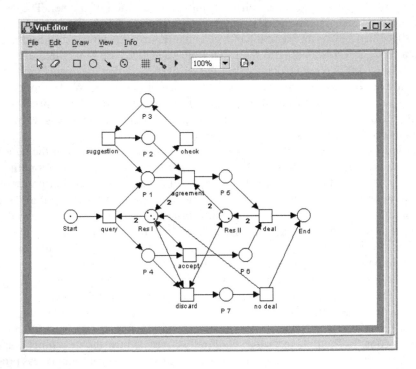

Fig. 1. Screenshot of the VipNetEditor, including an example net, which is explained later in a case study

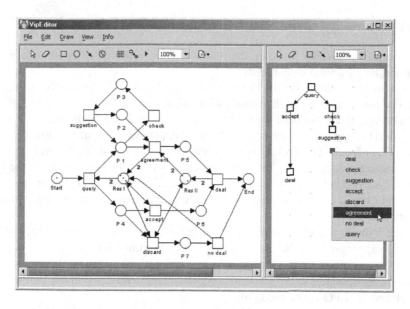

Fig. 2. Screenshot of the VipLpoEditor, showing the popup menu for adding a new node

nodes arranged on top of each other. For each added node, the related transition is chosen over a popup menu. For clarity, only the skeleton arcs of an LPO are drawn. Figure 2 shows a screen-shot of the VipEditor consisting of the VipNetEditor and the VipLpoEditor.

The last building block appears by starting the VipVerify Module to calculate if the LPO drawn in the VipLpoEditor is executable in the Petri net given in the VipNetEditor. The VipVerify Module distinguishes between the two following cases:

- If the LPO is executable, then a process corresponding to the LPO is computed by the VipVerify Module and is visualized in the VipProEditor. Moreover, then the VipVerify Module tests whether the LPO is minimal executable, i.e. whether there is another LPO with strictly less order between events which is also executable. If there is such another LPO, there are arcs in the given LPO representing an unnecessary ordering between events. Such arcs are highlighted.
- If the LPO is not executable, then the VipVerify Module computes a maximal executable prefix of the LPO and a process corresponding to this prefix. The process is visualized in the VipProEditor. In the VipLpoEditor, the prefix and the set of those events which are not concurrently enabled to occur after the occurrence of the prefix are highlighted by different colors.

Both cases will by described more precisely in the case study. The processes are visualized using the existing VipVisualizer module, which is based on the Sugiyama graph-drawing algorithm accommodated in [8]. Besides that, the objects of the visualized processes remain movable.

3 Functionality of the New VipTool Modules: A Case Study

In this section we briefly illustrate the functionality of the VipVerify Module and the VipLpoEditor by a simple case study.

The Petri net model of Figure 1 represents a possible business process of some company. The company handles their tasks with two resources (places *Res I* and *Res II*). After a prospective customer asks for a product (transition *query*) the business process divides into two concurrent sub-processes. In the upper one the company first checks on their offers or special conditions (transition *check*) and then makes offerings to the customer (transition *suggestion*). Yet they may agree (transition *agreement*). In the lower sub-process a parallel decision, for example on the solvency of the customer, has to take place. Only if that decision is positive, the customer and the company close a bargain (transition *deal*).

Figure 3 shows an LPO drawn in the VipLpoEditor, which represents a scenario that should be supported by the business process model. For this simple example it is easy to see that the specified scenario is minimal executable. By checking on the executability, a process corresponding to the scenario is computed and visualized in the VipProEditor. All nodes of the LPO are marked green.

Figure 4 shows another LPO which represents a scenario that should also be supported by the business process model. It is not an execution. In such a case, the VipVerify Module computes four other helpful contributions:

- A *maximal* executable prefix of the LPO. It is highlighted green and consists of the events *query*, *check* and *suggestion* in the example.

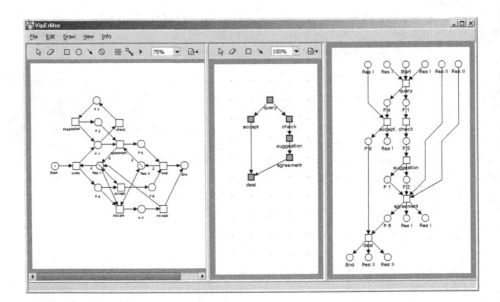

Fig. 3. The VipLpoEditor shows an executable LPO, and the VipProEditor shows a corresponding process of the Petri net

– The successor events which fail to be concurrently enabled after the occurrence of this prefix. They are highlighted red and are the events *agreement* and *discard* in the example.
– The place in the Petri net which does not carry enough tokens after the occurrence of the prefix to enable the red highlighted transitions. It is the place *Res II* in the example and highlighted red.
– A process corresponding to the executable prefix, which is visualized in the Vip-ProEditor.

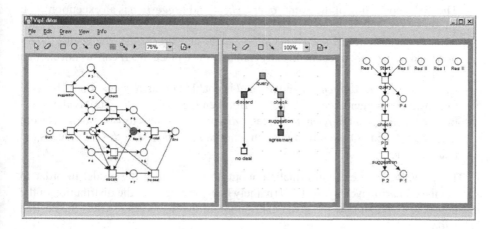

Fig. 4. Screenshot showing an LPO which is not executable in the given Petri net

For the interpretation of this information we have to describe the verification algorithm more precisely. For this, we say an LPO *is an execution w.r.t. a place p*, if this LPO is an execution of the given Petri net restricted to the place set containing only p. An inductive procedure verifies separately for every place p of the Petri net model, if the given LPO is an execution w.r.t. p. This is done by considering prefixes of the LPO increasing according to a calculated order over the set of nodes respecting the LPO. If for a place p this procedure stops before all nodes were considered, a prefix of the LPO is computed

– which is an execution w.r.t. p, and
– whose extension by the set of its direct successor nodes is not an execution w.r.t. p.

This prefix serves as input-LPO of this procedure for the next place. Thus, if there are more places preventing the execution of the LPO, the VipVerify Module computes that place (highlighted red in the VipNetEditor) with the smallest corresponding executable prefix among all places (resp. one of them). Those direct successor nodes of the prefix representing transitions which consume tokens from p are highlighted red in the VipLpoEditor (and therewith all events preventing the executability of the LPO w.r.t. p). Those direct successor nodes of the prefix representing transitions which do not consume tokens from p are highlighted yellow. Observe that possibly several of the red highlighted events are enabled after the occurrence of the prefix, but not all of them concurrently. Extending the prefix by such events would result in a bigger executable prefix, but with less information about the non-executability. The executable prefix corresponding to a place p depends on the calculated total ordering of the LPO-nodes. Nevertheless it is maximal w.r.t. the red highlighted nodes: A prefix corresponding to p computed w.r.t. another total ordering, which contains a red highlighted node, can not include the first prefix. Finally, a process corresponding to the prefix is visualized, showing a possible distribution of tokens among the pre- and post-conditions of the events of the prefix. In the example, that means:

- The prefix consisting of the events *query*, *check* and *suggestion* is an execution w.r.t. all places.
- After the occurrence of the events *query*, *check* and *suggestion*, the events *agreement* and *discard* are not concurrently enabled, since place *Res II* carries not enough tokens.
- Each of the events *agreement* and *discard* is enabled on its own and could be used to construct a bigger executable prefix. But then the user would get only the information that the event *agreement* (or, resp. *discard*) consumes too much tokens from place *Res II*, and not the information that the *combination of both events* needs too many tokens.

This gives the user clear information about how to change the model in order to support the scenario given by the LPO (namely how to reorganize the distribution of the resources), resp. about how the given distribution of the resources restricts the desired behavior.

4   Relating Old and New Functionalities of VipTool

In this section, we discuss, how the new implemented functionalities described in the previous two sections additionally fit into the existing validation and verification concept supported by VipTool. For this we briefly introduce this concept, but omit a detailed motivation, discussion and comparison to other approaches (here we refer to several publications from the last years ([3, 6, 7, 8]).

VipTool was originally developed at the University of Karlsruhe within the research project VIP [1] as a tool for modeling, simulation, validation and verification of business processes using Petri nets. It was implemented in the scripting language Python [8]. In [3] we presented a completely new and modular implementation in Java (using standard object oriented design) that allows to add extensions in a more flexible way.

The paper [7] proposes the following iterative validation procedure of Petri net models:

1. A requirement to be implemented is identified and formalized in terms of the graphical language of the model.
2. This formal specification is validated by distinction of those process nets that satisfy the specification from all other process nets. This way, the question *"what behavior is excluded by the specification?"* gets a clear and intuitive answer. The specification is changed until it precisely matches the intended property.

---

[1] Verification of Information systems by evaluation of Partially ordered runs.

3. The valid specification is implemented, i.e. new elements are added to the model such that the extended model matches all previous and the new specifications. Obviously this step requires creativity and cannot be automated. However, again by generation and analysis of process nets it can be tested whether the extended model satisfies the specifications (actually, when all runs are constructed, this test can be viewed as a verification). At this stage, other verification methods can be applied as well.

4. If some requirements are still missing, we start again with the first item, until all specifications are validated and hold for the designed model.

VipTool supports all these four steps. In particular, process nets representing partially ordered runs of Petri net models are generated. They are visualized, employing particularly adopted graph-drawing algorithms. Specifications can be expressed on the system level by graphical means. Process nets are analyzed w.r.t. these specified properties. The distinction of process nets that satisfy a specification is supported. For the test phase, the simulation stops when an error was detected.

The new functionalities now complement the second and third step of the above described validation procedure as follows: First, for complex Petri nets it can be (too) time consuming to construct all processes. In such cases it is helpful to have the possibility not to check on all the processes by unfolding the Petri net, but to directly test a particular scenario to be a possible execution of the Petri net or not. Second, the user now can specify concurrency of events. If an LPO representing a desired behavior turns out to be not executable, then the user gets detailed information about the reasons by visualizing the first state of the system which does not enable a set of concurrent events of the LPO. This facilitates the creative step of changing the specification in the second step as well as changing the model in the third step. Finally, in the third step the user now can verify particular concurrent runs directly.

5 Experimental Results

In this Section we test the performance of the presented algorithms by means of experimental results for the example instances $(N_{1,n}, lpo_{1,n})$ and $(N_{2,n}, lpo_{2,n})$ shown

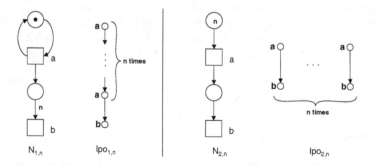

Fig. 5. Two place/transition Petri nets $N_{i,n}$, each together with an executable LPO $lpo_{i,n}$, $i = 1, 2$, dependent on the parameter $n \in \mathbb{N}$

Table 1. Results of the executability test

n	$(N_{1,n}, lpo_{1,n})$				$(N_{2,n}, lpo_{2,n})$			
	10	**50**	**100**	**500**	**10**	**50**	**100**	**500**
(A)	0.001	0.008	0.037	3.139	0.001	0.001	0.001	0.013
(B)	0.004	0.679	8.564	959.205	0.001	0.008	0.031	0.534
(C)	0.027	5.063	55.356	–	0.039	0.198	1.273	78.413
(D)	0.476	170.719	–	–	0.008	0.297	0.591	7.396

in Figure 5 with $n \in \mathbb{N}$ increasing. All experiments were performed on a Windows PC with 256 MByte of RAM and 1 GHz Intel Pentium III CPU. The times are measured in seconds. Table 1 shows the results for the following four procedures executed by the VipVerify Module: (A) Translation into the associated flow network, (B) Test whether $lpo_{i,n}$ is an execution of $N_{i,n}$, $i \in \{1, 2\}$, (C) Test whether $lpo_{i,n}$ is a minimal execution of $N_{i,n}$, $i \in \{1, 2\}$, and (D) Computation and Visualization of the corresponding process. The results show that the algorithms work better for LPOs with much concurrency between events. The weaker performance for LPOs with little concurrency is partially due to the quite general construction of the associated flow network, which could be further optimized.

6   Conclusion

We have presented the new VipVerify Module of the VipTool supporting scenario based verification of Petri net models of distributed systems and described its functionality within a small case study. VipVerify fits in the existing functionalities of the VipTool of supporting the step-wise design of business process models, employing validation of non-sequential specifications and verification of the model in each step. The further development of VipTool includes the following tasks:

- We plan to implement more efficient maximal flow algorithms underlying the presented verification algorithms.
- At present, VipTool only supports low level Petri nets. We plan to extend its functionalities to an appropriate restricted kind of predicate/transition nets.
- We are currently working on the synthesis of place/transition nets from given sets of LPOs. In [1] we present the first theoretical results which we plan to adapt for practical use and then to implement into VipTool.

We acknowledge the work of all other members of the VipTool development team, namely Niko Switek and Sebastian Mauser.

References

1. Robert Lorenz and Gabriel Juhás. *Towards Synthesis of Petri nets from Scenarios*. Accepted for ICATPN 2006.
2. Gabriel Juhás, Robert Lorenz and Jörg Desel. *Can I Execute my Scenario in your Net?*. LNCS 3536, pages 289-308, 2005.

3. Jörg Desel, Gabriel Juhás, Robert Lorenz and Christian Neumair. *Modeling and Validation with VipTool*. LNCS 2678, pages 380-389, 2003.
4. VipTool-Homepage. *http://www.informatik.ku-eichstaett.de/projekte/vip/*.
5. W.M.P. van der Aalst, J. Desel and A. Oberweis (Eds.). *Business Process Management*. Springer, LNCS 1806, 2000.
6. J. Desel. *Validation of Process Models by Construction of Process Nets*. In [5], pages 110-128.
7. J. Desel. *Model Validation - A Theoretical Issue?*. LNCS 2360, pages 23-42, 2002.
8. T. Freytag. *Softwarevalidierung durch Auswertung von Petrinetz-Abläufen*. Dissertation, Karlsruhe, 2001.
9. L.R. Ford, Jr. and D.R. Fulkerson. *Maximal Flow Through a Network*. Canadian Journal of Mathematics 8, pages 399–404, 1955.
10. A.V. Karzanov. *Determining the Maximal Flow in a Network by the Method of Preflows*. Soviet Math. Doc. 15, pages 434–437, 1974.
11. A. Goldberg and S. Rao. *Beyond the Flow Decomposition Barrier*. Journal of the ACM 45/5, pages 783–797, 1998.
12. U. Goltz and W. Reisig. *The Non-Sequential Behaviour of Petri Nets*. Information and Control 57(2-3), pages 125-147, 1983.
13. U. Goltz and W. Reisig. *Processes of Place/Transition Nets*. LNCS 154, pages 264-277, 1983.
14. C. Schröter, S. Schwoon and J. Esparza *The Model-Checking Kit*. LNCS 2676, pages 463-472, 2003.
15. http://theoretica.informatik.uni-oldenburg.de/ pep/
16. http://www.fmi.uni-stuttgart.de/szs/tools/mckit/
17. D. Harel, H. Kugler and A. Pnueli. *Synthesis Revisited: Generating Statechart Models from Scenario-Based Requirements*. LNCS 3393, pages 309-324, 2005.
18. http://www.wisdom.weizmann.ac.il/ playbook/
19. http://www.inf.uni-konstanz.de/soft/tools_en.php?sys=1
20. H. Ben-Abdallah and S. Leue. *MESA: Support for Scenario-Based Design of Concurrent Systems*. LNCS 1384, pages 118-135, 1998.
21. M. Lettrari and J. Klose. *Scenario-Based Monitoring and Testing of Real-Time UML Models*. LNCS 2185, pages 317-328, 2001.
22. http://www.osc-es.de/products/en/testconductor.php
23. F. Khendek and X.J. Zhang. *From MSC to SDL: Overview and an Application to the Autonomous Shuttle Transport System*. LNCS 3466, pages 228-254, 2005.
24. N. Mansurov. *Automatic synthesis of SDL from MSC and its applications in forward and reverse engineering*. Comput. Lang. 27/1, pages 115-136, 2001.
25. W. Damm and J. Klose. *Verification of a Radio-Based Signaling System Using the STATE-MATE Verification Environment*. Formal Methods in System Design 19/2, pages 121-141, 2001.
26. J. Klose and H. Wittke. *An Automata Based Interpretation of Live Sequence Charts*. LNCS 2031, pages 512-527, 2001.

EXHOST-PIPE:
PIPE Extended for Two Classes of Monitoring Petri Nets

Olivier Bonnet-Torrès[1,2], Patrice Domenech[3], Charles Lesire[1,2],
and Catherine Tessier[1]

[1] ONERA-CERT/DCSD, 2, avenue Édouard Belin, 31055 Toulouse cedex 4, France
[2] ENSAE-Supaero, 10, avenue Édouard Belin, 31055 Toulouse cedex 4, France
[3] at Onera-DCSD for a training period Feb-June 2005
{olivier.bonnet, charles.lesire, catherine.tessier}@onera.fr

Abstract. This paper presents the EXHOST-PIPE software: an extension of PIPE
– Platform Independent Petri net Editor – that complies with our own specifications. The new version supports the two kinds of Petri nets we have developed: particle Petri nets and plan Petri nets. EXHOST-PIPE also supports colours,
modularity, time, guards, differential equations... The Petri net player has been
extended and specific algorithms (such as an estimation process and specific reduction rules) have been implemented as plug-in modules.

1   Introduction

In the context of multiagent system monitoring, e.g. a crew-aircraft system or a mul-
tirobot system, we have designed two Petri net-based frameworks for (i) tracking and
predicting the (possibly conflicting) states of a human-controlled system and (ii) super-
vising and controlling a team of autonomous robots. As the question of the implementation arose, it was decided not to develop yet another tool from scratch, but to start from
an existing open-source software. A review of the tools available on the Internet led us
to select PIPE – Platform Independent Petri net Editor – for its clear and intuitive interface, its simulation module, its comprehensive UML documentation and the possibility
to modify the Java code quite easily.

The original PIPE software[1] [1] manages ordinary Petri nets. It applies the object-
oriented Model-Controller-View architecture pattern to implement several Petri net
analysis plug-in modules. The architecture separates the graphical rendering of the Petri
nets (the View), the processing of user requests (the Controller) and the Petri net structure itself including state and data (the Model). The model layer is also responsible
for PNML input/output management. With little modification the graphical interface
manages multiple Petri nets. As for the animation the original player is very simple and
fires either once at user click or a user-specified number of randomly chosen transitions.

The paper is organised as follows:

1. an explanation of the specifications will be provided first; the tool must be able to
   deal with the two special kinds of Petri nets that we have designed;

---

[1] http://pipe2.sourceforge.net

S. Donatelli and P.S. Thiagarajan (Eds.): ICATPN 2006, LNCS 4024, pp. 391–400, 2006.
© Springer-Verlag Berlin Heidelberg 2006

2. a presentation of the architecture; the tool is evolutive thanks to the distribution of object behaviour knowledge;
3. a description of some specific operations, embedded in the tool as modules; they provide specific algorithms to deal with special nets.

2   Specifications

The high level specifications were very simple: the PIPE-derived tool had to allow us to represent and play the two kinds of Petri nets we have designed for our applications and implement specific algorithms on these nets. The specifications are detailed hereafter.

2.1   Particle Petri Nets

Particle Petri nets [2] have been designed for hybrid numerical-symbolic situation monitoring and conflict detection in human-supervised systems.

The particle Petri net model (Fig. 1) is based on differential Petri nets and features: (1) *numerical places and transitions* that model the numerical behaviour of the system (using differential equations and guards) and (2) *symbolic places and transitions* that model the symbolic behaviour of the system (events and external actions).

The estimated state of a hybrid system is represented thanks to multiple tokens: *particles* $\pi^{(i)}_{k+1|k}$ – meaning particle number i estimated at time $k+1$ knowing the observation at time k – carry the numerical state vector of the system and evolve within the numerical places; *configurations* $\delta^{(j)}_{k+1|k}$ – meaning configuration number j estimated at time $k+1$ knowing the observation at time k – carry the symbolic state vector of the system and evolve within the symbolic places. A possible state of the system of Fig. 1

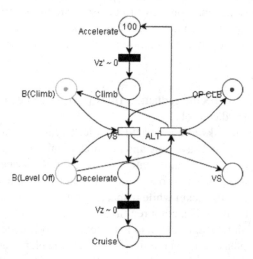

Fig. 1. The particle Petri net of a flight phase. It represents both the behaviour of the aircraft and the interactions between the crew and the autopilot.

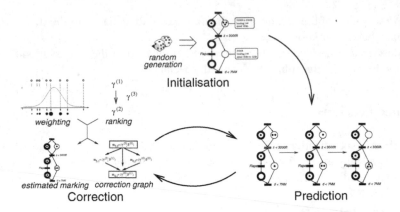

Fig. 2. The particle Petri net-based estimation process

is represented by marking $m_{i,j,l} = (\pi^{(i)}, \delta^{(j)}, \delta^{(l)})$, where $\pi^{(i)}$ is a possible state of the aircraft, $\delta^{(j)}$ is a possible mode of the autopilot and $\delta^{(l)}$ is a possible belief of the crew.

The particle Petri net-based estimation (Fig. 2) is a two-step process: a *prediction* of the system state by having the estimated marking evolve within the net according to specific firing rules and differential equations, and a *correction* of the predicted states according to an observation of the system. The correction consists in weighting/ranking the tokens to update the estimated marking of the Petri net and construct a *correction graph* used to detect inconsistent behaviours of the system [3].

Specifications: the PIPE-derived tool has to allow two kinds of places to be created *via* the GUI and dealt with with the Petri net player: (1) the symbolic places and (2) the numerical places that are linked to differential equations and whose tokens are numerical vectors. Receptivities and conditions have to be associated to transitions. As far as the simulation is concerned, the Petri net player has to take into account the specific firing rules for prediction and the whole estimation process has to be implemented.

2.2   Plan Petri Nets

Plan Petri nets and their specific reduction rules [4, 5] have been designed for robot team plan supervision and control.

A plan is a course of tasks to be realised in order to carry out a mission. Tasks are associated with the places and have time and resource requirements. Time requirements are start-time, finish-time and duration intervals. A team of service-performing robots is allocated to the mission. A plan is one possible way to complete the mission through actions with the given robotic team while respecting time and resource constraints. The mission plan Petri net (Fig. 3) is a coloured time Petri net[2] transcription of the plan.

For a complex mission the mission plan Petri net might be too detailed to provide an efficient interface for plan execution monitoring. Hence the plan is abstracted into a layered plan. The transformation is carried out by reducing some structures in the mission

---

[2] Start-times and finish-times correspond to a *t*-time Petri net. Durations are associated to the places and modify the behaviour by adding some properties of the *p*-timed Petri nets.

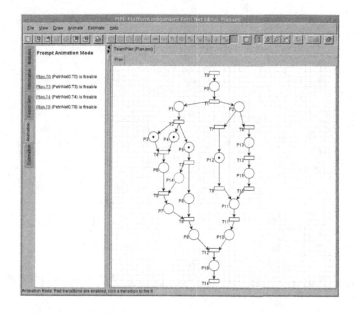

Fig. 3. A mission plan. Places correspond to tasks.

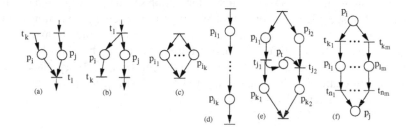

Fig. 4. Reduceable structures: (a) `source`, (b) `sink`, (c) `parallel`, (d) `sequence`, (e) `transfer`, (f) `choice`

plan Petri net (Fig. 4): `source` (agent arrival), `sink` (agent withdrawal), `sequence` (sequential activities), `parallel` (concurrent activities), `transfer` (transfer of a robot from a task to another) and `choice` (alternate activities).

The reduction yields a plan organised as a hierarchy and represented as a modular coloured time Petri net (Fig. 5). Each reduction step structures a part of the team: a reduced place corresponds to a subteam whose members are the robots or subteams corresponding to the places that have been reduced. Hence the reduction provides a team organised as a hierarchy that changes with the marking (Fig. 5, lower left corner). The dynamic team organisation allows the mission execution to be monitored at different levels of abstraction.

The control of individual robots must be performed at the local level. The plan Petri net is hence projected onto individual teammates: the leaf-places that involve the concerned robot are kept as well as all their ancestors in the hierarchy of places (Fig. 6).

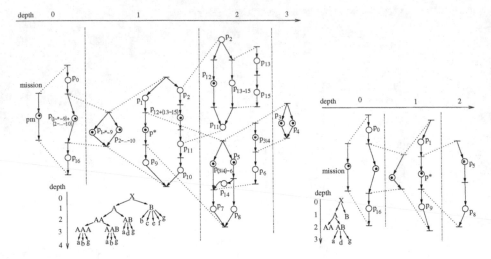

Fig. 5. Team plan Petri net and hierarchy associated with the current marking

Fig. 6. Team plan projected on agent d

Specifications: the PIPE-derived tool has to allow several nets to be drawn *via* the GUI and played at the same time. Defining transition fusion sets and place fusion sets has to be possible. The specific reduction rules have to be implemented: transition and place fusion sets have to be created automatically, special places (as the "transfer" places) have to be created and the different reduced Petri nets have to be played at the same time. Moreover, consistent and semantically valid names for the fused places have to be assigned automatically. The tool also has to be able to project a plan organised as a hierarchy onto any given individual.

3 Architecture

As the original version of Pipe [1], EXHOST-PIPE is organised according to a Model-View-Controller architecture. In this section we will focus on the model itself. Due to the very different natures of the Petri nets to deal with, some changes have been made. In particular the firing rules associated to our models are more complicated than in ordinary Petri nets and have been relocated within the transition objects themselves. For the same reasons the animator has been modified as well.

3.1 Tokens, Places and Transitions

The basic Petri net object in EXHOST-PIPE models a modular time Petri net. It decomposes into a set of coloured time Petri nets and a set of fusion sets (see Sect. 3.2). Each coloured Petri net has a set of places and a set of transitions that are specialised according to the type of Petri nets it models. The tokens allowed in the places are also specialised as they bear various types of information. The features that have been designed are the following:

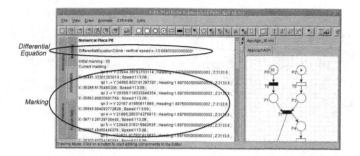

Fig. 7. A particle Petri net and the information attached to numerical place p_0: the differential equation is a Descent and the marking is detailed (the values of the vectors are given)

1. Ordinary token: the token corresponds to a coloured token. The allowed colour sets do not convey complex information yet.
2. Configuration: the token corresponds to a vector of discrete state variables. It models a configuration of the system under monitoring.
3. Particle: the token corresponds to a vector of continuous state variables. It models a numerical state of the system under monitoring.
4. Ordinary place: the place comes with a colour function and time specifications. The time specifications comprise a minimum duration, a maximum duration and a most-likely duration. If it pertains to a place fusion set, its marking is common with all places in the fusion set.
5. Symbolic place: the place represents the discrete state of the monitored system. It accepts configuration tokens.
6. Numerical place (Fig. 7): the place represents the way the particle parameters evolve. It is attached a differential equation.
7. Transfer place: the place models the transfer of an agent from one branch to another in a plan Petri net. It behaves exactly the same way as an ordinary place.
8. Ordinary transition: the transition is responsible for firing in a coloured time Petri net. The time specifications are given in the form of a firing time interval. If it pertains to a transition fusion set, it has to check whether the fusion set is enabled, *i.e.* if all transitions in the fusion set are enabled.
9. Symbolic transition: the transition fires to acknowledge event occurrences or to predict the possible external actions applied to the system.
10. Numerical transition: the transition is guarded with some particle parameter threshold values.

These elements when combined constitute a Petri net layer. As described above several layers may be linked in a modular Petri net.

3.2 Modular Petri Nets

In a modular Petri net [6, 7], each Petri net layer is a module. Several modules may be linked by sharing transitions and/or places. Each shared node constitutes a fusion set.

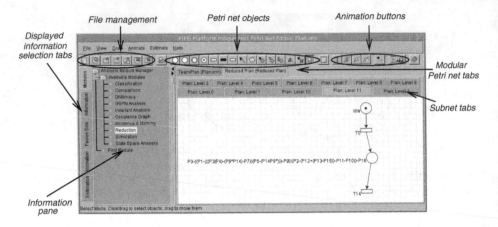

Fig. 8. Screenshot presenting the different parts of the interface. The modular Petri net shown is obtained from Fig. 3 after reduction.

The functioning rules are the following:

1. **Place fusion set:** if one place in the fusion set is marked, its marking is shared with the other fused places, *i.e.* all transitions downstream the places in the fusion set are enabled. These transitions enter in a conflict between modules. A place may appear in different fusion sets, therefore:
   (a) if it is marked, every place in all fusion sets can use the marking;
   (b) if it is not marked and only one other place is marked, only the places in the corresponding fusion set share the marking; it is not shared from one fusion set to another.
2. **Transition fusion set:** the transitions in the fusion set may fire only if all transitions in the set are enabled. The firing marks all downstream places in all modules and decreases the marking in all upstream places. A transition may belong to several fusion sets, in which case:
   (a) if two or more fusion sets are enabled, there is a conflict;
   (b) if only one fusion set is enabled, there is no conflict: it fires.

Hierarchical Petri nets [8] are represented by modular Petri nets [7]. A master place in a hierarchical Petri net is equivalent to a place whose upstream and downstream transitions are fused with the corresponding transitions within the slave subnet.

The user interface has been modified to take into account the modular Petri nets. The basic interface element is a modular Petri net, whose selection tabs call the different layers. All places and transitions can be added into fusion sets and a side tab in the left panel shows the properties of the fusion sets (Fig. 8).

3.3   Animation: Playing the Nets

EXHOST-PIPE features both Petri net models and manages fusion sets of places and transitions. Consequently, the original animator had to be modified: specific enabling conditions and firing rules are contained in the transitions themselves and fusion sets are defined within the modular Petri net object.

The modified animation process now:

1. checks whether each transition is enabled: when a transition is possibly enabled (according to the Petri net fundamental equation), the associated specific condition is checked (colours, incoming events, numerical condition...),
2. checks whether the fusion sets are enabled,
3. and finally follows one of the four implemented firing behaviours:
   (a) *prompt* mode: the animation asks the user to choose the transition (or fusion set) to fire;
   (b) *prompt on conflict* mode: if there is only one enabled transition (or fusion set), the animation makes it fire, otherwise the user is asked to choose the transition to fire;
   (c) *random* mode: the animation randomly draws an enabled transition (or fusion set) and makes it fire;
   (d) *timed* mode: the animation takes into account the time specifications: a time-stamp is randomly drawn for each transition firing or place duration interval, making use of a maximum likelihood parameter and a dilution factor. The transitions are drawn in ascending order of drawn firing times. If a firing time interval is violated the animation exits with a time violation message.

3.4 Estimation of Particle Petri Nets

All the algorithms/operations that are useful for the estimation process have been implemented in dedicated classes: prediction, weighting, resampling, ranking... The core of the estimation process is to perform a prediction of the Petri net marking while waiting for an observation to come.

The estimation process has thus been implemented as a multi-threading architecture. When a new observation comes, the Observer thread interrupts the Estimator thread to have the correction performed. Then the Estimator uses the time between two observations to predict the Petri net markings.

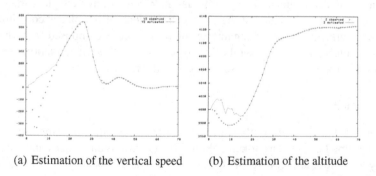

(a) Estimation of the vertical speed   (b) Estimation of the altitude

Fig. 9. Aircraft state estimation: the dots are the observations; the lines are the means of particles

Additionally, a frame has been added to display all the information and results that concern the estimation process. On this frame, one can read the current time of estimation, the latest observation, the particle equivalence classes, the correction graph, and the possible detection of a conflict.

The PIPE interface has been modified to allow observation files and connection sockets to be opened. Moreover an option has been created to log all the predicted and corrected tokens in a file so that the data could be analysed and plotted using an external tool (spreadsheets, plotters...) Figure 9 shows the result of estimating the altitude of an aircraft, from the Petri net of Fig. 1 and observations made on an aircraft simulator.

4   New Modules

In PIPE several plug-in modules offer services for Petri net analysis and comparison. Other modules have been implemented in EXHOST-PIPE for performing some operations on the new Petri net types we have introduced.

4.1   Marking Graph

A module has been created to compute the occurrence graph of a Petri net. The `Occurrence Graph` module is quite simple as the Petri nets involved in the applications have specific properties: (1) the estimation principle uses the `OccurrenceGraph` module to compute the set of reachable markings of the safe Petri net associated to the particle Petri net, and (2) the plan Petri nets do not contain loops and are bounded.

Therefore the occurrence graph is generally bounded and the algorithm is the classical one (no coverability tree method has been implemented yet).

4.2   Reduction

The reduction module implements the reduction rules as described in [4, 5]. It organises a coloured Petri net as a modular Petri net. The root layer features an *"idle"* place and another place that summarises the whole net (Fig. 8). At each algorithm step the reduced structure is placed on a different layer and its input and output nodes are linked with the corresponding nodes in the initial layer by fusion. When required the place that replaces the reduced structure is located at the geometrical mean of the places constituting the reduced structure. Meanwhile the name of the new place is composed from the names of the initial places by adding symbols to represent the type of reduction performed. This property produces a complete semantics for the reduced places.

4.3   Projection

The projection module is currently being implemented. It will enable the user to project the modular time Petri net on a token colour. The transitions and places that are not concerned by the colour are removed as well as the tokens that bear other colours (Fig. 6).

5   Conclusion

In this paper we have presented EXHOST-PIPE, an extension of PIPE for designing, analysing and playing ordinary, modular, coloured and time Petri nets as well as two types of nets we have developed: particle Petri nets and plan Petri nets. The tool interfaces according to PNML standards. Several plug-in modules implement analysis or transformation algorithms and allow operations on both new kinds of Petri nets. The user interface has been modified so as to help customising the displayed information.

A projection algorithm for coloured Petri nets is currently being developed. In the near future the tool will be modified to interface in real time with external systems (i.e. an aircraft simulator and a robot real-time executive) through socket connections. EXHOST-PIPE will be published as an open-source software afterwards.

References

1. Bloom, J., Clark, C., Clifford, C., Duncan, A., Khan, H., Papantoniou, M.: Platform Independent Petri-net Editor – final report. Technical report, Imperial College, London, UK (2003) *Under the supervision of W. Knottenbelt.*
2. Lesire, C., Tessier, C.: Particle Petri nets for aircraft procedure monitoring under uncertainty. In: ATPN'05, 26 th International Conference On Application and Theory of Petri Nets and Other Models of Concurrency, Miami, FL (2005)
3. Lesire, C., Tessier, C.: Estimation and conflict detection in human controlled systems. In: HSCC'06, 9th Workshop on Hybrid Systems: Computation and Control, Santa Barbara, CA (2006)
4. Bonnet-Torrès, O., Tessier, C.: From teamplan to individual plans: a Petri net-based approach. In: AAMAS'05, 4th International Joint Conference on Autonomous Agents and Multiagent Systems, Utrecht, The Netherlands (2005)
5. Bonnet-Torrès, O., Tessier, C.: From multiagent plan to individual agent plans. In: ICAPS'05, International Conference on Automated Planning & Scheduling, Workshop on Multiagent Planning and Scheduling, Monterey, CA (2005)
6. Christensen, S., Petrucci, L.: Towards a modular analysis of coloured Petri nets. In: ATPN'92, 13th International Conference on Application and Theory of Petri Nets, Sheffield, UK (1992)
7. Lakos, C.: From coloured Petri nets to object Petri nets. In: ATPN'95, 16th International Conference on Application and Theory of Petri Nets, Turin, Italy (1995)
8. Huber, P., Jensen, K., Shapiro, R.: Hierarchies in coloured Petri nets. In: ATPN'89, 10th International Conference on Application and Theory of Petri Nets, Bonn, Germany (1989)

ArgoSPE: Model-Based Software Performance Engineering*

Elena Gómez-Martínez and José Merseguer

Dpto. de Informática e Ingeniería de Sistemas, Universidad de Zaragoza
C/María de Luna, 1 50018 Zaragoza, Spain
{megomez, jmerse}@unizar.es

Abstract. Stochastic Petri nets (SPNs) have been proved useful for the quantitative analysis of systems. This paper introduces ArgoSPE, a tool for the performance evaluation of software systems in the first stages of the life-cycle. ArgoSPE implements a performance evaluation process that builds on the principles of the software performance engineering (SPE). The theory behind the tool, i.e. the underlying SPE process, has been presented in previous papers and consists in translating some performance annotated UML diagrams into SPN models. Therefore, ArgoSPE prevents software engineers to model with SPN since they are obtained as a by-product of their UML models. The design of the tool follows the architecture proposed by OMG in the UML Profile for Schedulability, Performance and Time specification.

Keywords: GSPN, UML, UML-SPT, software performance evaluation.

1 Introduction

Performance evaluation focusses on the analysis of the dynamic behavior of systems and the prediction of indices or measures such as their throughput, utilization or response time. Among the different formalisms used in this field, stochastic Petri nets (SPN) [1] have been proved as a very powerful one.

Concerning the performance evaluation of software systems, Software Performance Engineering (SPE) [2] proposes methods to evaluate them in the early stages of the development process. Being the Unified Modelling Language (UML) [3] a standard *de facto* for software engineers, the SPE community has adopted it to specify performance parameters in software designs, then defining the *UML Profile for Schedulability, Performance and Time Specification* (UML-SPT) [4].

Many approaches, see [5] for a survey, have arisen to derive from UML-SPT specifications, performance models based on a given modelling formalism, such as SPN. A step forward for the application of these approaches, and to contrast its maturity, should be the development of tools that support their proposals.

---

* This work has been developed within the projects: TIC2003-05226 of the Spanish Ministry of Science and Technology; and IBE2005-TEC-10 of the University of Zaragoza.

S. Donatelli and P.S. Thiagarajan (Eds.): ICATPN 2006, LNCS 4024, pp. 401–410, 2006.

The availability of these tools is a necessary condition for the industry in order to apply these proposals and to discover the feasibility of the SPE research field. The UML-SPT, being concerned about the problem, has proposed the main steps and modules that any SPE tool should follow, see Figure 1.

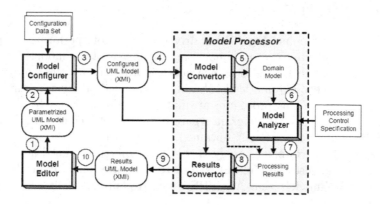

Fig. 1. Architecture proposed in the UML-SPT of OMG

Nevertheless, to the best of our knowledge, there not exist tools that fully implement the remarked SPE methodologies and follow the UML-SPT proposals. We guess that it is mainly motivated because the field is young. As a consequence, software performance prediction is still accomplished by means of *classical* performance evaluation tools such as [6, 7, 8], that introduce a gap between their methods and notations and those commonly used by software engineers. Good news are that prototypes and first attempts to implement parts of these approaches are arising in these last years.

In this paper, we present ArgoSPE, a tool for the performance evaluation of software systems in the first stages of the development process. It is based and implements most of the features given in our previous works [9, 10, 11] and gathered in [12] as a software performance modelling process. So, the system is modeled as a set of UML diagrams, annotated according to the UML-SPT, which are translated into Generalized Stochastic Petri Nets (GSPN) [13]. The UML diagrams used to obtain a performance model by means of ArgoSPE are those considered in our process: statemachines, activity diagrams and interaction diagrams. The use case diagram is taken into account in the process, but it has not been considered in ArgoSPE yet. The class and the implementation diagrams (components and deployment) are used to collect some system parameters (system population or network speed).

ArgoSPE has been implemented as a set of Java modules, that are plugged into the open source ArgoUML CASE tool [14]. It follows the software architecture proposed in the UML-SPT, see Figure 1. ArgoSPE has been used to model and analyze software fault tolerant systems [15] and mobile agents software [16].

The rest of the article is organized as follows. Section 2 presents the most interesting features of ArgoSPE, while section 3 focusses on its software architecture.

Section 4 surveys those tools prototypes, developed to analyze performance of software systems, based on the UML-SPT. Finally, conclusions and further works are exhibited in section 5.

2   ArgoSPE Features

From the user viewpoint, ArgoSPE is driven by a set of "performance queries" that s/he can execute to get the quantitative analysis of the modeled system.

We understand that a performance query is a procedure whereby the UML model is analyzed to automatically obtain a predefined performance index. The steps carried out in this procedure are hidden to the user. Each performance query is related to a UML diagram where it is interpreted, but it is computed in a GSPN model automatically obtained by ArgoSPE.

Moreover, the performance analyst, that has expertise in Petri net modelling and analysis, can use the GreatSPN tool to compute domain specific metrics using the GSPN models, that ArgoSPE generates automatically.

2.1   Queries in the Statechart Diagram

A statechart models the behavior of a class. Figure 2 depicts an example of statechart that models a Consumer class and a very simplified version of its GSPN translation.

- **State population.** This query computes the percentage of objects in each state. For example, in Figure 2, the 40% of the objects could be `Consuming` and the others `WaitingForProducer`.

  The query can be useful to detect saturated software processes or to know how an agent shares out its execution among different tasks (states). The state population is obtained by dividing the number of objects in the state among the mean number of objects that populate the class.

  For instance, in the state `WaitingForProducer`, the *State population* is computed by dividing the mean marking of place **p9** among the initial marking of the net in place **p1**.

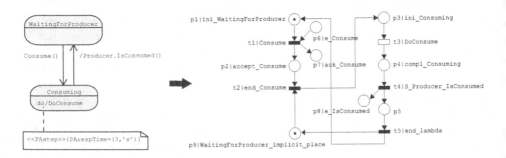

Fig. 2. Statechart and its corresponding GSPN

- **Stay time.** Represents the mean time that the objects of a class spend in each state. For each state, this value is computed by dividing the mean number of objects in it among its throughput, therefore, applying the Little's Law.

    In the example of Figure 2, the *Stay time* of the state WaitingFor-Producer represents the mean time that a Consumer spends waiting for consuming, and it is computed by dividing the mean marking of place p9 among the throughput of the transition t5.
- **Message delay.** When the sender and the receiver of a message reside in different physical nodes, this query calculates the time spent by the message to reach the receiver's node. This value is straightforward calculated by dividing the size of the message among the network delay (see section 2.2).

2.2   Queries in the Deployment and Collaboration Diagrams

The deployment diagram specifies the execution architecture of a system. Hardware resources are represented as nodes where software components can be deployed. Moreover, the physical network connections are modeled as relationships between nodes.

- **Network delay.** Calculates the network delay (bit rate) between two non adjacent hardware resources (nodes). Given a system configuration, the network delay is useful to find the node where a new software component could be deployed, i.e. the node that minimizes the delay of the component's messages.

The collaboration diagram is an interaction diagram that focusses on how objects exchange messages. It describes the behavior of the system in a specific context (scenario).

- **Response time.** For a given collaboration diagram (system scenario), this query computes its mean response time, i.e., the mean duration of a certain system execution.

2.3   Performance Annotations

ArgoSPE uses as input a UML-SPT annotated model, i.e. UML models have to explicitly include performance characteristics. These performance annotations, defined in the UML-SPT, are made by means of the UML extension mechanisms: *stereotypes* and *tagged values*.

The *stereotypes* specify the main performance characteristics of the UML model elements, while the *tagged values* specify the attributes of the stereotypes. As an example, see the performance annotation in Figure 2, where the stereotype PAstep means that DoConsume is a computation step, while the tagged value PArespTime models its response time.

The UML-SPT defines a *Tag Value Language* (TVL), a subset of the Perl language, that allows to specify complex and parameterized expressions in the tagged values.

The annotations supported by ArgoSPE are those necessary to compute the proposed performance queries, see Table 1.

Table 1. Performance annotations in ArgoSPE

Annotation	Stereotype	Tagged value	Unit	Model elements and Diagrams
Activity duration	PAstep	PArespTime	ms, s, m, h	Activities in the SC and AD
Probability	PAstep	PAprob	-	Transitions in SC & AD Messages in the Coll.
Size	PAstep	PAsize	b, B, kb, kB, Mb, MB	Messages in the SC and Coll.
Network speed	PAcommu- nication	PAspeed	bps, Bps, kbps, kBps, mbps, MBps	Deployment
Population	PAclosedLoad	PApopulation	-	Class in Class diagram
Initial state	PAinitial- Condition	PAinitialState	$true or $false	State in the SC and AD
Resident classes	GRMcode	-	-	Deployment

3   Software Architecture

ArgoSPE follows the architecture proposed in the UML-SPT [4], see Figure 1. The ArgoUML [14] CASE tool works as the Model Editor, while the ArgoSPE modules implement and coordinate the Model Configurer and the Model Processor (Model Convertor, Model Analyzer and Results Convertor) functions. Figure 3 depicts the ArgoSPE menu inside the menu bar of ArgoUML.

3.1   Model Editor

The Model Editor is used to create and modify performance-annotated UML diagrams.

ArgoUML allows to model and to annotate the UML diagrams involved in the translation process [9, 10, 11]. ArgoUML, as most CASE tools, exports UML models into XMI [17] files, allowing the standard exchange of information with another tools.

From the performance-annotated UML diagrams, ArgoSPE creates a parameterized XMI file, that will be an input for the Model Configurer. This XMI file contains the modeling and performance information of: the statecharts, describing the behavior of the classes in the system; the activity diagrams, specifying the activities of the statecharts; the deployment diagram, that gathers information about physical nodes location and network transmission speed; the class diagram with information about the system workload and the collaboration diagram.

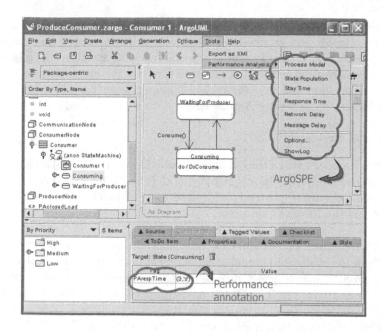

Fig. 3. ArgoSPE menu inside ArgoUML

3.2   Model Configurer

The Model Configurer functionality, see Figure 1, consists in converting a parameterized UML model in XMI format, into a configured UML model using a configuration data set. The main target is to substitute in the XMI file, the tagged values written in TVL with parameterized expressions that represent the performance annotations, for the equivalent evaluated expressions.

The first task is to parse the XMI file, obtaining a tree structure, called Document Object Model (DOM) [18]. This one is visited recursively from its root node into their children nodes to search for performance annotations. So, a list of XMI identifiers with known stereotypes is extracted. For each element in that list, a TVL expression is obtained, some of them with variables, that will be evaluated, then modifying the tree. At the same time, a symbol table is created containing the performance annotations. Finally, the tree is serialized to an XMI file.

Since the TVL expressions can contain variables, ArgoSPE prompts the user to choose a configuration file containing a configuration data set. An example of this kind of file is depicted in Figure 4.

```
#A very simple configuration file written in Perl
$value=5;
$value2=10;
$value3=($value<40)?100-$value2:100;
```

Fig. 4. A configuration file

ArgoSPE evaluates this file by invoking a Perl interpreter to get the actual value for variables and expressions. Then, for a performance annotation like `<<PAstep>>{PArespTime=($value,'s')}` the variable `$value` will be evaluated and replaced according to the configuration file (e.g. the one in Figure 4), so the evaluated expression will be `<<PAstep>>{PArespTime=(5,'s')}`. Multi-valued expressions are supported by ArgoSPE.

3.3   Model Processor

The Model Processor turns the configured model, obtained from the Model Configurer, into an analyzable model (GSPN model), analyzes it and returns the results. These tasks are respectively addressed by the Model Convertor, the Model Analyzer and the Results Convertor.

Model Convertor. The Model Convertor module encapsulates the translation process from the configured model into the target performance formalism. Therefore, in ArgoSPE it is a heavy process that implements the translation theory proposed in [9, 10, 11, 12]. The GSPN models are obtained in the GreatSPN file format [7].

The high-level algorithm implemented in the tool for the Model Convertor is illustrated in Algorithm 1. Note that it needs as inputs not only the configured XMI, as proposed in the UML-SPT, also the symbol table, that allows to speed up the translation process, but it increases the module coupling.

The lines 2 to 14 correspond to the translation process of the statecharts and its associated activity diagrams. Then, a GSPN, called `SysGSPN`, that models the whole system behavior is obtained by merging the Petri nets of the classes (line 26). The translation of the collaboration diagrams is described from lines 17 to 24. The result is a set of GSPNs, $Scenario_iGSPN$, each one modeling the scenario specified by the collaboration diagram i.

Model Analyzer. The Model Analyzer implements the performance queries described in sections 2.1 and 2.2. Concretely, the queries for the statecharts are computed using the `SysGSPN` net. While the *Response time* query for a collaboration diagram i is computed in the $Scenario_iGSPN$ net.

The Model Analyzer invokes the GreatSPN programs to get the answers to the queries. ArgoSPE currently uses the GreatSPN programs that implement analytical/numerical techniques. The use of GreatSPN simulation techniques will be considered to complement the results.

Results Convertor. The main function of the Results Convertor is to convert the results of the analysis back to the UML Model Editor, in a way that a software engineer can interpret them easily.

In the current version of ArgoSPE, the returned results are directly displayed by the Model Editor in a simple message window, then not directly in the UML models.

Algorithm 1. Model Convertor

Require: A configured XMI file and a symbol table
Ensure: A set of GreatSPN models (SysGSPN, $Scenario_i$ GSPN)
 1: UML diagrams ← XMI file
 2: **for all** class c ∈ Class diagram **do**
 3:     **if** c has Statechart **then**
 4:         node ← Locate node in Deployment diagram(c)
 5:         A ← Activity diagrams associated with c
 6:         **for all** Activity ac ∈ A **do**
 7:             pa ← Annotations associated with ac ∈ symbol table
 8:             acGSPN[j] ← TranslateToGSPN(ac,pa)
 9:         **end for**
10:         sc ← Statechart associated with c
11:         pa ← Annotations associated with sc ∈ symbol table
12:         scGSPN ← TranslateToGSPN(sc,pa,node)
13:         clGSPN[k] ← Merge(scGSPN, acGSPN[])
14:     **end if**
15: **end for**
16: SysGSPN ← Merge(clGSPN[])
17: **for all** class c ∈ Class diagram **do**
18:     C ← Collaboration diagrams associated with c
19:     **for all** Collaboration co ∈ C **do**
20:         pa ← Annotations associated with co ∈ symbol table
21:         coGSPN[i] ← TranslateToGSPN(co,pa)
22:         $Scenario_i$ GSPN ← Merge(SysGSPN,coGSPN[i])
23:     **end for**
24: **end for**

---

4   Related Work

A number of performance evaluation tools based on Petri nets have been developed in the last decade, such as Möbius [6], GreatSPN [7] or TimeNET [8]. But in this work, we only revise and compare those tools that focus in the SPE field.

DSPNexpress-NG [19], proposed by Lindemann et al., constitutes a framework that can evaluate both discrete-event systems specified as Petri nets and UML system models. It uses UML statecharts which are not annotated according to the UML-SPT, and transforms them into deterministic and stochastic Petri nets (DSPNs) to obtain numerical solutions.

Distefano et al. [20] developed a performance plug-in for ArgoUML. Following the UML-SPT, they focuss on use cases, deployment and activity diagrams and introduce an intermediate model, which is used to gather performance information. This intermediate model is transformed into SPN and analyzed with a web-based non-markovian Petri net tool.

Using formalisms different from Petri nets, Petriu and Shen [21] propose an algorithm to transform activity diagrams into LQN models. They obtain the XML files from existing UML tools, and change them by hand in order to add performance annotations to the different model elements. The tool of Gilmore and Kloul [22] uses ArgoUML to compile statecharts and collaboration diagrams through a process algebra language. D'Ambrogio [23] introduces a framework to automatically translate LQN models from annotated activity and deployment diagrams. Cortellessa et al. [24] propose a tool that in two phases gets a parameterized QN from use cases, sequence diagrams and deployment diagrams. Marzolla and Balsamo [25] transform annotated use case, deployment and activity diagrams into a discrete-event simulation model.

5 Conclusion and Further Work

Petri nets are recognized as a useful modeling paradigm for the performance evaluation of a wide range of systems. Nevertheless, most software engineers do not feel comfortable far from their pragmatic (non formal) modeling languages, such as UML. Moreover, engineers find easier and more productive to use only one modeling paradigm for all the project stages. Since ArgoSPE obtains GSPNs as a by-product of the software life-cycle, software engineers can use their UML models to assess system performance properties.

A number of new features can improve the tool: First, the more system properties assessed the more useful the tool become. New performance queries have to be implemented. Second, a standard format, PNML [26], could be the target file format, then gaining the possibility to use other Petri net analyzers.

Acknowledgments. The authors would like to thank Aitor Acedo, Borja Fernández, Luis Carlos Gallego, Álvaro Iradier, Juan Pablo López-Grao and Isaac Trigo for their work in the development of this tool. Finally, thanks to Simona Bernardi for her useful paper corrections and for her work testing the tool.

Availability. ArgoSPE is GNU soft. Download at: http://argospe.tigris.org.

References

1. Molloy, M.K.: Performance analysis using stochastic Petri nets. IEEE Transactions on Computers **31**(9) (1982) 913–917
2. Smith, C.U.: Perf. Engineering of Software Systems. Addison–Wesley (1990)
3. Unified Modeling Language Specification. (http://www.uml.org) Version 1.4.
4. UML Profile for Schedulabibity, Performance and Time Specification. (http://www.uml.org) Version 1.1.
5. Balsamo, S., Marco, A.D., Inverardi, P., Simeoni, M.: Model-Based Performance Prediction in Software Development: A Survey. IEEE Trans. Software Eng. **30**(5) (2004) 295–310
6. The Möbius tool (http://www.mobius.uiuc.edu/)
7. The GreatSPN tool (http://www.di.unito.it/~greatspn)
8. The TimeNET tool (http://pdv.cs.tu-berlin.de/~timenet/)
9. Merseguer, J., Bernardi, S., Campos, J., Donatelli, S.: A compositional semantics for UML state machines aimed at performance evaluation. (In: IEEE WODES'02) 295–302

410 E. Gómez-Martínez and J. Merseguer

10. Bernardi, S., Donatelli, S., Merseguer, J.: From UML sequence diagrams and statecharts to analysable Petri Net models. (In: ACM WOSP'02.) 35–45
11. López-Grao, J.P., Merseguer, J., Campos, J.: From UML Activity Diagrams to Stochastic Petri Nets: Application to Software Performance Engineering. (In: ACM WOSP'04) 25–36
12. Merseguer, J.: Software Performance Engineering based on UML and Petri nets. PhD thesis, University of Zaragoza, Spain (2003)
13. Ajmone Marsan, M., Balbo, G., Conte, G., Donatelli, S., Franceschinis, G.: Modelling with Generalized Stochastic Petri Nets. John Wiley Series (1995)
14. The ArgoUML project (http://argouml.tigris.org)
15. Bernardi, S., Merseguer, J.: QoS assesment of fault tolerant applications via stochastics analysis. IEEE Internet Computing (2006) Accepted for publication.
16. Merseguer, J., Campos, J., Mena, E.: Analysing internet software retrieval systems: Modeling and performance comparison. Wireless Networks 9(3) (2003) 223–238
17. XML Metadata Interchange (XMI) (http://www.omg.org)
18. Java Tecnology (http://java.sun.com)
19. DSPNexpressNG (http://www.dspnexpress.de)
20. Distefano, S., Paci, D., Puliafito, A., Scarpa, M.: UML Design and Software Performance Modeling. (In: ISCIS'04, vol. 3280 of LNCS.) 564–573
21. Petriu, D., Shen, H.: Applying the UML performance profile: Graph grammar-based derivation of LQN models from UML specifications. (In: TOOLS'02, vol. 2324 of LNCS.) 159–177
22. Gilmore, S., Kloul, L.: A unified tool for performance modelling and predicition. (In: SAFECOMP'03, vol. 2788 of LNCS.) 179–192
23. D'Ambrogio, A.: A model transformation framework for the automated building of performance models from UML models. (In: ACM WOSP'05.) 75–86
24. Cortellessa, V., et al.: XPRIT: An XML-Based Tool to Translate UML Diagrams into Execution Graphs and Queueing Networks. (In: IEEE QEST'04.) 342–343
25. Marzolla, M., Balsamo, S.: UML-PSI: The UML Performance Simulator. (In: IEEE QEST'04.) 340–341
26. PNML (http://www.informatik.hu-berlin.de/top/pnml/about.html)

Petriweb: A Repository for Petri Nets

R. Goud, Kees M. van Hee, R.D.J. Post, and J.M.E.M. van der Werf

Department of Mathematics and Computer Science
Eindhoven University of Technology
P.O. Box 513, 5600 MB Eindhoven, The Netherlands
rg@petriweb.org, k.m.v.hee@tue.nl, r.d.j.post@tue.nl, jmw@petriweb.org

Abstract. This paper describes Petriweb, a web application for managing collections of Petri nets.

When a collection of nets is large or has multiple users, it becomes difficult for users to survey the collection and to find specific nets. Petriweb addresses this issue by supporting arbitrary content-based filtering. Nets can be assigned properties with values of arbitrary types. Properties can be used in searching and are displayed in search results. Their values can be manually assigned by users or derived automatically by applying a tool. This allows server-side integration of Petri net analysis tools. Properties can also define translations to output formats, on which the user can invoke client-side viewers and analyzers. Petriweb supports communities: members submit nets and property definitions, community moderators approve them.

The paper discusses Petriweb's features and architecture, and how it relies on the proper application of a common document format for Petri nets, the Petri Net Markup Language (PNML).

1 Needs for Model Repositories

Many tools exist to support the modeling process. One tool may be better suited for designing models, another for analysis. Therefore, in a typical development or research environment, multiple tools are used in combination. This creates the need to work on the same models with different tools. This can be addressed by defining a standard file format that all tools can use.

Models need not only be shared by tools, but also by different users. For instance, a model may need to be reviewed by a colleague of the designer. We can send the model, and the colleague can open and use the model, but as soon as new versions appear, it becomes hard to make sure the right version is always used. Here, a shared location for the models is needed.

A shared collection of models also encourages users to reuse existing models or parts of them. This can be useful for different kinds of users. Designers, who employ modeling to describe and design systems, can use this to streamline the modeling process. Researchers and educators can build up collections of models used as illustrations, e.g. examples or counterexamples in proofs.

Most collections of models will be assembled in the context of a specific project, with a small group of participants. But collections can also be turned into

S. Donatelli and P.S. Thiagarajan (Eds.): ICATPN 2006, LNCS 4024, pp. 411–420, 2006.

company-wide or world-wide resources. In such cases, users will rarely be familiar with all the models in the collection, and collections can grow quite large.

Various solutions can be considered to support these needs. The simplest approach is to use a shared (web) location for models, but this does not offer any help in organizing them. An improvement is to use standard version control software, such as CVS or Subversion. However, these systems still require the users to be familiar with the organization of the material in terms of file names and directory structure. For larger collections, or an open-ended user base, this does not suffice. Additional facilities for searching and browsing will be required that employ knowledge of the model contents. Specialized repository software is needed that combines general file management with domain specific knowledge.

A specific domain to support at our university is process modeling with Petri nets. Many examples in course material and exercises are reused over the years; they are often recreated from memory or from paper. It is attractive to make such examples available in a shared repository, accessible by both students and teachers. Here, the need for both browsing and searching facilities is evident. Since users recognize Petri nets by their graphical representation, browsing the collection can only be supported with a graphical browser. In larger collections, users also need to filter the collection based on properties of the content. For instance, users may want to find examples of Petri nets that are bounded and contain a deadlock.

Petriweb is a web application for maintaining repositories of Petri nets. A Petriweb installation can host many different collections, each with their own administrators and users. The software is portable; different Petriweb installations can be easily installed.

2   Petriweb

2.1   Model Retrieval

Knowledge about the contents. The main focus of Petriweb is how to retrieve a specific Petri net in a collection. To be able to search for Petri nets by giving criteria, Petriweb needs to have knowledge about the Petri net.

An example query Petriweb needs to support: "Find the smallest net present that is live, unbounded and free-choice". Another example, from a developer's perspective: "Find a component with this interface and performing task P". Queries must also include metadata, e.g. the author or original publication of the net. We see that Petriweb must support different kinds of properties: structural properties, e.g. the number of places; behavioral properties, such as boundedness or liveness; and metadata.

Properties. Giving search criteria in terms of properties is a powerful manner of retrieving the correct Petri net. Petriweb supports different kinds of properties on Petri nets. These are:

- Metadata
- Application characteristics

– Structural and behavioral properties
– Transformations

By allowing metadata as properties in Petriweb, the user uploading the net can specify related information about it, such as when it was created, by whom, etc. This kind of property can narrow search results.

Application characteristics describe what the model is about: in what domain, for what purposes, etc.

Simple structural properties can be determined by simple programs, while more complex structural and behavioral properties such as free choice, liveness, boundedness, can be determined with existing Petri net analysis tools. Many such tools can be called as filters that take a net as input and produce results as output. Petriweb can incorporate this through *automated* properties. These are not specified by the user while uploading the net, but instead, computed automatically by invoking an XSLT stylesheet or an external command. This allows non-trivial structural and behavioral properties to be determined automatically; they can even be used in search criteria.

This mechanism can also be used to automate conversions from PNML to other file formats, e.g. the TPN format of Woflan [11]. Calling an analysis tool is often preceded by a transformation, but the results can also be presented to the user, e.g., as the input for a client-side tool. In supporting transformations, Petriweb becomes more than a repository: it functions as a mediator between different tools and formats.

View Petri Nets. Not only searching on properties is important, the graphical representation of a Petri net is also a great help in finding a specific Petri net. It is hard to describe a Petri net in words, such that others can find it; the graphical display of a net is much easier to recognize. Therefore, search results do not only list the Petri nets' names and properties, but also display them as diagrams. This allows a combination of searching and browsing.

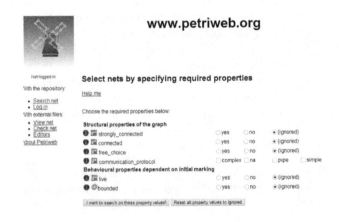

Fig. 1. Property based searching

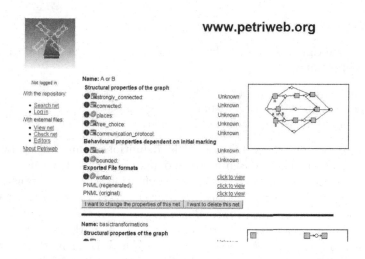

Fig. 2. Search results

2.2 Model Management

Addition. Petriweb has been designed for public and private use. Petri nets can be shared within a community. Anyone can register at Petriweb and upload their own Petri nets. To allow this, a standard file format is used for uploading Petri nets: the Petri Net Markup Language (PNML). Before adding the Petri net, Petriweb first checks its syntax against a PNML syntax definition [12]. The net is then parsed and stored.

Approval. While anyone can be registered at Petriweb and upload Petri nets to it, nets go through a built-in approval mechanism. A net can be in three possible states: uploaded, approved, or deleted. After a user uploads a Petri net for a community, the community moderator receives a notification and can decide whether the Petri net is approved or denied. In this way, collections remain manageable.

Retrieval. Sharing the Petri nets also means that it is possible to download and use the Petri nets from Petriweb. Petriweb supports this in two ways, by allowing downloading the original uploaded file, or a file generated from the parsed information. In this way it is possible to serve as many tools as possible.

2.3 Users and Communities

Petriweb is primarily intended to serve the public community by sharing Petri nets. Users may not want to share their Petri nets with the total public community, but only with a small group of users. Therefore, Petriweb features a built-in mechanism to support communities. Each registered user can request a community. After the Petriweb administrator has approved the request, the user becomes moderator of the private community and can invite users to join. In this way restricted project areas can be defined and used, with the full search and properties functionality of Petriweb.

2.4   Architecture

Storing knowledge. Petriweb is based on the data schema given in Fig. 3. It consists of three parts:

- the structure and marking of the Petri net,
- the graphical information to display the Petri net, and
- the properties associated with the Petri net.

Petriweb supports component based models: definitions define the behavior and structure, and can be instantiated in other definitions. Each Petri net definition is either a transition definition or a (sub)net definition. For every transition and subnet, its definition and its instance are stored. Every instance is a part of a (sub)net. The relation between *ProcessInstance* and *Subnet* depicts the hierarchy of the Petri net. Places are also part of a (sub)net. A marking is stored in the *MarkingInformation* table. A Petri net can have a trace of markings (a firing sequence); this information is stored in the *StateHistory* table. Multiple markings and traces per net are supported. Arcs are not considered as separate objects in the Petri net, but instead, both transitions and subnets have connectable pins. These pins connect instances with places or other pins. Due to the separation of definition and instance, a distinction is made between formal and actual pins.

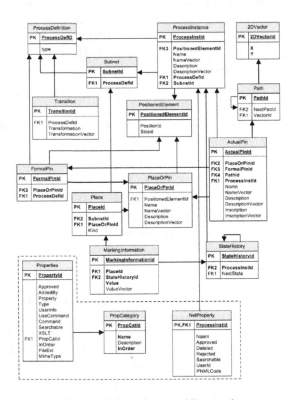

Fig. 3. Data schema of Petriweb

Fig. 4. Communities and their users

A formal pin is part of the definition, an actual pin is an instance of the formal pin. Graphical information is part of the data schema, although not all relations are drawn in the figure.

Properties. Properties in Petriweb are stored in the table *Properties*. Each property belongs to a specific category. The table *PropCategory* contains information about the different categories. The table *NetProperty* contains the values of properties of (sub)nets in the repository. The table contains some standard properties that should always be available for every Petri net, such as its name, the location of its file and whether the (sub)net is approved. For each property defined in Petriweb (thus an entry in the table *Properties*) a column is present in this table. Automated properties can be derived either by applying a stylesheet (XSLT) or by calling a tool on the command line. The output is parsed into the database.

Communities. To support communities, it is of importance to separate the different communities from each other, in such a way that there is no connection between them. An advantage of this approach is that only properties needed in the community are stored and shown. To support this, each community receives its own database to fill. In order to administer the different communities, the data schema of Fig. 4 is used.

Implementation. The design of Petriweb is component based (Fig. 5). It is divided into three main components: repository, viewer and checker. The repository uses the viewer and checker. All components are designed and implemented as stand-alone applications. It was a goal to provide easily installation on different machines.[1]. The implementation in PHP with MySQL meets this goal.

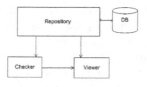

Fig. 5. Component model

---

[1] Petriweb has been tested on two platforms: Linux and Microsoft Windows.

3   PNML: The Petri Net Markup Language

Petriweb was designed to be used with as many different tools as possible. Hence it relies on a standard file format, and Petriweb is based on the Petri Net Markup Language, PNML [1][4][12], a draft ISO standard interchange format for Petri nets. Many tools already support PNML, and it is well-described, which makes it easy to write translators for tools that do not support it. Petriweb employs such translations automatically by means of automated properties.

PNML defines a basic syntax for flat colorless Petri nets, and an extension mechanism, PNTDs [13], to define syntax for additional features. Some extensions, e.g. structured PNML to denote hierarchy, are part of the standard. The PNML supported by Petriweb is defined by a specific PNTD we call EPNML 1.1. It encompasses basic Petri nets, as defined in [13], and most structural nets[2]. Further, it defines special arc and transition types[3]. These are employed by some of our tools, such as the Yasper editor [5][6].

3.1   PNML Viewer

PNML offers a way to store graphical information in the net. Therefore, Petriweb contains a basic Petri net viewer, that displays the Petri net as a Portable Network Graphics (PNG) image. The viewer supports hierarchical Petri nets; the user can browse through the different levels of a Petri net by clicking in the images. The viewer is implemented as a stand-alone web application. It can also animate firing sequences. Traces have to be uploaded in a special-purpose file format.

3.2   PNML Checker

Petriweb ensures that only correct PNML files are uploaded, by running them through a checker. This checker validates the file against the PNTD of EPNML 1.1. A PNTD is a RELAX NG [2] specification that extends the standard PNML definitions. The checker can also be used as a stand-alone application.

3.3   Issues with PNML

Despite the power and clarity of the PNML standard, interoperability problems remain. Generally they are due to the following issues:

PNML is a moving target: The PNML standard still changes in details. At present there are three versions to cope with: the current ISO draft [1], the currently published RELAX NG syntax definitions, and the original PNML as used by the Petri Net Kernel [8].

The differences between the first two are minor; for instance, in structured PNML, the ISO draft requires that all elements are included in a *page* element, which the syntax definitions do not.

---

[2] Some extra restrictions on reference nodes apply.

[3] The fully documented specification can be found in [14].

The last two have major incompatibilities, e.g. in the handling of hierarchy. We have found that even new tools sometimes produce PNK-compatible PNML instead of standard-conformant PNML.

Tools implement PNML incorrectly: Sometimes, nets being uploaded fail to validate for petty technical reasons, e.g. a failure to use XML namespaces properly.

PNML is open to interpretation: Some details left open in the PNML specification may lead to practical interoperability issues. For instance, the unit of measurement of graphical coordinates is not specified.

PNTDs are too liberal: In order to maximize interoperability between tools, PNML extensions (PNTDs) should be very conservative:

- all PNML documents that conform to the base definition should also conform to the extended definition, whenever this makes sense
- all such documents should continue to have the same interpretation

The first rule is broken by the present ISO draft, which, for structured PNML, requires every element to be in a *page*. With this requirement, basic (flat) PNML and structured PNML become disjoint sets of documents. In order to support compatibility with tools for flat Petri nets, designers of hierarchical Petri net tools will now need to support both basic and structured PNML, and take care that a net that happens to not use hierarchy is always written as basic PNML.

The second rule was broken by our own extension, EPNML. It adds a type label to arcs that determined their directions. The choice of source and target used to be defined as arbitrary. This breaks compatibility with tools that do not recognize our arc types. In EPNML 1.1, the arc type and the choice of source and target must agree.

These examples demonstrate how easy it is to accidentally break interoperability when designing Petri net extensions in PNML. If the utmost care is not taken, PNML will not achieve its intended purpose; it will just be a convenient mechanism for defining toolspecific file formats for Petri net tools.

No support for constraints: Precise syntax checking is crucial to a tool such as Petriweb. However, both standard PNML and our own extensions impose constraints on the use of PNML that cannot be expressed in RELAX NG (or, for that matter, in XML Schema). More accurate syntax checking can be attained by adding a mechanism to express constraints, e.g., Schematron [7].

4   Future Work

Petriweb is installed at `http://www.petriweb.org/`; we are populating it with collections of Petri nets. Specifically, we collect a large set of 'interesting' nets for researchers, as well as well-known standard Petri nets, e.g. from [3], [9], and [10].

Automated properties are a useful feature. They provide for the invocation of file format conversions and analysis tools in a batch run. However, the communication between Petriweb and tools leaves room for improvement.

A facility for batch uploading and downloading would be very convenient.

Support for communities could be further improved. At present, there is only one community with public access; it would be useful if there could be more. Further, it would be very useful if nets and property definitions could be shared among communities.

Adding support for version control will greatly aid the gradual development of models.

Another possible improvement is the unification of Petriweb properties and PNML labels. When retrieving Petri nets from Petriweb, the properties defined in Petriweb would be included into in the PNML. Conversely, when a net is uploaded, any of its attributes already present in PNML would show up as Petriweb properties. This can be taken one step further, by automatically generating PNTDs corresponding to properties defined in Petriweb. Thus, Petriweb can be turned into a tool for creating and managing PNTDs.

At the moment, only static Petri nets are stored with a single initial marking. It is not yet possible to add occurrence sequences to a Petri net. The datamodel supports this extension, but it is not yet integrated in Petriweb.

5 Conclusions

The use of repositories to handle large collections of models helps the researcher and designer in his or her work. Petriweb is a repository to handle such collections. Multiple collections are supported by a single Petriweb installation.

A stable version of the Petriweb software is installed at www.petriweb.org. It is a fully functional prototype, and demonstrates the benefits of such repositories. The manual and automated property definition mechanisms allow for powerful search facilities. Automated properties also provide an elegant mechanism for batch processing on nets, used for structural and behavioral Petri net analysis and for file format conversions. A graphical Petri net viewer is integrated to support browsing.

Petriweb would be useless without a good standard file format for Petri nets. The qualities of PNML make it a good choice, but it can be improved in various respects.

References

1. Software and Systems Engineering – High-level Petri Nets, Part 2: Transfer Format – ISO/IEC WD 15909-2:2005(E). June 2005.
2. James Clark and Makoto Murata. RELAX NG Specification, 2001. http://www.oasis-open.org/committees/relax-ng/spec-20011203.html.
3. J. Desel and J. Esparza. *Free Choice Petri Nets*, volume 40 of *Cambridge Tracts in Theoretical Computer Science*. Cambridge University Press, Cambridge, 1995.
4. J. Billington et al. The Petri Net Markup Language: Concepts, Technology, and Tools. http://citeseer.nj.nec.com/billington03petri.html.
5. R. Post et al. Yasper, Yet Another Smart Process EditoR, 2004-2006. www.yasper.org.

6. K. van Hee, R. Post, and L. Somers. Yet Another Smart Process Editor. In *Proc. of European Simulation and Modelling Conference (ESM 2005), EUROSIS, Porto, Portugal,* 2005.
7. Rick Jeliffe. The Schematron. An XML Structure Validation Language using Patterns in Trees, 2003. `http://www.ascc.net/xml/resource/schematron/schematron.html`.
8. Ekkart Kindler and Michael Weber. The Petri Net Kernel - An infrastructure for building Petri net tools. *International Journal on Software Tools for Technology Transfer,* 3(4):486–497, 2001.
9. T. Murata. Petri Nets: Properties, Analysis and Applications. *Proceedings of the IEEE,* 77(4):541–580, April 1989.
10. W. Reisig. *Petri Nets: An Introduction,* volume 4 of *Monographs in Theoretical Computer Science: An EATCS Series.* Springer-Verlag, Berlin, 1985.
11. Eric Verbeek and Wil M.P. van der Aalst. Woflan 2.0, A Petri-Net-Based Workflow Diagnosis Tool. In *Lecture Notes in Computer Science,* volume 1825, page 475, 2000.
12. Michael Weber and Ekkart Kindler. The Petri Net Markup Language, April 2002. `http://www.informatik.hu-berlin.de/top/pnml/`.
13. Michael Weber and Ekkart Kindler. Petri Net Markup Language schema RELAX NG implementation of structured PNML, 2004. `http://www.informatik.hu-berlin.de/top/pnml/1.3.2/structuredPNML.rng`.
14. J.M.E.M. van der Werf and R.D.J. Post. EPNML 1.1 - an XML format for Petri nets, March 2004. `http://www.petriweb.org/specs/epnml11/pnmldef.pdf`.

Geist3D, a Simulation Tool for Geometry-Driven Petri Nets

Jochen Stier, Jens Jahnke, and Hausi Müller

University of Victoria, Victoria

Abstract. Petri Nets have proven useful as a language for expressing distributed control logic. This paper presents a tool that integrates the formalism with virtual reality technology in order to model functioning mechatronic systems in 3D. A virtual environment generates sensor telemetry and reflects the state of actuators by computing the geometric and physical properties of a system and the surrounding environment. Petri Nets, combined with the Python programming language, model control systems in terms of virtual sensors and actuators. This methodology simulates the interactions between the structure and logic of mechatronic systems, allowing for an early verification of designs.

1 Introduction

Physical prototypes generate realistic input when testing control systems and therefore play an important role in the development process. However, functioning components are often not readily available and opportunities to evaluate a system are lacking. Yet, the costs of correcting control system errors at a later stage of development can be orders of magnitude higher than had they been discovered earlier on. In the absence of physical prototypes, the opportunity to analyze a model for flaws should therefore always be taken [1]. Software-based simulation is an effective tool to project designs forward in time and explore a greater solution space more safely.

This paper introduces a method for simulating the control of distributed mechatronic systems using a combination of virtual reality technology and Petri Nets [6]. The Petri Net constructs model distributed control logic and a computer generated virtual environment simulates the natural artifacts which lead to sensor telemetry. The goal is to re-create the feedback loop between the logic and environment, which physical prototypes close implicitly through natural processes. A modelling tool called Geist3D was specially constructed to support Geometry-driven Petri Nets by combining a Petri Net editor and kernel with the Python programming language and a virtual reality component into a single development environment.

The title of this paper refers to Geometry-driven Petri Nets, and the focus was to present the research from that viewpoint. However, the virtual reality environment is also an important component which encapsulates the results of much research and development effort. Yet, only a small subset of its functionality is introduced in this paper. The development environment including a number

S. Donatelli and P.S. Thiagarajan (Eds.): ICATPN 2006, LNCS 4024, pp. 421–430, 2006.

of models is available for download at www.geist3d.org. The following sections discuss related work, define the modeling language, present an application of the language and discuss future research.

2   Related Work

Industry routinely uses factory simulation tools such as Flexsim or Delmia to model repetitive assembly processes in 3D [2] [4]. Many types of manufacturing systems are inherently designed so that components are always in the proper position during manufacture. Devices mostly traverse predetermined paths through space and there is little need for sensor telemetry to monitor the surrounding environment. The corresponding control systems can therefore be modelled using a collection of fixed animation sequences.

Research efforts now focus on methods for simulating autonomic [3] or adaptive [5] mechatronic systems. A growing number of groups are developing simulation tools which are characterized by integrating feedback from the virtual environment with the logic that defines the animation [7] [8] [16]. All of the tools in this domain have to represent an environment capable of generating realistic telemetry. However, most do not include a process modelling language such as Petri Nets to develop distributed control systems more effectively than by using only a procedural programming language.

The most relevant related work with respect to Petri Nets is the *CPN Tools* [9] development environment for Colored Petri Nets. *CPN Tools* appears to be a more elaborate Petri Net simulation tool than Geist3D, because it also includes methods to analyze the Petri Net graph and underlying process models. However, *CPN Tools* does not include a representation from which to derive meaningful physical quantities such as sensor telemetry. Assumptions therefore have to be made about parameters representing physical artifacts and many of the variables in a mechatronics simulation have to be randomized or approximated.

Petri Nets have also been used to control animations before [20, 21]. The basic concepts among these approaches and the one presented here are the same – tokens represent artifacts in the virtual environment. However, in the literature presented so far, Petri Nets have only been used to control the animation and none of the internal variables depended on the state of the 3D environment.

3   Overview

Geometry-driven Petri Nets are a hybrid modelling language based on timed hierarchical Colored Petri Nets and the Python programming language [10]. The Petri Net constructs express concurrency and synchronization, and the Python scripts define discrete time step algorithms that control actuators and interpret sensors. The scripts are attached to individual Petri Net constructs and evaluated as it executes. At the same time, the virtual reality component generates sensor telemetry and reflects the state of actuators. Most of the artifacts in the virtual environment can become tokens or variables in the Petri Net. This section provides an overview of the virtual reality and Petri Net components, respectively.

3.1 Virtual Reality

Virtual reality technology is based on software that simulates artifacts of the physical world in real time. A virtual environment can therefore reflect the effects of mechanical actuators and produce sensor telemetry. Geist3D contains a growing number of algorithms that model a variety of artifacts, including terrains and buildings as well as mechanical systems in terms of rigid body dynamics. Gradient fields and real-time Delaunay triangulations are also supported to annotate the environment with contextual information and maintain proximity among objects. A number of different types of sensors are also available, including range finders, collision detectors, radio transmitters and receivers, force sensors and gradient field sensor.

A scene tree is a common data structure to organize the artifacts of a virtual reality system. The recursive hierarchical decomposition of trees supports efficient divide and conquer algorithms for rendering, collision detection and methods to synthesize sensor telemetry [11]. The scene tree nodes contain information such as geometric description of surfaces, material properties, light sources and viewpoints. The rendering process generates a single image by applying each node to the current drawing state in the order that they are encountered during a traversal. The contents of the nodes as well as the structure of the tree therefore determine the visual properties of the scene. Animations are produced by changing the structure or the content between rendering steps.

The Petri Net constructs in Geist3D are also treated as scene tree nodes. As a result, transitions and places contain additional contextual information such as a position and orientation in 3D space. Although the benefits of these new properties have not yet been explored, it appears that relating the Petri Net constructs with objects in the virtual environment provides good cognitive support when constructing and maintaining models.

3.2 Geometry-Driven Petri Nets

Geometry-driven Petri Nets are a derivative of timed Colored Petri Nets. The color sets can represent abstract data types which integrate well with a procedural programming language [13]. Timed transition occurrences have traditionally been designed to describe and analyze timed dynamical systems more effectively [12]. In Geometry-driven Petri Nets, the durations are also used to execute discrete time-step functions which animate the virtual environment.

A color set consists of tokens and variables. The token types include scene tree nodes and integers, and the variables include sensors. All Petri Net constructs can be annotated with inscriptions, or event handlers. Transitions contain pre-fire, post-fire and in-fire handlers and places contain different sets of handlers for each color type. Token colors, for example, contain *OnDeposit* and *OnRemove* handlers which are called when the marking changes. The handlers of variable colors depend on the types of sensors. For example, a range finder contains

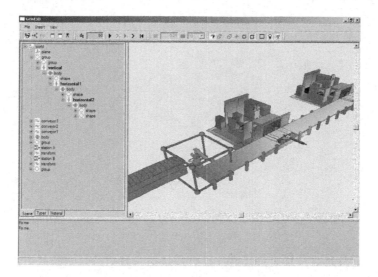

Fig. 1. Geist 3D application

OnChange and *OnSample* handlers which are invoked when the distance changes or every time when the sensor samples.

The policies of Geometry-driven Petri Nets are close to timed Colored Petri Nets. However, the tokens in Geometry-driven Petri Nets are still treated as discrete entities. The $<, >, +, -$ operators are therefore arithmetic functions that add and remove tokens from the color sets and compare their size. In the future, however, the operators may have different meanings. For example, adding a token containing a triangle mesh to a place, adds the corresponding node as a discrete entity to the set of tokens already on that place. Yet, it is also conceivable that this operation merges the triangles of the mesh with another node. In this case, only one node remains containing the triangles of both meshes.

In order to avoid redundancy and add structure to large scale systems, hierarchal nesting of Petri Nets is supported in the form of place refinement [17]. An entire Petri Net including a subtree of the scene are encapsulated by a *page*. Pages can be instantiated and deleted, and Geist3D supports mechanisms to propagate type changes to all instantiations. In the context of a mechatronics simulation, pages represent different types of mechanical devices which encapsulate control logic and mechanical structure into a single entity.

4   Development Environment

Geist3D was specifically designed to integrate a 3D graphics engine and Petri Net simulation tool into a single development environment. All of the components, except the Python runtime environment and *Open Dynamics Engine* physics library have been custom developed for this purpose. The main challenges included developing and testing all of the algorithms and user interface widgets

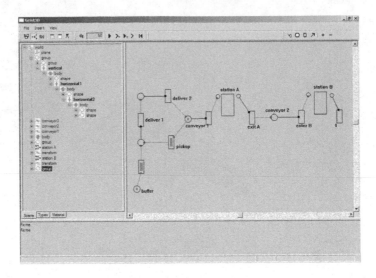

Fig. 2. Geist 3D application

required to support a comprehensive environment, as well as determining how to synchronize the frame rate with the execution of the Petri Net.

Geist3D includes numerous 3D graphical user interface widgets to construct elaborate virtual environments. User can design systems in terms of shapes and mechanical joints, including parameters such friction, masses and forces. Devices can be placed in an environment constructed of elevation data sets or general triangles meshes imported from 3D Studio Max. The laws of rigid body dynamics implicitly hold true for all of the objects, which leads to simplified animation control and realistic sensor telemetry.

Geist3D also includes a hierarchical Petri Net editor and a runtime kernel which integrates the Python interpreter. The editor allows user to define hierarchical Petri Nets including color sets using simple drag and drop operations, and to develop and debug event handlers. Each scene tree node type publishes a procedural programming interface consisting of Python objects, methods and members. All parameters of the simulation are therefore accessible from the scripts embedded within the Petri Net.

Figures 1 and 2 depicts the Geist3D development environment containing a simulation of a material handling systems. Only the virtual reality and Petri Net views are shown. The application also contains an additional view which displays the color set and event handlers of the currently selected Petri Net construct.

5   Example

This section illustrates an application of Geometry-driven Petri Nets by simulating a 4-way stop intersection. Cars arrive at the intersection randomly and then either turn or continue on straight. If other cars are currently passing through the

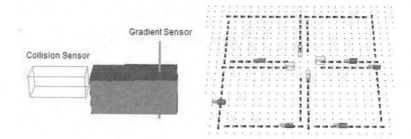

Fig. 3. 3D model of 4-way stop and car

intersection, a driver has to stop and wait. The following paragraphs illustrate selected aspects of the model in terms of the Geist3D user interface.

Figure 3 depicts the 3D model of a vehicle and the intersection. A gradient field defines a road grid of four city blocks with the intersection in the center. A vehicle consists of a grey box and a collision and gradient field sensor. The sensors are displayed as a transparent red box in front of the vehicle and as a vertical red line. The gradient sensor continuously samples the field below and sets the velocity of the car to the value of the gradient. As a result, the cars move along the paths defined by the topology of the field. A car stops when it moves on top of a gradient with magnitude 0 or when the collision sensor detects an object that is too close to the front of the car.

The intersection is controlled by the Petri Net shown in Figure 4. Each of the four transitions represents a direction from which cars can arrive. The place in the center contains the gradient field as a token. It acts as a mutual exclusion ensuring that only one transition fires at a time and that only one car enters the intersection.

Four additional collision sensors represent the stop signs at which cars have to wait before entering the intersection. If a collision with the corresponding *stop sign* sensor is detected and if the exclusion token is available then a transition

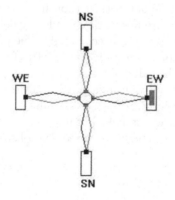

Fig. 4. Petri Net of 4-way stop

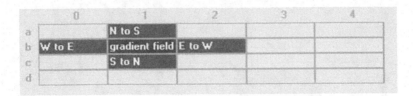

Fig. 5. Color set of place

	0	1	2	3	4
a					
b		gradient field	E to W		
c					
d					

Fig. 6. Color set of transition EW

becomes enabled and fires. At this point, it consumes the token and rewrites the gradient field at the intersection to guide the car into one of the three directions. Until the transitions deposits the token, no other transition can fire and all other cars will stop when they reach the gradient of magnitude 0 at the intersection.

Figure 5 shows the color set of the only place. It consist of four sensor variables and one integer token arranged in a 2D grid of slots. Depending on the color type, each slot has different properties such as background color and programming interface. The letters in front of each row and the numbers above each column uniquely identify a slot in an inscription function. For example, $a1.name$ refers to the name of the collision sensor in slot a1 of Figure 5. The API of slots and tokens depends on their respective types.

Figure 6 shows the color set of the transition which controls the traffic flow from East to West. The color set consist of the gradient field and the collision sensor in the East-West direction. When a transition fires then it executes the scripts shown in Figure 5. The pre-fire inscription randomly determines the direction of a car and then changes the gradients accordingly. The post-fire inscription resets the grid back to gradients of magnitude 0. The API of the gradient field includes a *set* function that alters the value of the gradient at a particular grid point.

A place containing the collision and gradient field sensors in its color set is used to model a car. The inscriptions set the linear velocity of the box to the value of the underlying gradient field. Simple checks are performed to ensure that that the collision sensor currently detects no obstacles.

6   Conclusion

This paper has outlined a new development environment for simulating mecha-tronic systems using Petri Nets and virtual reality technology. This methodology

Pre-fire script	Post-fire script

import random
 	
rand = random.uniform(0, 1)
 	
if rand < 0.33: # turn left
    self.b1.set(23,15,1,0)
    self.b1.set(24,15,1,0)
    self.b1.set(25,15,0,1)
    self.b1.set(25,16,0,1)
    self.b1.set(25,17,0,1)
elif rand < 0.66: # go straight
    self.b1.set(23,15,1,0)
    self.b1.set(24,15,1,0)
    self.b1.set(25,15,1,0)
    self.b1.set(26,15,1,0)
elif rand < 1.00: # turn right
    self.b1.set(23,15,1,0)
    self.b1.set(24,15,0,−1)
    self.b1.set(24,14,0,−1)
	self.b1.set(23,15,0,0)
	self.b1.set(24,15,0,0)
	
	self.b1.set(25,15,0,0)
	self.b1.set(25,16,0,0)
	self.b1.set(25,17,0,0)

Fig. 7. Inscription of Transition **EW**

helps to synthesis more *correct* Petri Nets, because parameters that otherwise have to be approximated or randomized, are bound to a close approximation of the structural and physical properties of a system. For example, the firing durations in the model above express the amount of time that a car can spend traveling through the intersection. Incorrect firing durations which may cause cars to collide can be detected by observing the virtual environment. Since the roads are expressed in terms of physical dimensions and since the cars move in terms of velocities, it is likely that the final durations arc close to realistic. More accurate analysis such as cycles times, are therefore to be expected as well.

7   Future Research

The types of systems which Geist3D can currently express are limited by the types of supported sensors. An important research goal is therefore to provide a more diverse set of telemetry by computing other natural processes. The scope of this task includes areas such as computational biology and physics. Some of the future plans include simulating plant growth using L-Systems [15] and articulated characters using neural network based controllers [14].

Geist3D should also support hybrid Petri Nets which can model fluid dynamics in terms of ordinary first-order differential equations [18]. Most of the necessary infrastructure and graphical widgets already exist and little additional development would be required. While Geist3D already supports hierarchical

Petri Nets, it may further be useful to support higher-order Petri Nets in which entire pages can become tokens [19]. This extension, however, requires a new form of *port place* that provides an interface to the pages which occupy places. It would also be useful to support analysis methods that automatically warn users of undesirable or faulty designs.

Another important area of research is determining how to synthesize control logic from a Geist3D simulation by mapping a Geometry-driven Petri Net to embedded hardware. Webots, for example, uses the same programming interface for the virtual sensors and actuators as for the ones in the corresponding physical system. The control logic of the simulation is therefore interchangeable between virtual and real robots. The Petri Net places in Geist3D already seem to correspond to processors, and the arcs to communication links. The color slots represent registers which are connected to networking or I/O devices and the place inscriptions are interrupt handlers.

References

1. Daniel Jackson and Martin Rinard : Software analysis: a Roadmap. Proceedings of the Conference on The Future of Software Engineering. (2000) 133–145
2. Daniel Williams and Daniel Finke and D. J. Medeiros and Mark Traband : Discrete simulation development for a proposed shipyard steel processing facility. Proceedings of the 33nd conference on Winter simulation. **1** (2001) 882–887
3. Jeffrey Kephart and David Chess : The Vision of Autonomic Computing. Computer Journal. **1** (2003) 41–50
4. Deogratias Kibira and Chuck McLean : Manufacturing modeling methods: virtual reality simulation of a mechanical assembly production line. Proceedings of the Winter Simulation Conference. (2002) 1130–1137
5. Pattie Maes : Modeling adaptive autonomous agents. Artificial Life. **1** (1994) 9
6. Olivier Michel : Fundamentals of a Theory of Asynchronous Information Flow. Proc. of IFIP Congress 62. (1963) 386–390
7. Olivier Michel : Webots: Symbiosis Between Virtual and Real Mobile Robots. Proceedings of the International Conference on Virtual Worlds. (1998) 254–263
8. Nathan Koenig and Andrew Howard : Design and Use Paradigms for Gazebo, An Open-Source Multi-Robot Simulator. International Conference on Intelligent Robots and Systems. (2004) 2149-2154
9. A. V. Ratzer and L. Wells and H. Michael Lassen and M. Laursen and S. Christensen and K. Jensen : CPN Tools for Editing, Simulating, and Analysing Coloured Petri Nets. ICATPN. (2003) 450-462
10. P. Antsaklis and X. Koutsoukos and J. Zaytoon : On hybrid control of complex systems: A survey. European Journal of Automation. **32** (1985) 1023–1045
11. Alfred Aho and John Hopcroft : The Design and Analysis of Computer Algorithms. Addison-Wesley Longman Publishing Co., Inc. (1974)
12. C. Ramchandani : Analysis of asynchronous concurrent systems by timed Petri Nets. Massachusetts Institute of Technology. (1974)
13. Kurt Jensen : Coloured Petri Nets and the Invariant Method. Theoretical Computer Science. **14** (1981) 317–336

14. Radek Grzeszczuk and Demetri Terzopoulos and Geoffrey Hinton : NeuroAnimator: fast neural network emulation and control of physics-based models. SIGGRAPH '98: Proceedings of the 25th annual conference on Computer graphics and interactive techniques. (1998) 9–20
15. Aristid Lindenmayer : Mathematical model for cellular interaction in development, Parts I and II. Journal of Theoretical Biology. **18** (1968) 280–315
16. G. A. Kaminka and M. M. Veloso and S. Schaffer and C. Sollitto and R. Adobbati and A. N. Marshal and A. Scholer and S. Tejada : GameBots: The ever-challenging multi-agent research test-bed. ACM Transactions on Computer Graphics. (2002) 280–315
17. Wilfried Brauer and Robert Gold and Walter Vogler : A Survey of Behaviour and Equivalence Preserving Refinements of Petri Nets. Lecture Notes in Computer Science; Advances in Petri Nets. **483** (1991) 1–46
18. Fabio Balduzzi and Alessandro Giua and Carla Seatzu : Modeling and simulation of manufacturing systems with first-order hybrid Petri Nets. International Journal of Production Research. **483** (2001) 255–282
19. Jörn Janneck and Robert Esser : Higher-order Petri net modeling— techniques and applications. Workshop on Software Engineering and Formal Methods. (2001)
20. E. Kindler and C. Pales : 3D-Visualization of Petri Net models: Concept and Realization. Application and Theory of Petri Nets, ATPN 2004. 464–473
21. F. Baldini and G. Bucci and E. Vicario: A Tool Set for Modeling and Simulation of Robotic Workcells. Workshop on Techniques, Methodologies and Tools for Performance Evaluation of Complex Systems. (2005) 106–114

The BRITNeY Suite Animation Tool

Michael Westergaard and Kristian Bisgaard Lassen

Department of Computer Science, University of Aarhus,
IT-parken, Aabogade 34, DK-8200 Aarhus N, Denmark
{mw, k.b.lassen}@daimi.au.dk

Abstract. This paper describes the BRITNeY suite, a tool which enables users to create visualizations of formal models. BRITNeY suite is integrated with CPN Tools, and we give an example of how to extend a simple stop-and-wait protocol with a visualization in the form of message sequence charts. We also show examples of animations created during industrial projects to give an impression of what is possible with the BRITNeY suite.

1 Introduction

Colored Petri nets (CP-nets or CPN) [7] have proved their usefulness in modeling and understanding complex systems [2,10,12,19], e.g., for verification of existing behavior or requirements engineering of needed behavior.

However, when using CP-nets, only people familiar with the formalism are able to truly understand the model of the system. A domain expert may understand a CP-net, when introduced to CP-nets in general and when the particular CP-net is explained by the model developer, but the domain expert is seldom able to talk back, say precisely what is wrong with the model, and offer suggestions to fix the model of the system, because of lack of technical expertise with the formalism. CP-net models of systems are prone to errors if they can not be fully understood and validated by someone with domain knowledge. The contribution of the BRITNeY[1] suite animation tool is to give a visualization of the state and actions of a CP-net so the domain expert can validate the model.

In this paper we present the BRITNeY suite [23] which introduces an animation layer for CP-nets. BRITNeY suite provides a uniform way to implement, integrate, and deploy visualizations of CP-nets and has a pluggable architecture which makes it possible to write customized plug-ins to animate the model in addition to more than a dozen predefined plug-ins. The BRITNeY suite has already been used successfully to animate a network protocol [11], to animate a workflow process in a bank for the purpose of requirements engineering [8] and to visualize how patient, nurse and doctor work together with a system that dispenses sedatives, again for the purpose of requirements engineering [13].

Even though BRITNeY suite is designed with CPN Tools in mind, it is possible to integrate the tool with any executable formalism as the interface to CPN Tools is based on well-known public standards. For example the tool has

---

[1] An abbreviation for **B**asic **R**eal-time **I**nteractive **T**ool for **Net**-based animation.

S. Donatelli and P.S. Thiagarajan (Eds.): ICATPN 2006, LNCS 4024, pp. 431–440, 2006.
© Springer-Verlag Berlin Heidelberg 2006

been used successfully to visualize the execution of a timed automaton [1] model
as well as the reachability graphs of systems created using a subset of the π-
calculus [17], bigraphical reactive systems [15], finite and timed automata, and
Coloured Petri nets. Also goto-graphs of Java programs have been visualized
using the BRITNeY suite.

The paper is structured as follows. In Sect. 2 we give a brief overview of the
architecture of the BRITNeY suite. In Sect. 3 we demonstrate how to add a
message sequence chart visualization to a CP-net of a simple stop-and-wait protocol. Sect. 4 contains some example visualizations created as part of industrial
projects. In Sect. 5 we mention related work and outline some of the new features
planned for BRITNeY suite.

2   Architectural Overview

A well-known design pattern from the object-oriented world is the model-view-
controller (MVC) design pattern [6]. In the MVC design pattern, three participants collaborate to provide the implementation of an application, namely a
model, a view, and a controller, see Fig. 1. The model contains the state of the
system, the view is a (graphical) representation of the current state of the model,
and the controller implements the behavior of the system. The view may initiate
actions in the controller.

The idea behind the BRITNeY suite is to use a CP-net (or any other formal
executable model) to model the state and behavior of the system (the model
and controller), and use BRITNeY suite for visualizing the system (view). This
division is natural as places of CP-nets are used to model the state of a system
and transitions the behavior.

In Fig. 2, we see how BRITNeY suite is integrated with CPN Tools [3] to
provide simulation-based visualizations and animations. CPN Tools itself is split
into two components, an editor and a simulator. The animation tool, in the
right part of the figure, communicates with CPN Tools using a standard Remote
Procedure Call protocol, called XML-RPC [25], in order to allow vendors of other
tools to directly integrate their tools with BRITNeY suite. BRITNeY suite uses
plug-ins to make the actual visualizations, which makes it easy to create your
own animations. 15 plug-ins are currently available in the tool. Table 1 lists each
plug-in with a short description. Over time, more plug-ins will be added.

BRITNeY does not contain a fixed set of plug-ins as plug-ins can be added
and removed, so stubs are generated on-the-fly as needed by using the reflection

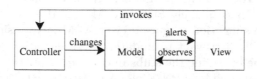

Fig. 1. Architectural overview of the model-view-controller design pattern

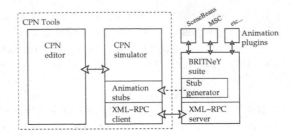

Fig. 2. A more detailed view of the integration of the animation tool with CPN Tools

Table 1. Plug-ins for the BRITNeY suite. The first group of plug-ins is for creating various charts, the second group is for displaying directed graphs, the third group is for interacting with a user and the final group contain plug-ins that do not fit in any group.

Name	Description
AreaChart	For visualizing data values by filling the area below them
GanttChart	For drawing Gantt charts
Histogram	For drawing histograms
MSC	For drawing message sequence charts
PieChart	For drawing pie charts
PieChart3D	For drawing 3D pie charts
StepChart	Similar to a histogram
XYChart	For visualizing data values as points
Graph	For drawing 2D graphs
Graph3D	For drawing 3D graphs
GetString	For getting short text-messages from the user
ShowString	Display short text-messages to the user
DataStore	Storage for simple data-types
Report	Nice presentation of data
SceneBeans	For displaying and interacting with a SceneBeans [20] animation

mechanism in Java to inspect the signatures of the plug-ins. The stubs make sure that values are passed correctly to the appropriate Java object's method and takes care of passing the return value back to the caller. Stubs are generated automatically by the stub generator component of BRITNeY suite. The stubs are injected into CPN Tools and are available as regular functions in the inscription language of CPN Tools, namely Standard ML (SML) [16], which allows the modeler to use the animation plugins anywhere SML expressions are allowed.

The modeler will often want to update the visualization when a transition occurs. This is done by calling the stubs in code segments that are special transition inscriptions allowed by CPN Tools. A code segment is executed when the transitions it belongs to occurs. It consists of input, output, and action parts. The input and output parts make it possible to receive input from the model and to provide situmli back to the model respectively. This makes it possible to, e.g., invoke a stub with values dictated by tokens and to generate new tokens from the result of executing the stub.

3   Using BRITNeY to Generate Message Sequence Charts

In this section we will describe how to show a simulation of a CP-net as a message sequence chart (MSC), i.e. generate a chart which displays the simulation of the CP-net in terms of events being passed between processes. This is instead of, e.g., in CPN Tools where simulation is shown as enabling of transitions, and tokens being consumed and generated when transitions occur. The description is fairly high-level, and a more detailed and technical description can be found in [24], but the reader is assumed to have basic knowledge of object oriented programming and an ability to read Java and SML code.

3.1   Model

The model that we will use in this paper is a very simple stop-and-wait protocol as seen in Fig. 3. The model consists of three parts: 1) A sender who can Send Data from the Out Buffer with a packet number from Next Id. Also the sender can Receive Ack thereby updating the token on Next Id. 2) A network that can Drop packets that are sent to the receiver from place Network 1. Network 2 contains acknowledgments that the receiver is sending back to the sender. 3) The receiver can Receive Data and update the Receive Id that the next packet must have.

3.2   Adding the MSC Primitives in CPN Tools

MSCs are well-known to protocol engineers, and it is therefore a good idea to be able to present the execution of a CP-net as an MSC. The first part of an MSC that is generated from the model in Fig. 3 can be seen in Fig. 4. The Sender process corresponds to the sender part of the CP-net, Network process to the network part of the CP-net and Receiver process to the receiver part of the

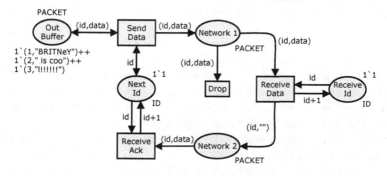

Fig. 3. CP-net of a stop-and-wait protocol

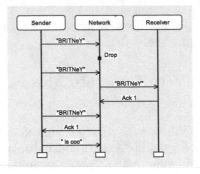

Fig. 4. First part of an MSC generated from the model in Fig. 3

CP-net. In the following we will describe how to extend the model in Fig. 3 with primitives to draw this MSC.

In Listing 1 we show the signature of the Java plug-in for the MSC class. It contains functions for adding new process, adding events between processes, and adding events internal to a single process. This will, as explained in Sect. 2, be translated, by the stub generator, to a corresponding SML representation. In the following we show how to apply SML primitives to the CP-net to call these methods.

Listing 1. Java signature of the MSC object.

```
1   void addProcess(String name);
2   void addEvent(String from, String to, String name);
3   void addInternalEvent(String process, String name);
```

Listing 2. Initialization of the MSC view.

```
1   structure msc = MSC(val name = "Stop-and-Wait Protocol");
2   val _ = msc.addProcess("Sender");
3   val _ = msc.addProcess("Net");
4   val _ = msc.addProcess("Receiver");
```

To set up the MSC view we need to add some declarations to the CP-net. In CPN Tools we add declarations as in Listing 2. Line 1 initializes an MSC object with the name "Stop-and-Wait Protocol". Lines 2–4 creates the three processes as seen in Fig. 4; i.e. Sender, Network, and Receiver.

Next we need to extend our model from Fig. 3 to generate the events that correspond to those in Fig. 4.

In Fig. 5 we see how the methods from Listing 1 are incorporated into the CP-net. The idea is that we want to generate an event in the MSC when one of

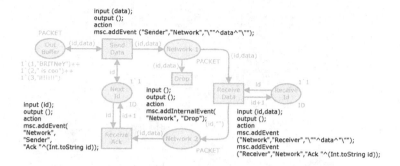

Fig. 5. Model from Fig. 3 with MSC primitives

the transitions in the model occurs. We did this as follows: When Send Data in the CP-net occurs we add an event from Sender to Network in the MSC, where the label is the same as the data being sent, i.e. "data" where data is bound from the string in the packet from Out Buffer. When Drop in the CP-net occurs we add the internal event Drop on the process Network in the MSC. When Receive Data in the CP-net occurs, an event is added from Network to Receiver in the MSC, with label stating what data is received (the label is "data", where data is bound from the string in the packet from Network 1) and also, an event from Receiver to Network in the MSC, with an acknowledgment with the received packet number as label; the label is Ack i where i is the integer in the packet bound in the occurence of Receive Data. Finally, when Receive Ack occurs, an event is sent from Network to Sender in the MSC with the acknowledgment as the label; here the label is again Ack i.

4   Visualization Examples

This section will give a number of examples of practical use of BRITNeY suite. We will not describe the examples in detail, but just refer to papers with detailed descriptions.

In Fig. 6, we see an animation created to visualize an interoperability protocol for mobile ad-hoc networks [11]. The protocol is used to ensure that the mobile ad-hoc nodes (the laptops) can communicate with the stationary host, even when on the move. The domain-specific GUI makes it possible for the user to observe the behavior of the system as packets, visualized by colored dots, flow along the network and to provide stimuli to the protocol by dragging and dropping the laptops to indicate the node movements. The use of an underlying formal model can be completely hidden when experimenting with the prototype. The domain-specific GUI has been used in the project both internally during protocol design and externally when presenting the designed protocol to management and protocol engineers not familiar with CPN modeling.

In Fig. 7 we see the domain specific animation based on the SceneBeans plug-in. This was used in [8] for requirements engineering of a new workflow

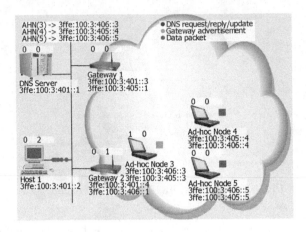

Fig. 6. A visualization of an interoperability protocol for mobile ad-hoc networks

Fig. 7. SceneBeans animation used for requirements engineering

system. The goal of the workflow was to support the handling of a blanc loan applications.

The animation is constructred as follows: There are always two bank assistants, Ann and Bill. Up to two customers can be present, in the figure only Mr. Smith is present. A bank manager, Mr. Banks, is always present. The balls represent blanc loan requests and the position of it shows who is responsible for the request. Whenever a transfer of responsibility occurs in the CP-net the ball is moved from one person to another in the animation. One ball has a P on it. This means that it is suspended, or parked, but can be picked up by one of the bank employees when they have the time. The square is part of the animation interface.

Once in a while the user can interact with the animation by e.g., setting up a loan for the customer when he wants to make a loan request, or setting the status of a loan request on behalf of Ann, Bill or Mr. Banks to e.g. granted or rejected. By not making the animation look like a normal prototype with windows, menues etc., the focus of the user was on the workflow and not on how the interface of the future system should be like.

5   Related Work and Future Improvements

BRITNeY suite supports adding animations to CPN models by annotating transitions with function calls, which are executed whenever the transition occurs. In the following, we outline how a number of other modeling tools facilitate visualization.

ExSpect [21], a tool for modeling based on CP-nets, allows the user to view the state by associating widgets with the state of the model, and to asynchronously interact with the model, also using simple widgets. In this way, it is easy to create simple user interfaces that support displaying information, but support for creating more elaborate animations is not readily available.

MIMIC/CPN [18] makes it possible to animate models within DESIGN/CPN [4], which is another tool for modeling using CP-nets. CPN models are animated by MIMIC/CPN by using function calls that are executed whenever a transition of the CP-net occurs. The animations are drawn using an application that resembles traditional drawing programs. Input from the user is possible by showing a modal dialog, where the simulation of the model is stopped while the user is expected to input information. It is also possible to make click-able regions, and the model can then query if one of these has been clicked. Another approach, which is taken by the COMMS/CPN [5] library for DESIGN/CPN and CPN Tools, is to provide a TCP/IP abstraction, allowing the user to code the user interface in any language and use RPC to communicate with it.

LTSA [14], a tool for modeling using timed labeled transition systems, allows users to animate models using the SceneBeans library. In LTSA animations are tied to the models by associating each animation activity with a clock; resetting a clock corresponds to starting an animation sequence. The animation sequence or a user with his mouse can then send events which correspond to the progress of the timer.

PNVis [9] is an add-on for the Petri Net Kernel [22], a highly modular tool for editing Petri nets. PNVis associates tokens with 3D objects and certain places with locations in a 3D world. Moving tokens corresponds to moving the associated object in the 3D world. PNVis is suitable for modeling physical systems, but not so applicable for creating prototypes of software or requirements engineering.

Using some of these animation tools/libraries, animation is integrated with the modeling formalism, such as the use of timers in LTSA or the ability to view or change the marking of places in ExSpect. Some libraries are easy to extend, such as animations in LTSA, as the SceneBeans library allows users to easily extend it with new animation primitives. Also, animations created using COMMS/CPN can easily be extended, as the "animation" is just a custom (e.g. Java) application. Some libraries make it easy to design animations, such as ExSpect and MIMIC/CPN, which both provide a graphical user interface to design animations. The approach of the current version of BRITNeY suite resembles a combination of MIMIC/CPN and COMMS/CPN, as the animation is driven by function calls associated with transitions to an external application. The main feature offered by BRITNeY suite from a user point of view is thus compatibility with CPN Tools (rather than the discontinued DESIGN/CPN) and platform-independence. BRITNeY suite also makes it easy to extend the tool using simple Java classes. From a developer point of view, BRITNeY provides good foundations for allowing closer integration with the model by allowing parts of the animation to inspect and modify tokens on fusion places of the CPN model, much like how widgets are associated with places in ExSpect. This is an important part of future work.

An important new feature of BRITNeY suite is that it is possible to deploy animations in a way that allows even non-technical users to download and experiment with the animation. Another part of the future work is to make this process even easier by adding a wizard to take care of all the details.

BRITNeY suite has already proven itself useful in real projects, and has already been used in several industrial projects.

References

1. J. Bengtsson and W. Yi. Timed Automata: Semantics, Algorithms and Tools. In *Lectures on Concurrency and Petri Nets*, volume 3098 of *LNCS*, pages 87–124. Springer-Verlag, 2004.
2. C. Bossen and J.B. Jørgensen. Context-descriptive prototypes and their application to medicine administration. In *DIS '04: Proc. of the 2004 conference on Designing interactive systems*, pages 297–306, Boston, MA, USA, 2004. ACM Press.
3. CPN Tools. Online www.daimi.au.dk/CPNTools/.
4. Design/CPN. Online www.daimi.au.dk/designCPN/.
5. G. Gallasch and L.M. Kristensen. A Communication Infrastructure for External Communication with Design/CPN. In *Proc. of Third CPN Workshop*, volume PB-554 of *DAIMI*, pages 79–93, 2001.
6. E. Gamma, R. Helm, R. Johnson, and J. Vlissides. *Design Patterns: Elements of Reusable Object-Oriented Software*. Addison-Wesley, 1995.
7. K. Jensen. *Coloured Petri Nets—Basic Concepts, Analysis Methods and Practical Use. Volume 1: Basic Concepts*. Springer-Verlag, 1992.
8. J.B. Jørgensen and K.B. Lassen. Aligning Work Processes and the Adviser Portal Bank System. In *REBNITA05*, 2005.
9. E. Kindler and C. Páles. 3D-Visualization of Petri Net Models: Concept and Realization. In *Proc. of ICATPN 2004*, volume 3099 of *LNCS*, pages 464–473. Springer-Verlag, 2003.
10. L.M. Kristensen and K. Jensen. Specification and Validation of an Edge Router Discovery Protocol for Mobile Ad-hoc Networks. In *Integration of Software Specification Techniques for Application in Engineering*, volume 3147 of *LNCS*, pages 248–269. Springer-Verlag, 2004.
11. L.M. Kristensen, M. Westergaard, and P.C. Nørgaard. Model-based Prototyping of an Interoperability Protocol for Mobile Ad-hoc Networks. Accepted for Fifth International Conference on Integrated Formal Methods, 2005.
12. L. Lorentsen, A-P Tuovinen, and J. Xu. Modelling Features and Feature Interactions of Nokia Mobile Phones Using Coloured Petri Nets. In *Proc. of ICATPN 2002*, volume 2360 of *LNCS*, pages 294–313, 2002.
13. R.J. Machado, K.B. Lassen, S. Oliveira, M. Couto, and P. Pinto. Execution of UML Models with CPN Tools for Workflow Requirements Validation. In *Proc. of Sixth CPN Workshop*, volume PB-576 of *DAIMI*, pages 231–250, 2005.

14. J. Magee and J. Kramer. *Concurrency – State Models and Java Programs*. John Wiley & Sons, 1999.
15. R. Milner. Bigraphical Reactive Systems. In K.G. Larsen and M. Nielsen, editors, *Proc. of CONCUR 2001*, volume 2154 of *LNCS*, pages 16–35. Springer-Verlag, 2001.
16. R. Milner, R. Harper, and M. Tofte. *The Definition of Standard ML*. MIT Press, 1990.
17. R. Milner, J. Parrow, and D. Walker. A Calculus of Mobile Processes. *Information and Computation*, 100(1):1–77, 1992.
18. J.L. Rasmussen and M. Singh. *Mimic/CPN. A Graphical Simulation Utility for Design/CPN. User's Manual.* www.daimi.au.dk/designCPN.
19. J.L. Rasmussen and M. Singh. Designing a Security System by Means of Coloured Petri Nets. In *Proc. ICATPN 1996*, volume 1091 of *LNCS*, pages 400–419. Springer-Verlag, 1996.
20. SceneBeans. Online www-dse.doc.ic.ac.uk/Software/SceneBeans.
21. The ExSpect tool. www.exspect.com.
22. M. Weber and E. Kindler. The Petri Net Kernel. In *Petri Net Technologies for Modeling Communication Based Systems*, volume 2472 of *LNCS*, pages 109–123. Springer-Verlag, 2003.
23. M. Westergaard. BRITNeY suite website. Online wiki.daimi.au.dk/tincpn/.
24. M. Westergaard and K.B. Lassen. Building and Deploying Visualizations of Coloured Petri Net Models Using BRITNeY animation and CPN Tools. In *Proc. of Sixth CPN Workshop*, volume PB-576 of *DAIMI*, pages 119–136, 2005.
25. D. Winer. XML-RPC Specification. www.xmlrpc.org/spec.

Author Index